IIT JEE Objective Chemistry

Question Papers and Model Solutions

For IIT Objective Type Joint Entrance Examination

Second Edition

Arun Syamal
M.Sc., Ph.D., D.Sc.
Formerly Chairman,
Professor of Chemistry and Director,
National Institute of Technology
(Deemed University), Kurukshetra 136119
Haryana

ATLANTIC
PUBLISHERS & DISTRIBUTORS (P) LTD

Published by
ATLANTIC
PUBLISHERS & DISTRIBUTORS (P) LTD
7/22, Ansari Road, Darya Ganj,
New Delhi-110002
Phones : +91-11-40775252, 23273880, 23275880, 23280451
Fax : +91-11-23285873
Web : www.atlanticbooks.com
E-mail : orders@atlanticbooks.com

Branch Office
5, Nallathambi Street, Wallajah Road,
Chennai-600002
Phones : +91-44-64611085, 32413319
E-mail : chennai@atlanticbooks.com

Printed in India at Nice Printing Press, A-33/3A, Site-IV,
Industrial Area, Sahibabad, Ghaziabad, U.P.

Preface

The book contains IIT JEE previous years' objective type chemistry question papers from 1999 onwards and their model solutions. The book will be extremely useful to the students appearing at IIT JEE examination. The book will also serve the needs of the students appearing for AIEEE and other engineering entrance examinations. It is very important for the aspiring students appearing at IIT JEE to solve the objective type questions of previous years of IIT joint entrance examinations. A further advantage of the book is that it contains very rich Appendices.

Being a teacher at the National Institute of Technology, Kurukshetra, I have vast experience of dealing with engineering entrance tests and setting question papers for such tests. In addition, I have long experience of teaching chemistry to aspiring students for IIT JEE and other engineering entrance examinations. This book will be extremely useful to all those students who do not have the facility to receive extra guidance for the preparation to appear at such examinations. I wish to thank Dr. M. M. Singh, Senior Model School, Kurukshetra, Dr. Dinesh Kumar, Professor of chemistry, National Institute of Technology, Kurukshetra and Mr. R. K. Tomar, Govt. High School, Pehowa, Haryana for making useful suggestions. I will welcome constructive criticisms and suggestions from teachers and students for improvement of the book.

Arun Syamal

E-mail: arunsyamal@yahoo.co.in

Contents

I.I.T. JEE Objective Question Paper—1999

1. The electrons, identified by quantum numbers n and l, (i) $n = 4, l = 1$, (ii) $n = 4, l = 0$, (iii) $n = 3, l = 2$, (iv) $n = 3, l = 1$ can be placed in order of increasing energy, from the lowest to highest, as
   (A) (iv) < (ii) < (iii) < (i)
   (B) (ii) < (iv) < (i) < (iii)
   (C) (i) < (iii) < (ii) < (iv)
   (D) (iii) < (i) < (iv) < (ii)
2. The number of neutrons accompanying the formation of $^{139}_{54}Xe$ and $^{94}_{38}Sr$ from the absorption of a slow neutron by $^{235}_{92}U$, followed by nuclear fission is
   (A) 0
   (B) 2
   (C) 1
   (D) 3
3. The correct order of increasing C—O bond length of CO, CO_3^{2-}, CO_2 is
   (A) $CO_3^{2-} < CO_2 < CO$
   (B) $CO_2 < CO_3^{2-} < CO$
   (C) $CO < CO_3^{2-} < CO_2$
   (D) $CO < CO_2 < CO_3^{2-}$
4. A gas will approach ideal behaviour at
   (A) low temperature and low pressure
   (B) low temperature and high pressure
   (C) high temperature and low pressure
   (D) high temperature and high pressure.
5. The normality of $0.3M$ phosphorous acid (H_3PO_3) is
   (A) 0.1
   (B) 0.9
   (C) 0.3
   (D) 0.6
6. The coordination number of a metal crystallising in a hexagonal close-packed structure is
   (A) 12
   (B) 4
   (C) 8
   (D) 6
7. A gas X at 1 atm is bubbled through a solution containing a mixture of 1 MY^- and 1 MZ^- at 25 °C. If the reduction potential of Z > Y > X, then
   (A) Y will oxidise X and not Z
   (B) Y will oxidise Z and not X
   (C) Y will oxidise both X and Z
   (D) Y will reduce both X and Z.
8. The pH of $0.1M$ solution of the following salts increases in the order
   (A) $NaCl < NH_4Cl < NaCN < HCl$
   (B) $HCl < NH_4Cl < NaCl < NaCN$
   (C) $NaCN < NH_4Cl < NaCl < HCl$
   (D) $HCl < NaCl < NaCN < NH_4Cl$
9. For the chemical reaction $3X(g) + Y(g) \rightleftharpoons X_3Y(g)$, the amount of X_3Y at equilibrium is affected by
   (A) temperature and pressure
   (B) temperature only
   (C) pressure only
   (D) temperature, pressure and catalyst.

10. In the dichromate dianion,

(A) 4 Cr—O bonds are equivalent (B) 6 Cr—O bonds are equivalent

(C) all Cr—O bonds are equivalent (D) all Cr—O bonds are non-equivalent.

11. One mole of calcium phosphide on reaction with excess water gives

(A) one mole of phosphine (B) two moles of phosphoric acid

(C) two moles of phosphine (D) one mole of phosphorus pentaoxide.

12. The oxidation number of sulphur in S_8, S_2F_2, H_2S respectively, is

(A) 0, + 1 and – 2 (B) + 2, + 1 and – 2

(C) 0, + 1 and + 2 (D) – 2, + 1 and – 2

13. On heating ammonium dichromate, the gas evolved is

(A) oxygen (B) ammonia (C) nitrous oxide (D) nitrogen

14. In the commercial electrochemical process for aluminium extraction the electrolyte used is

(A) $Al(OH)_3$ in NaOH solution

(B) an aqueous solution of $Al_2(SO_4)_3$

(C) a molten mixture of Al_2O_3 and Na_3AlF_6

(D) a molten mixture of AlO(OH) and $Al(OH)_3$

15. The geometry of H_2S and its dipole moment are

(A) angular and non-zero (B) angular and zero

(C) linear and non-zero (D) linear and zero.

16. The geometry of $Ni(CO)_4$ and $Ni(PPh_3)_2Cl_2$ is

(A) both square planar

(B) tetrahedral and square planar respectively

(C) both tetrahedral

(D) square planar and tetrahedral respectively.

17. In compounds of type ECl_3, where E = B, P, As or Bi, the angles Cl—E—Cl for different E are in the order

(A) B > P = As = Bi (B) B > P > As > Bi

(C) B < P = As = Bi (D) B < P < As < Bi

18. In the compound $CH_2{=}CH{-}CH_2{-}CH_2{-}C{\equiv}CH$, the C_2—C_3 bond is of the type,

(A) *sp*-sp^2 (B) sp^3-sp^3 (C) *sp*-sp^3 (D) sp^2-sp^3

19. When propanoic acid is treated with aqueous sodium bicarbonate, CO_2 is liberated. The C of CO_2 comes from

(A) methyl group (B) carboxylic acid group

(C) methylene group (D) bicarbonate.

20. The enol form of acetone, after treatment with D_2O, gives

(A) $CH_3{-}C(OD){=}CH_2$ (B) $CD_3{-}C(=O){-}CD_3$

(C) $CH_2{=}C(OH){-}CH_2D$ (D) $CD_2{=}C(O){-}CD_2$

21. A positive carbylamine test is given by

(A) N,N-dimethylaniline (B) 2,4-dimethylaniline
(C) N-methyl-*o*-methylaniline (D) *p*-methylbenzylamine

22. The optically active tartaric acid is named as D-(+)-tartaric acid because it has a positive

(A) optical rotation and is derived from D-glucose
(B) *p*H in organic solvent
(C) optical rotation and is derived from D-(+)-glyceraldehye
(D) optical rotation only when substituted by deuterium.

23. A solution of (+)-2-chloro-2-phenylethene in toluene racemises slowly in the presence of small amount of $SbCl_5$ due to the formation of

(A) carbanion (B) carbene (C) free radical (D) carbocation.

24. The product(s) obtained via oxymercuration ($HgSO_4 + H_2SO_4$) of but-l-yne would be

(A) $CH_3-CH_2-\overset{\overset{O}{\|}}{C}-CH_3$ (B) $CH_3-CH_2-CH_2-CHO$
(C) $CH_3-CH_2-CHO + HCHO$ (D) $CH_3CH_2COOH + HCOOH$

25. The most unlikely representation of resonance structures of *p*-nitrophenoxide ion is:

(A) 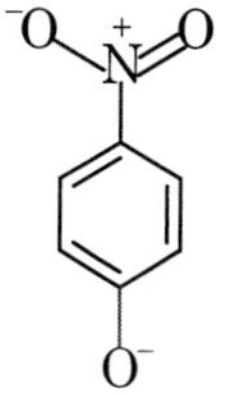 (B) 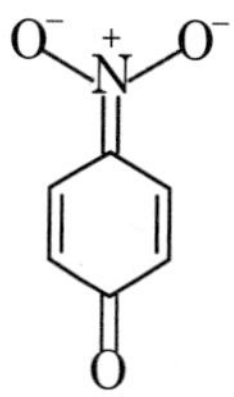 (C) 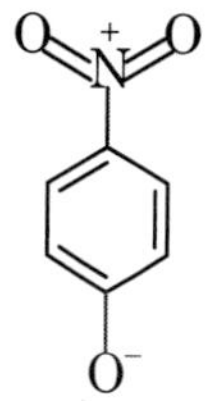 (D) 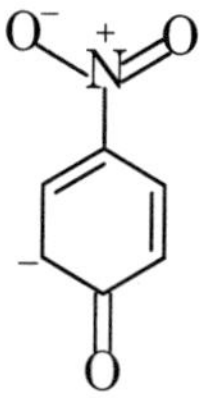

Directions: Question numbers 26–35 carry 3 marks each and may have more than one correct answer. All correct answers must be marked to get any credit in these questions.

26. The ether 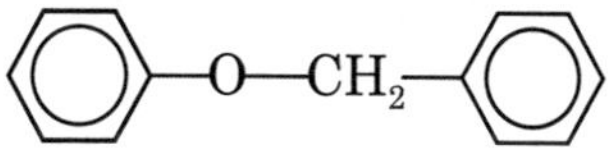 when treated with HI produces

(A) 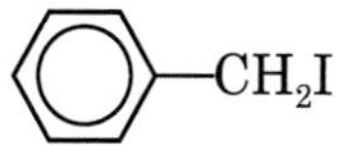 (B) 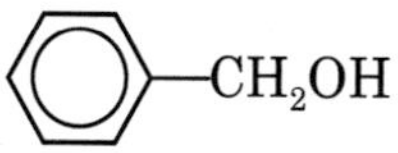 (C) 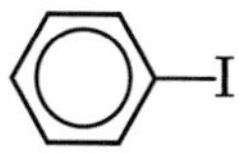 (D) 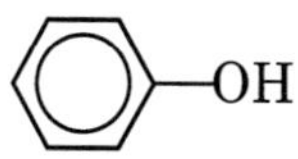

27. Toluene, when treated with Br_2/Fe, gives *p*-bromotoluene as the major product because the CH_3 group

(A) is *para* directing (B) is *meta* directing
(C) deactivates the ring (D) activates the ring by hyperconjugation.

28. The following statements(s) is(are) correct

(A) A plot of log K_p versus $1/T$ is linear
(B) A plot of log [X] *versus* time is linear for a first order reaction, $X \longrightarrow P$
(C) A plot of P *versus* $1/T$ is linear at constant volume
(D) A plot of P *versus* $1/V$ is linear at constant temperature.

29. The following is(are) endothermic reaction(s)

(A) Combustion of methane (B) Decomposition of water
(C) Dehydrogenation of ethane to ethylene (D) Conversion of graphite to diamond.

30. Ground state electronic configuration of nitrogen atom can be represented by

(A) [↓↑] [↓↑] [↑|↑|↑] (B) [↑↓] [↑↓] [↑|↓|↑] (C) [↓↑] [↓↑] [↑|↓|↓] (D) [↑↓] [↑↓] [↓|↓|↓]

31. In the depression of freezing point experiment, it is found that the
(A) vapour pressure of the solution is less than that of pure solvent
(B) vapour pressure of the solution is more than that of pure solvent
(C) only solute molecules solidify at the freezing point
(D) only solvent molecules solidify at the freezing point.

32. Ionic radii of
(A) $Ti^{4+} < Mn^{7+}$ (B) $^{35}Cl^- < {}^{37}Cl^-$ (C) $K^+ > Cl^-$ (D) $P^{3+} > P^{5+}$

33. Ammonia, on reaction with hypochlorite anion, can form
(A) NO (B) NH_4Cl (C) N_2H_4 (D) HNO_2

34. A buffer solution can be prepared from a mixture of
(A) sodium acetate and acetic acid in water
(B) sodium acetate and hydrochloric acid in water
(C) ammonia and ammonium chloride in water
(D) ammonia and sodium hydroxide in water.

35. An aromatic molecule will
(A) have $4n$ π electrons (B) have $(4n + 2)$ π electrons
(C) be planar (D) be cyclic.

ANSWER KEYS

1. A	2. D	3. D	4. C	5. D	6. A
7. A	8. B	9. A	10. B	11. C	12. A
13. D	14. C	15. A	16. C	17. B	18. D
19. D	20. B	21. B, D	22. C	23. D	24. A
25. C	26. A, D	27. A, D	28. A, B, D	29. B, C, D	30. A, D
31. A, D	32. D	33. C	34. A, C	35. B, C, D	

SOLUTIONS

1. (A)

(i) value of $(n + l) = 4 + 1 = 5$ (ii) value of $(n + l) = 4 + 0 = 4$

(iii) value of $(n + l) = 3 + 2 = 5$ (iv) value of $(n + l) = 3 + 1 = 4$

Thus $(n + l)$ value is lower for (ii) and (iv) but is equal in both. In (iv) the n value is lower than that in (ii) and hence (iv) is lower energy than (ii). Similarly, (iii) is lower energy than (i). Hence the order of increasing energy is: (iv) < (ii) < (iii) < (i).

2. (D)

$$^{235}_{92}U + ^{1}_{0}n \longrightarrow ^{139}_{54}Xe + ^{94}_{38}Sr + 3^{1}_{0}n$$

3. (D)

Bond order in CO_2 = 3 (from molecular orbital theory)

$$\text{Bond order in } CO_2 = \frac{\text{No. of bonds in all possible sides}}{\text{No. of resonating structures}} \text{ (from resonance)}$$

$$= \frac{4}{2} = 2$$

$$\text{Bond order in } CO_3^{2-} = \frac{4}{3} = 1.33 \text{ (from resonance)}$$

$$\text{Bond length} \propto \frac{1}{\text{Bond order}}$$

Hence the order of increasing bond length: $CO < CO_2 < CO_3^{2-}$

4. (C)

A gas will approach ideal behaviour at high temperature and low pressure.

5. (D)

Phosphorous acid (H_3PO_3) is a dibasic acid and hence its molecular weight = 2 × Eq. wt. Thus for H_3PO_3, $1M = 2N$. Hence $0.3M = 0.6N$.

6. (A)

The coordination number of a metal crystallising in a hexagonal close-packed structure is 12.

7. (A)

The reduction potential of X, Y and Z is in the order: Z > Y > X.

On the basis of reduction potential, the following spontaneous reactions can be predicted:

(a) Z is reduced and X is oxidised. (b) Y is reduced and X is oxidised.

(c) Z is reduced and Y is oxidised.

Hence Y will oxidise X and not Z.

8. (B)

Among the given substances, HCl is the strongest acid. An aqueous solution of NH_4Cl is slightly acidic due to hydrolysis. NH_4Cl is a salt of a strong acid and a weak base.

$$NH_4Cl + H_2O \rightleftharpoons NH_4OH + HCl \qquad ([OH^+] > [OH^-])$$

An aqueous solution of NaCN is basic due to hydrolysis. NaCN is salt of a weak acid and a strong base.

$$NaCN + H_2O \rightleftharpoons NaOH + HCN \qquad ([OH^-] > [H^+])$$

An aqueous solution of NaCl is neutral since it is a salt of a strong acid and a strong base. Hence the increasing order of pH is:

$$HCl < NH_4Cl < NaCl < NaCN$$

9. (A)

The equilibrium is affected by temperature and pressure due to change in heat and change in volume of substance, respectively.

10. (B)

The structure of dichromate ion $\left(Cr_2O_7^{2-}\right)$ is as follows:

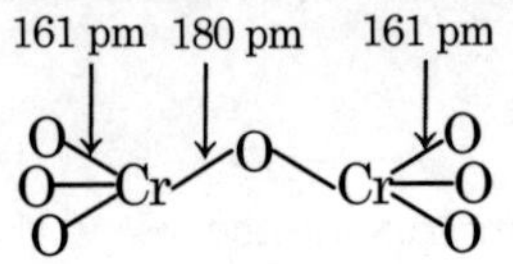

In dichromate ion, all the six non-bridging Cr—O bonds are equivalent and two bridged Cr—O bonds are equivalent.

11. (C)

$$Ca_3P_2 + 6H_2O \longrightarrow 3Ca(OH)_2 + \underset{\text{two moles}}{2PH_3}$$

12. (A)

The oxidation number of S in $S_8 = 0$

The oxidation number of S in $S_2F_2 = +1$

The oxidation number of S in $H_2S = -2$

13. (D)

$$(NH_4)_2Cr_2O_7 \xrightarrow{\Delta} Cr_2O_3 + 4H_2O + N_2$$

14. (C)

Na_3AlF_6 is used to decrease the melting point of alumina and to increase the conductivity of molten mass.

15. (A)

The S atom in H_2S is sp^3 hybridised and H_2S molecule is angular (V-shaped) giving rise to non-zero value of dipole moment.

16. (C)

Both $Ni(CO)_4$ and $Ni(PPh_3)_2Cl_2$ have sp^3 hybridisation and hence are tetrahedral.

$_{28}Ni: 1s^2 2s^2 2p^6 3s^2 3p^6 3d^8 4s^2 \qquad Ni^{2+}: 1s^2 2s^2 2p^6 3s^2 3p^6 3d^8$

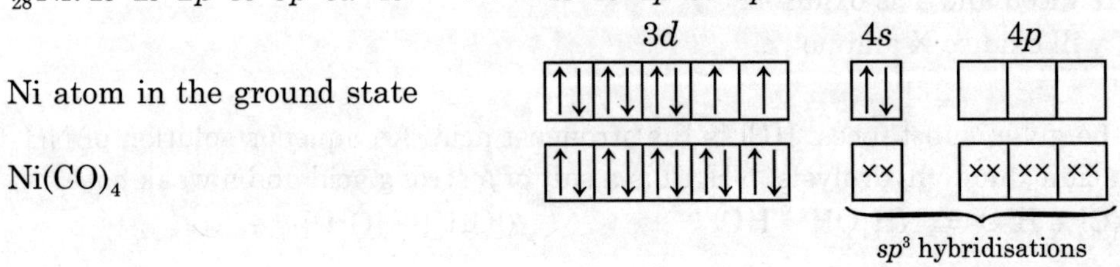

Ni^{2+} ion

Ni(PPh$_3$)$_2$Cl$_2$

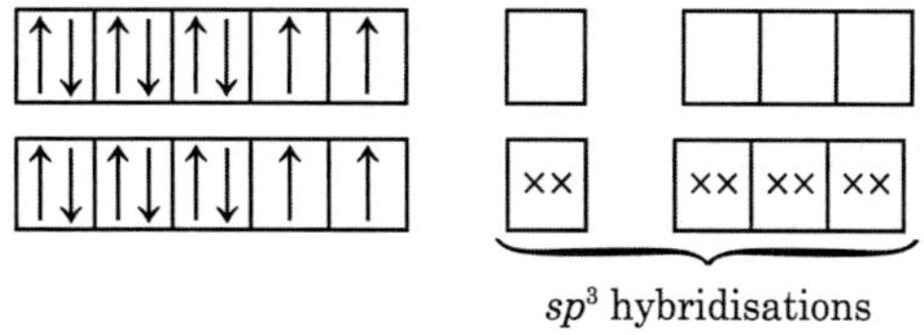

17. (B)

The structure, hybridisation and bond angles of BCl_3, PCl_3, $AsCl_3$ $BiCl_3$ are as follows:

Molecule	Structure	Hybridisation	Bond angle
BCl_3	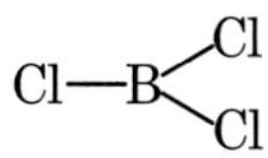 Triangular planar	sp^2	120°
PCl_3	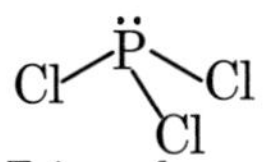 Trigonal pyramid	sp^3	100°
$AsCl_3$	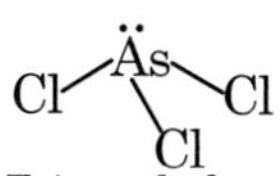 Trigonal planar	sp^3	98°
$BiCl_3$	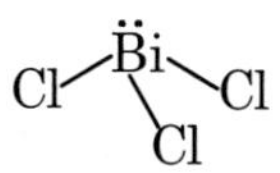	sp^3	96°

The electronegativity value of P, As and Bi is in the order Bi < As < P. As the electronegativity of the central atom increases, there occurs more repulsion between two bonding pairs (*bp-bp*) of electrons and this leads to greater bond angle in PCl_3 in comparison to $AsCl_3$ and $BiCl_3$. Hence the order of bond angle ($\angle$Cl—E—Cl) is $BCl_3 > PCl_3 > AsCl_3 > BiCl_3$.

18. (D)

$$\overset{1}{\underset{sp^2}{CH_2}}=\overset{2}{\underset{sp^2}{CH}}-\overset{3}{\underset{sp^3}{CH_2}}-\overset{4}{\underset{sp^3}{CH_2}}-\overset{5}{\underset{sp}{C}}\equiv\overset{6}{\underset{sp}{CH}} \quad \rightleftharpoons$$

Hence the C_2—C_3 bond is of the type sp^2-sp^3.

19. (D)

$$C_2H_5COOH + NaHCO_3 \longrightarrow C_2H_5COONa + H_2O + CO_2$$

20. (B)

$$CH_3-\underset{OH}{\underset{|}{C}}=CH_2 \xrightarrow{D_2O} CD_3-\underset{OD}{\underset{|}{C}}=CD_2 \qquad CD_3-\underset{O}{\underset{\|}{C}}-CD_3$$

21. (B, D)

A positive carbylamine test is given by primary amines.

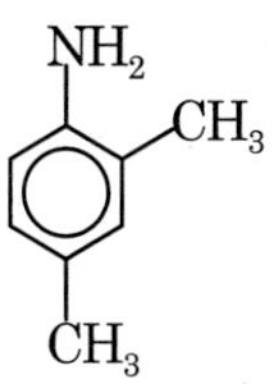

2,4-dimethylaniline

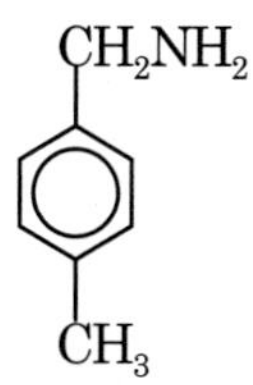

p-methylbenzylamine

22. (C)

D-(+)-tartaric acid has a positive optical rotation and its name is derived from D-(+)-glyceraldehye. The symbol D is assigned to the arrangement of —OH group on the right and the hydrogen atom on the left.

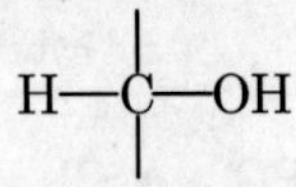

The + sign is used to represent the isomer which rotates the plane of polarised light towards right. D-(+)-tartaric acid is derived from D-(+)-glyceraldehye as shown below.

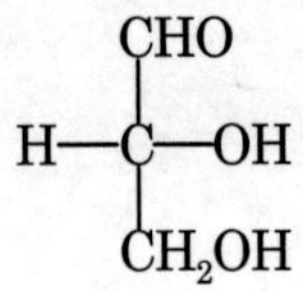

D-(+)-glyceraldehye

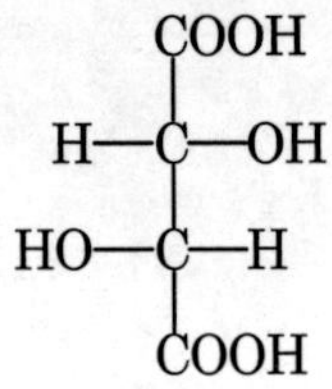

D-(+)-tartaric acid

23. (D)

$SbCl_5$ is used for the formation of intermediate carbocation.

24. (A)

$$CH_3-CH_2-C\equiv CH + H_2O \xrightarrow[\text{Markownikoff addition}]{HgSO_4 + H_2SO_4} \left[CH_3-CH_2-\underset{OH}{\underset{|}{C}}=CH_2\right] \xrightarrow{\text{tautomerisation}}$$

$$CH_3-CH_2-\underset{O}{\underset{\|}{C}}-CH_3$$

The enol form tautomerises because the keto form is more stable than the enol form.

25. (C)

The structure C is most unlikely because N atom has five bonds and a positive charge.

26. (A, D)

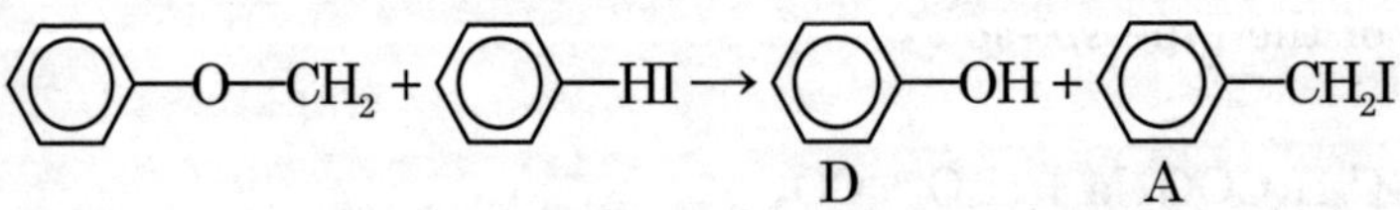

27. (A, D)

The —CH_3 group being an electron-releasing group activates the benzene ring by hyperconjugation. The electron-releasing or activating group —CH_3 is *ortho* and *para* directing.

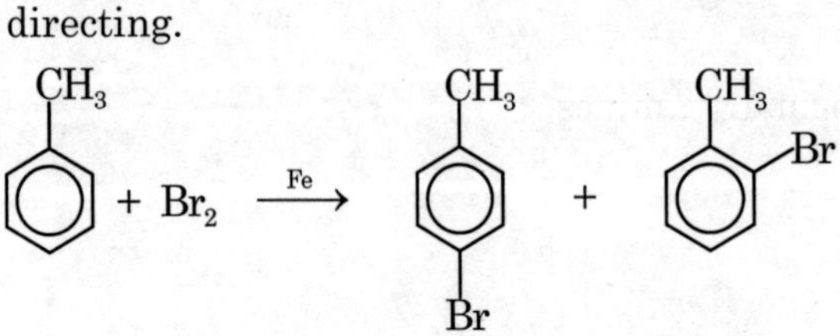

p-bromotoluene major product　*o*-bromotoluene minor product

28. (A, B, D)

The relevant expressions are as follows:

(A) $\log K_p = -\dfrac{\Delta H}{R}\cdot\dfrac{1}{T} + 1$

(B) $\log [X] = \log [X_0] + kt$

(C) $\dfrac{P}{T}$ = constant (at constant V)

(D) PV = constant (at constant T)

29. (B, C, D)

All combustion reactions are exothermic. The other three reactions are endothermic.

30. (A, D)

A and D follow Hund's rules. B and C do not follow Hund's rule.

31. (A, D)

32. (D)

Larger the effective nuclear charge, smaller is the radius.

33. (C)

$2NH_3 + OCl^- \longrightarrow NH_2{-}NH_2 + H_2O + Cl^-$

34. (A, C)

The mixture given in A and C can constitute buffer solution because a buffer solution is prepared by mixing a weak acid/weak base with salt of its conjugate base/acid.

35. (B, C, D)

An aromatic molecule will have:

(B) ($4n\ \pi$) electrons (by Huckel's rule).

(C) planar structure (due to resonance).

(D) cyclic structure (due to the presence of sp^2 hybridised carbon atoms).

I.I.T. JEE Objective Question Paper—2000

1. Amongst the following identify the species with an atom in + 6 oxidation state.
   (A) MnO_4^- (B) $Cr(CN)_6^{3-}$ (C) NiF_6^{2-} (D) CrO_2Cl_2
2. The correct order of acidic strength is
   (A) $Cl_2O_7 > SO_2 > P_4O_{10}$ (B) $CO_2 > N_2O_5 > SO_3$
   (C) $Na_2O > MgO > Al_2O_3$ (D) $K_2O > CaO > MgO$
3. The correct order of radii is
   (A) $N < Be < B$ (B) $F^- < O^{2-} < N^{3-}$
   (C) $Na < Li < K$ (D) $Fe^{3+} < Fe^{2+} < Fe^{4+}$
4. The number of P—O—P bonds in cyclic metaphosphoric acid is
   (A) zero (B) two (C) three (D) four
5. Ammonia can be dried by
   (A) conc. H_2SO_4 (B) P_4O_{10}
   (C) CaO (D) anhydrous $CaCl_2$
6. Molecular shapes of SF_4, CF_4, and XeF_4 are
   (A) the same, with 2, 0 and 1 lone pairs of electrons respectively
   (B) the same, with 1, 1 and 1 lone pairs of electrons respectively
   (C) different, with 0, 1, and 2 lone pairs of electrons respectively
   (D) different, with 1, 0 and 2 lone pairs of electrons respectively.
7. The chemical processes in the production of steel from haematite ore involve
   (A) reduction (B) oxidation
   (C) reduction followed by oxidation (D) oxidation followed by reduction.
8. The hybridisations of atomic orbitals of nitrogen in NO_2^+, NO_2^- and NH_4^+ are
   (A) *sp*, sp^3 and sp^2 respectively (B) *sp*, sp^2 and sp^3 respectively
   (C) sp^2, *sp* and sp^3 respectively (D) sp^2, sp^3 and *sp* respectively.
9. Amongst H_2O, H_2S, H_2Se and H_2Te, the one with the highest boiling point is
   (A) H_2O because of hydrogen bonding
   (B) H_2Te because of higher molecular weight
   (C) H_2S because of hydrogen bonding
   (D) H_2Se because of lower molecular weight.

10. Electrolytic reduction of alumina to aluminium by Hall–Heroult process is carried out
(A) in the presence of NaCl
(B) in the presence of fluorite
(C) in the presence of cryolite which forms a melt with lower melting temperature
(D) in the presence of cryolite which forms a melt with higher melting temperature.

11. The ΔH_f^0 for $CO_2(g)$, $CO(g)$ and $H_2O(g)$ is – 393.5, – 110.5 and – 241.8 kJ mol^{-1} respectively. The standard enthalpy change (in kJ) for the reaction $CO_2(g) + H_2(g) \longrightarrow CO(g) + H_2O(g)$ is
(A) 524.1 (B) 41.2 (C) – 262.5 (D) – 41.2

12. The rate constant for the reaction $2N_2O_5 \longrightarrow 4NO_2 + O_2$, is 3.0×10^{-5} s^{-1}. If the rate is 2.4×10^{-5} mol L^{-1} s^{-1}, then the concentration of N_2O_5 (in mol L^{-1}) is
(A) 1.4 (B) 1.2 (C) 0.04 (D) 0.8

13. For the electrohemical cell, $M | M^+ || X^- | X$, E^0 (M^+/M) = 0.44 V and E^0 (X/X^-) = 0.33 V. From this data one can deduce that
(A) $M + X \longrightarrow M^+ + X^-$ is the spontaneous reaction
(B) $M^+ + X^- \longrightarrow M + X$ is the spontaneous reaction
(C) $E_{cell} = 0.77$ V
(D) $E_{cell} = -0.77$ V

14. For the reversible reaction, $N_2(g) + 3H_2(g) \rightleftharpoons 2NH_3(g)$ at 500°C, the value of K_p is 1.44×10^{-5} when partial pressure is measured in atmospheres. The corresponding value of K_c with concentration in mole $litre^{-1}$ is
(A) $1.44 \times 10^{-5}/(0.082 \times 500)^{-2}$ (B) $1.44 \times 10^{-5}/(8.314 \times 773)^{-2}$
(C) $1.44 \times 10^{-5}/(0.082 \times 773)^{2}$ (D) $1.44 \times 10^{-5}/(0.082 \times 773)^{-2}$

15. The rms velocity of hydrogen is $\sqrt{7}$ times the rms velocity of nitrogen. If T is the temperature of the gas,
(A) $T(H_2) = T(N_2)$ (B) $T(H_2) > T(N_2)$
(C) $T(H_2) < T(N_2)$ (D) $T(H_2) = \sqrt{7}\, T(N_2)$

16. The compressibility factor of a gas is less than unity at STP. Therefore,
(A) $V_m > 22.4$ litres (B) $V_m < 22.4$ litres
(C) $V_m = 22.4$ litres (D) $V_m = 44.8$ litres.

17. At 100 °C and 1 atm, if the density of liquid water is 1.0 g cm^{-3} and that of water vapours is 0.0006 g cm^{-3}, then the volume occupied by water molecules in 1 L of steam at that temperature is
(A) 6 cm^3 (B) 60 cm^3 (C) 0.6 cm^3 (D) 0.06 cm^3

18. The electronic configuration of an element is $1s^2 2s^2 2p^6 3s^2 3p^6 3d^5 4s^1$. This represents its
(A) excited state (B) ground state (C) cationic form (D) anionic form.

19. The number of nodal planes in a p_x orbital is
(A) one (B) two (C) three (D) zero

20. When two reactants, A and B are mixed to give products C and D, the reaction quotient Q, at the initial stages of the reaction
(A) is zero (B) decreases with time
(C) is independent of time (D) increases with time.

21. Which of the following compounds will exhibit geometrical isomerism?
(A) 1-phenylbut-2-ene (B) 3-phenylbut-1-ene
(C) 2-phenylbut-1-ene (D) 1,1-diphenylprop-1-ene

22. Which of the following has the most acidic hydrogen?
(A) hexan-3-one (B) hexane-2,4-dione
(C) hexane-2,5-dione (D) hexane-2,3-dione

23. Among the following, the strongest base is
(A) $C_6H_5NH_2$ (B) p-NO_2—$C_6H_4NH_2$
(C) m-NO_2—$C_6H_4NH_2$ (D) $C_6H_5CH_2NH_2$

24. Which one of the following will most readily be dehydrated in acidic condition?
(A) 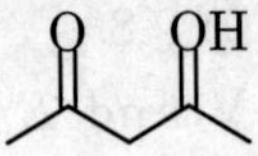 (B) 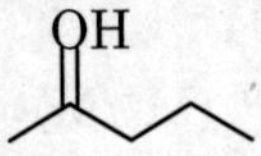 (C) 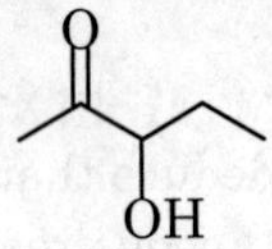 (D) 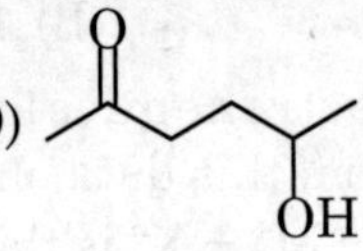

25. Which of the following has the highest nucleophilicity?
(A) F^- (B) OH^- (C) CH_3^- (D) NH_2^-

26. Propyne and propene can be distinguished by
(A) conc. H_2SO_4 (B) Br_2 in CCl_4
(C) dil. $KMnO_4$ (D) $AgNO_3$ in ammonia.

27. The order of reactivities of the following alkyl halides for an S_{N^2} reaction is
(A) RF > RCl > RBr > RI (B) RF > RBr > RCl > RI
(C) RCl > RBr > RF > RI (D) RI > RBr > RCl > RF

28. The appropriate reagent for the following transformation is

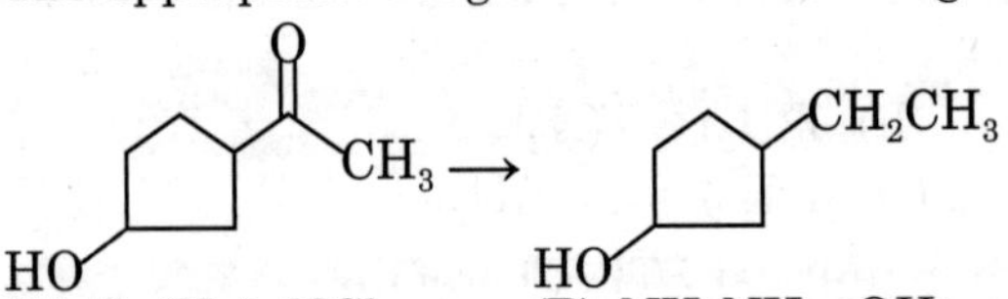

(A) Zn(Hg), HCl (B) NH_2NH_2, OH^- (C) H_2/Ni (D) $NaBH_4$

29. Benzoyl chloride is prepared from benzoic acid by
(A) Cl_2, hv (B) SO_2Cl_2 (C) $SOCl_2$ (D) Cl_2, H_2O

30. Which one of the following alkenes will react fastest with H_2 under catalytic hydrogenation condition?
(A) 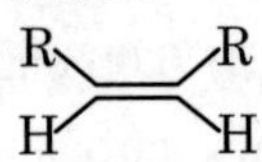 (B) 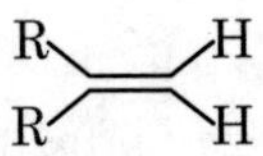 (C) 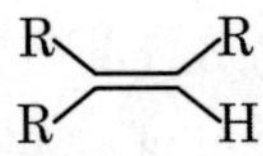 (D) 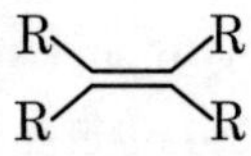

The questions below (31 to 35) consist of an *Assertion* and the *Reason*. Use the following key to choose the appropriate answer.

(A) If both *assertion* and *reason* are CORRECT, and *reason* is the CORRECT explanation of the *assertion*.

(B) If both *assertion* and *reason* are CORRECT, but reason is NOT the CORRECT explanation of the *assertion*.

(C) If *assertion* is CORRECT, but *reason* is INCORRECT.
(D) If *assertion* is INCORRECT, but *reason* is CORRECT.

31. *Assertion*: The pressure of a fixed amount of an ideal gas is proportional to its temperature.
Reason: Frequency of collisions and their impact both increase in proportion to the square root of temperature.

32. *Assertion*: The heat absorbed during the isothermal expansion of an ideal gas against vacuum is zero.
Reason: The volume occupied by the molecules of an ideal gas is zero.

33. *Assertion*: But-1-ene on reaction with HBr in the presence of a peroxide produces 1-bromobutane.
Reason: It involves the formation of a primary radical.

34. *Assertion*: Phenol is more reactive than benzene towards electrophilic substitution reaction.
Reason: In the case of phenol, the intermediate carbocation is more resonance stabilised.

35. *Assertion*: The first ionization energy of Be is greater than that of B.
Reason: $2p$ orbital is lower in energy than $2s$.

ANSWER KEYS

1. D	2. A	3. B	4. C	5. C	6. D
7. D	8. B	9. A	10. C	11. B	12. D
13. B	14. D	15. C	16. B	17. C	18. B
19. A	20. D	21. A	22. B	23. D	24. A
25. C	26. D	27. D	28. B	29. C	30. A
31. B	32. B	33. C	34. C	35. C	

SOLUTIONS

1. (D)

The oxidation states of the central atom in the given species are as follows:

$\overset{+7}{Mn}O_4^-$, $\overset{+3}{Cr}(CN)_6^{3-}$, $\overset{+4}{Ni}F_6^{2-}$, $\overset{+6}{Cr}O_2Cl_2$

In CrO_2Cl_2, Cr has + 6 oxidation state.

2. (A)

The oxides Cl_2O_7, SO_2 and P_4O_{10} are the anhydrides of $HClO_4$, H_2SO_3 and H_3PO_3 respectively. The acidic character of acids increases along a period from left to right:

$HClO_4 > H_2SO_3 > H_3PO_3$

The corresponding anhydrides also follow the same order: $Cl_2O_7 > SO_2 > P_4O_{10}$.

3. (B)

F^-, O^{2-} and N^{3-} are isoelectronic species and in these ions the radius decreases with increase in atomic number due to increase in effective nuclear charge (Z_{eff}).

$Z_{eff} = Z - \sigma$

where Z = atomic number and σ = screening constant. Due to th presence of equal number of electrons, σ is constant in F^-, O^{2-} and N^{3-}. Hence the decreasing order of Z_{eff} is: $F^- > O^{2-} > N^{3-}$. Since radius $\propto \dfrac{1}{Z_{eff}}$, the increasing order of radius is: $F^- < O^{2-} < N^{3-}$.

4. (C)

The structure of metaphosphoric acid is

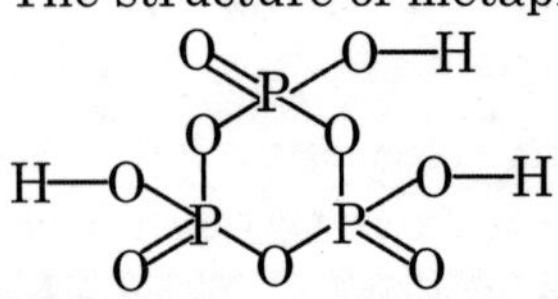

Hence the number of P—O—P bonds is three.

5. (C)

6. (D)

The hybridisation of the central atom in SF_4, CF_4 and XeF_4 is sp^3d, sp^3 and sp^3d^2 respectively. Hence the structure of SF_4, CF_4 and XeF_4 is trigonal bipyramidal, tetrahedral and octahedral respectively. The structures of these molecules are as follows:

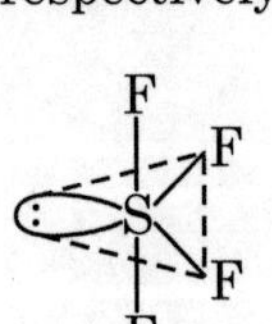

sp^3*d*, trigonal pyramidal, one lone pair

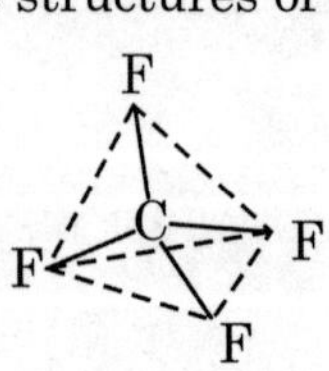

sp^3*d*, tetrahedral no lone pair

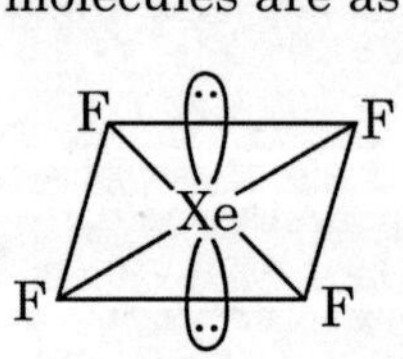

sp^3*d*^{2}, octahedral two lone pairs

7. (D)
Carbon which is added to haematite is first oxidised to CO and CO_2. The CO formed acts as a reducing agent for the reduction of haematite to Fe.

8. (B)
The structures of NO_2^+, NO_3^- and NH_4^+ ions are linear, triangular planar and tetrahedral respectively. Hence the hybridisation of atomic orbitals of nitrogen in these ions is sp, sp^2 and sp^3 respectively,

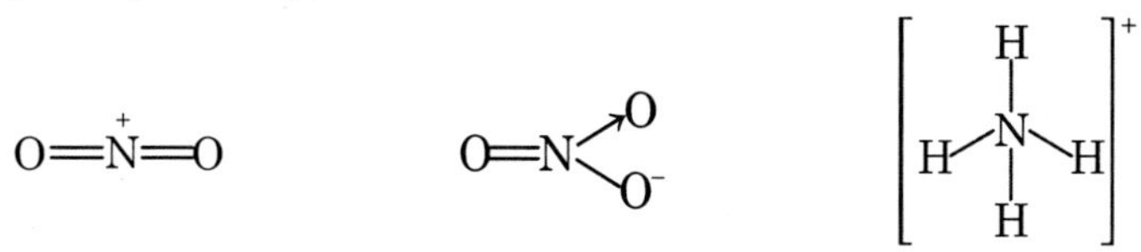

linear, sp triangular planar, sp^2 tetrahedral, sp^3

9. (A)
In spite of lower molecular mass, H_2O exhibits the highest boiling point due to the presence of strong intermolecular hydrogen bonding. H_2S, H_2Se and H_2Te do not form hydrogen bond.

10. (C)
Cryolite (Na_3AlF_6) is used in Hall–Heroult process to lower the melting point of alumina as well as to make the molten mass a good conductor of electric current.

11. (B)
$CO_2(g) + H_2(g) \longrightarrow CO(g) + H_2O(g),\ \Delta H^0 = ?$

$$\Delta H^0 = \Delta H_f^0(CO, g) + \Delta H_f^0(H_2O, g) - \Delta H_f^0(CO_2, g)$$
$$= [-110.5 - 241.8 - (-393.5)]\ \text{kJ mol}^{-1} = 41.2\ \text{kJ mol}^{-1}$$

12. (D)
$2N_2O_5(g) \longrightarrow 4NO_2(g) + O_2(g)$
The rate expression is given by
$r = k[N_2O_5]$
$2.40 \times 10^{-5}\ \text{mol L}^{-1}\ \text{s}^{-1} = (3.0 \times 10^{-5}\ \text{s}^{-1})\ [N_2O_5]$

$$\text{or } [N_2O_5] = \frac{2.40 \times 10^{-5}}{3.0 \times 10^{-5}}\ \text{mol L}^{-1} = 0.8\ \text{mol L}^{-1}$$

13. (B)
For the electrochemical cell M | M^+ | | X^- | X, the cell reaction is derived as follows:

$$\begin{array}{lll} & X^- \longrightarrow X + e^- & E^0 = -0.33\ \text{V} \\ & M^+ + e^- \longrightarrow M & E^0 = 0.44\ \text{V} \\ \hline \text{Add} & M^+ + X^- \longrightarrow M + X & E_{cell} = 0.11\ \text{V} \end{array}$$

Since E^0_{cell} is positive, the cell reaction is spontaneous.

14. (D)
$N_2(g) + 3H_2(g) \rightleftharpoons 2NH_3(g)$
K_p and K_c are related by the equation,
$K_p = K_c(RT)^{\Delta n}$

where Δn is the change in stoichiometric number of gaseous speices in the equation and it is 2 – 4 = – 2. Hence

$$K_p = K_c(RT)^{-2}$$

$$\text{or} \quad K_c = \frac{K_p}{(RT)^{-2}} = \frac{1.44 \times 10^{-5}}{(0.082 \times 773)^{-2}}$$

15. (C)

Given that

$$u_{\text{rms}}(\text{H}_2) = \sqrt{7}\, u_{\text{rms}}(\text{N}_2)$$

$$\sqrt{\frac{3RT_{\text{H}_2}}{M_{\text{H}_2}}} = \sqrt{7}\sqrt{\frac{3RT_{\text{N}_2}}{M_{\text{N}_2}}}$$

$$\text{or} \quad \frac{T_{\text{H}_2}}{M_{\text{H}_2}} = 7 \times \frac{T_{\text{N}_2}}{M_{\text{N}_2}} \qquad \text{or } T_{\text{H}_2} = 7 \times \frac{M_{\text{H}_2}}{M_{\text{N}_2}} \times T_{\text{N}_2} = 7 \times \frac{2}{28} \times T_{\text{N}_2} = \frac{1}{2} T_{\text{N}_2}$$

Hence $T_{\text{H}_2} < T_{\text{N}_2}$

16. (B)

The compressibility factor of a gas,

$$Z = \frac{V_{\text{m, real}}}{V_{\text{m, ideal}}}$$

It is given that $Z < 1$ at STP. Hence

$V_{\text{m, real}} < V_{\text{m, ideal}}$

But for an ideal gas at STP, $V_{\text{m, ideal}} = 22.4$ L

Hence $V_{\text{m, real}} < 22.4$ L

17. (C)

$$\text{Density } (\rho) = \frac{\text{Mass}}{\text{Volume}} = \frac{m}{V}$$

The ratio of denisty of liquid and vapour having same mass is

$$\frac{\rho_l}{\rho_v} = \frac{V_v}{V_l}$$

$$\text{or} \quad V_l = \frac{\rho_v \times V_v}{\rho_l} = \frac{0.0006 \times 1000 \text{ cm}^3}{1} = 0.6 \text{ cm}^3$$

18. (B)

The electronic configuration represents the ground state of chromium. The half-filled orbital (d^5) is more stable.

19. (A)

Number of nodal planes in p_x orbital is one in the yz plane.

20. (D)

$$\text{Reaction quotient, Q} = \frac{[\text{C}][\text{D}]}{[\text{A}][\text{B}]}$$

In the initial stages of the reaction, the concentrations of C and D increase and the concentrations of A and B decrease. Hence the value of Q increases with time.

21. (A)

Only 1-phenylbut-2-ene will exhibit geometrical isomerism.

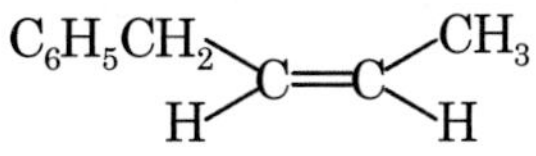

cis-1-phenylbut-2-ene

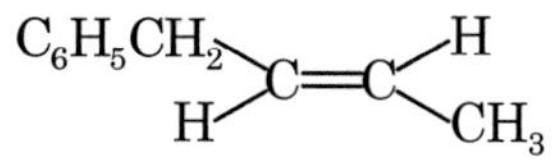

trans-1-phenylbut-2-ene

22. (B)

Only in hexane-2,4-dione, a methylene group is surrounded by two electron-withdrawing carbonyl groups. Hence its both hydrogen atoms are acidic in character.

23. (D)

Aliphatic amine is a stronger base than aromatic amine which is stabilised due to resonance. The electron-withdrawing benzene ring in aniline reduces the availability of lone pair of electron on nitrogen atom for protonation.

24. (A)

The dehydration of the alcohol (A) takes places readily due to the acidity of the α-hydrogen atom and also the product is stabilised due to the presence of conjugate double bonds.

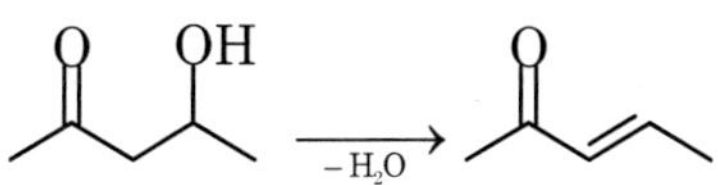

25. (C)

The increasing order of basic strength (nucleophilicity) is:

$F^- < OH^- < NH_3^- < CH_3^-$

Hence CH_3^- has the highest nucleophilicity.

26. (D)

Propyne forms white pecipitate of silver methyl acetylide on reaction with ammoniacal silver nitrate. Propene does not form acetylide due to the absence of acidic hydrogen atom.

$CH_3—C{\equiv}CH + NH_3 + AgNO_3 \rightarrow CH_3—C{\equiv}CAg\downarrow + NH_4NO_3$

$CH_3—CH{=}CH_2 + NH_3 + AgNO_3 \rightarrow$ No reaction

27. (D)

Of the halide ions, the iodide ion (I^-) is the best leaving group and the fluoride ion (F^-) is the poorest.

28. (B)

Wolff–Kishner reduction occurs in strongly basic medium and can be used for reduction of compounds which are sensitive to acid.

29. (C)

Thionyl chloride readily replaces —OH group of an acid by —Cl group producing acid chloride.

$$\underset{\text{Benzoic acid}}{C_6H_5COOH} + SOCl_2 \rightarrow \underset{\text{Benzoyl chloride}}{C_6H_5COCl} + HCl + SO_2$$

30. (A)

A *cis*-alkene is more reactive than a *trans*-alkene due to greater strain at the double bond and least steric hindrance. The smaller the number of R substituents, lesser is the steric hindrance and faster is the rate of hydrogenation.

31. (B)

Both assertion and reason are correct but reason is not the correct explanation of assertion.

32. (B)

Both assertion and reason are correct but reason is not the correct explanation of assertion. For an isothermal expansion, the work done by the gas is zero ($w = 0$) and internal energy of an ideal gas depends only on temperature.

33. (C)

Assertion is correct but reason is incorrect. In this reaction, a secondary free radical is formed as an intermediate.

$$(C_6H_5CO)_2O_2 \rightarrow 2C_6H_5COO\cdot$$

$$C_6H_5COO\cdot + HBr \rightarrow C_6H_5COOH + Br\cdot$$

$$CH_3—CH_2—CH{=}CH_2 + Br\cdot \rightarrow \underset{\text{Secondary free radical}}{CH_3—CH_2—\dot{C}H—CH_2Br}$$

$$CH_3—CH_2—\dot{C}H—CH_2Br + HBr \rightarrow CH_3—CH_2—CH_2—CH_2Br + Br$$

34. (C)

Assertion is correct but reason is incorrect. In the electrophilic substitution reaction of phenol, the intermediates formed are oxonium ions, instead of carbocations.

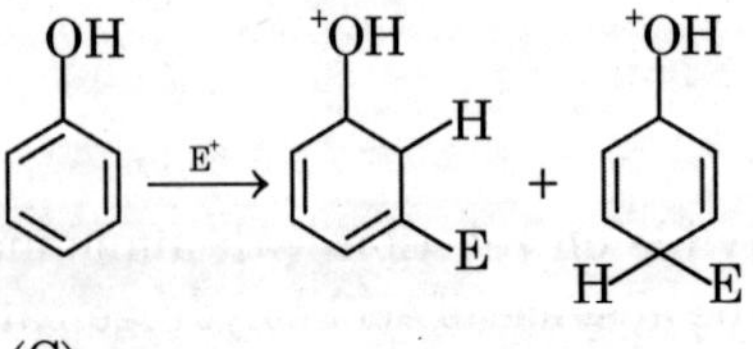

35. (C)

Assertion is correct but reason is incorrect. Boron ($1s^2 2s^2 2p^1$) contains one extra electron in $2p$ orbital in comparison to beryllium ($1s^2 2s^2$). The removal of $2p$ electron requires less energy in comparison to the removal of an electron from the more stable completely filled $2s$ orbital.

I.I.T. JEE Objective Question Paper—2001

Useful Data

Planck's constant, $h = 9.2626 \times 10^{-34}$ J s

Atomic numbers: Cr = 24, Mn = 25, Fe = 26, Co = 27, Pt = 78

1. In the presence of peroxide, hydrogen chloride and hydrogen iodide do not give anti-Markownikoff addition to alkenes because

(A) both are highly ionic

(B) one is oxidising and the other is reducing

(C) one of the steps is endothermic in both the cases

(D) all the steps are exothermic in both the cases.

2. The correct order of basicities of the following compounds

1. $CH_3—C(=NH)—NH_2$ 2. $CH_3—CH_2—NH_2$ 3. $(CH_3)_2NH$ 4. CH_3CONH_2 is

(A) 2 > 1 > 3 > 4 (B) 1 > 3 > 2 > 4 (C) 3 > 1 > 2 > 4 (D) 1 > 2 > 3 > 4

3. The compound that will react most readily with NaOH to form methanol is

(A) $(CH_3)_4N^+I^-$ (B) CH_3OCH_3 (C) $(CH_3)_3S^+I^-$ (D) $(CH_3)_3CCl$

4. The number of isomers for the compound with molecular formula $C_2BrClFI$ is

(A) 3 (B) 4 (C) 5 (D) 6

5. In thermodynamics, a process is called reversible when

(A) surroundings and system change into each other

(B) there is no boundary between system and surroundings

(C) the surroundings are always in equilibrium with the system

(D) the system changes into the surroundings spontaneously.

6. In a solid AB having the NaCl structure, A atoms occupy the corners of the cubic unit cell. If all the face-centred atoms along one of the axes are removed, then the resultant stoichiometry of the solid is

(A) AB_2 (B) A_2B (C) A_4B_3 (D) A_3B_4

7. The wavelength associated with a golf ball weighing 200 g and moving at a speed of 5 m/h is of the order

(A) 10^{-10} m (B) 10^{-20} m (C) 10^{-30} m (D) 10^{-40} m

8. The quantum numbers + 1/2 and – 1/2 for the electron spin represent
(A) rotation of the electron in clockwise and anticlockwise direction respectively
(B) rotation of the electron in anticlockwise and clockwise direction respectively
(C) magnetic moment of the electron pointing up and down respectively
(D) two quantum mechanical spin states which have no classical analogue.

9. Which one of the following statements is false?
(A) Work is a state function
(B) Temperature is a state function
(C) Change in the state is completely defined when the initial and final states are specified
(D) Work appears at the boundary of the system.

10. The RMS velocity of an ideal gas at constant pressure varies with density (d) as
(A) d^2 (B) d (C) $\sqrt{d}$ (D) $1\sqrt{d}$

11. The reaction, $3ClO^-(aq) \longrightarrow ClO_3^-(aq) + 2Cl^-(aq)$, is an example of
(A) oxidation reaction (B) reduction reaction
(C) disproportionation reaction (D) decomposition reaction.

12. The set with correct order of acidity is
(A) $HClO < HClO_2 < HClO_3 < HClO_4$ (B) $HClO_4 < HClO_3 < HClO_2 < HClO$
(C) $HClO < HClO_4 < HClO_3 < HClO_2$ (D) $HClO_4 < HClO_2 < HClO_3 < HClO$

13. An aqueous solution of 6.3 g oxalic acid dihydrate is made up to 250 mL. The volume of $0.1N$ NaOH required to completely neutralise 10 mL of this solution is
(A) 40 mL (B) 20 mL (C) 10 mL (D) 4 mL

14. The number of S—S bonds in sulphur trioxide trimer (S_3O_9) is
(A) three (B) two (C) one (D) zero

15. The chemical composition of slag formed during the smelting process in the extraction of copper is
(A) $Cu_2O + FeS$ (B) $FeSiO_3$ (C) $CuFeS_2$ (D) $Cu_2S + FeO$

16. The common features among the species CN^-, CO and NO^+ are
(A) bond order three and isoelectronic
(B) bond order two and π-acceptors
(C) isoelectronic and weak field ligands
(D) bond order three and weak field ligands.

The questions below (17–21) consist of an *Assertion* in column 1 and the *Reason* in column 2. Use the following key to choose the appropriate answer.
(A) If both *assertion* and *reason* are correct, and *reason* is the correct explanation of the *assertion*
(B) If both *assertion* and *reason* are correct, but *reason* is not the correct explanation of the *assertion*
(C) If *assertion* is correct, but *reason* is incorrect
(D) If *assertion* is incorrect, but *reason* is correct.

Assertion (column 1)	*Reason* (column 2)
17. Addition of bromine to *trans* but-2-ene yields meso-2,3-dibromobutane.	Bromine addition to an alkene is an electrophilic addition.
18. Dimethylsulphide is commonly used for the reduction of an ozonide of an alkene to get the carbonyl compounds.	It reduces the ozonide giving water soluble dimethyl sulphoxide and excess of it evaporates.
19. In any ionic solid [MX] with Schottky defects, the number of positive and negative ions are same.	Equal number of cation and anion vacancies are present.
20. In strongly acidic solutions, aniline becomes more reactive towards electrophilic reagents.	The amino group being completely protonated in strongly acidic solution, the lone pair of electrons on the nitrogen is no longer available for resonance.
21. Between $SiCl_4$ and CCl_4, only $SiCl_4$ reacts with water.	$SiCl_4$ is ionic and CCl_4 is covalent.

22. In the standardisation of $Na_2S_2O_3$ using $K_2Cr_2O_7$ by iodometry, the equivalent weight of $K_2Cr_2O_7$ is
(A) (molecular weight)/2
(B) (molecular weight)/6
(C) (molecular weight)/3
(D) same as molecular weight.

23. The complex ion which has no *d* electron in the central metal atom is
(A) $[MnO_4]^-$
(B) $[Co(NH_3)_6]^{3+}$
(C) $[Fe(CN)_6]^{3-}$
(D) $[Cr(H_2O)_6]^{3+}$

24. The set representing the correct order of first ionisation energy is
(A) K > Na > Li
(B) Be > Mg > Ca
(C) B > C > N
(D) Ge > Si > C

25. The order of hybridisation of the central atom in the species NH_3, $[PtCl_4]^{2-}$, PCl_5 and BCl_3 is
(A) dsp^2, dsp^3, sp^2, sp^3
(B) sp^3, dsp^2, dsp^3, sp^2
(C) dsp^2, sp^2, sp^3, dsp^3
(D) dsp^2, sp^3, sp^2, dsp^3

26. For a sparingly soluble salt A_pB_q, the relationship of its solubility product (L_s) with its solubility (S) is
(A) $L_s = s^{p+q} \cdot p^q \cdot q^q$
(B) $L_s = s^{p+q} \cdot p^q \cdot q^p$
(C) $L_s = s^{pq} \cdot p^p \cdot q^p$
(D) $L_s = s^{pq} \cdot (pq)^{p+q}$

27. Saturated solution of KNO_3 is used to make salt-bridge because
(A) velocity of K^+ is greater than that of NO_3^-
(B) velocity of NO_3^- is greater than that of K^+
(C) velocity of K^+ and NO_3^- is nearly the same
(D) KNO_3 is highly soluble in water.

28. The order of equivalent conductance at infinite dilution of LiCl, NaCl and KCl is
(A) LiCl > NaCl > KCl
(B) KCl > NaCl > LiCl
(C) NaCl > KCl > LiCl
(D) LiCl > KCl > NaCl

29. If I is the intensity of absorbed light and C is the concentration of AB for the photochemical process $AB + h\nu \longrightarrow AB\cdot$, the rate of formation of $AB\cdot$ is directly proportional to
(A) C
(B) I
(C) I^2
(D) $C.I$

30. At constant temperature, the equilibrium constant (K_p) for the decomposition reaction $N_2O_4 \rightleftharpoons 2NO_2$ is expressed by $K_p = (4x^2P)/(1 - x^2)$, where P = pressure, x = extent of decomposition. Which of the following statements is true?

(A) K_p increases with increase of P
(B) K_p increases with increase of x
(C) K_p increases with decrease of x
(D) K_p remains constant with change in P and x.

31. A mixture of benzaldehyde and formaldehyde on heating with aqueous NaOH solution gives

(A) benzyl alcohol and sodium formate
(B) sodium benzoate and methyl alcohol
(C) sodium benzoate and sodium formate
(D) benzyl alcohol and methyl alcohol.

32. An S_{N^2} reaction at an asymmetric carbon of a compound always gives

(A) an enantiomer of the substrate
(B) a product with opposite optical rotation
(C) a mixture of diastereomers
(D) a single stereoisomer.

33.

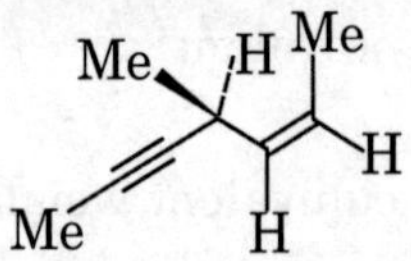

Hydrogenation of the above compound in the presence of poisoned palladium catalyst gives

(A) an optically active compound
(B) an optically inactive compound
(C) a racemic mixture
(D) a diastereomeric mixture.

34. The reaction of propene with HOCl proceeds via the addition of

(A) H^+ in the first step
(B) Cl^+ in the first step
(C) OH^- in the first step
(D) Cl^- and OH^- in a single step.

35. 1-Propanol and 2-propanol can be best distinguished by

(A) oxidation with alkaline $KMnO_4$ followed by reaction with Fehling solution
(B) oxidation with acidic dichromate followed by reaction with Fehling solution
(C) oxidation by heating with copper followed by reaction with Fehling solution
(D) oxidation with concentrated H_2SO_4 followed by reaction with Fehling solution.

ANSWER KEYS

1. C	2. B	3. A	4. D	5. C	6. D
7. C	8. D	9. A	10. D	11. C	12. A
13. A	14. D	15. B	16. A	17. B	18. A
19. A	20. D	21. C	22. B	23. A	24. B
25. B	26. A	27. C	28. B	29. B	30. D
31. A	32. D	33. B	34. B	35. C	

SOLUTIONS

1. (C)
One of the steps in the reaction is endothermic and hence sufficient energy is not available for anti-Markownikoff addition of HCl or HI to alkenes in the presence of peroxide.

2. (B)
The amine 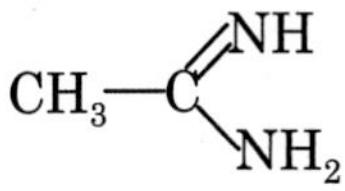 is most basic as its protonation leads to resonating cation having a large delocalisation energy.

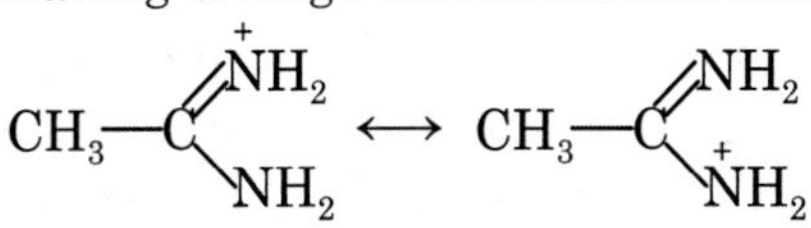

Secondary amine is more basic than primary amine. Amides are less basic than primary amines. Hence the order of decreasing basicity is: 1 > 3 > 2 > 4.

3. (A)
The positive change on nitrogen makes $(CH_3)_4N^+I^-$ most reactive towards nucleophilic attack.

$(CH_3)_4N^+I^- + NaOH \longrightarrow (CH_3)_4\overset{+}{N}\overset{-}{O}H + NaI$

$(CH_3)_4\overset{+}{N}\overset{-}{O}H \xrightarrow{\Delta} (CH_3)_3N + CH_3OH$

4. (D)
The six isomers are:

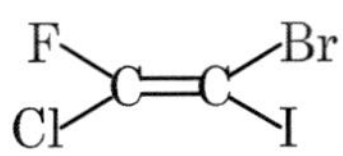
1. Z isomer

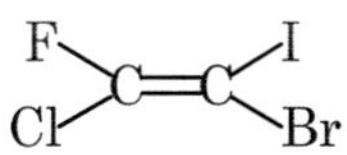
2. E isomer

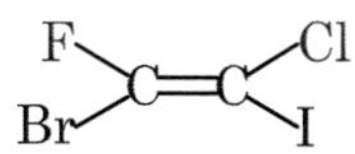
3. Z isomer

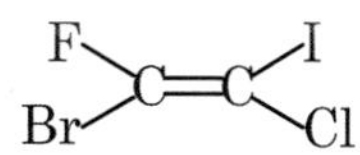
4. E isomer

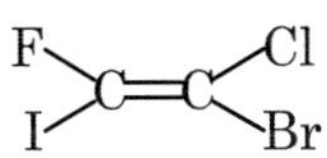
5. Z isomer

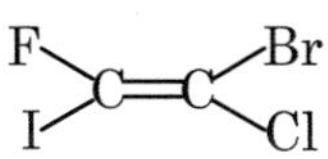
6. E isomer

5. (C)
In thermodynamics, a process is called reversible when the surroundings are always in equilibrium with the system.

6. (D)
The structure of AB after the removal of all the face-central atoms along one of the axis is:

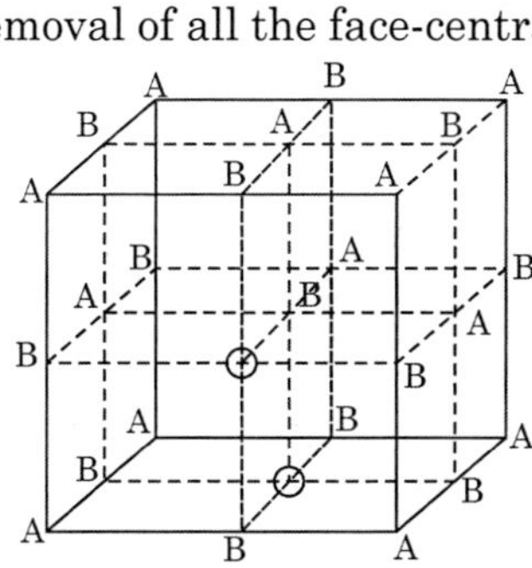

Number of A atoms per unit cell $= \underset{\text{from corners}}{8 \times \frac{1}{8}} + \underset{\text{from face-centre}}{4 \times \frac{1}{2}} = 1 + 2 = 3$

Number of B atoms per unit cell $= \underset{\text{from edges}}{12 \times \frac{1}{4}} + \underset{\text{from body-centre}}{1} = 3 + 1 = 4$

Hence the stoichiometry of the solid is A_3B_4.

7. (C)
de Broglie wavelength, $\lambda = \frac{h}{mv} = \frac{6.626 \times 10^{-34}\,\text{J s}}{(0.200\,\text{kg})[5\,\text{m}/(60 \times 60.5)]} = 2.4 \times 10^{-30}$ m

8. (D)
The magnetic spin quantum numbers $m_s = +\frac{1}{2}$ and $m_s = -\frac{1}{2}$ for the electron spin represent two quantum mechanical spin states which have no classical analogues. Classically, the two spin states are considered as the rotation of an electron on its own axis either in clockwise or anticlockwise direction. But spin is a purely quantum mechanical property and differs from its classical namesake.

9. (A)
Work is not a state function. Work has to be performed.

10. (D)

$$u_{rms} = \sqrt{\frac{3RT}{M}} = \sqrt{\frac{3PV}{M}} = \sqrt{\frac{3P}{M/V}} = \sqrt{\frac{3P}{d}}$$

Hence at constant pressure,

$$u_{rms} \propto \frac{1}{\sqrt{d}}$$

11. (C)
The oxidation number of Cl in ClO^-, ClO_3^-, and Cl^- is + 1, + 5 and – 1 respectively. In the reaction $3ClO^- \rightarrow ClO_3^-, + 2Cl^-$, Cl in ClO^- undergoes oxidation as well as reduction reaction as follows:
oxidation: $ClO^- + 4OH^- \rightarrow ClO_3^-, + 2H_2O + 4e^-$
reduction: $ClO^- + H_2O + 2e^- \rightarrow Cl^- + 2OH^-$
Hence it is a disproportionation reaction.

12. (A)
The larger the oxidation number of Cl atom, the larger is the acidity. Hence the order of acidity is:
$HClO < HClO_2 < HClO_3 < HClO_4$

13. (A)

$$\text{Normality of oxalic acid} = \frac{6.3\,\text{g}/63\,\text{g equiv.}^{-1}}{0.25\,\text{L}} = 0.4N$$

$$N_1V_1 = N_2V_2$$
$$0.1 \times V_1 = 0.4 \times 10$$
or $$V_1 = 40\text{ mL}$$

14. (D)

Number of S—S bonds = 0

The structure of sulphur trioxide trimer (S_3O_9) is:

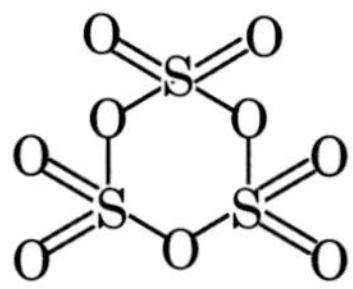

15. (B)

$FeO + SiO_2 \longrightarrow FeSiO_3$ (slag)

16. (A)

All the three species contain 14 electrons. Their molecular electronic configuration is:

$$\left(\sigma_{1s}^{b}\right)^2\left(\sigma_{1s}^{*}\right)^2\left(\sigma_{2s}^{b}\right)^2\left(\sigma_{2s}^{*}\right)^2\left(\pi_{p_x}^{b}\right)^2\left(\pi_{p_y}^{b}\right)^2\left(\pi_{p_z}^{b}\right)^2$$

$$\text{Bond order} = \frac{\text{No. of bonding electrons} - \text{No. of antibonding electrons}}{2} = \frac{10-4}{2} = 3$$

17. (B)

Both assertion and reason are correct and reason is not the correct explanation of assertion.

18. (A)

Both assertion and reason are correct and reason is the correct explanation of assertion.

19. (A)

Both assertion and reason are correct and reason is the correct explanation of assertion.

20. (D)

Assertion is incorrect but reason is correct. Due to protonation of the $-NH_2$ group, the resultant $-NH_3^+$ becomes *meta* director. The reactivity of aniline decreases in concentrated acidic solution due to the protonation of the lone pair of electrons by H^+ ions and the lone pair of electrons is no longer available for resonance. Resonance is responsible for activity of a compound.

21. (C)

Assertion is correct but reason is incorrect. Hydrolysis of $SiCl_4$ is due to the presence of *d*-orbitals.

22. (B)

The reaction undergone by $Cr_2O_7^{2-}$ ion is:

$$Cr_2O_7^{2-} + 14H^+ + 6e^- \longrightarrow 2Cr^{3+} + 7H_2O$$

$$\text{Equivalent weight of } K_2Cr_2O_7 = \frac{\text{Mol. wt.}}{\text{Decrease in O. N.} \times n}$$

where n = number of atoms of Cr.

$$= \frac{\text{Mol. wt.}}{3 \times 2} = \frac{\text{Mol. wt.}}{6}$$

23. (A)

Central atom	*Oxidation state*	*Number of electrons*	*Electronic Configuration*	*No. of d-electrons*
Mn	+ 7	25 – 7 = 18	$1s^22s^22p^63s^23p^6$	0
Co	+ 3	27 – 3 = 24	$1s^22s^22p^63s^23p^63d^6$	6
Fe	+ 3	26 – 3 = 23	$1s^22s^22p^63s^23p^63d^5$	5
Cr	+ 3	24 – 3 = 21	$1s^22s^22p^63s^23p^63d^3$	3

Hence MnO_4^- has no *d*-electron.

24. (B)
Ionisation energy decreases on moving down a group from top to bottom. Hence the order of ionisation energy is: Be > Mg > Ca.

25. (B)
The structures of NH_3, $[PtCl_4]^{2-}$, PCl_5 and BCl_3 are tetrahedral, square planar, trigonal bipyramid and triangular planar respectivily. Hence the hybridisation of the central atoms in the above compounds is sp^3, dsp^2, dsp^3, sp^2 respectively.

26. (A)
$A_pB_q \rightleftharpoons pA^{q+} \;\; + qB^{p-}$
If *s* is the solubility of A_pB_q, then $[A^{q+}] = ps$ and $[B^{p-}] = qs$
$K_{sp} = [A^{q+}]^p\,[B^{p-}]^q = (ps)^p(qs)^q = s^{p+q} \cdot p^p \cdot q^q$

27. (C)
Saturated solution of KNO_3 is used to make salt bridge because the velocities of K^+ and NO_3^- ions are nearly the same. Hence the conductivity in the cathodic and anodic compartments is not affected.

28. (B)
The size of hydrated ions in solution is as follows:
$Li^+ > Na^+ > K^+$
Hence the conductivity of these ions follows the reverse order:
KCl > NaCl > LiCl

29. (B)
Rate $\propto I$

30. (D)
K_p is not dependent on pressure (*P*) and concentration (*x*) of substances.

31. (A)
Formaldehyde is always converted into formic acid.
$C_6H_5CHO + HCHO + NaOH \longrightarrow C_6H_5CH_2OH + HCOONa$
It is a crossed Cannizzaro reaction.

32. (D)
In an S_{N^2} reaction, optical inversion takes place. Hence an S_{N^2} reaction always gives a product with opposite optical rotation and the product is a single stereoisomer of the reactant optically active compound.

33. (B)

An optically inactive alkene is formed.

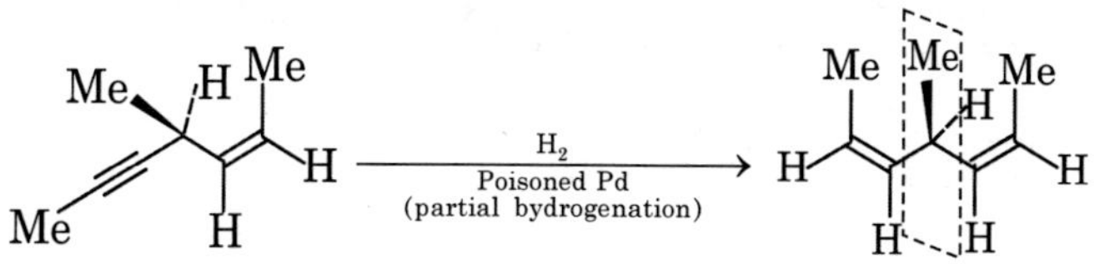

Optically active — Optically inactive

Due to the *cis*-addition of hydrogen to the triple bond, the reduction product has a plane of symmetry and hence is optically inactive.

34. (B)

It is an electrophilic addition reaction (Markownikoff addition).

$$CH_3—CH{=}CH_2 + \overset{+}{Cl} \xrightarrow{\text{slow}} CH_3—\overset{+}{C}H—\underset{\underset{Cl}{|}}{C}H_2 \xrightarrow[\text{fast}]{HO^-} CH_3—\underset{\underset{OH}{|}}{C}H—\underset{\underset{Cl}{|}}{C}H_2$$

35. (C)

Heating of an alcohol with copper causes dehydration.

1-Propanol on dehydration gives propanol while 2-propanol on dehydration gives propanone. Only propanol reacts with Fehling solution.

$$\underset{\text{1-Propanol}}{CH_3—CH_2—CH_2—OH} \xrightarrow{\text{Cu, 300°C}} CH_3—CH_2—CHO \xrightarrow[\text{Fehling solution}]{2CuO} \underset{\text{Red}}{Cu_2O\downarrow}$$

$$\underset{\text{2-Propanol}}{CH_3—\underset{\underset{OH}{|}}{C}H—CH_3} \xrightarrow{\text{Cu, 300°C}} \underset{\text{Propanone}}{CH_3—CO—CH_3} \xrightarrow[\text{Fehling solution}]{2CuO} \text{No reaction}$$

I.I.T. JEE Objective Question Paper—2002

1. How many moles of electron weigh one kilogram?
   (A) 6.023×10^{23} (B) $(1/9.108) \times 10^{31}$
   (C) $(6.023/9.108) \times 10^{54}$ (D) $(1/9.108 \times 6.023) \times 10^{8}$

2. Which of the following volume (V)-temperature (T) plots represents the behaviour of one mole of an ideal gas at one atmospheric pressure?

(A) 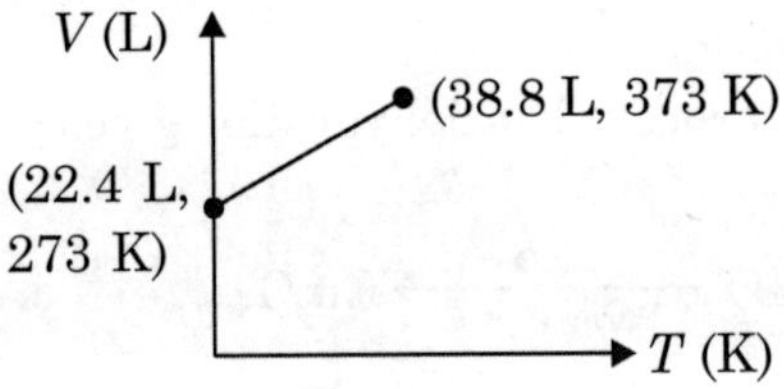

(B) 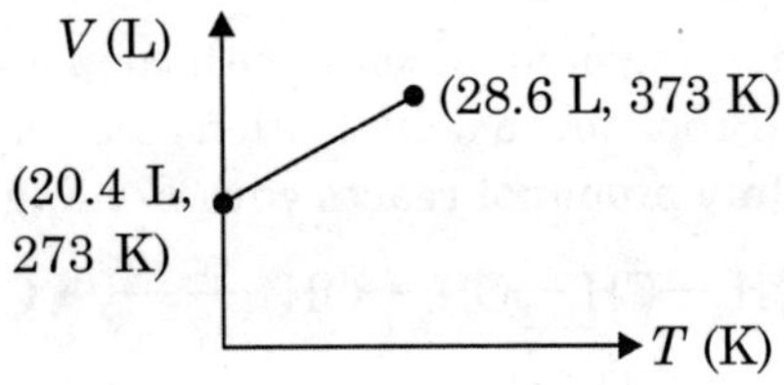

(C) 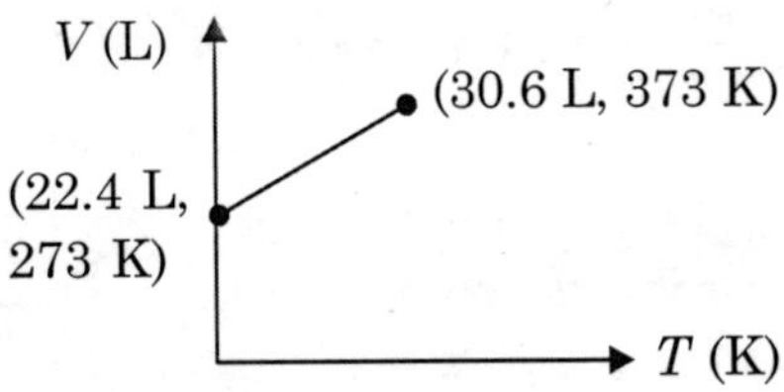

(D) 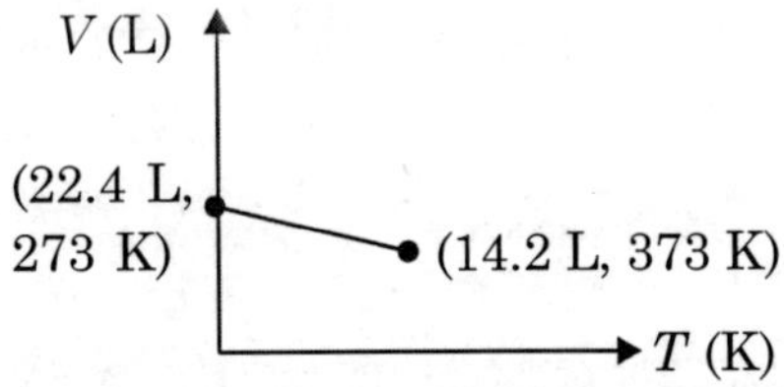

3. A substance A_xB_y crystallises in a face centred cubic (*fcc*) lattice in which atoms A occupy each corner of the cube and atoms B occupy the centres of each face of the cube. Identify the correct composition of the substance A_xB_y.
   (A) AB_3 (B) A_4B_3
   (C) A_3B (D) Composition cannot be specified.

4. When the temperature is increased, surface tension of water
   (A) increases (B) decreases
   (C) remains constant (D) shows irregular behaviour.

5. Rutherford's experiment, which established the nuclear model of the atom, used a beam of
   (A) α-particles, which impinged on a metal foil and got absorbed
   (B) γ-rays, which impinged on a metal foil and ejected electrons
   (C) helium atoms, which impinged in a metal foil and got scattered
   (D) helium nuclei, which impinged on a metal foil and got scattered.

6. If the nitrogen atom had electronic configuration $1s^7$, it would have energy lower than that of the normal ground state configuration $1s^2 2s^2 2p^3$, because the electrons would be closer to the nucleus. Yet $1s^7$ is not observed because it violates
(A) Heisenberg uncertainty principle
(B) Hund's rule
(C) Pauli exclusion principle
(D) Bohr postulate of stationary orbits.

7. Specify the coordination geometry around and hybridisation of N and B atoms in a 1 : 1 complex of BH_3 and NH_3.
(A) N: tetrahedral, sp^3, B: tetrahedral, sp^3
(B) N: pyramidal, sp^3, B: pyramidal, sp^3
(C) N: pyramidal, sp^3, B: planar, sp^2
(D) N: pyramidal, sp^3, B: tetrahedral, sp^3

8. One mole of a non-ideal gas undergoes a change of state (2.0 atm, 3.0 L, 95 K) $\longrightarrow$ (4.0 atm, 5.0 L, 245 K) with a change in internal energy, $\Delta U = 30.0$ L atm. The change in enthalpy (ΔH) of the process in L atm is
(A) 40.0
(B) 42.3
(C) 44.0
(D) not defined, because pressure is not constant.

9. Consider the following equilibrium in a closed container: $N_2O_4(g) \rightleftharpoons 2NO_2(g)$. At a fixed temperature, the volume of the reaction container is halved. For the change, which of the following statements holds true regarding the equilibrium constant (K_p) and degree of dissociation (α)?
(A) neither K_p nor a changes
(B) both K_p and α change
(C) K_p changes, but α does not change
(D) K_p does not change, but α changes.

10. Standard electrode potential data are useful for understanding the suitability of an oxidant in a redox titration. Some half-cell reactions and their standard potentials are given below:

$MnO_4^-(aq) + 8H^+(aq) + 5e^- \longrightarrow Mn^{2+}(aq) + 4H_2O(l)$ $\quad E^0 = 1.51$ V

$Cr_2O_7^{2-}(aq) + 14H^+(aq) + 6e^- \longrightarrow 2Cr^{3+}(aq) + 7H_2O(l)$ $\quad E^0 = 1.38$ V

$Fe^{3+}(aq) + e^- \longrightarrow Fe^{2+}(aq)$ $\quad E^0 = 0.77$ V

$Cl_2(g) + 2e^- \longrightarrow 2Cl^-(aq)$ $\quad E^0 = 1.40$ V

Identify the only incorrect statement regarding the quantitative estimation of aqueous $Fe(NO_3)_2$.
(A) MnO_4^- can be used in aqueous HCl
(B) $Cr_2O_7^{2-}$ can be used in aqueous HCl
(C) MnO_4^- can be used in aqueous H_2SO_4
(D) $Cr_2O_7^{2-}$ can be used in aqueous H_2SO_4

11. Consider the chemical reaction, $N_2(g) + 3H_2(g) \longrightarrow 2NH_3(g)$. The rate of this reaction can be expressed in terms of time derivative of concentration of $N_2(g)$, $H_2(g)$ or $NH_3(g)$. Identify the correct relationship amongst the rate expressions.
(A) Rate $= -d[N_2]/dt = -1/3d[H_2]/dt = 1/2d[NH_3]/dt$
(B) Rate $= -d[N_2]/dt = -3d[H_2]/dt = 2d[NH_3]/dt$
(C) Rate $= d[N_2]/dt = 1/3d[H_2]/dt = 1/2d[NH_3]/dt$
(D) Rate $= -d[N_2]/dt = -d[H_2]/dt = d[NH_3]/dt$

12. Anhydrous ferric chloride is prepared by
(A) heating hydrated ferric chloride at a high temperature in a stream of air
(B) heating metallic iron in a stream of dry chlorine gas
(C) reaction of metallic iron with hydrochloric acid (*aq*)
(D) reaction of metallic iron with hydrochloric acid (*g*).

13. Identify the correct order of solubility of Na_2S, CuS and ZnS in aqueous medium.
(A) CuS > ZnS > Na_2S
(B) ZnS > Na_2S > CuS
(C) Na_2S > CuS > ZnS
(D) Na_2S > ZnS > CuS

14. Which of the following processes is used in the extractive metallurgy of magnesium?
(A) fused salt electrolysis
(B) self-reduction
(C) aqueous solution electrolysis
(D) thermite reduction.

15. An aqueous solution of a substance gives a white precipitate on treatment with dilute hydrochloric acid, which dissolves on heating. When hydrogen sulfide is passed through the hot acidic solution, a black precipitate is obtained. The substance is a
(A) Hg_2^{2+} salt
(B) Cu^{2+} salt
(C) Ag^+ salt
(D) Pb^{2+} salt

16. A gas X is passed through water to form a saturated solution. The aqueous solution on treatment with silver nitrate gives a white precipitate. The saturated aqueous solution also dissolves magnesium ribbon with evolution of a colourless gas Y. Identify X and Y.
(A) X = CO_2, Y = O_2
(B) X = Cl_2, Y = CO_2
(C) X = Cl_2, Y = H_2
(D) X = H_2, Y = Cl_2

17. Polyphosphates are used as water softening agents because they
(A) form soluble complexes with anionic species
(B) precipitate anionic species
(C) form soluble complexes with cationic species
(D) precipitate cationic species.

18. Identify the correct order of acidic strengths of CO_2, CuO, CaO, H_2O.
(A) CaO < CuO < H_2O < CO_2
(B) H_2O < CuO < CaO < CO_2
(C) CaO < H_2O < CuO < CO_2
(D) H_2O < CO_2 < CaO < CuO

19. Identify the least stable ion amongst the following.
(A) Li^-
(B) Be^-
(C) B^-
(D) C^-

20. Which of the following molecular species has unpaired electron(s)?
(A) N_2
(B) F_2
(C) O_2^-
(D) O_2^{2-}

21. The nodal plane in the π-bond of ethene is located in
(A) the molecular plane
(B) a plane parallel to the molecular plane
(C) a plane perpendicular to the molecular plane which bisects the carbon-carbon σ-bond at right angle
(D) a plane perpendicular to the molecular plane which contains the carbon-carbon σ-bond.

22. Which of the following compounds exhibits stereoisomerism?
(A) 2-methylbut-1-ene
(B) 3-methylbut-1-yne
(C) 3-methylbutanoic acid
(D) 2-methylbutanoic acid

23. Which of the following acids has the smallest dissociation constant?
(A) $CH_3CHFCOOH$ (B) FCH_2CH_2COOH (C) $BrCH_2CH_2COOH$ (D) $CH_3CHBrCOOH$

24. Consider the reaction: $H_3C-CH(D)-CH(CH_3)-CH_3 + Br\cdot \rightarrow X + HBr$. Identify the structure of the major product X.

(A) $H_3C-CH(D)-CH(CH_3)-\dot{C}H_2$
(B) $H_3C-CH(D)-\dot{C}(CH_3)-CH_3$
(C) $H_3C-\dot{C}(D)-CH(CH_3)-CH_3$
(D) $H_3C-\dot{C}H-CH(CH_3)-CH_3$

25. Identify the reagent from the following list which can easily distinguish between but-1-yne and but-2-yne.
(A) bromine, CCl_4
(B) H_2, Lindlar catalyst
(C) dilute H_2SO_4, $HgSO_4$
(D) ammoniacal Cu_2Cl_2 solution.

26. Identify the correct order of boiling points of the following compounds:
$CH_3CH_2CH_2OH$ (1) $CH_3CH_2CH_2CHO$ (2) $CH_3CH_2CH_2COOH$ (3)
(A) $1 > 2 > 3$ (B) $3 > 1 > 2$ (C) $1 > 3 > 2$ (D) $3 > 2 > 1$

27. Identify the set of reagent/reaction conditions X and Y in the following set of transformations:
$CH_3-CH_2-CH_2Br \xrightarrow{X} \text{Product} \xrightarrow{Y} CH_3-CH(Br)-CH_3$
(A) X = dilute aqueous NaOH, 20°C; Y = HBr/acetic acid, 20°C
(B) X = concentrated alcoholic NaOH, 80°C; Y = HBr/acetic acid, 20°C
(C) X = dilute aqueous NaOH, 20°C; Y = $Br_2/CHCl_3$, 0°C
(D) X = concentrated alcoholic NaOH, 80°C; Y = $Br_2/CHCl_3$, 0°C

28. Identify the correct order of reactivity in electrophilic substitution reactions of the following compounds.

(1) 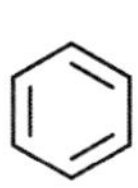
(2) 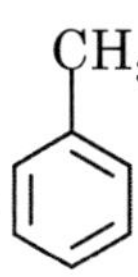

(3) 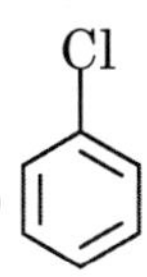

(4) 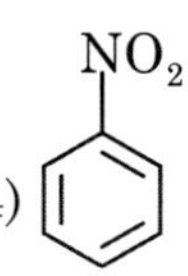

(A) $1 > 2 > 3 > 4$ (B) $4 > 3 > 2 > 1$ (C) $2 > 1 > 3 > 4$ (D) $2 > 3 > 1 > 4$

29. Compound *A* (molecular formula C_3H_8O) is treated with acidified potassium dichromate to form a product *B* (molecular formula C_3H_6O). *B* forms a shining silver mirror on warming with ammoniacal silver nitrate. *B* when treated with an aqueous solution of $H_2NCONHNH_2.HCl$ and sodium acetate gives a product *C*. Identify the structure of *C*.
(A) $CH_3CH_2CH{=}NNHCONH_2$
(B) $CH_3-C(CH_3){=}NNHCONH_2$
(C) $CH_3-C(CH_3){=}NCONHNH_2$
(D) $CH_3CH_2CH{=}NCONHNH_2$

30. Which of the following hydrocarbons has the lowest dipole moment?

(A) 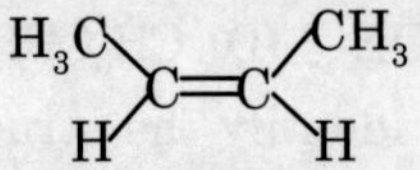

(B) $CH_3C{\equiv}CCH_3$

(C) $CH_3CH_2C{\equiv}CH$

(D) $CH_2{=}CH{-}C{\equiv}CH$

ANSWER KEYS

1. D	2. C	3. A	4. B	5. D	6. C
7. A	8. D	9. D	10. A	11. A	12. B
13. D	14. A	15. D	16. C	17. C	18. A
19. B	20. C	21. A	22. D	23. C	24. B
25. D	26. B	27. B	28. C	29. A	30. B

SOLUTIONS

1. (D)

Mass of electron = 9.108×10^{31} kg

9.108×10^{31} kg of electron contain $= \dfrac{1}{6.023 \times 10^{23}}$ moles

1 kg of electron contains $= \dfrac{1}{6.023 \times 10^{23} \times 9.108 \times 10^{-31}}$ moles

$= \dfrac{1}{6.023 \times 9.108} \times 10^{8}$ moles

2. (C)

At constant pressure

$$\frac{V_1}{T_1} = \frac{V_2}{T_2} \quad \Rightarrow \quad \frac{224}{273} = \frac{V_2}{373}$$

or $$V_2 = \frac{224 \times 373}{273} = 20.6 \text{ L}$$

3. (A)

A atoms occupy eight corners of the face centred cube. Hence the number of A atoms in the unit cell = $8 \times \dfrac{1}{8} = 1$. B atoms occupy the centres of six faces of the cube. Hence the number of B atoms sharing the unit cell = $6 \times \dfrac{1}{2} = 3$. Then the composition of the substance A_xB_y is AB_3.

4. (B)

When the temperature is increased the ionic mobility of ions increases. Hence the surface tension increases.

5. (D)

Rutherford's experiment used a beam of α-particles and an alpha particle is represented by doubly charged helium nucleus containing two protons and two neutrons. The scattering of α-particles established the nuclear model of atom according to which an atom consists of positively charged nucleus.

6. (C)

According to the Pauli exclusion principle not more than two electrons can be placed in one and the same ortital and the two electrons in an ortital must have opposite spins. Hence placing of seven electrons in an ortital ($1s^7$) violates the Pauli exclusion principle.

7. (A)

The 1 : 1 complex of BF_3 and NH_3 is represented as

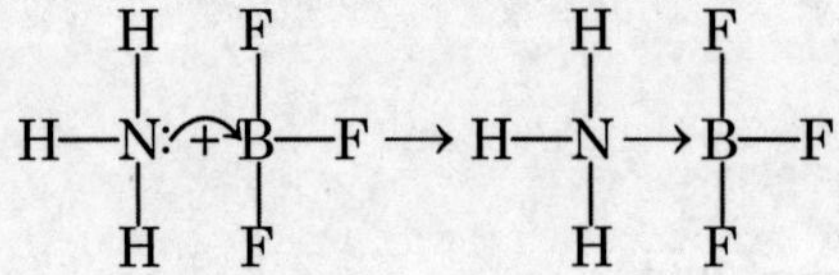

In this coordination compound, both B and N attain tetrahedral geometry and sp^3 hybridisation.

8. (D)

To calculate ΔH, pressure should remain constant.

9. (D)

$$N_2O_4(g) \rightleftharpoons 2NO_2(g)$$

Initial moles	1	0
Moles at equi.	$1-\alpha$	2α

Total number of moles at equilibrium = $1-\alpha+2\alpha = 1+\alpha$

If P is the total pressure at equilibrium,

$$p_{N_2O_4} = \frac{1-\alpha}{1-\alpha}\times P \text{ and } p_{NO_2} = \frac{2\alpha}{1+\alpha}\times P$$

$$K_p = \frac{(p_{NO_2})^2}{p_{N_2O_4}} = \frac{4\alpha^2\times P}{(1-\alpha)(1+\alpha)}$$

If volume is halved, the pressure will be doubled. Since K_p is constant, α will be lowered in order to maintain the constant value of K_p. Hence K_p does not change but α changes.

10. (A)

The ionic equiation for the reaction of MnO_4^- and HCl is as follows:

$2MnO_4^- + 16H^+ + 10Cl^- \longrightarrow 2Mn^{2+} + 8H_2O + 5Cl_2$

The electrochemical cell is represented as

$Pt|Cl_2(g)$ (1 atm) $Pt|Cl^-(aq)||MnO_4^-(aq)|Mn^{2+}(aq)$

$$E^0_{cell} = E^0_{cathode} - E^0_{anode} = 1.51\text{ V} - 1.4\text{ V} = 0.11\text{ V}$$

Since E^0_{cell} is positive, ΔH^0 is negative. Hence the above cell reation is feasible. But MnO_4^- is an oxidising agent and it can ozidise Fe^{2+} to Fe^{3+} and Cl^- to Cl_2. Hence MnO_4^-/HCl is not a suitable reagent for quantitative estimation of $Fe(NO_3)_2$.

11. (A)

For the reaction,

$N_2(g) + 3H_2(g) \longrightarrow 2NH_3(g)$

the rate of reaction is expressed as

$$\text{rate} = -\frac{d[N_2]}{dt} = -\frac{1}{3}\frac{d[H_2]}{dt} = \frac{1}{2}\frac{d[NH_3]}{dt}$$

The negative sign is due to the decrease in concentration of N_2 and H_2 with time and the positive rign is due to the increase in the concentration with time.

12. (B)

Anhydrous iron(III) chloride is prepared by heating metallic iron in a stream of dry chlorine gas.

$2Fe + 3Cl_2 \xrightarrow{heat} 2FeCl_3$

13. (D)

Sodium sulphide is the strongest electrolyte among Na_2S, ZnS and CuS, and sodium being an alkali metal, Na_2S is highly soluble in water. The solubility product of ZnS and CuS is in the order: CuS < ZnS. Hence CuS is less soluble than ZnS. Thus the order of solubility is: Na_2S > ZnS > CuS.

14. (A)

Magnesium being a reactive metal and its position being higher in the electrochemical series, it is extracted by the electrolysis of fused oxide or fused anhydrous chloride.

$MgCl_2 \longrightarrow Mg^{2+} + 2Cl^-$

At anode: $2Cl^- \longrightarrow Cl_2 + 2e^-$

At cathode: $M^{2+} + 2e^- \longrightarrow Mg$

15. (D)

$Pb^{2+} + HCl\ (dil.) \longrightarrow \underset{White}{PbCl_2\downarrow} + 2H^+$

The white precipitate of $PbCl_2$ dissolves on heating. When H_2S is passed through the hot acidic solution of $PbCl_2$, a black precipitate of PbS is formed.

$PbCl_2 + H_2S \xrightarrow[medium]{Acidic} \underset{Black}{PbS\downarrow} + 2HCl.$

Hence the substance is a Pb^{2+} salt.

16. (C)

$\underset{X}{Cl_2} + H_2O \longrightarrow HCl + HOCl$

$HCl + AgNO_3 \longrightarrow \underset{White}{AgCl\downarrow} + HNO_3$

$2HCl + Mg \longrightarrow MgCl_2\downarrow + H_2$

Hence X = Cl_2, Y = H_2

17. (C)

Polyphosphates are used as water softeners because they form soluble complexes with cationic species (Ca^{2+}, Mg^{2+}) present in hard water.

$\underset{\text{Sodium polyhexametasphosphate}}{Na_2[Na_4(PO_3)_6]} + CaCl_2 \longrightarrow \underset{\text{Soluble complex}}{Na_2[CaNa_2)PO_3]_6} + 2NaCl$

18. (A)

The acidic/basic properties of CO_2, CuO, CaO, H_2O are as follows: CO_2—acidic, CuO—weakly basic, CaO—strongly basic, H_2O—neutral. CO_2 is an anhydride of carbonic acid (H_2CO_3) and is an acidic oxide. Calcium is a group 2 metal and its oxide (CaO) is strongly basic. Copper is a transition metal (group 11) and its oxide is weakly basic. Hence the order of acidic strength is: CaO < CuO < H_2O < CO_2.

19. (B)

The electronic configuration of the ions are as follows:

Li^- : $1s^22s^2$ Stable due to the presence of completely-filled $2s$ subshell

Be^- : $1s^22s^22p^1$ Least stable due to the presence of only one electron

B^- : $1s^22s^22p^2$ Less stable due to the presence of two electrons in $2p$ subshell

C^- : $1s^22s^22p^3$ Stable due to the presence of half-filled $2p$ subshell

20. (C)

The species O_2^- has one unpaired electron.

Molecular orbital configuration of O_2^-:

$$(\sigma^b_{1s})^2(\sigma^*_{1s})^2(\sigma^b_{2s})^2(\sigma^*_{2s})^2(\sigma^b_{p_z})^2(\pi^b_{2p_x})^2(\pi^b_{2p_y})^2(\pi^*_{2p_x})^2(\pi^*_{2p_y})^1$$

21. (A)

The carbon atoms in ethane is sp^2 bybridised. The unhybridised $2p_x$ orbital of each carbon atom houses one unpaired eletron. A π-bond is formed by the $2p_x$(C)–$2p_x$(C) overlap as shown below:

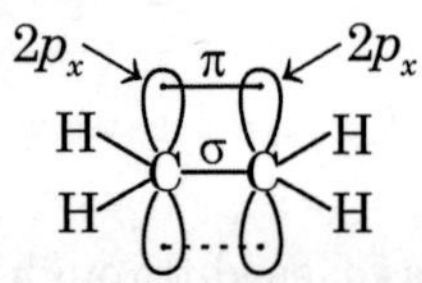

The $2p_x$ orbitals are pendicular to the plane containing ethane molecule. Hence the molecular plane does not contain any π-electron density and the nodal plane lies along the molecular plane.

22. (D)

The structures of the compounds are as follows:

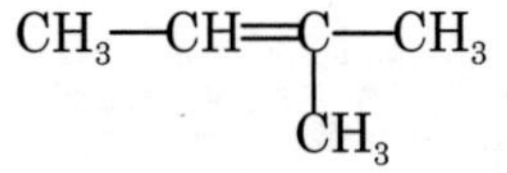

(A)

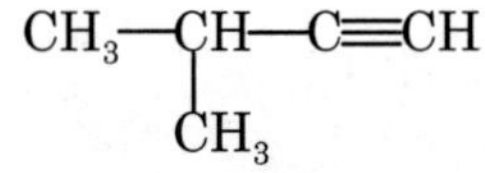

(B)

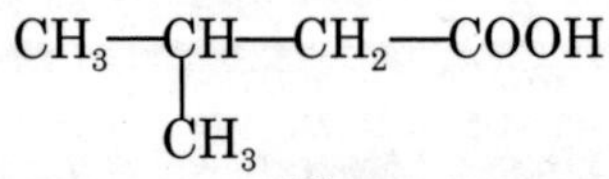

(C)

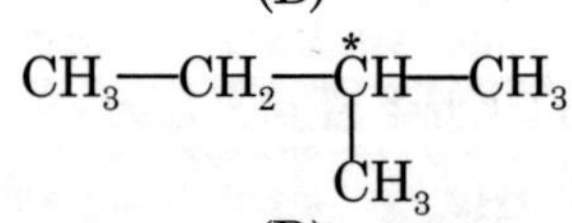

(D)

In D, there is one asymmetric carbon atom. Hence it exhibits the property of stereoisomerism. The compound A is of the type ab$C{=}Ca_2$ and it does not exhibit the property of geometrical isomerism since 180º rotation about C=C produces the same compound.

23. (C)

Acidic strength is proportional to dissociation constant (K_a) of acid. In the given acids, the order of increasing acidic strength is as follows:

$BrCH_2CH_2COOH < FCH_2CH_2COOH < CH_3CHBrCOOH < CH_3CHFCOOH$

Hence the acid $BrCH_2CH_2COOH$ has the smallest dissociation constant.

24. (B)

The tertiary free radical $CH_3-\underset{\displaystyle D}{\underset{|}{CH}}-\underset{\displaystyle CH_3}{\underset{|}{\dot{C}}}-CH_3$ is more stable among all other free radicals because the order of stability of free radicals is: primary < secondary < terticary.

25. (D)

Acidic hydrogen is present in but-1-yne and is absent in but-2-yne. Hence but-1-yne gives a red precipitate with ammoniacal Cu_2Cl_2 solution while but-2-yne does not form such a precipitate.

$$\underset{\text{But-1-yne}}{2CH_3-CH_2-C\equiv CH} + Cu_2Cl_2 + 2NH_4OH \longrightarrow \underset{\text{Red}}{2CH_3-CH_2-C\equiv CCu\downarrow} + 2NH_4Cl + 2H_2O$$

$$CH_3-C\equiv C-CH_3 + Cu_2Cl_2 + 2NH_4OH \longrightarrow \text{No reaction}$$

26. (B)

The compound 3 forms intermolecular hydrogen bonding and exists as a dimer. The compound 1 forms intermolecular hydrogen bonds. The compound 2 is not involved in hydrogen bonding. Hence the decreasing order of molecular mass of these compounds is: 3 > 2 > 1. So they follow the same order in their boiling point: 3 > 2 > 1.

27. (B)

$$CH_3-CH_2-CH_2Br \xrightarrow[80°C]{\text{alcoholic NaOH}} CH_3-CH=CH_2 \xrightarrow[20°C]{HBr/CH_3COOH} CH_3-\underset{\displaystyle Br}{\underset{|}{CH}}-CH_3$$

Hence X = alcoholic NaOH, 80°C and Y = HBr/CH_3COOH, 20°C.

28. (C)

The $-NO_2$ group is an electron-withdrawing group and hence it deactivates the benzene ring: The —Cl group also deactivates the benzene ring but its deactivation of the benzene ring is less than that of the $-NO_2$ group. The $-CH_3$ group is an electron-releasing group and hence it activates the benzene ring. Thus on the basis of activation and deactivation of the benzne ring, the order of reactivity in electrophilic substitution reaction is: 2 > 1 > 3 > 4.

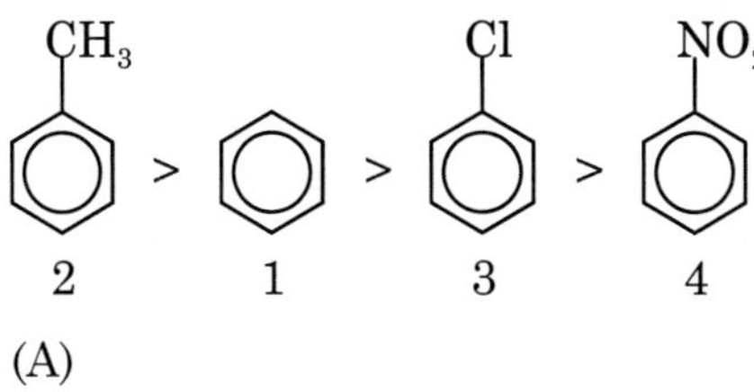

2 1 3 4

29. (A)

Since compound B forms a shining mirror on warming with ammoniacal silver nitrate, B is an aldehyde. Since the compound A on oxidation gives an aldehyde B, the compound A is a primary alcohol. Hence B (C_3H_6O) is CH_3CH_2CHO and A (C_3H_8O) is $CH_3CH_2CH_2OH$. The compound B reacts with an aqueous solution of $H_2NCONHNH_2$ and sodium acctate to form C. The compound C is $CH_3CH_2CH{=}NNHCONH_2$. The involved reactions are as follows:

$$\underset{A}{CH_3CH_2CH_2OH} \xrightarrow{K_2Cr_2O_7/H^+} \underset{B}{CH_3CH_2CHO} \xrightarrow[AgNO_3]{Ammoniacal} Ag\downarrow + CH_3CH_2COOH$$

$$\underset{B}{CH_3CH_2CHO} \xrightarrow{H_2NCONHNH_2} \underset{C}{CH_3CH_2CH{=}NNHCONH_2}$$

30. (B)

The dipole moment of $CH_3-C\equiv C-CH_3$ is zero because the molecule has a symmetrical planar structure.

I.I.T. JEE Objective Question Paper—2003

1. Among the following, the molecule with the highest dipole moment is
   (A) CH_3Cl (B) CH_2Cl_2 (C) $CHCl_3$ (D) CCl_4

2. Which of the following are isoelectronic and isostructural?
   NO_3^-, CO_3^{2-}, ClO_3^-, SO_3
   (A) NO_3^-, CO_3^{2-} (B) SO_3, NO_3^- (C) ClO_3^-, CO_3^{2-} (D) CO_3^{2-}, SO_3

3. In the reaction, 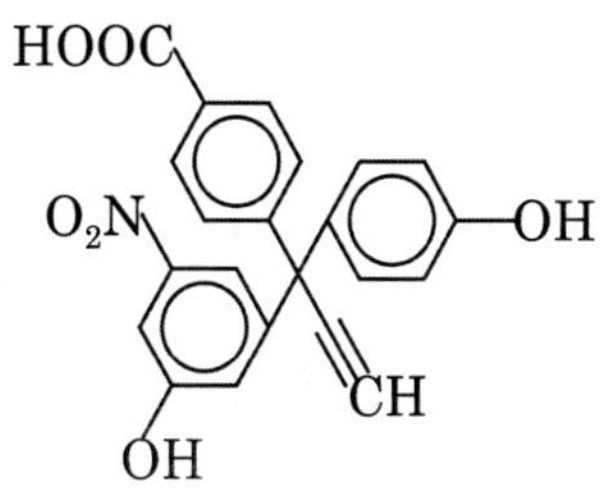 $\xrightarrow{\text{2 moles NaNH}_2}$ A, the product is

(A) 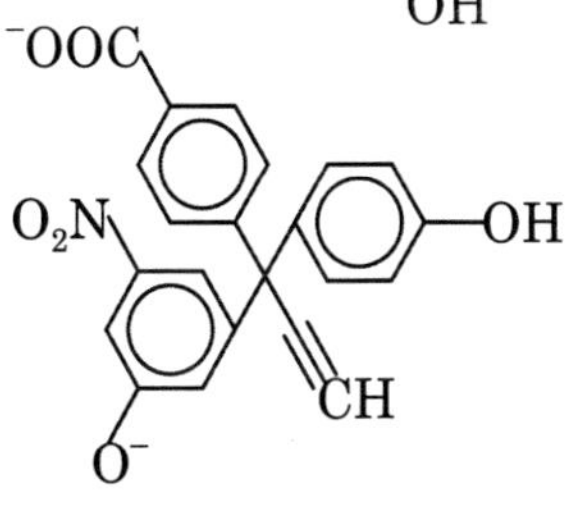

(B) 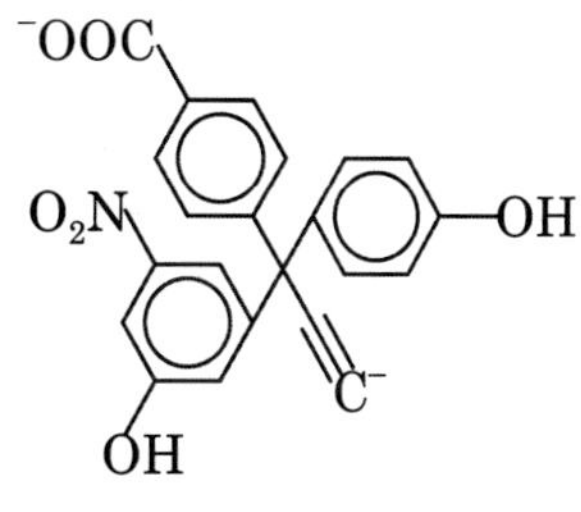

(C) 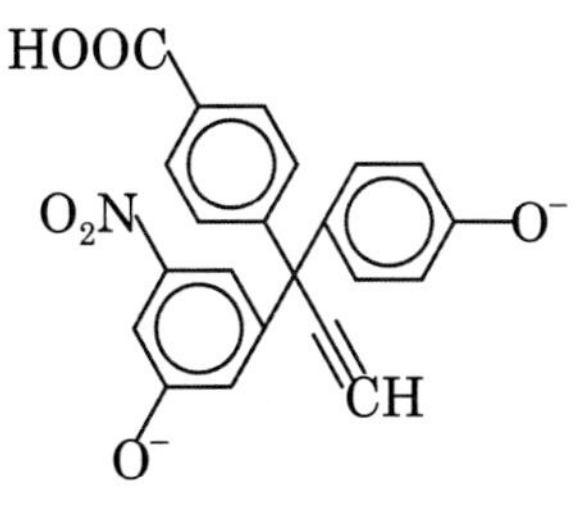

(D) 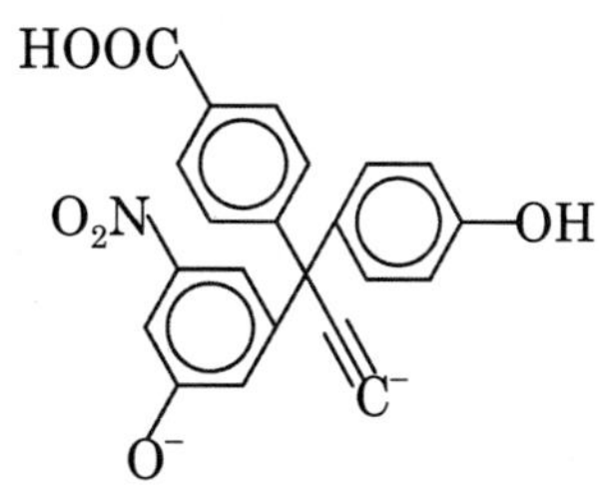

4. [X] + $H_2SO_4 \longrightarrow$ [Y] a colourless gas with irritating smell; [Y] + $K_2Cr_2O_7$ + $H_2SO_4 \longrightarrow$ green solution. [X] and [Y] are
   (A) SO_3^{2-}, SO_2 (B) Cl^-, HCl (C) S^{2-}, H_2S (D) CO_3^{2-}, CO_2

5. For H_3PO_3 and H_3PO_4 the correct choice is
(A) H_3PO_3 is dibasic and reducing
(B) H_3PO_3 is dibasic and non-reducing
(C) H_3PO_4 is tribasic and reducing
(D) H_3PO_3 is tribasic and non-reducing.

6. When MnO_2 is fused with KOH, a coloured compound is formed. The product and colour are
(A) Mn_3O_4, black
(B) $KMnO_4$, purple
(C) Mn_2O_3, brown
(D) K_2MnO_4, purple green

7. Rate of physisorption increases with
(A) decrease in temperature
(B) increase in temperature
(C) decrease in pressure
(D) decrease in surface area.

8. Which of the following represents the given mode of hybridisation *sp*^{2}-*sp*^{2}-*sp*-*sp* from left to right?
(A) $H_2C{=}CH{-}C{\equiv}N$
(B) $HC{\equiv}C{-}C{\equiv}CH$
(C) $H_2C{=}C{=}C{=}CH_2$
(D) $H_2C{=}CH{-}CH{=}CH_2$

9. In the sequence of reactions, 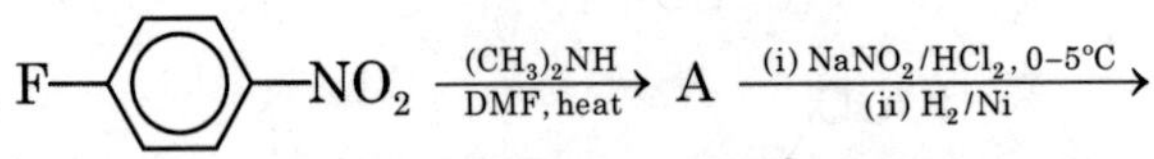 B, B is

(A) 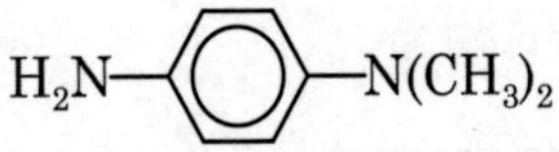
(B) 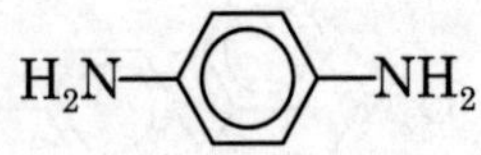
(C) 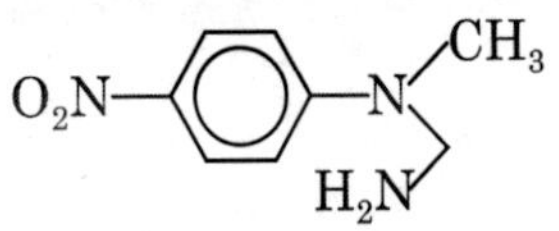
(D) 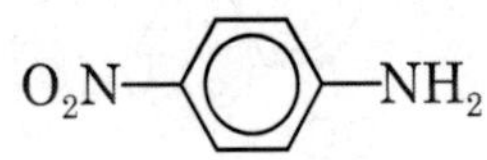

10. The product of acid hydrolysis of P and Q can be distinguished by
P = 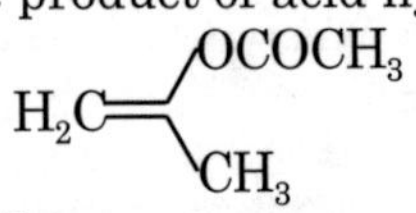
Q = 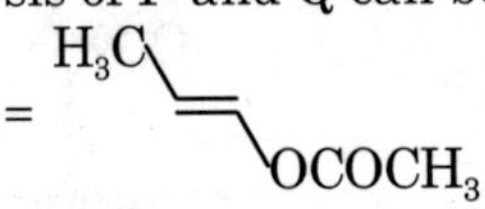
(A) Lucas reagent
(B) 2,4-DNP
(C) Fehling's solution
(D) $NaHSO_3$

11. 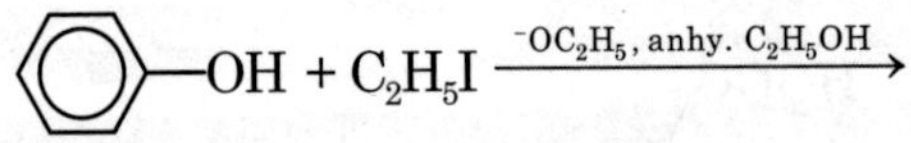
(A) $C_6H_5OC_2H_5$
(B) $C_2H_5OC_2H_5$
(C) $C_6H_5OC_6H_5$
(D) C_6H_5I

12. Which has the maximum number of atoms?
(A) 24 g of C (12)
(B) 56 g of Fe (56)
(C) 27 g of Al (27)
(D) 108 g of Ag (108)

13. In the reaction, 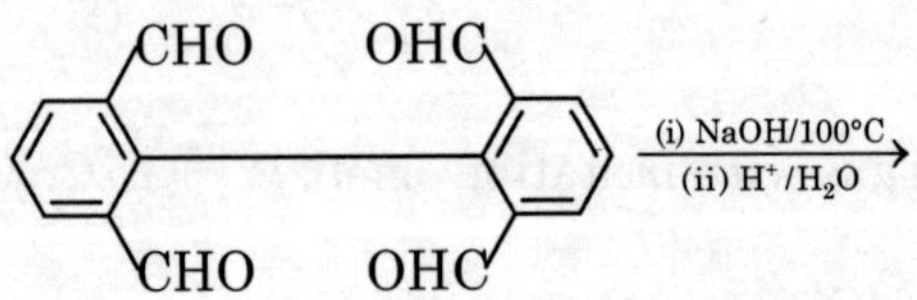 the major product is

(A) 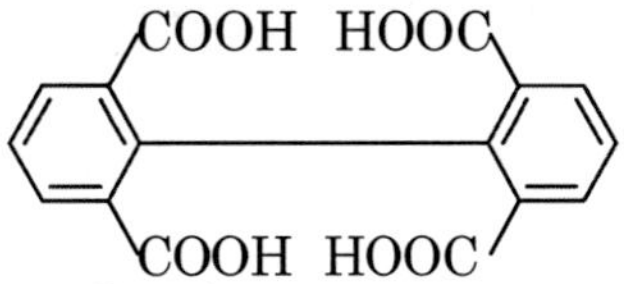

(B) 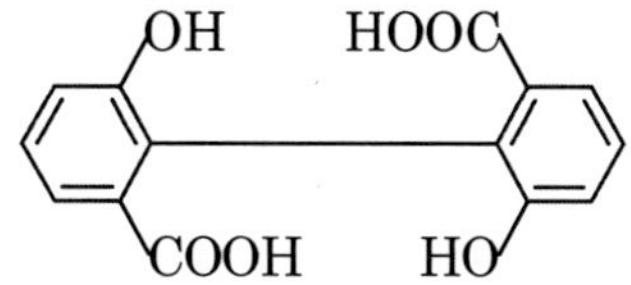

(C) 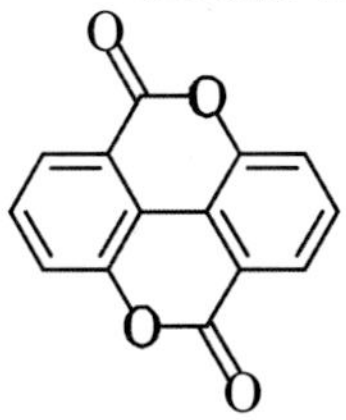

(D) 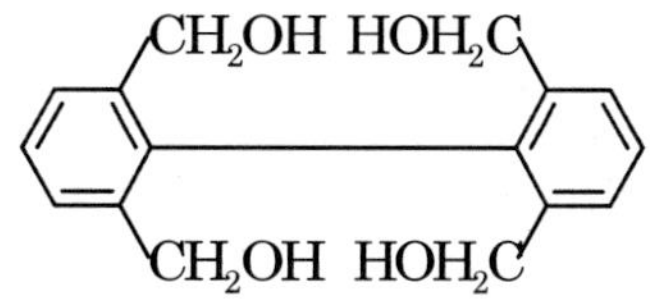

14. In the electrolytic cell, the flow of electrons is from
(A) cathode to anode in solution
(B) cathode to anode through external supply
(C) cathode to anode through internal supply
(D) anode to cathode through internal supply.

15. In a first order reaction, the concentration of reactant decreases from 800 mol/dm^3 to 50 mol/dm^3 in 2×10^4 sec. The rate constant of reaction in sec^{-1} is
(A) 2×10^4 (B) 3.45×10^{-5} (C) 1.386×10^{-4} (D) 2×10^{-4}

16. During depression of freezing point in a solution, the following are in equilibrium
(A) liquid solvent, solid solvent (B) liquid solvent, solid state
(C) liquid solute, solid solute (D) liquid solute, solid solvent

17. H_3BO_3 is
(A) monobasic and weak Lewis acid (B) monobasic and weak Bronsted acid
(C) monobasic and strong Lewis acid (D) tribasic and weak Bronsted acid.

18. In the reaction, $Ph-C\equiv C-CH_3 \xrightarrow{Hg^{2+}/H^+} A$, A is

(A) 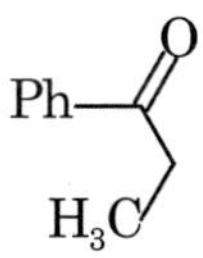 (B) 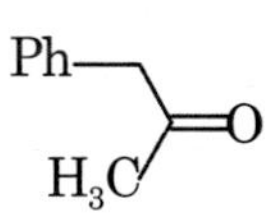 (C) 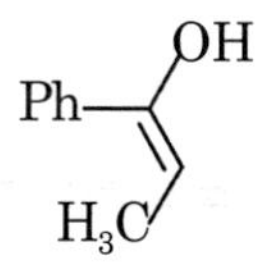 (D) 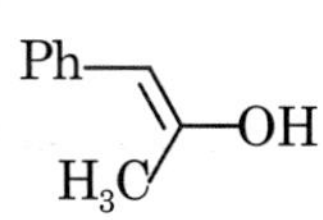

19. In the reaction, ethyl ester $\xrightarrow{\text{excess } CH_3MgBr}$ P, the product P is

(A)  (B) 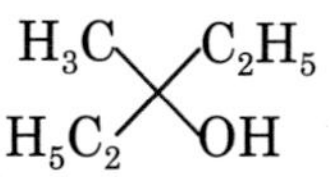 (C) 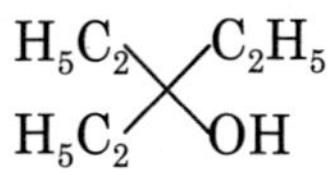 (D) 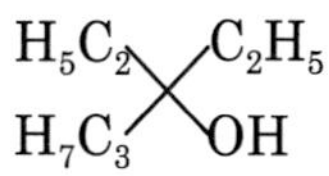

20. Mixture X = 0.02 mol of $[Co(NH_3)_5SO_4]Br$ and 0.02 mol of $[Co(NH_3)_5Br]SO_4$ was prepared in 2 L of solution. 1 L of mixture X + excess $AgNO_3 \longrightarrow$ Y. 1 L of mixture X + excess of $BaCl_2 \longrightarrow$ Z. Number of moles of Y and Z are
(A) 0.01, 0.01 (B) 0.02, 0.01 (C) 0.01, 0.02 (D) 0.02, 0.02

21. Which of the reaction defines ΔH_f^0?
(A) $C_{diamond} + O_2(g) \longrightarrow CO_2(g)$ (B) $1/2H_2(g) + 1/2F_2(g) \longrightarrow HF(g)$
(C) $N_2(g) + 3H_2(g) \longrightarrow 2NH_3(g)$ (D) $CO(g) + 1/2O_2(g) \longrightarrow CO_2(g)$.

22. ^{23}Na is more stable isotope of Na. Find out the process by which $^{24}_{11}Na$ can undergo radioactive decay.
(A) β^- emission (B) α emission (C) β^+ emission (D) K electron capture

23. $(Me)_2SiCl_2$ on hydrolysis will produce
(A) $(Me)_2Si(OH)_2$ (B) $(Me)_2Si{=}O$
(C) $-[O-(Me)_2Si-O]_n-$ (D) $Me_2SiCl(OH)$

24. A solution which is 10^{-3} M in Mn^{2+}, Fe^{2+}, Zn^{2+} and Hg^{2+} is treated with 10^{-16} M sulphide solution. If K_{sp} of MnS, FeS, ZnS and HgS is 10^{-15}, 10^{-23}, 10^{-20} and 10^{-54} respectively, which one will precipitate first?
(A) FeS (B) MnS (C) HgS (D) ZnS

25. In the process of extraction of gold, roasted gold ore + CN^- + $H_2O \xrightarrow{O_2}$ [X] + OH^-. [X] + Zn $\longrightarrow$ [Y] + Au, identify the complexes [X] and [Y].
(A) $[X] = [Au(CN)_2]^-$, $Y = [Zn(CN)_4]^{2-}$ (B) $X = [Au(CN)_4]^{2-}$, $Y = [Zn(CN)_4]^{2-}$
(C) $X = [Au(CN)_2]^-$, $Y = [Zn(CN)_6]^{4-}$ (D) $X = [Au(CN)_4]^-$, $Y = [Zn(CN)_4]^{2-}$

26. Positive deviation from ideal behaviour takes place because of
(A) molecular interaction between atoms and $PV/nRT > 1$
(B) molecular interaction between atoms and $PV/nRT < 1$
(C) Finite size of atoms and $PV/nRT > 1$
(D) Finite size of atoms and $PV/nRT < 1$

27. 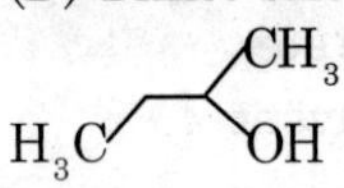 $\xrightarrow{H^+/H_2O}$ [F] $\xrightarrow{Br_2/CCl_4}$ C_4H_8Br (5 such products are possible). How many structures of F are possible?
(A) 2 (B) 5 (C) 6 (D) 3

28. An enantiomerically pure acid is treaterd with racemic mixture of an alcohol having one chiral carbon. The ester formed will be
(A) optically active mixture (B) pure enantiomer
(C) meso compound (D) racemic mixture.

ANSWER KEYS

1. A	2. A	3. A	4. A	5. A	6. D
7. A	8. A	9. A	10. C	11. B	12. A
13. B	14. C	15. C	16. A	17. A	18. A
19. A	20. A	21. B	22. A	23. C	24. C
25. A	26. A	27. D	28. A		

SOLUTIONS

1. (A)

Methane is a symmetrical molecule and hence its dipole moment is zero. The replacement of one of the hydrogen atoms by a chlorine atom increases the dipole moment in CH_3Cl. This is because the bond dipole moment of C—H bond and that of C—Cl bond reinforces one another and the resultant of two C—H dipoles add to the resultant of C—H and C—Cl dipoles.

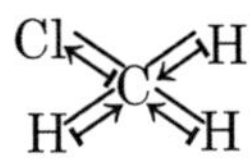

CH_3Cl (μ = 1.86 D)

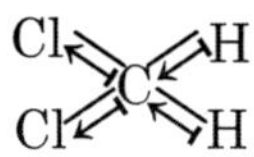

CH_2Cl_2 (μ = 1.62 D)

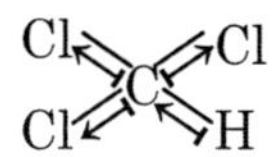

$CHCl_3$ (μ = 1.03 D)

The replacement of another hydrogen atom by a chlorine atom causes the resultant of two C—H dipoles to add to the resultant of two C—Cl dipoles. But due to bond pair-bond pair repulsion between C—Cl bonds, the $\angle$Cl—C—Cl increases leading to reduction of dipole moment of the molecule. In $CHCl_3$, the resultant of C—H and C—Cl dipoles opposes the resultant of two C—Cl dipoles and increase in bond angles is there due to the introduction of the third chlorine atom. On introduction of the fourth chlorine atom, the molecule (CCl_4) again becomes symmetric and dipole moment reduces to zero. Hence CH_3Cl has the maximum dipole moment.

2. (A)

Both NO_3^- and CO_3^{2-} ions have the same number of electrons (32) and the central atom in each is sp^2 hybridised. Hence NO_3^- and CO_3^{2-} ions are isoelectronic and isostructural.

3. (A)

The order of acidic strength of the various groups present in the compound is:

—COOH > —OH (phenolic, deactivated by NO_2 group) > —OH (phenolic) > alkynic H.

Two moles of NH_2^- ions abstract two moles of most acidic hydrogen out of the four moles of hydrogen present in one mole of the compound. This would lead to the formation of A.

4. (A)

Since both SO_2 and H_2S being reducing agents, can turn acidified dischromate solution green, [Y] may be SO_2 or H_2S. SO_2 is produced by the action of acid on sulphites while H_2S is produced by the action of acid in sulphides. But SO_2 has an odour of burning sulphur, which is irritating. H_2S has the odour of rotten egg. Hence [Y] is SO_2.

$$\underset{[X]}{SO_3^{2-}} + H_2S \longrightarrow SO_4^{2-} + H_2O + \underset{[Y]}{SO_2}$$

$$SO_3^{2-} + K_2Cr_2O_7 + H_2SO_4 \longrightarrow K_2SO_4 + \underset{\text{Green solution}}{Cr_2(SO_4)_3} + H_2O$$

5. (A)

The structure of H_3PO_3 indicates that there are two —OH groups and hence it is a dibasic acid. The oxidisation number of P in H_3PO_3 is + 3. We know that phosphorus exhibits the highest oxidation state of + 5. Hence H_3PO_3 can be oxidised and this means that H_3PO_3 is a reducing agent.

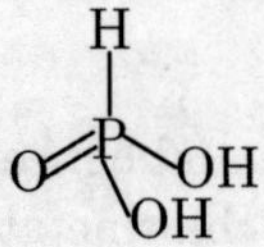

6. (A)

In alkaline medium, the stable oxidation state of manganese is + 6. Hence MnO_2 is oxidised to K_2MnO_4 by atmospheric oxygen in KOH medium.

$2MnO_2 + 4KOH + O_2 \rightarrow 2K_2MnO_4 + 2H_2O$

7. (A)

Physisorption is an exothermic process. Hence according to the Le Chatelier's principle, a decrease in temperature will favour the direction in which heat is evolved. So the rate of physisorption increases with decrease of temperature.

8. (A)

$H_2C{=}CH{-}C{\equiv}N$

sp^2 sp^2 sp sp

9. (A)

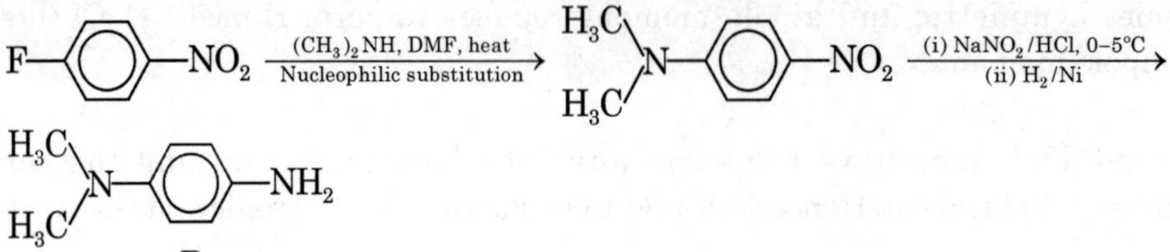

B

10. (C)

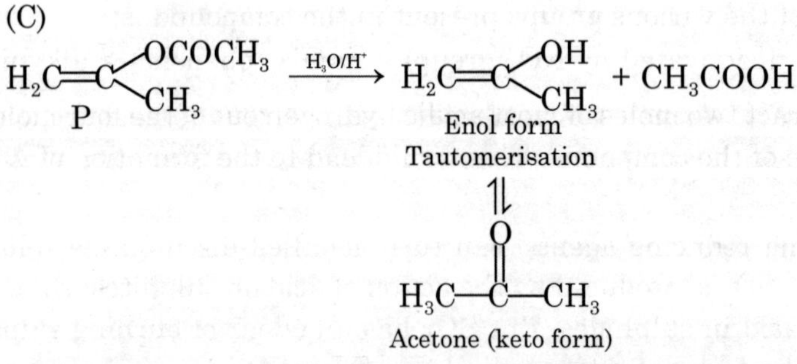

Acetone does not reduce Fehling solution.

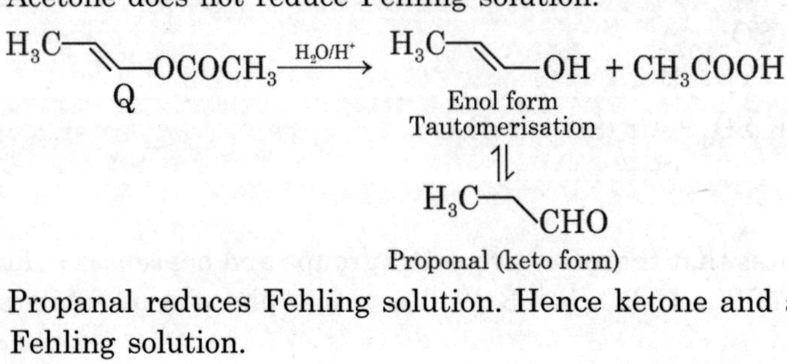

Propanal reduces Fehling solution. Hence ketone and aldehyde can be distinguished by Fehling solution.

11. (B)

$C_2H_5O^-$ abstracts the proton from phenol and converts phenol to phenoxide ion. The phenoxide ion then makes nucleophilic attack on the methylene carbon of iodoethane. But $C_2H_5O^-$ ion is in excess. It is a better nucleophile than $C_6C_5O^-$ ion since in $C_2H_5O^-$ the negative charge is localised over oxygen and in $C_6H_5O^-$ the negative charge is delocalised over the whole molecular framework. Hence the $C_2H_5O^-$ ion makes the nucleophilic attack at iodoethane to give ethoxyethane (Williamson's synthesis).

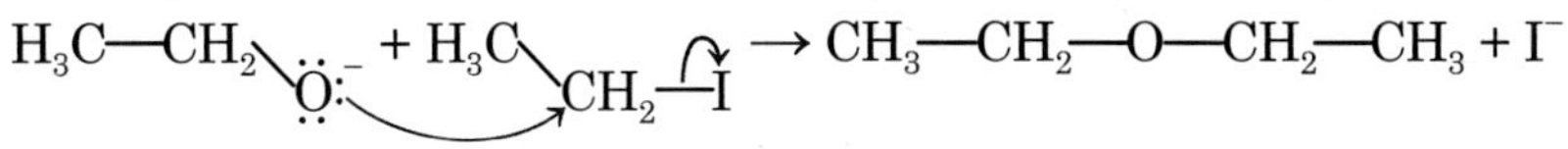

12. (A)

$$\text{Number of atoms in 24 g of C} = \frac{24}{12} \times 6.023 \times 10^{23} = 2 \times 6.023 \times 10^{23}$$

$$\text{Number of atoms in 56 g of Fe} = \frac{56}{56} \times 6.023 \times 10^{23} = 1 \times 6.023 \times 10^{23}$$

$$\text{Number of atoms in 27 g of Al} = \frac{27}{27} \times 6.023 \times 10^{23} = 1 \times 1.6023 \times 10^{23}$$

$$\text{Number of atoms in 108 g of Ag} = \frac{108}{108} \times 6.023 \times 10^{23} = 1 \times 6.023 \times 10^{23}$$

Hence the maximum number of atoms is present in 24 g of carbon.

13. (B)

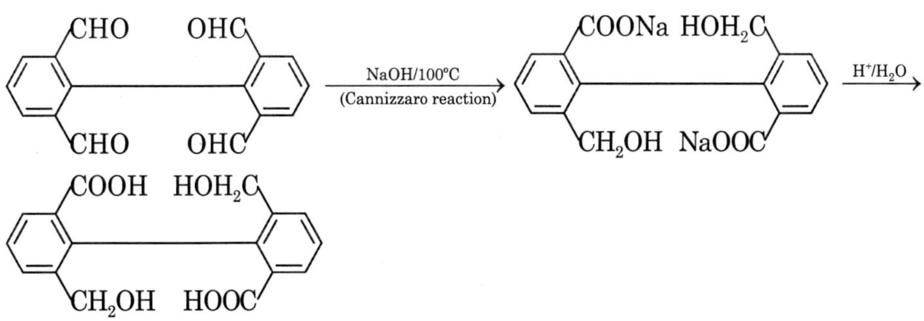

14. (C)

In an electrochemical cell, the flow of electron occurs from cathode to anode through internal supply of cations.

15. (C)

For a first order reaction,

$$k = \frac{2.303}{t}\log\frac{[A]_0}{[A]_t} = \frac{2.303}{2\times10^4}\log\frac{800}{50} = 1.386 \times 10^4\ s^{-1}$$

16. (A)

At the freezing point liquid and solid are in equilibrium. When a solution of a non-volatile solute is cooled to a temperature below the freezing point of a solution, some of liquid solvent separates as a solid solvent. Hence the concentration of solution increases.

17. (A)

The boron atom in H_3BO_3 is electron-deficient with one vacant *p*-orbital. Hence boric acid behaves as Lewis acid. Boron does not have a *d*-orbital to accept electrons from a Lewis base. Hence it can accommodate only one additional pair of electrons in its outermost shell. Thus it behaves as a monobasic weak Lewis acid.

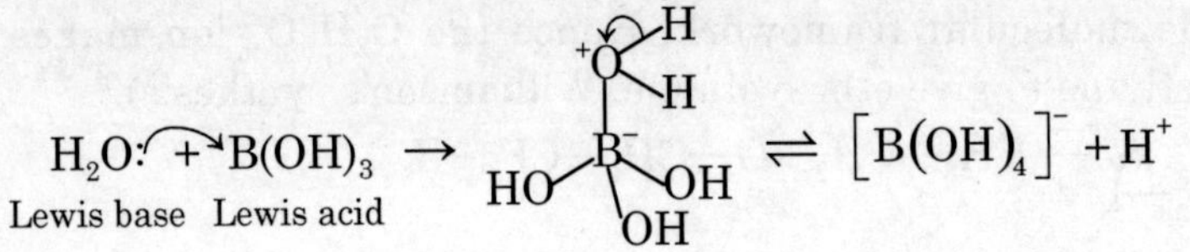

18. (A)

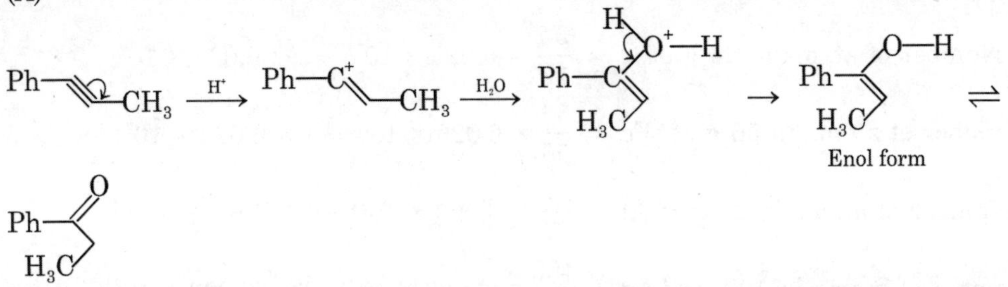

19. (A)

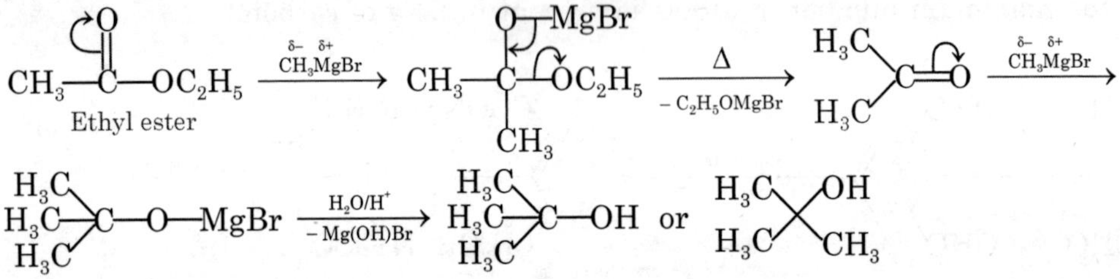

20. (A)

A mixture containing 0.02 mol of $[Co(NH_3)_5SO_4]Br$ and 0.02 mol of $[Co(NH_3)_5Br]SO_4$ was prepared in 2 L of solution. Hence the concentrations of $[Co(NH_3)_5SO_4]Br$ and $[Co(NH_3)_5Br]SO_4$ in solution are 0.01 mol L^{-1} and 0.01 mol L^{-1} respectively. The following reaction occurs when excess of $AgNO_3$ is added to the mixture X:

$$\underset{0.01\ \text{mol L}^{-1}}{[Co(NH_3)_5SO_4]Br} + \underset{\text{Excess}}{AgNO_3} \rightarrow \underset{\text{Y (soluble)}}{[Co(NH_3)_5SO_4]NO_3} + \underset{0.01\ \text{mol L}^{-1}}{AgBr\downarrow}$$

Hence the number of moles of Y = 0.01.

The following reaction occurs when excess of $BaCl_2$ is added to the mixture X.

$$\underset{0.01\ \text{mol L}^{-1}}{[Co(NH_3)_5Br]SO_4} + \underset{\text{Excess}}{BaCl_2} \rightarrow \underset{\text{Z (soluble)}}{[Co(NH_3)_5Br]Cl_2} + \underset{0.01\ \text{mol L}^{-1}}{BaSO_4\downarrow}$$

Hence the number of moles of Z = 0.01.

21. (B)

The standard enthalpy of formation (ΔH_f^0) is the amount of heat absorbed or evolved at constant pressure when 1 mole of a compound is formed from its elements in their standard state at 1 atom and 298 K. The correct equation which defines (ΔH_f^0) is (B). At 298 K and 1 atm, the standard state of carbon is graphite and hence (A) is incorrect. In (C), two moles of $NH_3(g)$ are formed and hence (C) is incorrect. In (D), $CO_2(g)$ gas is formed from a compound $CO(g)$ and hence (D) is incorrect.

22. (A)

The n/p ratio in ^{24}Na is 13/11 which is greater than 1 and hence ^{24}Na is radioactive. In order to achieve stability, ^{24}Na adjusts its n/p ratio to the proper value of unity. It is done by breaking a neutron to a proton and electron $_0^1n \longrightarrow {}_1^1p + {}_{-1}^{0}e$ or β^-. Hence ^{24}Na is changed into ^{23}Na by means of β^- emission and proton emission.

23. (C)

Me_2SiCl_2 on hydrolysis gives an unstable compound $Me_2Si(OH)_2$. Due to very large size of silicon in comparison to oxygen, silicon cannot form π-bond with oxygen atom $(>Si{=}O)$. Hence $Me_2Si(OH)_2$ attains stabilisation through loss of water and polymerisation.

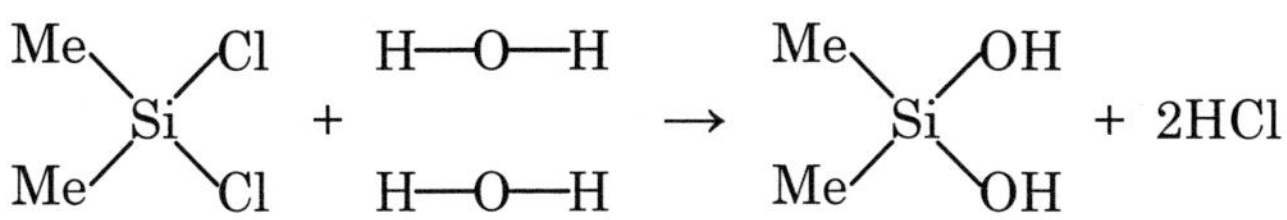

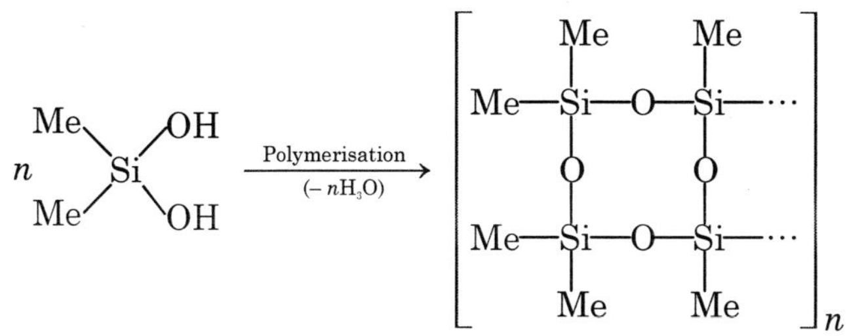

24. (C)

The precipitation of a salt occurs when the ionic product exceeds the solubility product. Hence HgS having the lowest K_{sp} among the given salts will precipitate first.

25. (A)

$2Au + 4CN^- + H_2O + \frac{1}{2}O_2 \longrightarrow 2[Au(CN)_2]^- + 2OH^-$

$2[Au(CN)_2]^- + Zn \longrightarrow [Zn(CN)_4]^{2-} + 2Au$

Hence $X = [Au(CN)_2]^-$ and $Y = [Zn(CN)_4]^{2-}$

26. (A)

For an ideal gas the compressibility factor, $Z = \dfrac{PV}{nRT} = 1$. If a gas deviates from ideal behaviour, $Z > 1$. Hence for positive deviation from ideal behaviour, $Z > 1$.

For positive deviation of gas from ideal behaviour,

$PV = nRT + nPb$

where b is the effective volume of the molecule. The factor nPb is responsible for increasing the value of PV above the ideal value. Hence it is the finite size of the molecules that is responsible for origin of b and positive deviation at high pressure.

27. (D)

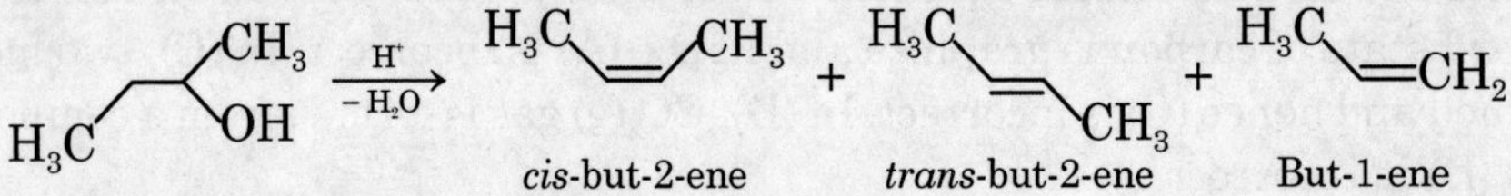

cis-but-2-ene *trans*-but-2-ene But-1-ene

These there alkenes on reaction with Br_2/CCl_4 give addition products having the formula $C_4H_8Br_2$. The five dibromo derivatives are:

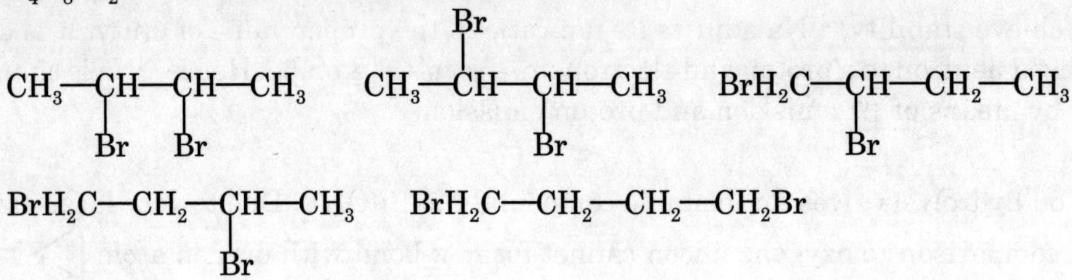

28. (A)

The optically active acid reacts with racemic mixture of alcohol to form two types of isomeric esters. In each the configuration of the chiral centre of acid remains the same. Hence the mixture will be optically active.

I.I.T. JEE Objective Question Paper—2004

1. 2-Phenylpropene on acidic hydration gives

(A) 2-phenylpropan-2-ol (B) 2-phenylpropan-1-ol

(C) 3-phenylpropan-1-ol (D) 1-phenylpropan-2-ol

2. 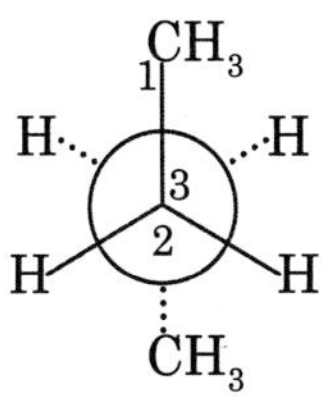

C_2 is rotated anticlockwise 120° about $C_2—C_3$ bond. The resulting conformer is

(A) staggered (B) eclipsed (C) gauche (D) partially eclipsed

3. Benzamide on treatment with $POCl_3$ gives

(A) aniline (B) benzonitrile (C) chlorobenzene (D) benzylamine

4. The methods chiefly used for the extraction of lead and tin from their ores are respectively

(A) self-reduction and carbon reduction (B) self-reduction and electrolytic reduction

(C) carbon reduction and self-reduction (D) cyanide process and carbon reduction.

5. 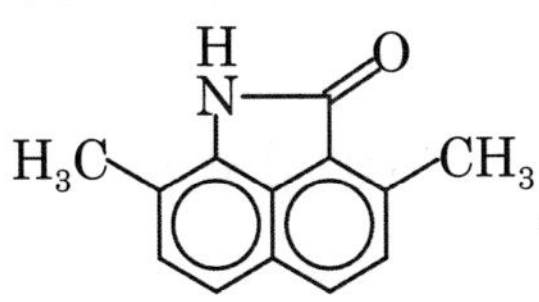 $\xrightarrow{Fe/Br_2}$. The product of monobromination of this compound is

(A) 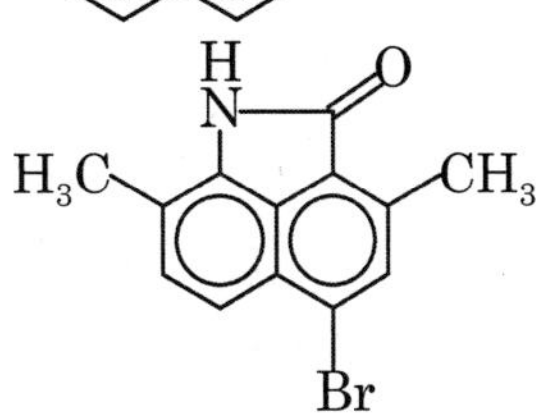

(B) 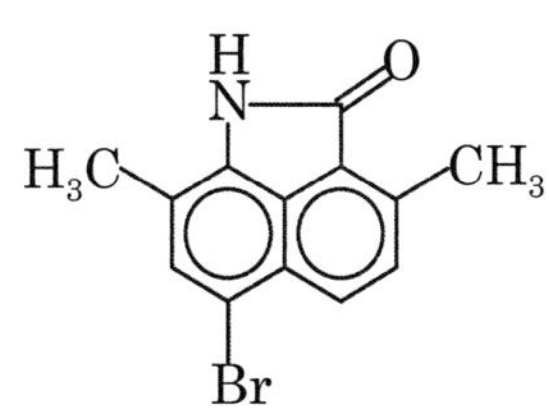

(C) 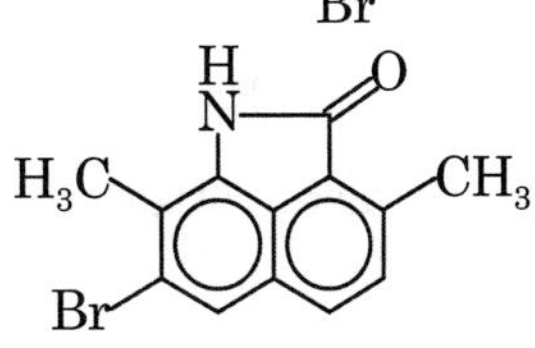

(D) 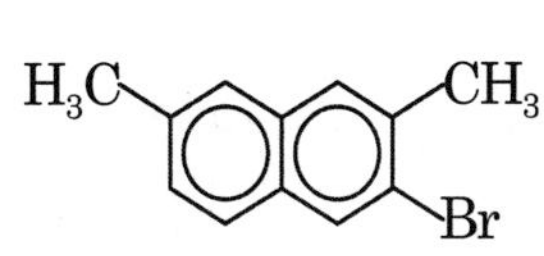

6. The order of reactivity of phenyl magnesium bromide with the following compounds is

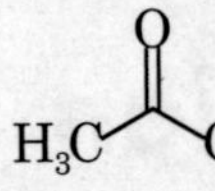
I

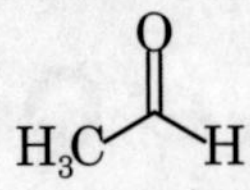
II

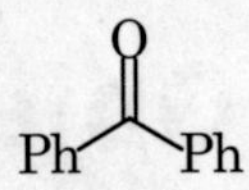
III

(A) II > III > I
(B) I > III > II
(C) II > I > III
(D) All react with same rate.

7. Ammonium dichromate on heating gives a gas which is also given by
(A) heating NH_4NO_2
(B) heating NH_4NO_3
(C) $Mg_3N_2 + H_2O$
(D) Na (comp.) + H_2O_2

8. The emf of the cell $Zn|Zn^{2+}$ (a = 0.1M) || Fe^{2+} (a = 0.01M) | Fe is 0.2905 V. Equilibrium constant for the cell reaction is
(A) $10^{0.32/0.591}$
(B) $10^{0.32/0.0295}$
(C) $10^{0.26/0.0295}$
(D) $e^{0.32/0.295}$

9. HX is a weak acid ($K_a = 10^{-5}$). It forms a salt NaX (0.1M) on reacting with sodium hydroxide. The degree of hydrolysis of NaX is
(A) 0.01%
(B) 0.001%
(C) 0.1%
(D) 0.5%

10. Spontaneous adsorption of a gas in solid surface is an exothermic process because
(A) ΔH increases for system
(B) ΔS increases for gas
(C) ΔS decreases for gas
(D) ΔG increases for gas

11. For a monoatomic gas, kinetic energy = E. The relation of E with rms velocity is
(A) $u = \left(\frac{2E}{m}\right)^{1/2}$
(B) $u = \left(\frac{3E}{2m}\right)^{1/2}$
(C) $u = \left(\frac{E}{2m}\right)^{1/2}$
(D) $u = \left(\frac{E}{3m}\right)^{1/2}$

12. The pair of compounds having metals in their highest oxidation state is
(A) MnO_2, $FeCl_3$
(B) MnO_4^-, CrO_2Cl_2
(C) $[Fe(CN)_6]^{3-}$, $[Co(CN)_3]$
(D) $[NiCl_4]^{2-}$, $[CoCl_4]^-$

13. The compound with tetrahedral geometry is
(A) $[Ni(CN)_4]^{2-}$
(B) $[Pd(CN)_4]^{2-}$
(C) $[PdCl_4]^{2-}$
(D) $[NiCl_4]^{2-}$

14. Spin-only magnetic moment of the compound $Hg[Co(NCS)_4]$ is
(A) $\sqrt{3}$
(B) $\sqrt{15}$
(C) $\sqrt{24}$
(D) $\sqrt{8}$

15. A sodium salt of an unknown anion when heated with $MgCl_2$ gives a white precipitate only on boiling. The anion is
(A) SO_4^{2-}
(B) HCO_3^-
(C) CO_3^{2-}
(D) NO_3^-

16. Which hydrogen-like species will have the same radius as that of Bohr orbit of hydrogen atom?
(A) $n = 2$, Li^{2+}
(B) $n = 2$, Be^{3+}
(C) $n = 2$, He^+
(D) $n = 3$, Li^{2+}

17. In 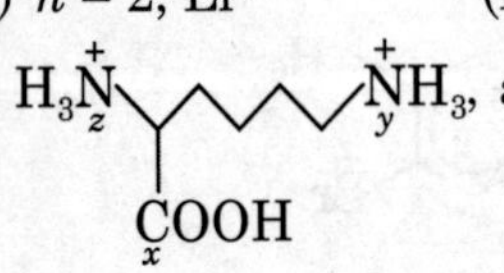, arrange x, y, z in the order of increasing acidic strength.
(A) $x > z > y$
(B) $z < x > y$
(C) $x > y > z$
(D) $z > x > y$

18. $0.004M$ Na_2SO_4 is isotonic with $0.01M$ glucose. Degree of dissociation of Na_2SO_4 is
(A) 75% (B) 50% (C) 25% (D) 85%

19. $\Delta H_{vap} = 30$ kJ mol^{-1} and $\Delta S_{vap} = 75$ J mol^{-1} K^{-1}. Find the temperature of vapour at 1 atmosphere.
(A) 400 K (B) 350 K (C) 298 K (D) 250 K

20. 2 mol of an ideal gas expanded isothermally and reversibly from 1 L to 10 L at 300 K. What is the enthalpy change?
(A) 4.98 kJ (B) 11.47 kJ (C) – 11.47 kJ (D) 0 kJ

21. (A) follows the first order reaction, (A) ⟶ Product. The concentration of A changes from $0.1M$ to $0.025M$ in 40 min. Find the rate of reaction of A when [A] = $0.01M$.
(A) 3.47×10^{-4} M min^{-1} (B) 3.47×10^{-5} M min^{-1}
(C) 1.73×10^{-4} M min^{-1} (D) 1.73×10^{-5} M min^{-1}

22. Hex-2-yne gives *trans*-hex-2-ene with
(A) Li/NH_3 (B) $Pd/BaSO_4$ (C) $LiAlH_4$ (D) Pt/H_2

23. How many chiral compounds are possible on mono chlorination of 2-methylbutane?
(A) 2 (B) 4 (C) 6 (D) 8

24. Which of the following pairs give positive Tollen's test?
(A) Glucose, sucrose (B) glucose, fructose
(C) Hexanal, acctophenone (D) Fructose, sucrose

25. Which of the following has —O—O— linkage?
(A) $H_2S_2O_6$ (B) $H_2S_2O_8$ (C) $H_2S_2O_3$ (D) $H_2S_4O_6$

26. When I^- ion is oxidised by MnO_4^- ion in alkaline medium, I^- ion converts to
(A) IO_3^- (B) I_2 (C) IO_4^- (D) IO^-

27. Number of lone pair(s) in $XeOF_4$ is
(A) 0 (B) 1 (C) 2 (D) 3

28. According to MO theory,
(A) O_2^+ is paramagnetic and bond order is greater than O_2
(B) O_2^+ is paramagnetic and bond order is less than O_2
(C) O_2^+ is diamagnetic and bond order is less than O_2
(D) O_2^+ is diamagnetic and bond order is more than O_2

ANSWER KEYS

1. A	2. C	3. B	4. A	5. B	6. C
7. A	8. B	9. A	10. C	11. A	12. B
13. D	14. B	15. B	16. B	17. A	18. A
19. A	20. D	21. A	22. A	23. B	24. B
25. B	26. A	27. B	28. A		

SOLUTIONS

1. (A)

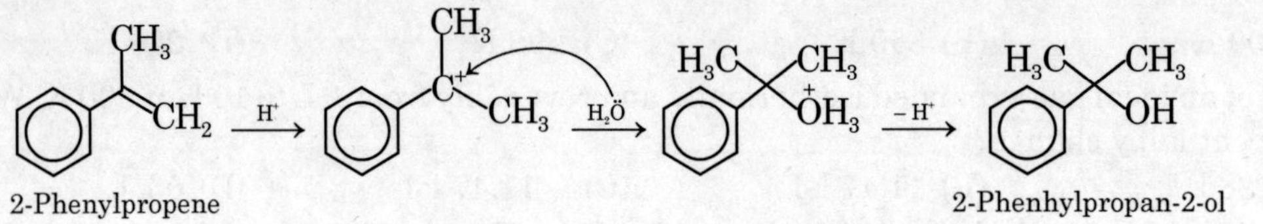

2. (C)

When C_2 is rotated anticlockwise 120° about C_2—C_3 bond, the resulting conformer is a gauche conformer.

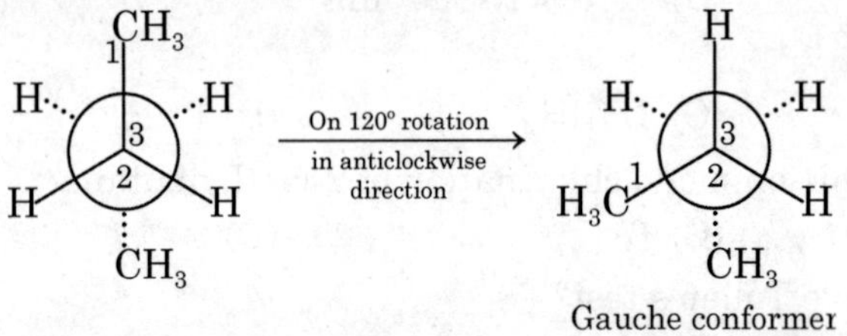

3. (B)

Benzamide on treatment with $POCl_3$ gives benzonitrile because dehydration occurs since $POCl_3$ is a dehydrating agent.

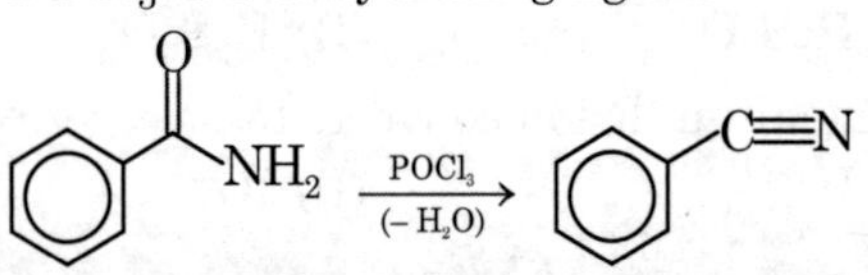

4. (A)

The methods chiefly used for the extraction of lead and tin from their ores are respectively self-reduction and carbon reduction.

(a) $2PbS + 3O_2 \longrightarrow 2PbO + 2SO_2$

$PbS + 2O_2 \longrightarrow PbSO_4$

$2PbO + PbS \longrightarrow 3Pb + SO_2$

$PbS + PbSO_4 \longrightarrow 2Pb + 2SO_2$

(b) $SnO_2 + 2C \longrightarrow Sn + 2CO$

5. (B)

The ring with maximum electron density will be substituted by an electrophile. The ring attached to —NH— will have rich electron density due to resonance. Since the *ortho* position is blocked, the electrophile attacks the *para* position.

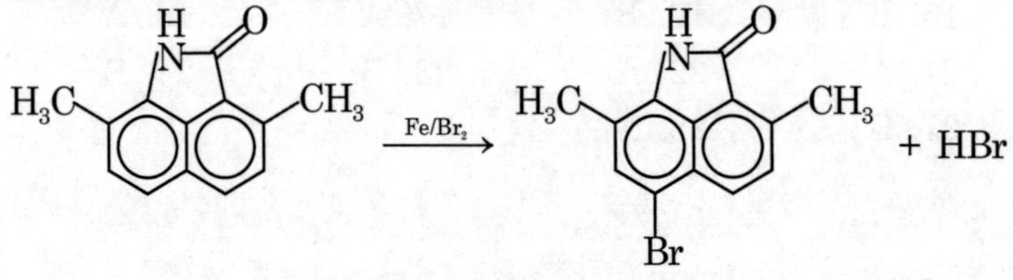

6. (C)

A nucleophile attacks the most electrophilic site first. Among aldehyde and ketone, the aldehyde is more electrophilic since in the ketone the $\delta+$ charge on carbonyl carbon is decreased by the + I-effect of both the alkyl groups.

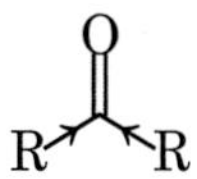
Ketone

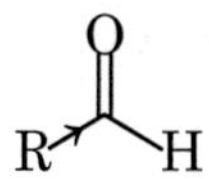
Aldehyde

Moreover, in the tetrahedral intermediate, the aldehyde has less steric repulsion than that in ketone. This facilitates the formation of intermediate in the aldehyde. Hence the order of reactivity is: II > I > III.

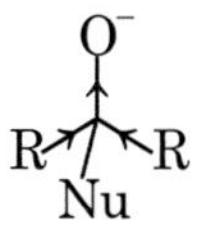
Ketone

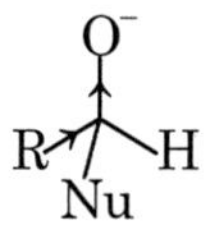
Aldehyde

7. (A)

Ammonium dichromate on heating gives a gas which is also given by heating ammonium nitrite.

$$(NH_4)_2Cr_2O_7 \xrightarrow{\Delta} Cr_2O_3 + 4H_2O + N_2\uparrow$$

$$NH_4NO_2 \xrightarrow{\Delta} 2H_2O + N_2\uparrow$$

8. (B)

For the cell Zn|Zn^{2+} (a = 0.1M)||Fe^{2+} (a = 0.01M)|Fe, the cell reaction is

$$Zn(s) \longrightarrow Zn^{2+}(aq) + 2e^-$$

$$\underline{Fe^{2+}(aq) + 2e^- \longrightarrow Fe(s)}$$

$$Zn(s) + Fe^{2+}(aq) \longrightarrow Zn^{2+}(aq) + Fe(s)$$

According to the Nernst equation,

$$E_{cell} = -\frac{0.0591}{n}\log\frac{[Zn^{2+}]}{Fe^{2+}}$$

$$0.2905 = E^0_{cell} - \frac{0.0591}{2}\log\frac{0.1}{0.01}$$

$$E^0_{cell} = 0.2905 + 0.0295 \log 10 = 0.2905 + 0.0295 = 0.32 \text{ V}$$

At equilibrium, $E_{cell} = 0$

$$E_{cell} = E^0_{cell} - \frac{0.0591}{n}\log K$$

$$0 = E^0_{cell} - \frac{0.0591}{n}\log K \quad \text{or} \quad E^0_{cell} = \frac{0.0591}{2}\log K$$

$$0.32 = \frac{0.0591}{2}\log K \quad \text{or} \quad K = 10^{0.32/0.0295}$$

9. (A)

$$NaX + H_2O \longrightarrow NaOH + HX$$

Since HX is a weak acid, NaX is a salt of a weak acid and a strong base. The hydrolysis constant (K_h) for such a salt is given by

$$K_h = \frac{K_w}{K_a} = \frac{1\times10^{-14}}{10^{-5}} = 1\times10^{-9}$$

For this type of salt, $K_h = Ch^2$ where h is the degree of hydrolysis.

Hence $1\times10^{-9} = 0.1\times h^2$

or $\quad h^2 = \dfrac{1\times10^{-9}}{0.1} = 1\times10^{-8} \quad$ or $\quad h = 1\times10^{-4}$

Hence % degree of hydrolysis $= 1\times10^{-4}\times100 = 0.01\%$

10. (C)

For a spontaneous adsorption, ΔG is negative. Since the degree of randomness of gas molecules on the surface of solid decreases, ΔS is negative. Hence $T\Delta S$ is negative. Since adsorption is an exothermic process, ΔH is negative.

$\Delta H = \Delta G + T\Delta S$

Hence during adsorption, ΔS decreases.

11. (A)

Root main square velocity, $u_{rms} = \sqrt{\dfrac{3PV}{m}}$

For 1 molecule of gas, $PV = kT$

Hence $u_{rms} = \sqrt{\dfrac{3kT}{m}}$. But K.E. $= \dfrac{3}{2}kT \quad$ or $\quad$ 2 K.E. $= 3kT$

Hence $u_{rms} = \sqrt{\dfrac{2\text{ K.E.}}{m}}$

12. (B)

In MnO_2 and $FeCl_3$, the oxidation states of Mn and Fe are + 4 and + 3 respectively.

In MnO_4^- and CrO_2Cl_2, the oxidation states of Mn and Cr are + 7 and + 6 respectively.

In $[Fe(CN)_6]^{3-}$ and $[Co(CN)_3]$, the oxidation states of Fe and Co are + 3 and + 3 respectively.

In $[NiCl_4]^{2-}$ and $[CoCl_4]^-$, the oxidation states of Ni and Co are + 2 and + 3 respectively.

Hence the pair of compounds having metals in their highest oxidation state is (B) MnO_4^-, CrO_2Cl_2.

13. (D)

Electronic configuration of Ni^{2+}: $1s^2 2s^2 2p^6 3s^2 3p^6 3d^8$

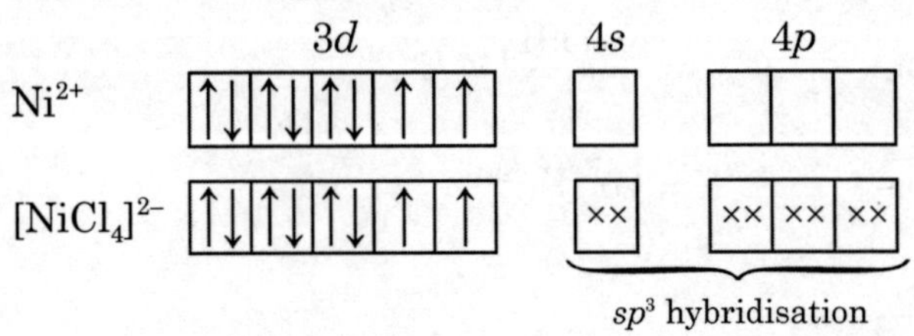

Since the hybridisation of $[NiCl_4]^{2-}$ is sp^3, the shape of $[NiCl_4]^{2-}$ is tetrahedral.

14. (A)

The oxidation state of Co in Hg[Co(NCS)$_4$] is + 2, Co^{2+} ion belongs to $3d^7$ system.

Co^{2+} 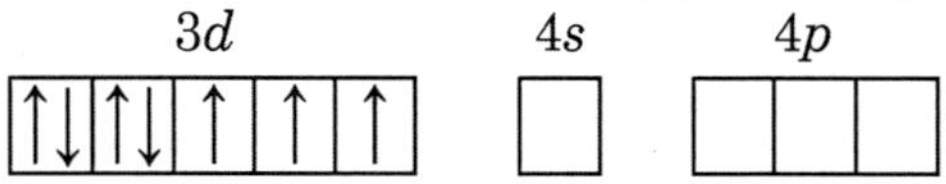

Hence $n = 3$.

Spin-only magnetic moment, $\mu_s = \sqrt{n(n+2)} = \sqrt{3(3+2)}$ B.M. $= \sqrt{15}$ B.M.

15. (B)

The anion is HCO_3^-.

$$MgCl_2 + 2NaHCO_3 \longrightarrow \underset{\text{Soluble}}{Mg(HCO_3)_2} + 2NaCl$$

$$Mg(HCO_3)_2 \xrightarrow{\Delta} \underset{\text{White}}{MgCO_3\downarrow} + H_2O + CO_2$$

16. (B)

We know $\quad r = \dfrac{n^2 r_{(H)}}{Z}$

Given that $\quad r = r_{(H)}$

Hence $\quad n^2 = Z$

For Be^{3+}, $\quad Z = 4$

Hence $\quad n = 2$

Thus $\quad r_{2\,(Be^{3+})} = r_{1(H)}$

17. (A)

The pK_a value of carboxylic group is less than the pK_a value of $—\overset{+}{N}H_3(y)$ in amino acid and $—\overset{+}{N}H_3(z)$ will have comparatively less pK_a value than $—\overset{+}{N}H_3(y)$ due to the – I-effect of the carboxylate group. Since the acid strength is inversely proportional to pK_a, the order of acid strength is

$$—COOH\ (x) > —\overset{+}{N}H_3\ (z) > —\overset{+}{N}H_3\ (y)$$

18. (A)

Since $0.004M$ Na_2SO_4 solution is isotonic with $0.01M$ solution of glucose, $\pi_{Na_2SO_4} = \pi_{glucose} = 0.01 \times RT$

	$Na_2SO_4 \rightleftharpoons$	$2Na^+$ +	SO_4^{2-}
At $t = 0$	1	0	0
At equilibrium	$1 - \alpha$	2α	α

(where α is the degree of dissociation of Na_2SO_4)

$(\pi_{cal})_{Na_2SO_4} = CRT = 0.004 \times R \times T$

By van't Hoff factor,

$$\frac{(\pi_{obs})_{Na_2SO_4}}{(\pi_{calc})_{Na_2SO_4}} = \frac{\text{No. of particles after dissociation}}{\text{No. of particles before dissociation}} = \frac{1 - \alpha + 2\alpha + \alpha}{1}$$

But $\quad (\pi_{obs})_{Na_2SO_4} = \pi_{glucose} = 0.01 \times RT$

Hence $\quad \dfrac{0.01 \times R \times T}{0.004 \times R \times T} = \dfrac{1 + 2\alpha}{1}$

or $\frac{10}{4} = \frac{1+2\alpha}{1}$ or $10 = 4 + 8\alpha$

or $\alpha = \frac{10-4}{8} = 0.75$ $\% \alpha = 0.75 \times 100 = 75$

19. (A)

$$\Delta G = \Delta H - T\Delta S$$

At equilibrium, $\Delta G = 0$. Hence $T\Delta S = \Delta H$

or $T = \frac{\Delta H}{\Delta S} = \frac{30 \times 10^3}{75} = 400 \text{ K}$ or $T = 400$ K

20. (D)

For an isothermal process, $\Delta T = 0$ and $\Delta E = 0$.

$$H = E + PV$$

$$\Delta H = \Delta E + \Delta(PV)$$

or $\Delta H = \Delta E + nR\Delta T = 0 + nR \times 0$ or $\Delta H = 0$

21. (A)

For first order reaction,

$$k = \frac{2.303}{t} \log \frac{[A]_0}{[A]_t}$$

At $t = 40$ minute,

$$k = \frac{2.303}{40} \log \frac{0.1}{0.025} = \frac{2.303}{40} \log 4 = \frac{2.303}{40} \times 2 \log 2$$

$$= \frac{2.303}{4} \times 2 \times 0.3010 = 0.0347 \text{ min}^{-1}$$

$$r = k[A]$$

$$= 0.0347 \times 0.01 = 3.47 \times 10^{-4} \text{ mol L}^{-1} \text{ min}^{-1}$$

22. (A)

Li/NH_3 brings about *trans* addition of H_2 to an alkyne.

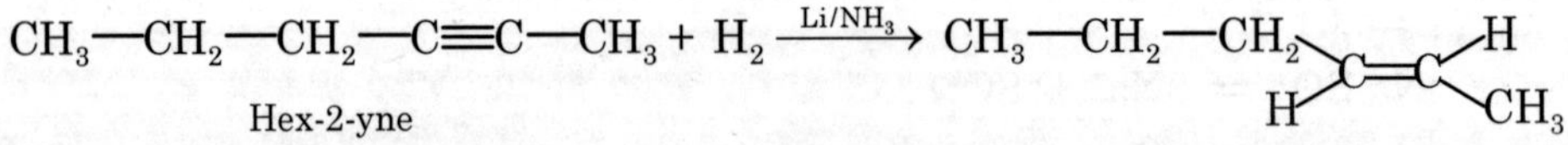

Hex-2-yne

trans-hex-2-ene

23. (B)

On chlorination of 2-methylbutane, four chiral compounds are formed.

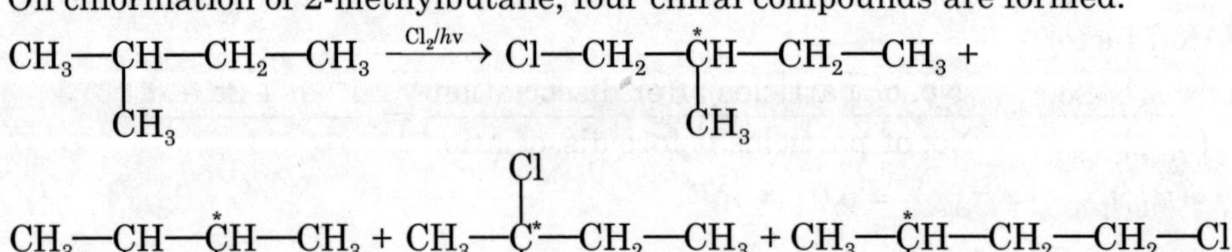

24. (B)

Aldehydes and α-hydroxyketones exhibit positive Tollen's test. Glucose has an aldehyde group and fructose is an α-hydroxyketone. Hence both glucose and fructose give positive Tollen's test.

25. (B)

$H_2S_2O_8$ has —O—O— linkage.

$H_2S_2O_8$: 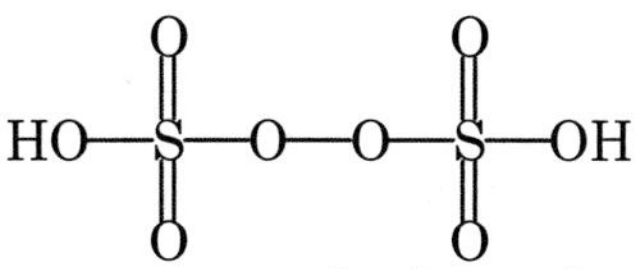

Other three compounds do not have —O—O— linkage.

$H_2S_2O_6$: 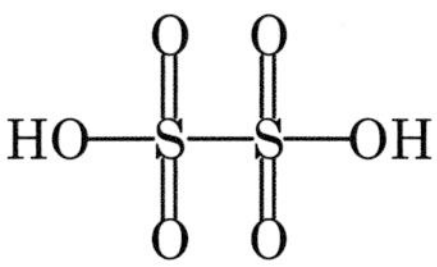 $H_2S_2O_3$: 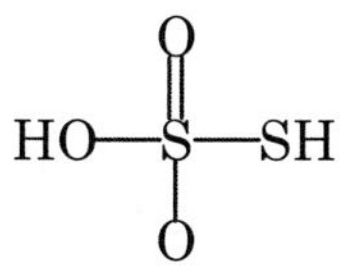 $H_2S_4O_6$: 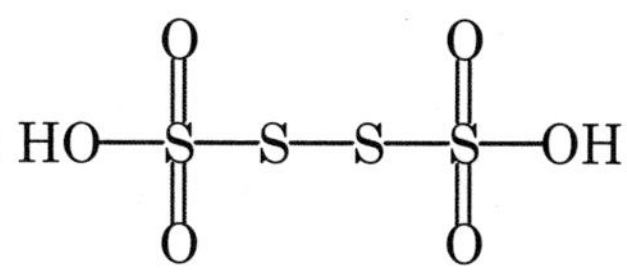

26. (A)

When I^- ion is oxidised by MnO_4^- ion in alkaline medium, I^- ion is converted into IO_3^- ion.

$KI + 2KMnO_4 + H_2O \longrightarrow 2MnO_2 + KIO_3 + 2KOH$

27. (B)

Number of lone pair on Xe in $XeOF_4$ is 1. The structure of $XeOF_4$ is as follows:

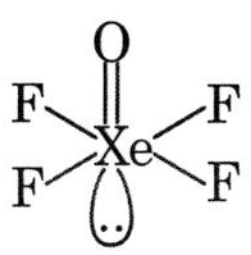

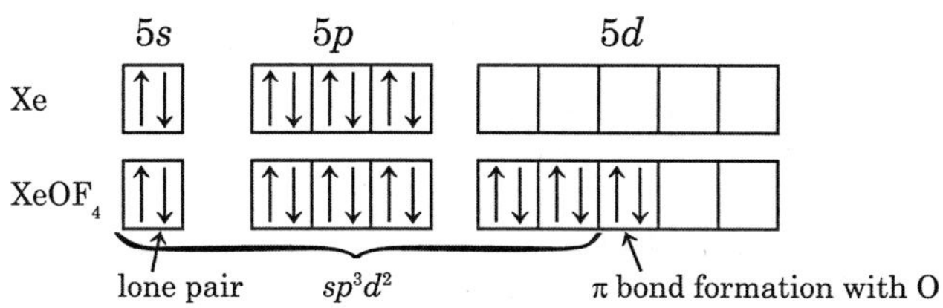

28. (A)

$O_2^+: (\sigma_{1s}^b)^2(\sigma_{1s}^*)^2(\sigma_{2s}^b)^2(\sigma_{2s}^*)^2(\sigma_{2p_z}^b)^2(\pi_{2p_x}^b)^2(\pi_{2p_y}^b)^2(\pi_{2p_x}^*)^1$

B. O. $= \frac{1}{2}$ (Number of bonding electrons − Number of antibonding electrons)

$= \frac{1}{2}(10 - 5) = 2.5$

$O_2: (\sigma_{1s}^b)^2(\sigma_{1s}^*)^2(\sigma_{2s}^b)^2(\sigma_{2s}^*)^2(\sigma_{2p_z}^b)^2(\pi_{2p_x}^b)^2(\pi_{2p_y}^b)^2(\pi_{2p_x}^*)^1(\pi_{2p_y}^*)^1$

B. O. $= \frac{1}{2}(10 - 6) = 2$

I.I.T. JEE Objective Question Paper—2005

1. When phenyl magnesium bromide reacts with tertiary butanol, which of the following is formed?

(A) Tertiary butyl methyl ether (B) Benzene

(C) Tertiary butyl benzene (D) Phenol

2. 1-Bromo-3-chlorocyclobutane when treated with two equivalents of Na in the presence of ether, which of the following will be formed?

(A) 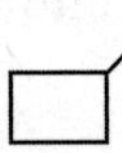

(B) 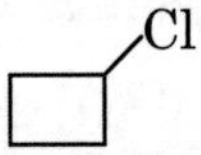

(C) 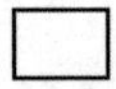
(D) 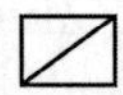

3. In the reaction,

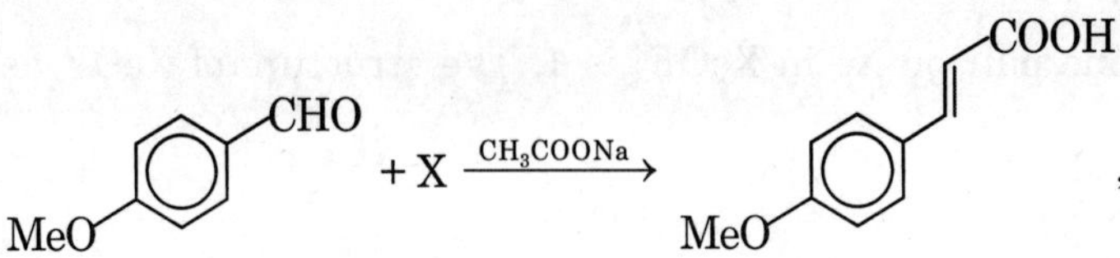

, what is X?

(A) CH_3COOH (B) $BrCH_2COOH$ (C) $(CH_3CO)_2O$ (D) OHC—COOH

4. Cyclohexene is best prepared from cyclohexanol by which of the following?

(A) Conc. H_3PO_4 (B) Conc. $HCl/ZnCl_2$ (C) Conc. HCl (D) Conc. HBr

5. 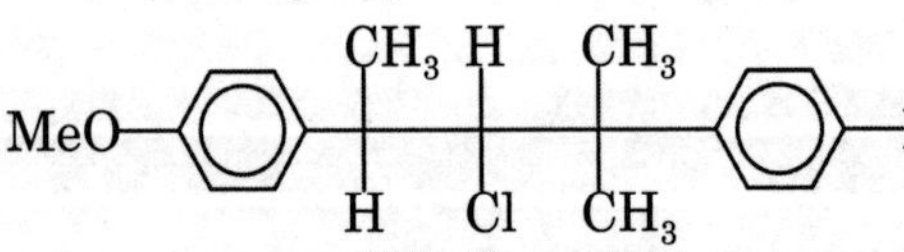

on hydrolysis in aqueous acetone gives

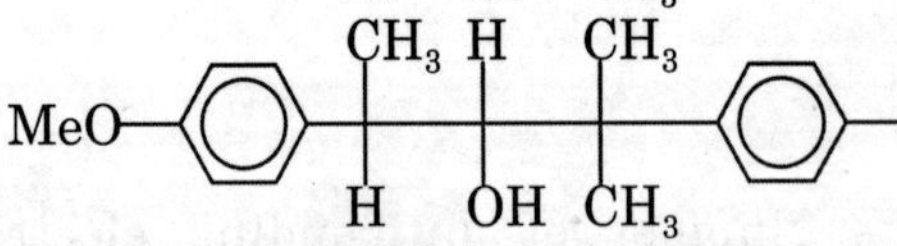

(K)

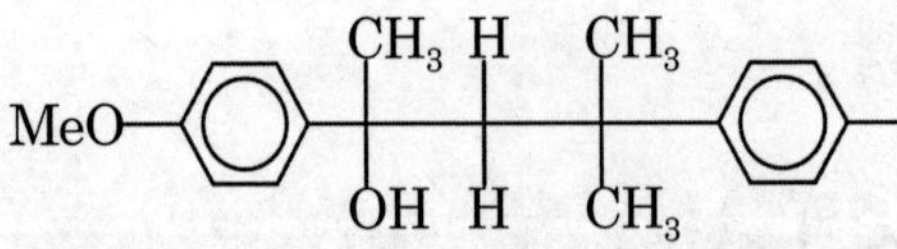

(L)

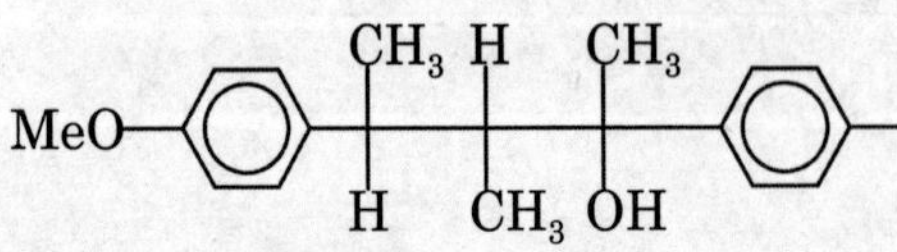

(M)

It mainly gives

(A) K and L (B) only K (C) L and M (D) only M

6. For 1-methoxy-1,3-butadiene, which of the following resonating structures is the least stable?
(A) $H_2\bar{C}-\overset{+}{C}H-CH=CH-O-CH_3$ (B) $H_2\bar{C}-CH=CH-CH=\overset{+}{O}-CH_3$
(C) $H_2C=CH-\overset{+}{C}H-\bar{C}H-O-CH_3$ (D) $H_2C=CH-\bar{C}H-CH=\overset{+}{O}-CH_3$

7. 2-Butanone can be converted into propanoic acid by which of the following?
(A) NaOH, NaI/H^+ (B) Fehling solution
(C) NaOH, I_2/H^+ (D) Tollen's reagent

8. Two forms of D-glucopyranose are called
(A) enantiomers (B) anomers (C) epimers (D) diastereoisomers

9. Which of the following pairs is expected to exhibit same colour in solution?
(A) $VOCl_2$; $FeCl_2$ (B) $CuCl_2$; $VOCl_2$ (C) $MnCl_2$; $FeCl_2$ (D) $FeCl_2$; $CuCl_2$

10. Which of the following allotropes of phosphorus is thermodynamically most stable?
(A) Red (B) White (C) Black (D) Yellow

11. A metal nitrate gives black precipitate with KI and on adding excess of KI gives orange colour. It is
(A) Hg^{2+} (B) Bi^{3+} (C) Sn^{2+} (D) Pb^{2+}

12. Which of the following will not be oxidised by O_3?
(A) KI (B) $FeSO_4$ (C) $KMnO_4$ (D) K_2MnO_4

13. Which type of isomerism is shown by $[Co(NH_3)_4Br_2]Cl$?
(A) Geometrical and ionisation (B) Optical and ionisation
(C) Geometrical and optical (D) Geometrical only.

14. Which of the following *fcc* structures contains cations in alternate tetrahedral voids?
(A) NaCl (B) ZnS (C) Na_2O (D) CaF_2

15. The elevation in boiling point, when 13.44 g of freshly prepared $CuCl_2$ are added to one kilogram of water, is (some useful data: $K_b = 0.52$ kg K mol^{-1}, molecular mass of $CuCl_2$ = 134 g)
(A) 0.05 (B) 0.1 (C) 0.16 (D) 0.21

16. The half-cell reactions for rusting of iron ore are:
$2H^+ + \frac{1}{2}O_2 + 2e^- \longrightarrow H_2O$; $E^0 = +1.23$ V, $Fe^{2+} + 2e^- \longrightarrow Fe(s)$; $E^0 = -0.44$ V ΔG^0 (in kJ) for the reaction is
(A) – 76 (B) – 322 (C) – 122 (D) – 176

17. The number of radial nodes in 3*s* and 2*p* orbitals respectively are
(A) 2 and 0 (B) 1 and 2 (C) 0 and 2 (D) 2 and 1

18. Which of the following ones contains both copper and iron?
(A) Cuprite (B) Chalcocite (C) Chalcopyrite (D) Malachite

19. A pale blue liquid which is obtained by equimolar mixture of two gases at – 30° C is
(A) N_2O (B) N_2O_3 (C) N_2O_4 (D) N_2O_5

20. Which of the following is obtained when 4-methylbenzenesulphonic acid is reacted with excess of sodium acetate solution?

(A) $CH_3-C_6H_4-COO^-\overset{+}{Na}$ (B) $CH_3-C_6H_5 + SO_3$

(C) $CH_3-C_6H_4-SO_3^-\overset{+}{Na} + CH_3COOH$ (D) $CH_3-C_6H_4-SO_2O-COCH_3 + NaOH$

21. CH_3NH_2 (0.1 mole, $K_b = 5 \times 10^{-4}$) is added to 0.08 mole of HCl and the solution is diluted to one litre. The resulting hydrogen ion concentration is

(A) 1.6×10^{-11} (B) 8×10^{-11} (C) 5×10^{-5} (D) 8×10^{-2}

22. Which silicates are formed from $[SiO_4]^{4-}$ tetrahedral units by sharing three oxygen atoms?

(A) Sheet silicates (B) Pyro silicates
(C) Linear chain silicates (D) 3-dimensional silicates.

23. Which gas is evolved when PbO_2 is treated with conc. HNO_3?

(A) NO_2 (B) O_2 (C) N_2 (D) N_2O

24. If helium and methane are allowed to diffuse out of the container under the similar conditions of temperature and pressure, then the ratio of rate of diffusion of helium to methane is

(A) 2.0 (B) 1.0 (C) 0.5 (D) 4.0

25. Which of the following contains maximum number of lone pairs on the central atom?

(A) ClO_3^- (B) XeF_4 (C) SF_4 (D) I_3^-

26. Which of the following is correct for lyophilic sols?

(A) They are irreversible
(B) They are formed by inorganic substances
(C) They are readily coagulated by addition of electrolytes
(D) They are self stabilised.

27. Which of the following statements is incorrect about order of reaction?

(A) Order of reaction is determined experimentally
(B) It is the sum of power of concentration terms in the rate law expression
(C) It does not necessarily depend on stoichiometric coefficients
(D) Order of reaction cannot have fractional value.

28. One mole of a monoatomic ideal gas expands adiabatically at initial temperature T against a constant external pressure of 1 atm from one litre to two litres. Find out the final temperature. ($R = 0.0821$ L atm K^{-1} mol^{-1})

(A) T (B) $\frac{T}{2^{[(5/3)-1]}}$ (C) $T - \frac{2}{3 \times 0.0821}$ (D) $T + \frac{2}{3 \times 0.0821}$

ANSWER KEYS

1. B	2. D	3. C	4. A	5. A	6. C
7. C	8. B	9. B	10. C	11. B	12. C
13. A	14. B	15. C	16. B	17. A	18. C
19. B	20. C	21. B	22. A	23. B	24. A
25. D	26. D	27. D	28. C		

SOLUTIONS

1. (B)
$PhMgBr + Me_3COH \rightarrow Ph—H + Me_3COMgBr$

2. (D)
It is an intermolecular Wurtz reaction.

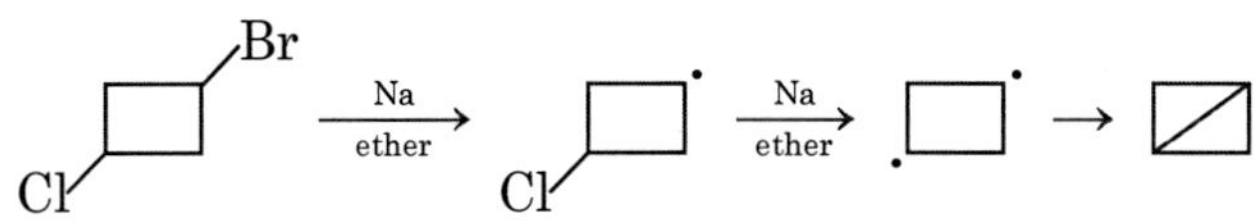

3. (C)

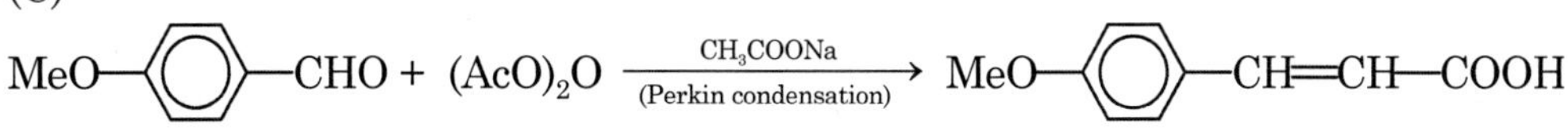

4. (A)

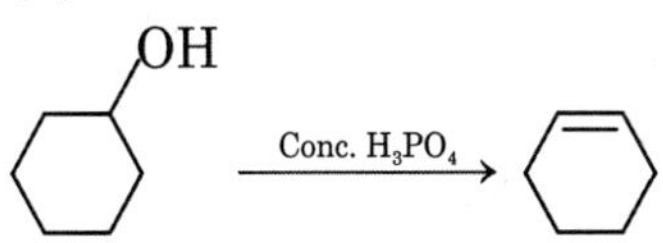

H_3PO_4 acts as a dehydrating agent.

5. (A)
Both S_{N^1} and S_{N^2} reactions are possible due to aqueous acetone solution.

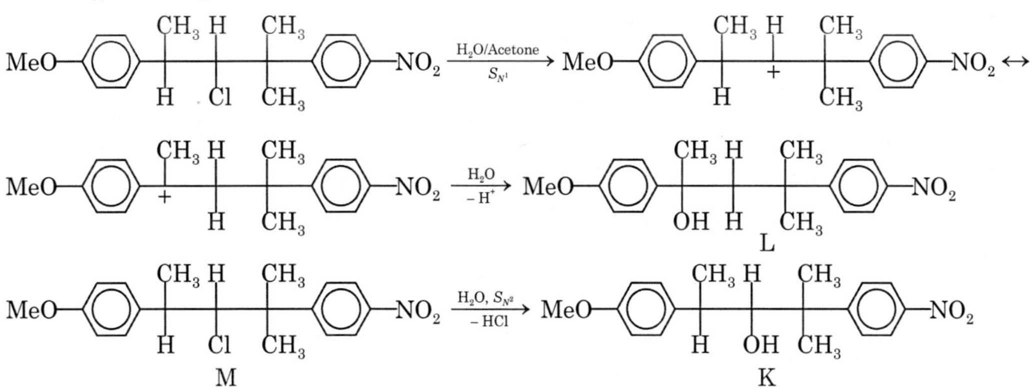

6. (C)
The point of difference is nature of anion. 2° carbanions are less stable than 1° carbanions.

7. (A)

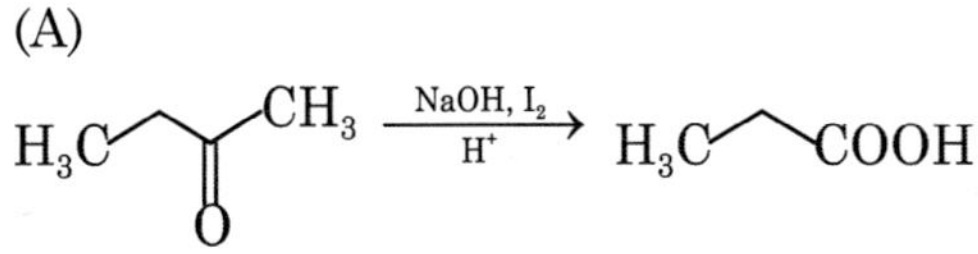

8. (B)
D-Glucopyranose is the cyclic form of glucose. Two isomers are observed around C_1 containing the chiral centre. These isomers are called α-anomer and β-anomer. Anomers have different configurations at the C_1 atom.

9. (B)

Both Cu^{2+} and V^{4+} ions have one unpaired electron and hence are coloured. Both $CuCl_2$ and $VOCl_2$ are green coloured compounds.

10. (C)

Black phosphorus is thermodynamically most stable due to the presence of layered structure. It is inert.

11. (B)

$$\underset{}{Bi^{3+}} + 3KI \longrightarrow \underset{\text{Black}}{BiI_3}$$

$$BiI_3 + KI \longrightarrow \underset{\text{Orange solution}}{K[BiI_4]}$$

12. (C)

Since manganese can exhibit the maximum possible oxidation state of + 7, $KMnO_4$ containing Mn(VII) cannot be oxidised by any oxidising agent.

13. (A)

$[Co(NH_3)_4Br_2]Cl$ can exhibit both geometrical isomerism (*cis*-$[Co(NH_3)_4Br_2]Cl$, *trans*-$[Co(NH_3)_4Br_2]Cl$) and ionisation isomerism ($[Co(NH_3)_4Br_2]Cl$, $[Co(NH_3)_4BrCl]Br$).

14. (B)

In ZnS, the anions (S^{2-}) are placed in *fcc* manner and the cations (Zn^{2+}) are placed in the alternate tetrahedral voids.

15. (C)

$$\Delta T_b = i \times K_b \times m = 3 \times 0.52 \times \left(\frac{13.44}{134.4} \times \frac{1000}{1000}\right) = 0.16$$

16. (B)

$$2H^+ + \frac{1}{2}O_2 + 2e^- \longrightarrow H_2O;\ E^0 = +\ 1.23\ V$$

$$Fe^{2+} + 2e^- \longrightarrow Fe(s);\ E^0 = -\ 0.44\ V$$

Subtract

$$Fe(s) + 2H^+ + \frac{1}{2}O_2 \longrightarrow Fe^{2+} + H_2O;\ E^0_{cell} = 1.67\ V$$

$$\Delta G^0 = -\ nFE^0_{cell} = -\ 2 \times 96500 + 1.67 = -\ 322\ kJ$$

17. (A)

Number of radial nodes = $n - l - 1$

For 3*s* orbital, the number of radial nodes = 3 – 0 – 1 = 2

For 2*p* orbital, the number of radial nodes = 2 – 1 – 1 = 0

18. (C)

Chalcopyrite ($CuFeS_2$) contains both copper and iron.

19. (B)

$$NO + NO_2 \xrightarrow{-30^\circ C} \underset{\text{Pale blue liquid}}{N_2O_3}$$

20. (C)

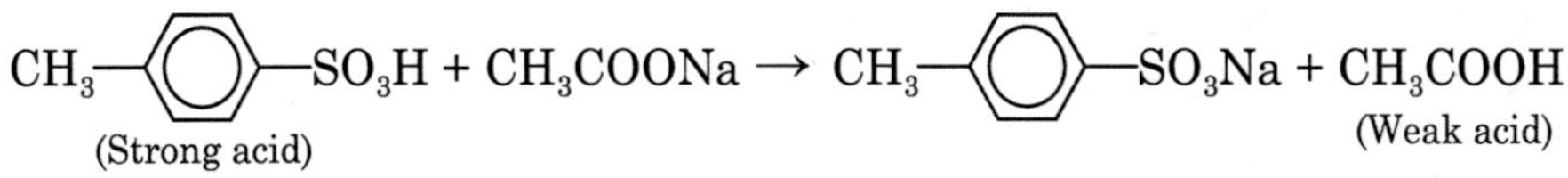

This reaction is an acid-base reaction and not a hydrolysis reaction.

21. (B)

$$CH_3NH_2 + HCl \longrightarrow CH_3\overset{+}{N}H_3 + Cl^-$$

	CH_3NH_2	HCl	$CH_3\overset{+}{N}H_3$
Initial mole	0.1	0.08	
Moles at equi.	0.02	–	0.08

$$[OH^-] = K_b \frac{[CH_3NH_2]}{[CH_3\overset{+}{N}H_3]} = \frac{5\times10^{-4}\times0.02}{0.08} = \frac{5}{4}\times10^{-4}$$

$$[H^+] = \frac{K_w}{[OH^-]} = \frac{10^{-14}\times4}{5\times10^{-4}} = 8\times10^{-11}$$

22. (A)

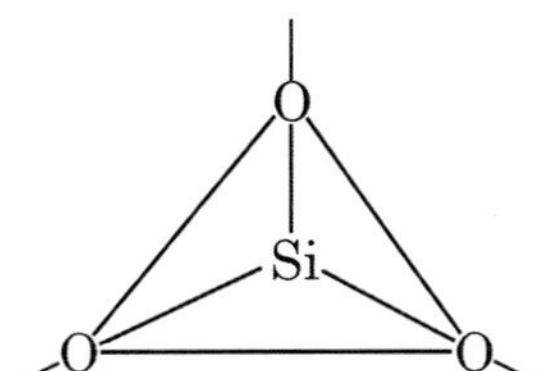

Sheet structure

23. (B)

$2PbO_2 + 4HNO_3 \longrightarrow 2Pb(NO_3)_2 + 2H_2O + O_2$

24. (A)

$$\frac{r_{He}}{r_{CH_4}} = \sqrt{\frac{16}{4}} = \sqrt{4} = 2$$

25. (D)

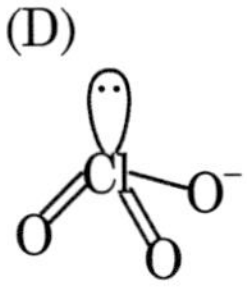

(One lone pair)

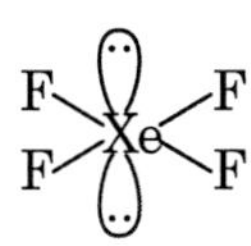

(Two lone pairs)

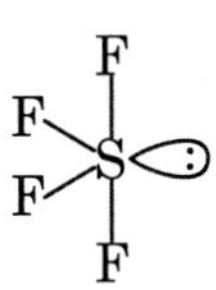

(One lone pair)

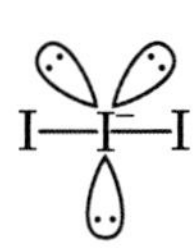

(Three lone pairs)

26. (D)

Lyophilic sols are solvent loving in nature and they are extensively solvated. Due to this property, the lyophilic sols are self-stabilised and they cannot be readily coagulated.

27. (D)

The order of reaction is determined experimentally and it may be fractional.

28. (C)

Work done against constant pressure = $-P_{ext}(V_2 - V_1)$

Under adiabatic condition, $\Delta q = 0$. Hence $w = \Delta U$.

$$\Delta U = nC_{v,m}(T_2 - T_1) = 1 \times \frac{3}{2}R(T_2 - T_1)$$

$$-P(V_2 - V_1) = \frac{3}{2}R(T_2 - T_1) \text{ (Expansion work is negative)}$$

$$\text{or } -1(2-1) = \frac{3}{2}R(T_2 - T_1)$$

$$\text{or } T_2 - T_1 = -\frac{2}{3R}$$

$$\text{or } T_2 = T_1 - \frac{2}{3R} = T_1 - \frac{2}{3 \times 0.0821}$$

I.I.T. JEE Objective Question Paper—2006

Section-I (MCQ with only one choice correct)

1. $B(OH)_3 + NaOH \rightleftharpoons Na[B(OH)_4]$
How can this reaction be made to proceed in forward direction?
(A) Addition of *cis*-1,2-diol (B) Addition of borax
(C) Addition of *trans*-1,2-diol (D) Addition of Na_2HPO_4

2. A solution when diluted with H_2O and boiled, it gives a white precipitate. On addition of excess NH_4Cl/NH_4OH, the volume of precipitate decreases leaving behind a white gelatinous precipitate. Identify the precipitate which dissolves in NH_4OH/NH_4Cl.
(A) $Zn(OH)_2$ (B) $Al(OH)_3$ (C) $Mg(OH)_2$ (D) $Ca(OH)_2$

3. When benzenesulphonic acid and *p*-nitrophenol are treated with $NaHCO_3$, the gases released respectively are
(A) SO_2, NO_2 (B) SO_2, NO (C) SO_2, CO_2 (D) CO_2, CO_2

4. A monoatomic ideal gas undergoes a process in which the ratio of P to V at any instant is constant and equals to 1. What is the molar heat capacity of the gas?
(A) $4R/2$ (B) $3R/2$ (C) $5R/2$ (D) 0

5. Arrange the following compounds according to decreasing boiling point.

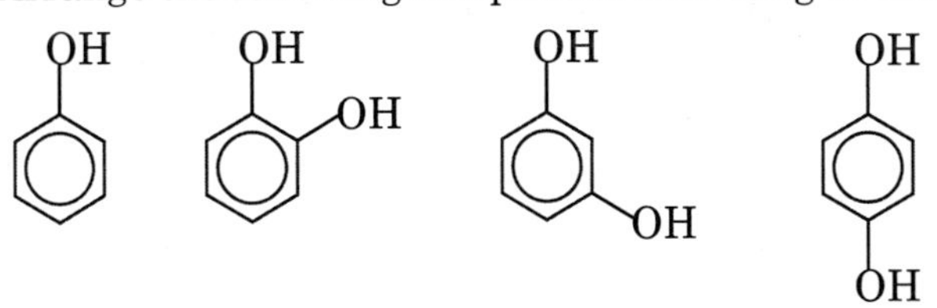

I II III IV

(A) IV > III > II > I (B) III > IV > II > I (C) I > II > III > IV (D) II > III > I > IV

6. $CH_3—CH=CH_2 + NOCl \rightarrow P$
Identify the product.
(A) $CH_3—CH(Cl)—CH_2(NO)$ (B) $CH_3—CH(NO)—CH_2(Cl)$
(C) $CH_3—CH_2—CH(NO)(NO)$ (D) $CH_2(NO)—CH_2—CH_2(Cl)$

7. The IUPAC name of C_6H_5COCl is
(A) Benzoyl chloride (B) Benzenechloroketone
(C) Benzenecarbonyl chloride (D) Chlorophenylketone

8. $CH_3NH_2 + CHCl_3 + 3KOH \longrightarrow (P) + 3KCl + H_2O$. The missing product is
(A) $CH_3{-}C{\equiv}N$ (B) $CH_3{-}NH{-}Cl$ (C) $CH_3{-}\bar{N}{\equiv}\overset{+}{C}$ (D) $CH_3{-}\overset{+}{N}{\equiv}\overset{-}{C}$

9. The blue solution of $CuSO_4$ is decolourised on addition of excess KCN. The product is
(A) $[Cu(CN)_4]^{2-}$ (B) Cu^{2+} (C) $Cu(OH)_2$
(D) CuCN is reduced by CN^- to Cu^+ which forms the complex $[Cu(CN)_4]^{3-}$.

10. The species present in solution when CO_2 is dissolved in water are
(A) CO_2, H_2CO_3, HCO_3, CO_3^{2-} (B) H_2CO_3, CO_3^{2-}
(C) CO_3^{2-}, HCO_3^- (D) CO_2, H_2CO_3

11. The direct converion of A to B is difficult. Hence it is carried out by the following shown path:

$A \longrightarrow C \longrightarrow D \longrightarrow B$

Given: $\Delta S_{(A \to C)} = 50$ eu, $\Delta S_{(C \to D)} = 30$ eu, $\Delta S_{(B \to D)} = 20$ eu, where eu is entropy unit. The value of $\Delta S_{(A \to B)}$ is
(A) + 100 eu (B) + 60 eu (C) – 100 eu (D) – 60 eu

12. Haber–Bosch process for the formation of NH_3 at 298 K is $N_2(g) + 3H_2(g) \rightleftharpoons 2NH_3(g)$; $\Delta H = -46.0$ kJ; $K_p = 14$
The correct statement is
(A) On adding N_2, the equilibrium will spontaneously shift to the right and ΔS will be positive.
(B) Even though at 298 K, the yield is more, still industrial manufacture of NH_3 is carried out at 400 K ($K_p = 41$). This is because at 298 K, the catalyst increases the rate by a factor of two for both forward and backward reaction. But at 400 K, the iron catalyst increases the rate of forward reaction by a factor of two and that of backward reactio by a factor of 1.7.
(C) At equilibrium, $2G_{NH_3} = G_{N_2} + 3G_{H_2}$ where G is the Gibbs free energy.
(D) Addition of catalyst does not alter K_p but changes ΔH.

Section-II (MCQ with one or more than one choice correct)

13. $Ag^+ + NH_3 \rightleftharpoons [Ag(NH_3)]^+$; $k_1 = 6.8 \times 10^3$
$[Ag(NH_3)]^+ + NH_3 \rightleftharpoons [Ag(NH_3)_2]^+$; $k_2 = 1.6 \times 10^3$
The overall formation constant of $[Ag(NH_3)_2]^+$ is
(A) 1.08×10^7 (B) 1.08×10^3 (C) 6.8×10^6 (D) 6.8×10^7

14. If the bond length of CO bond in carbon monoxide is 1.128 Å, then what is the value of CO bond length in $Fe(CO)_5$?
(A) 1.13 Å (B) 1.128 Å (C) 1.10 Å (D) 1.115 Å

15. Which of the following compounds on reaction with aqueous alkali followed by acidification gives the following lactone?

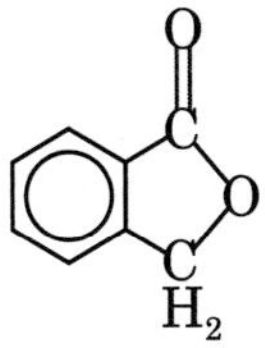

(A) 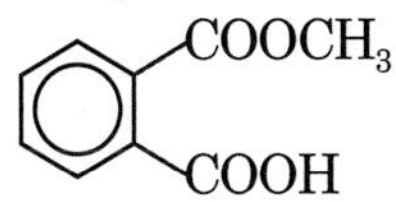 (B) 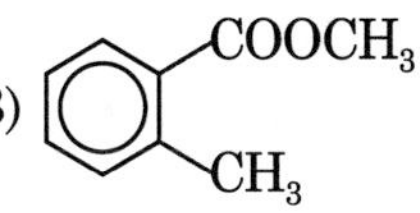 (C) 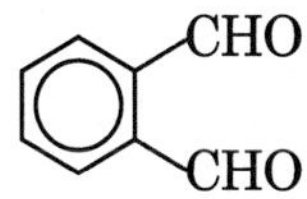 (D) 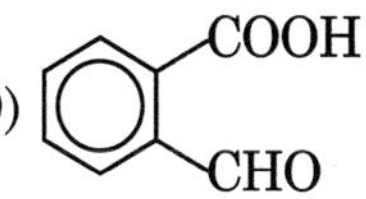

16.  $+ \text{Cl—CH}_2\text{—CH}_2\text{—CH}_3 \xrightarrow{\text{AlCl}_3} \text{P} \xrightarrow[\text{(ii) H}_3\text{O}^+]{\text{(i) O}_2/\Delta} \text{Q} + \text{Phenol}$

The major products P and Q are

(A) 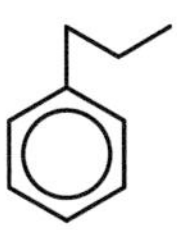 and CH_3CH_2CHO

(B) 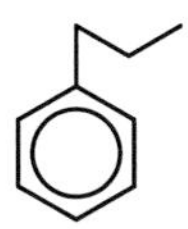 and CH_3COCH_3

(C) 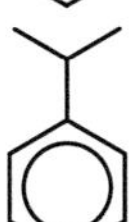 and CH_3COCH_3

(D) 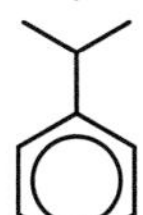 and CH_3CH_2CHO

17. The given graph represents the variation of Z (compressibility factor) *versus* P for three real gases A, B and C. Identify the incorrect statement.

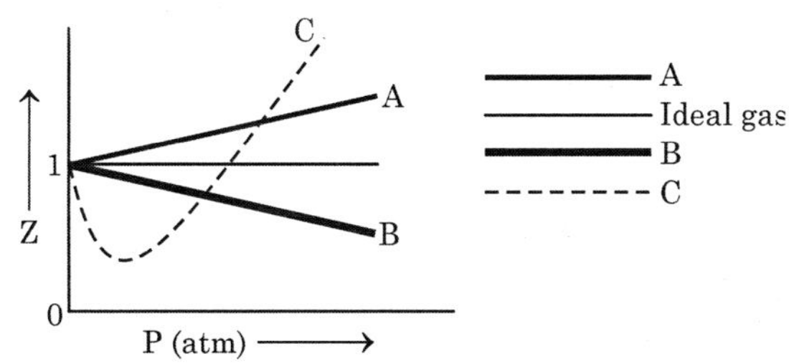

(A) For the gas A, $a = 0$ and the dependence of Z on P is linear

(B) For the gas B, $b = 0$ and the dependence of Z on P is linear

(C) Gas C is a real gas and we can find a and b if intersection data is given

(D) All van der Waals gases will behave like gas C and give positive slope at high pressure.

18. The lowest molecular mass ketone and its next homologue are reacted with NH_2OH to form oximes. Which of the following is/are correct?

(A) Two oximes are formed
(B) Three oximes are formed
(C) One oxime is optically active
(D) Two oximes are optically active

19. 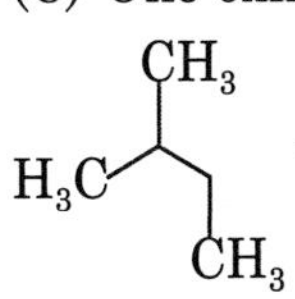 $\xrightarrow{\text{Cl}_2,\, h\nu}$ N isomers of $C_5H_{11}Cl$ $\xrightarrow{\text{fractional distillation}}$ M distilled products

The values of N and M are:

(A) 6, 6 (B) 6, 4 (C) 4, 4 (D) 3, 3

20. Ammoniacal solution of $MgSO_4$ in the presence of NH_4Cl is heated with Na_2HPO_4. A white crystalline precipitate is formed of

(A) $Mg(NH_4)PO_4$ (B) $Mg_3(PO_4)_2$ (C) $MgCl_2{\cdot}MgSO_4$ (D) $MgSO_4{\cdot}MgPO_4$

Section-III (Passage)

Passage I

$RCONH_2$ is converted into RNH_2 by means of Hofmann bromamide degradation.

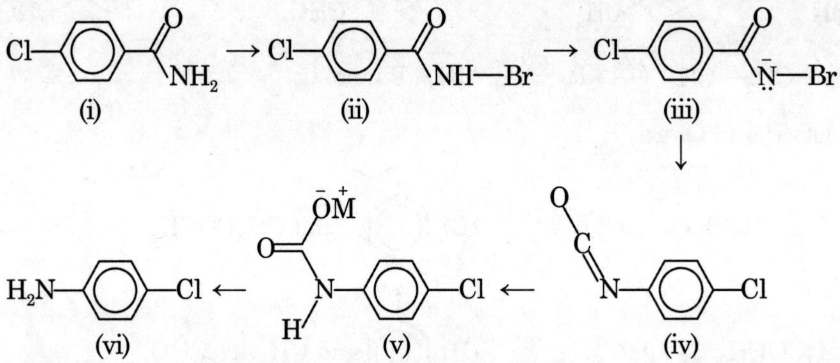

In this reaction, RCONHBr is formed from which this reaction has derived its name. Electron donating group at phenyl activates the reaction. Hofmann bromamide reaction is an intramolecular reaction.

21. How can the conversion of (i) to (ii) be brought about?

(A) NBS (B) KBr + CH_3ONa (C) KBr + NaOH (D) Br_2 + NaOH

22. Which is the rate determining step in Hofmann bromamide raction?

(A) Formation of (v) (B) Formation of (ii) (C) Formation of (iii) (D) Formation of (iv)

23. What are the constituent amines when the mixture of vii and viii undergoes Hofmann bromamide degradation?

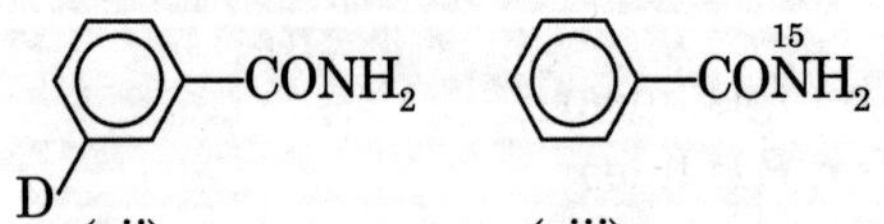

(vii) (viii)

(A) 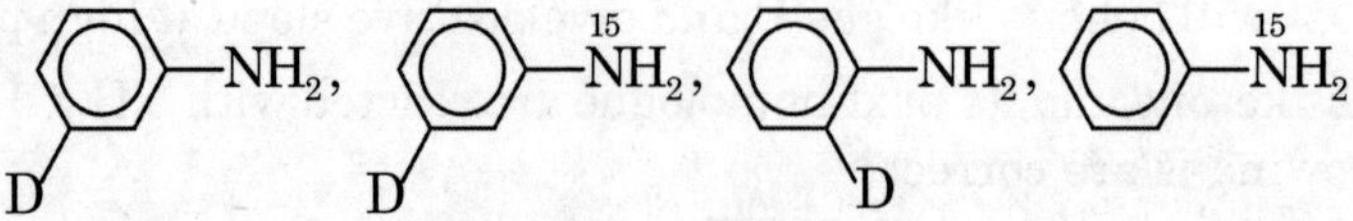

(B) 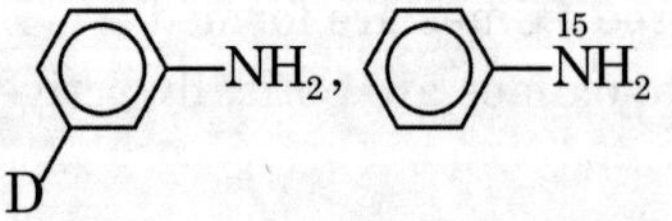

(C) 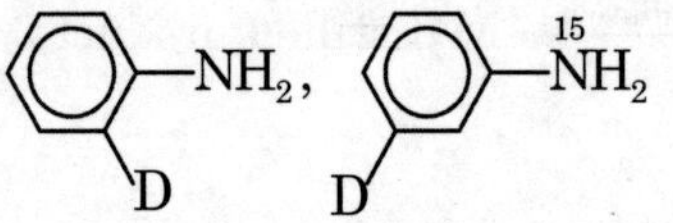 (D) 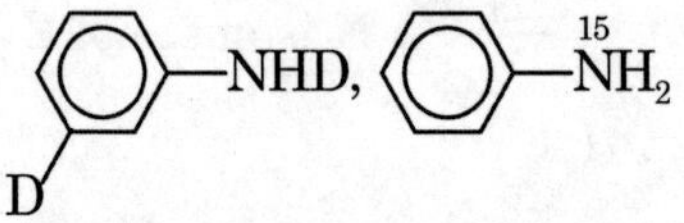

Passage II

Tollen's reagent is used for the detection of aldehyde. When a solution of $AgNO_3$ is added to glucose with NH_4OH, gluconic acid is formed.

$Ag^+ - e^- \longrightarrow Ag, E^0_{cell} = 0.8$ V

$C_6H_{12}O_6 + H_2O \longrightarrow$ Gluconic acid $(C_6H_{12}O_7) + 2H^+ + 2e^-$, $E^0_{ox} = -0.05$ V

$\left[Ag(NH_3)_2\right]^+ + e^- \longrightarrow Ag(s) + 2NH_3$, $E^0_{red} = 0.373$ V

$$\left[\text{Use } 2.303 \times \frac{RT}{F} = 0.0592 \text{ and } \frac{F}{RT} = 38.9\,\text{V}^{-1} \text{ at } 298\,\text{K}\right]$$

24. $2Ag^+ + C_6H_{12}O_6 + H_2O \longrightarrow 2Ag(s) + C_6H_{12}O_7 + 2H^+$

Find $\ln K$ of this reaction.

(A) 66.00 (B) 58.35 (C) 29.60 (D) 55.60

25. When ammonia is added to the solution, the pH increases to 11. Which half-cell reaction is affected by pH and by how much?

(A) E_{ox} will increase by a factor of 0.65 from E^0_{ox}

(B) E_{ox} will decrease by a factor of 0.65 from E^0_{ox}

(C) E_{red} will increase by a factor of 0.65 from E^0_{red}

(D) E_{red} will decrease by a factor of 0.65 from E^0_{red}

26. Ammonia is used in this reaction than any other base. What is the correct reason for this?

(A) $[Ag(NH_3)_2]^+$ is a weaker oxidising agent than Ag^+

(B) NH_3 prevents the decomposition of gluconic acid

(C) Ag^+ precipitates gluconic acid as its silver salt

(D) NH_3 changes the standard reduction potential of $[Ag(NH_3)_2]^+$.

Passage III

The coordination number of Ni^{2+} is 4 in A and B.

$NiCl_2$ + KCN (excess) $\longrightarrow$ A (cyano complex)

$NiCl_2$ + conc. HCl (excess) $\longrightarrow$ B (chloro complex)

27. The IUPAC names of A and B are respectively

(A) potassium tetracyanonickelate(II), potassium tetrachloronickelate(II)

(B) potassium tetracyanonickel(II), potassium tetrachloronickel(II)

(C) potassium tetracyanonickel, potassium tetrachloronickel

(D) potassium tetracyanonickelate(II), potassium tetrachloronickel(II)

28. Predict the magnetic nature of A and B.

(A) Both are diamagnetic (B) Both are paramagnetic

(C) A is paramagnetic and B is diamagnetic (D) A is diamagnetic and B is paramagnetic.

29. The hybridisations of A and B are respectively

(A) dsp^2, sp^3 (B) sp^3, sp^3 (C) dsp^2, dsp^2 (D) sp^3, dsp^2

Passage IV

Due to the neutrons present in cosmic rays, the following reaction takes place by the bombardment of nitrogen with neutrons.

$^{14}_{7}N + ^{1}_{0}n \rightarrow ^{14}_{6}C + ^{1}_{1}H;\ ^{14}_{6}C \rightarrow ^{14}_{7}N + ^{0}_{-1}e$

Cosmic rays form $^{14}_{6}C$ which gets circulated in the atmosphere as well as in living species. Whenever there is a nuclear explosion, the concentration of ^{14}C increases both in the atmosphere as will as in living species. When a species dies the ^{14}C concentration decreases and hence this decrease can be measured and the time estimated as to when the organism died. For this we require (1) half-life of ^{14}C (5760 years) (2) activity of ^{14}C in living species (3) activity of ^{14}C in fossil. Beyond 30000 years the age of fossil cannot be accurately determined as the activity then would be too low.

30. In radiocarbon dating for finding the age of fossils, the most appropriate statement is
(A) During the life-time ^{14}C assimilated by the human beings is in equilibrium with the ^{14}C that decomposes by β^- emission. At any time in a human life $^{14}C/^{12}C$ is always constant at a particular instant
(B) ^{14}C dating method is inappropriate for finding the life of a given sample because ^{14}C undergoes β^- emission and hence the ratio of $^{14}C/^{12}C$ in human beings in not constant
(C) Radiocarbon dating is inappropriate to determine the age because ^{14}C concentration varies from place to place
(D) None of these.

31. A fossil whose age can be determined using carbon dating must be
(A) 6 years old (B) 6000 years old (C) 60000 years old
(D) can be accurately calculated for any age of the sample.

32. Two organisms died on the same day. One died at a place where nuclear explosion had taken place while the other died at a place where no such explosion had occurred. The ratio of ^{14}C during life to that present in the fossil at an instant is C_1 for the former and C_2 for the latter. The age of the former was calculated as t_1 and for the latter as t_2. The correct answer is

(A) There is an increae in t_1 w.r.t. t_2 and $t_1 - t_2 = \frac{1}{\lambda}\ln\frac{C_1}{C_2}$

(B) There is a decreae in t_1 w.r.t. t_2 and $t_1 - t_2 = \frac{1}{\lambda}\ln\frac{C_1}{C_2}$

(C) $\frac{t_1}{t_2} = \frac{C_1}{C_2}$ (D) Cannot be determined.

33. Match the extraction processes listed in Column I with metals listed in Column II.

Column I	*Column II*
(A) Self-reduction	(i) Lead
(B) Carbon reduction	(ii) Silver
(C) Complex formation and displacement by metal	(iii) Copper
(D) Decomposition of iodide	(iv) Boron

34. Match the following:

Column I	Column II
(A) $Bi_2O_3 \longrightarrow (BiO)^+$	(i) Heating
(B) $[AlO_2]^- \longrightarrow Al(OH)_3$	(ii) Hydrolysis
(C) $SiO_4^{4-} \longrightarrow Si_2O_7^-$	(iii) Acidification
(D) $B_4O_7^{2-} \longrightarrow [B(OH)_3]$	(iv) Dilution with water

35. According to Bohr's theory of H-atom, E_n = total energy, K_n = kinetic energy of nth orbit, V_n = potential energy of nth orbit. r_n = radius of nth orbit. Match the following:

Column I	Column II
(A) $V_n/K_n = ?$	(i) 0
(B) If radius of nth orbit $\propto E_n^x$, x = ?	(ii) – 1
(C) Orbital angular momentum in lowest energy orbit	(iii) – 2
(D) $\frac{1}{V_n} \propto Z^y$, y = ?	(iv) 1

36. Match the following:

Column I	Column II
(A) $PhCH_2CD_2Br$ on reaction with $C_2H_5O^-$ gives $Ph{-}CH{=}CD_2$	(i) *E*1
	(ii) *E*2
(B) $PhCHBrCH_3$ and $PhCHBrCD_3$ both react with the same rate	
(C) $PhCHDCH_2Br$ on treatement with $C_2H_5O^-$ and C_2H_5OD gives $Ph{-}CD{=}CD_2$	(iii) *E*1CB
(D) $PhCH_2CH_2Br$ reacts faster than $PhCD_2CH_2Br$ on reaction with $C_2H_5O^-$ in ethanol	(iv) First order reaction

ANSWER KEYS

1. A	2. A	3. D	4. A	5. A	6. A
7. C	8. D	9. D	10. A	11. B	12. C
13. A	14. A	15. C	16. C	17. B	18. B
19. B	20. A	21. B	22. D	23. B	24. B
25. C	26. A	27. A	28. D	29. A	30. C
31. B	32. A				

Matching Type Qustions

33. A, i , iii; B, i, iii; C, ii; D, iv
34. A, ii; B, iv; C, i; D, ii, iii
35. A, iii; B, ii; C, i; D, iv
36. A, ii; B, i, iv; C, ii; D, ii

SOLUTIONS

1. (A)

$Na[B(OH)_4]$ is removed from the equilibrium due to the formation of chelate compound with *cis*-1,2-diol and the reaction moves in the forward direction.

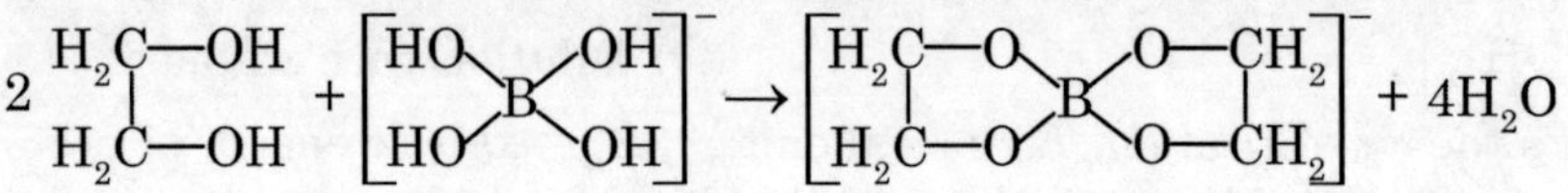

2. (A)

White precipitate of $Zn(OH)_2$ dissolves in NH_4OH/NH_4Cl to form tetraamminezinc(II) hydroxide $[Zn(NH_3)_4](OH)_2$: $Zn(OH)_2 + 4NH_4OH \longrightarrow [Zn(NH_3)_4](OH)_2 + 4H_2O$

3. (D)

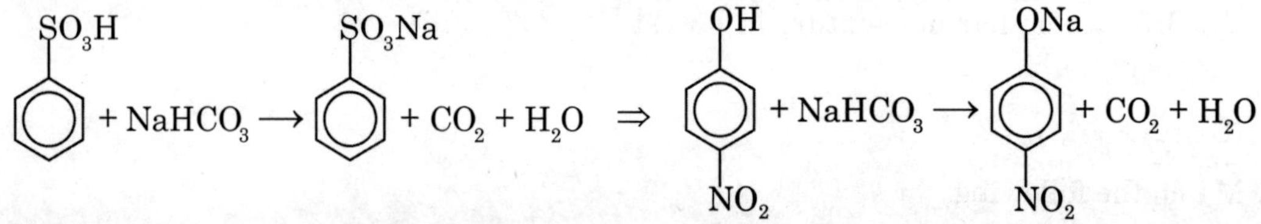

4. (A)

Equation for the process: PV^{-1} = constant

Molar heat capacity for the process $= \dfrac{R}{\gamma-1} + \dfrac{R}{1-(-1)} = \dfrac{R}{\frac{5}{3}-1} + \dfrac{R}{2} = \dfrac{3R}{2} + \dfrac{R}{2} = \dfrac{4R}{2}$

5. (A)

Increase in the number of hydroxyl groups increases the boiling point. Among the isomeric dihydroxybenzenes, as the distance between the hydroxyl groups increases, the repulsion between them decreases and hence the boiling point increases.

6. (A)

The reaction occurs through Markownikoff addition.

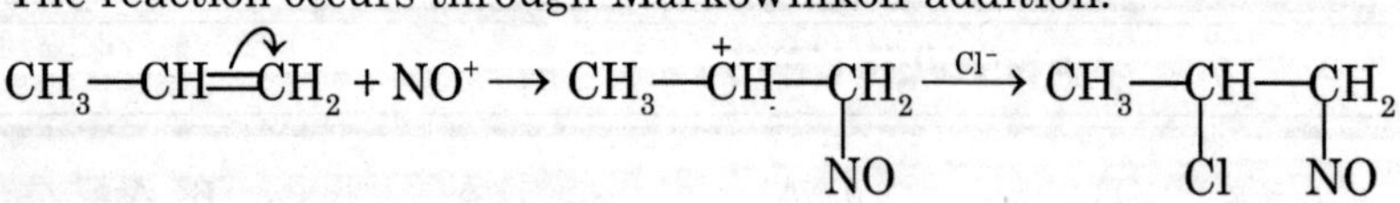

The 2° carbocation ($CH_3-\overset{+}{C}H-CH_2-NO$) formed is more stable than 1° carbocation ($CH_3-\underset{+}{C}H_2-CH_2$). The chloride ion attacks the more stable 2° carbocation to give the major product (A).

7. (C)

8. (D)

This is carlylamine reaction and alkyl isocyanide is formed.

9. (D)

$2CuSO_4 + 4KCN \longrightarrow 2CuCN + (CN)_2 + 2K_2SO_4$

$CuCN + 4KCN \longrightarrow [Cu(CN)_4]^{3-} + 4K^+ + CN^-$

10. (A)

$$CO_2 + H_2 \rightleftharpoons H_2CO_3 \rightleftharpoons H^+ + HCO_3^-; HCO_3^- \rightleftharpoons H^+ + CO_3^{2-}$$

11. (B)

$$\Delta S_{(A\to B)} = \Delta S_{(A\to C)} + \Delta S_{(C\to D)} - \Delta S_{(B\to D)} = 50 + 30 - 20 = 60 \text{ eu}$$

12. (C)

13. (A)

$$\beta_k = k_1 \times k_2 = 6.8 \times 10^3 \times 1.6 \times 10^3 = 1.08 \times 10^7$$

14. (A)

Due to the back donation of electrons from M to CO, the M—C bond becomes stronger and the C≡O bond becomes weaker leading to a shorter M—C bond and longer C—O bond as compared to M—C single bond and C≡O triple bond respectively.

15. (C)

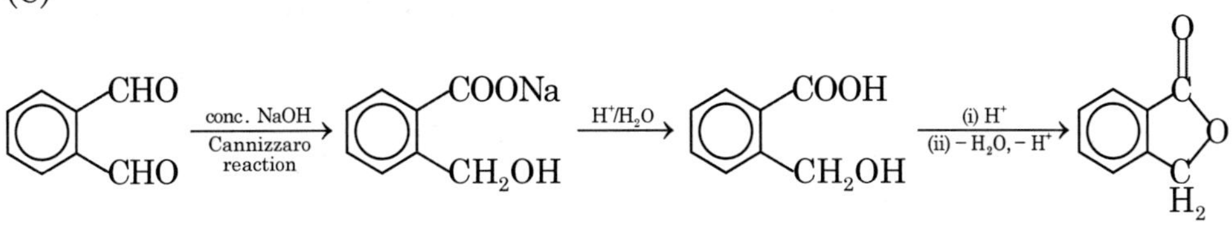

16. (C)

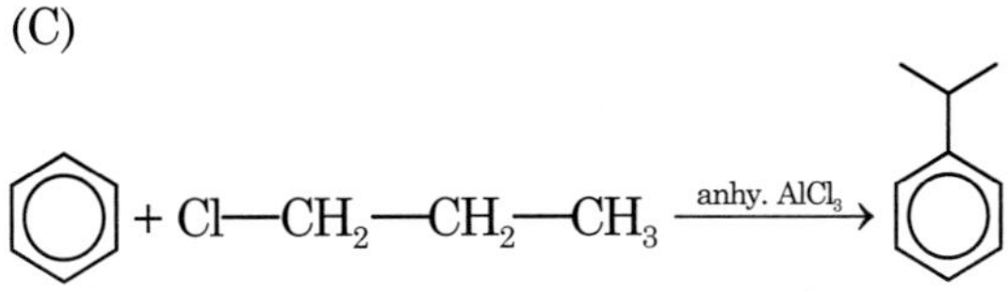

P (major product)

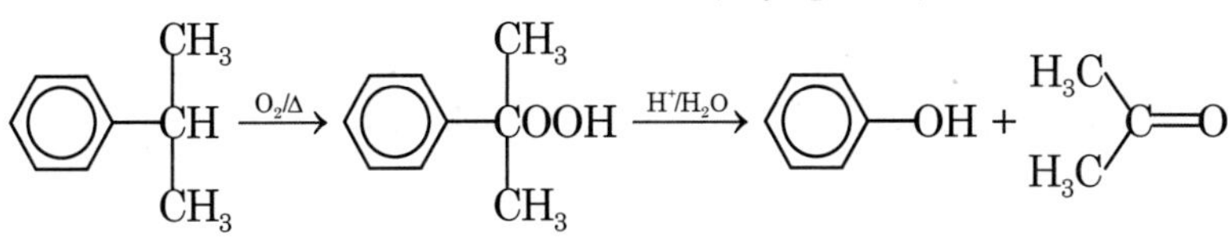

17. (B)

For gas B, $b = 0$, $Z = 1 - \dfrac{a}{VRT}$. Hence Z does not vary linearly with pressure. For gas A, $a = 0$, $Z = 1 + \dfrac{Pb}{RT}$ implies that Z varies linearly with pressure. Given the intersection data for gas C we can find the values of a and b. All van der Waals gases like gas C give positive slope at high pressure.

18. (B)

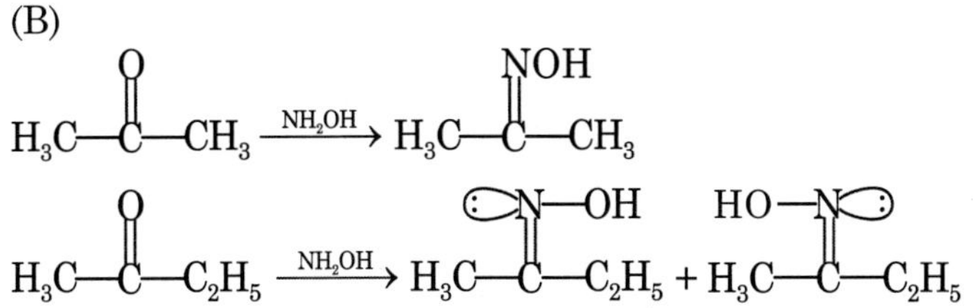

19. **(B)**

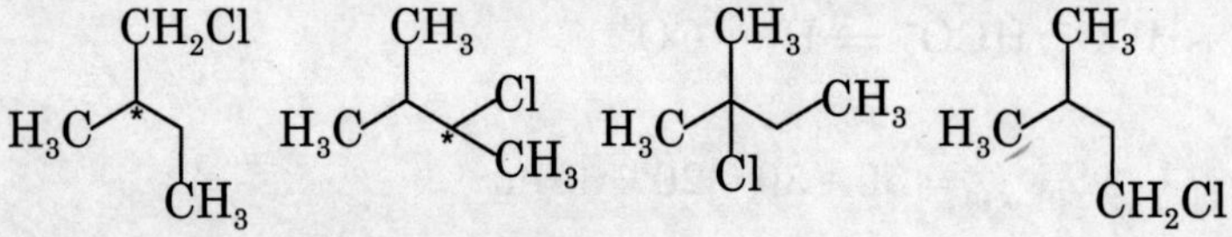

d, l (two isomers) *d, l* (two isomers)

M: Two *d, l* compounds (four isomers) cannot be separated by fractional distillation.

20. (A)

$Mg^{2+} + NH_3 + HPO_4^{2-} \longrightarrow Mg(NH_4)PO_4$

21. (B)

$(Br_2 + NaOH)$ is the required reagent.

22. (D)

Formation of compound (iv) will occur in the rate determining step.

23. (B)

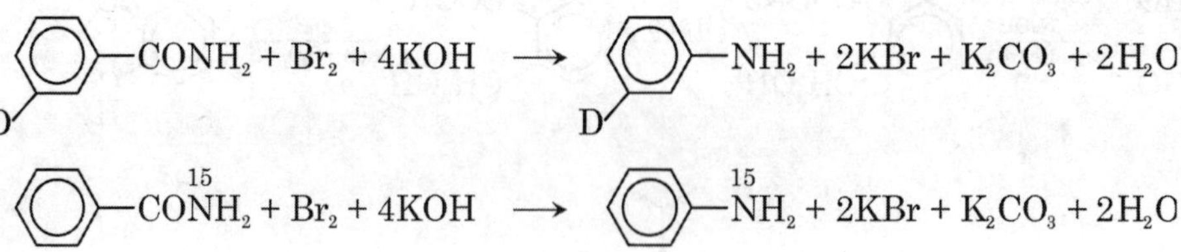

24. (B)

$$E^0_{cell} = \frac{RT}{nF}\ \ln K$$

$$(0.8 - 0.05) = \frac{1}{2} \times \frac{0.0592}{2.303} \ln K \text{ or } \ln K = \frac{0.75 \times 2 \times 2.303}{0.0592} = 58.35$$

25. (C)

On increasing the concentration of NH_3, the concentration of H^+ ion decreases. Hence E_{red} increases.

$C_6H_{12}O_6 + H_2O \rightarrow C_6H_{12}O_7 + 2H^+ + 2e^-$

$$E = E^0 - \frac{RT}{2F}\ln\left[H^+\right]^2 = E^0 - \frac{2}{2\times 38.9}\ln\left[H^+\right] = E^0 + \frac{2.303}{38.9}pH$$

$$= E^0 + \frac{2.303 \times 11}{38.9} = E^0 + \frac{1.302}{2} = E^0 + 0.65$$

$E - E^0 = 0.65$ V

26. (A)

Since $E^0_{Ag^+/Ag} > E^0_{[Ag(NH_3)_2]^+/Ag}$, the correct answer is (A).

27. (A)

$A = K_2[Ni(CN)_4]$, $B = K_2[NiCl_4]$

28. (D)

CN^- ion being a strong field ligand, it forms low-spin square planar (*dsp*^{2} hybridisation) $K_2[Ni(CN)_4]$ which is diamagnetic. Cl^- ion being a weak field ligand, it forms high-spin tetrahedral (*sp*^{3} hybridisation) $K_2[NiCl_4]$ which is paramagnetic with two unpaired electrons.

29. (A)

$K_2[Ni(CN)_4]$: 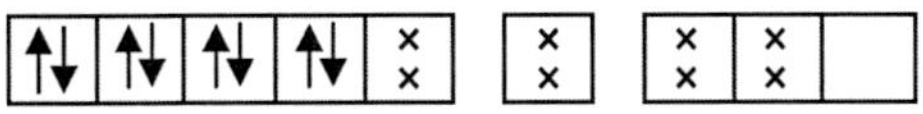 (*dsp*^{2} hybridisation)

$K_2[NiCl_4]$: 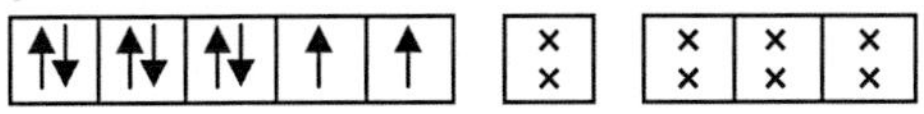 (*sp*^{3} hybridisation)

30. (C)

The correct statement is (C). (A) is incorrect because after nuclear explosion $^{14}C/^{12}C$ will not remain constant. (B) is incorrect because ^{14}C will always remain in equilibrium with the surroundings.

31. (B)

For effective carbon dating, the fossil must be minimum 1000 years old.

32. (A)

$t_1 = \frac{1}{\lambda}\ln C_1$; $t_2 = \frac{1}{\lambda}\ln C_2$. Hence $t_1 - t_2 = \frac{1}{\lambda}\ln\frac{C_1}{C_2}$. Since due to nuclear explosion the initial activity A_0 would be greater, it would take more time to reach the same value of activity A.

I.I.T. JEE Objective Question Paper—2007

PAPER—I

Section–I: Straight Objective Type

This section contains 9 multiple choice questions numbered 1 to 9. Each question has 4 choices (A), (B), (C) and (D), out of which ONLY ONE is correct.

1. Among the following, the least stable resonance structure is

(A) 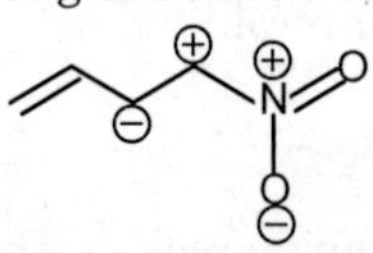 (B) 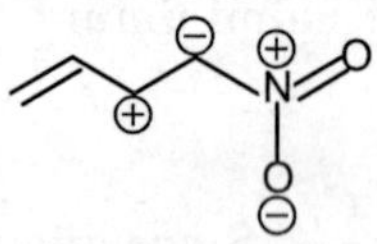 (C) 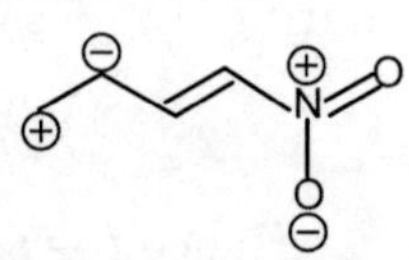 (D) 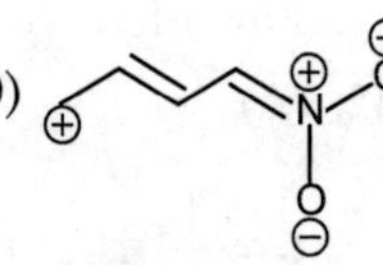

2. For the process $H_2O(l)$ (1 bar, 373 K) $\rightarrow$ $H_2O(g)$ (1 bar, 373 K), the correct set of thermodynamic parameters is

(A) $\Delta G = 0$, $\Delta S = +ve$ (B) $\Delta G = 0$, $\Delta S = -ve$

(C) $\Delta G = +ve$, $\Delta S = 0$ (D) $\Delta G = -ve$, $\Delta S = +ve$

3. Cyclohexene on ozonolysis followed by reaction with zinc dust and water gives compound E. Compound E on further treatment with aqueous KOH yields compound F. Compound F is

(A) 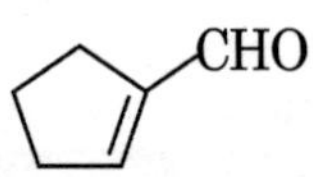 (B) 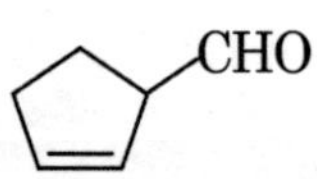 (C) 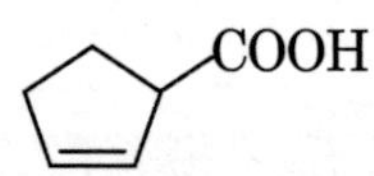 (D) 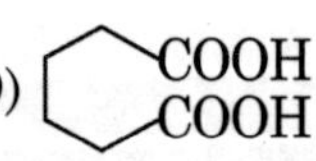

4. Consider a reaction $a\text{G} + b\text{H} \longrightarrow$ Products. When concentration of both the reactants G and H is doubled, the rate increases by eight times. However, when concentration of G is doubled keeping the concentration of H fixed, the rate is doubled. The overall order of the reaction is

(A) 0 (B) 1 (C) 2 (D) 3

5. Among the following metal carbonyls, the C—O bond order is lowest in

(A) $[Mn(CO)_6]^+$ (B) $Fe(CO)_5$ (C) $[Cr(CO)_6]$ (D) $[V(CO)_6]^-$

6. A positron is emitted from ${}^{23}_{11}Na$. The ratio of the atomic mass and atomic number of the resulting nuclide is

(A) 22/10 (B) 22/11 (C) 23/10 (D) 23/12

7. Consider a titration of potassium dichromate solution with acidified Mohr's salt solution using diphenylamine as indicator. The number of moles of Mohr's salt required per mole of dichromate is

(A) 3 (B) 4 (C) 5 (D) 6

8. The number of stereoisomers obtained by bromination of *trans*-but-2-ene is

(A) 1 (B) 2 (C) 3 (D) 4

9. A solution of metal ion when treated with KI gives a red precipitate which dissolves in excess KI to give a colourless solution. Moreover, the solution of metal ion on treatment with a solution of cobalt(II) thiocyanate gives rise to a deep blue crystalline precipitate. The metal ion is

(A) Pb^{2+} (B) Hg^{2+} (C) Cu^{2+} (D) Co^{2+}

Section–II: Assertion-Reason Type

This section contains 4 questions numbered 10 to 13. Each question contains Statement–1 (Assertion) and Statement–2 (Reason). Each question has four choices (A), (B), (C) and (D), out of which ONLY ONE is correct.

10. Statement–1: Molecules that are not superimposable on their mirror images are chiral.

because

Statement–2: All chiral molecules have chiral centres.

(A) Statement–1 is true, Statement–2 is true; Statement–2 is a correct explanation for Statement–1

(B) Statement–1 is true, Statement–2 is true; Statement–2 is not a correct explanation for Statement–1

(C) Statement–1 is true, Statement–2 is false

(D) Statement–1 is false, Statement–2 is true.

11. Statement–1: Alkali metals dissolve in liquid ammonia to give blue solution.

because

Statement–2: Alkali metals in liquid ammonia give solvated species of the type $[M(NH_3)_n]^+$ (M = alkali metals).

(A) Statement–1 is true, Statement–2 is true; Statement–2 is a correct explanation for Statement–1

(B) Statement–1 is true, Statement–2 is true; Statement–2 is not a correct explanation for Statement–1

(C) Statement–1 is true, Statement–2 is false

(D) Statement–1 is false, Statement–2 is true.

12. Statement–1: Band gap in germanium is small.

because

Statement–2: The energy spread of each germanium atomic energy level is infinitesimally small.

(A) Statement–1 is true, Statement–2 is true; Statement–2 is a correct explanation for Statement–1

(B) Statement–1 is true, Statement–2 is true; Statement–2 is not a correct explanation for Statement–1

(C) Statement–1 is true, Statement–2 is false

(D) Statement–1 is false, Statement–2 is true.

13. Statement–1: Glucose gives a reddish-brown precipitate with Fehling's solution.
because
Statement–2: Reaction of glucose with Fehling's solution gives CuO and gluconic acid.
(A) Statement–1 is true, Statement–2 is true; Statement–2 is a correct explanation for Statement–1
(B) Statement–1 is true, Statement–2 is true; Statement–2 is not a correct explanation for Statement–1
(C) Statement–1 is true, Statement–2 is false
(D) Statement–1 is false, Statement–2 is true.

Section–III: Linked Comprehensive Type

This section contains 2 paragraphs C_{14-16} and C_{17-19}. Based upon each paragraph, 3 multiple choice questions have to be answered. Each question has 4 choices (A), (B), (C) and (D), out of which ONLY ONE is correct.

C_{14-16}: Paragraph for Question Nos. 14 to 16

Reimer–Tiemann reaction introduces an aldehyde group, on to the aromatic ring of phenol, *ortho* to the hydroxyl group. This reaction involves electrophilic aromatic substitution. This is a general method for the synthesis of salicylaldehydes as depicted below:

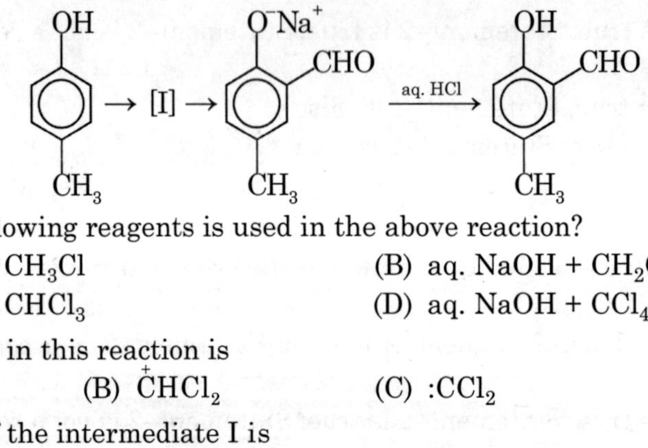

14. Which of the following reagents is used in the above reaction?
(A) aq. NaOH + CH_3Cl
(B) aq. NaOH + CH_2Cl_2
(C) aq. NaOH + $CHCl_3$
(D) aq. NaOH + CCl_4

15. The electrophile in this reaction is
(A) :CHCl
(B) $\overset{+}{C}HCl_2$
(C) $:CCl_2$
(D) $\cdot CCl_3$

16. The structure of the intermediate I is

(A) 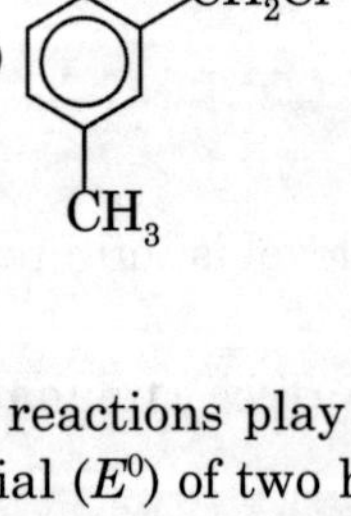

(B) 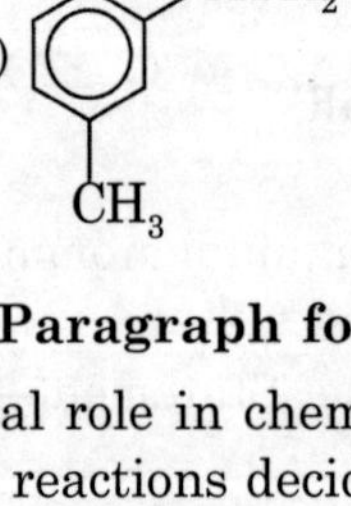

(C) 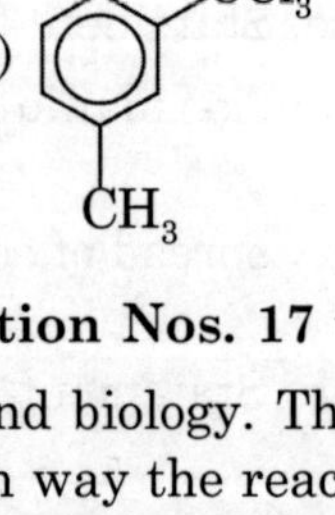

(D) 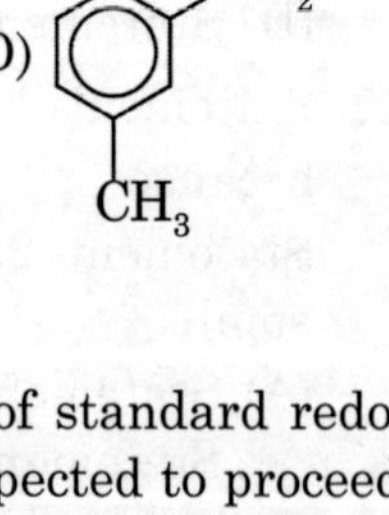

C_{17-19}: Paragraph for Question Nos. 17 to 19

Redox reactions play a pivotal role in chemistry and biology. The values of standard redox potential (E^0) of two half-cell reactions decide which way the reaction is expected to proceed. A simple example is a Daniel cell in which zinc goes into solution and copper gets deposited. Given below are a set of half-cell reactions (acidic medium) along with their E^0 (V with respect to normal hydrogen electrode) values. Using this data obtain correct explanations to

Questions 17 to 19. $I_2 + 2e^- \longrightarrow 2I^-$, $E^0 = 0.54$ V; $Cl_2 + 2e^- \longrightarrow 2Cl^-$, $E^0 = 1.36$ V; $Mn^{3+} + e^- \longrightarrow Mn^{2+}$, $E^0 = 1.50$ V; $Fe^{3+} + e^- \longrightarrow Fe^{2+}$, $E^0 = 0.77$ V; $O_2 + 4H^+ + 4e^- \longrightarrow 2H_2O$, $E^0 = 1.23$ V

17. Among the following, identify the correct statement.
(A) Chloride ion is oxidised by O_2
(B) Fe^{2+} is oxidised by iodine
(C) Iodide ion is oxidised by chlorine
(D) Mn^{2+} is oxidised by chlorine.

18. While Fe^{3+} is stable, Mn^{3+} is not stable in acid solution because
(A) O_2 oxidises Mn^{2+} to Mn^{3+}
(B) O_2 oxidises both Mn^{2+} and Fe^{2+} to Fe^{3+}
(C) Fe^{3+} oxidises H_2O to O_2
(D) Mn^{3+} oxidises H_2O and O_2

19. Sodium fusion extract obtained from aniline, on treatment with iron(II) sulphate and H_2SO_4 in the presence of air gives a prussian blue precipitate. The blue colour is due to the formation of
(A) $Fe_4[Fe(CN)_6]_3$
(B) $Fe_3[Fe(CN)_6]_2$
(C) $Fe_4[Fe(CN)_6]_2$
(D) $Fe_3[Fe(CN)_6]_3$

Section–IV: Matrix-Match Type

This section contains 3 questions. Each question contains statements given in two columns which have to be matched. Statements (A, B, C, D) in Column I have to be matched with statements (p, q, r, s) in Column II. The answers to these questions have to be appropriately bubbled as illustrated in the following example.

If the correct matches are A-p, A-s, B-q, B-r, C-p, C-q and D-s, then the correctly bubbled 4 × 4 matrix should be as follows:

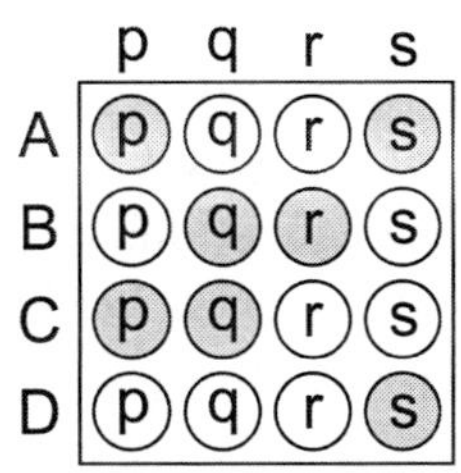

20. Match the reactions in Column I with nature of the reactions/type of the products in Column II. Indicate your answer by darkening the appropriate bubbles of the 4 × 4 matrix given in the ORS.

Column I	*Column II*
(A) $O_2^- \longrightarrow O_2 + O_2^{2-}$	(p) redox reaction
(B) $CrO_4^{2-} + H^+ \longrightarrow$	(q) one of the products has trigonal planar structure
(C) $MnO_4^- + NO_2^- + H^+ \longrightarrow$	(r) dimeric bridged tetrahedral metal ion
(D) $NO_3^- + H_2SO_4 + Fe^{2+} \longrightarrow$	(s) disproportionation

21. Match the compounds/ions in Column I with their properties/reactions in Column II. Indicate your answer by darkening the appropriate bubbles of the 4 × 4 matrix given in the ORS.

Column I	*Column II*
(A) C_6H_5CHO	(p) gives precipitate with 2,4-dinitrophenylhydrazine
(B) $CH_3C\equiv CH$	(q) gives precipitate with $AgNO_3$
(C) CN^-	(r) is a nucleophile
(D) I^-	(s) is involved in cyanohydrin formation

22. Match the crystal system/unit cells mentioned in Column I with their characteristic features mentioned in Column II. Indicate your answer by darkening appropriate bubbles of the 4×4 matrix given in the ORS.

Column I	*Column II*
(A) Simple cubic and face-centred cubic	(p) have these cell parameters: $a = b = c$ and $\alpha = \beta = \gamma$
(B) Cubic and rhombohedral	(q) are two crystal systems
(C) Cubic and tetragonal	(r) have only two crystallography angles of 90°
(D) Hexagonal and monoclinic	(s) belong to same crystal system

ANSWER KEYS

1. A	2. A	3. A	4. D	5. B	6. C
7. D	8. A	9. B	10. C	11. B	12. C
13. C	14. C	15. C	16. B	17. C	18. D
19. A					

20. A-p, s; B-r; C-p, q; D-p
21. A-p, q, s; B-q; C-q, r, s; D-q, r
22. A-p, s; B-p, q; C-q; D-q, r

PAPER—II

Section–I: Straight Objective Type

This section contains 9 multiple choice questions numbered 1 to 9. Each question has 4 choices (A), (B), (C) and (D), out of which ONLY ONE is correct.

1. The number of structural isomers for C_6H_{14} is

(A) 3 (B) 4 (C) 5 (D) 6

2. In the reaction, 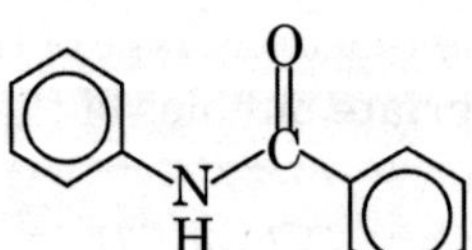 $\xrightarrow[\text{conc. } H_2SO_4]{\text{conc. } HNO_3}$ X, the structure of the major product X is

(A) 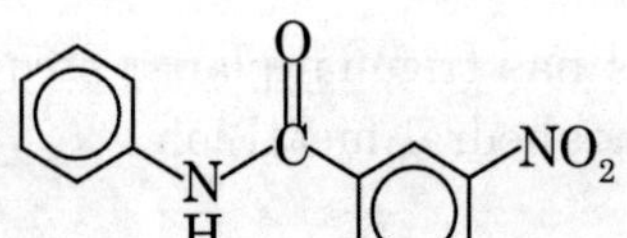

(B) 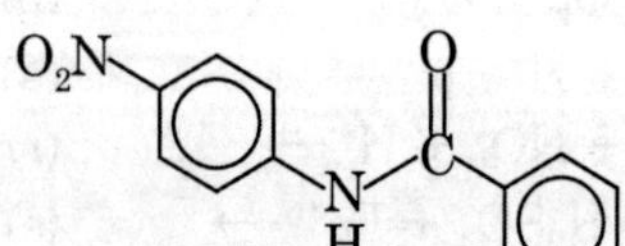

(C) 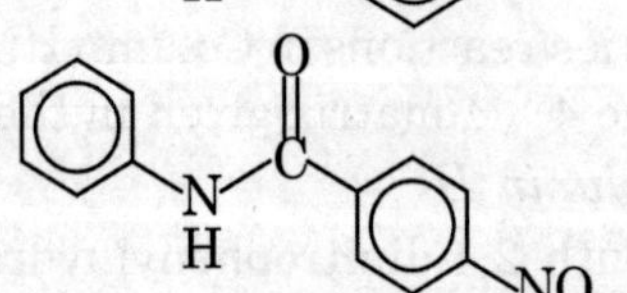

(D) 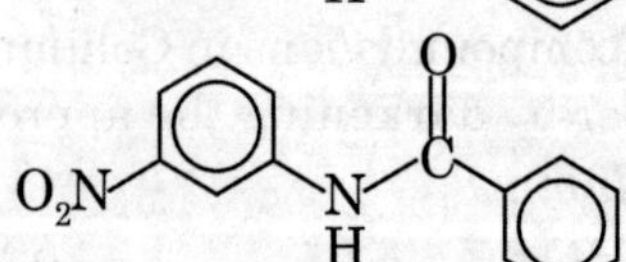

3. When 20 g of naphthoic acid ($C_{11}H_8O_2$) is dissolved in 50 g of benzene ($K_f = 1.72$ K kg mol^{-1}), a freezing point depression of 2 K is observed. The van't Hoff factor (i) is

(A) 0.5 (B) 1 (C) 2 (D) 3

4. Among the following the paramagnetic compound is
(A) Na_2O_2 (B) O_3 (C) N_2O (D) KO_2

5. The value of $\log_{10} K$ for a reaction $A \rightleftharpoons B$ is (Given: $\Delta H^0_{298\text{ K}} = -54.0$ kJ mol^{-1}, $\Delta S^0_{298\text{ K}} = 10$ J K^{-1} mol^{-1} and $R = 8.314$ J K^{-1} mol^{-1}; $2.303 \times 8.314 \times 298 = 5705$)
(A) 5 (B) 10 (C) 95 (D) 100

6. The species having bond order different from that in CO is
(A) NO^- (B) NO^+ (C) CN^- (D) N_2

7. The percentage of *p*-character in the orbitals forming P—P bonds in P_4 is
(A) 25 (B) 33 (C) 50 (D) 75

8. Extraction of zinc from zinc blende is achieved by
(A) electrolytic reduction
(B) roasting followed by reduction with carbon
(C) roasting followed by reduction with another metal
(D) roasting followed by self-reduction.

9. The reagent(s) for the conversion, 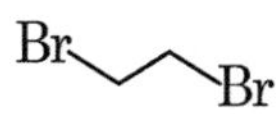 $\xrightarrow{?}$ H—≡—H is/are
(A) alcoholic KOH (B) alcoholic KOH followed by $NaNH_2$
(C) aqueous KOH followed by $NaNH_2$ (D) Zn/CH_3OH

Section–II: Assertion-Reason Type

This section contains 4 questions numbered 10 to 13. Each question contains Statement–1 (Assertion) and Statement–2 (Reason). Each question has 4 choices (A), (B), (C) and (D), out of which ONLY ONE is correct.

10. Statement–1: *p*-Hydroxybenzoic acid has a lower boiling point than *o*-hydroxybenzoic acid.
because
Statement–2: *o*-Hydroxybenzoic acid has intramolecular hydrogen bonding.
(A) Statement–1 is true, Statement–2 is true; Statement–2 is a correct explanation for Statement–1
(B) Statement–1 is true, Statement–2 is true; Statement–2 is not a correct explanation for Statement–1
(C) Statement–1 is true, Statement–2 is false
(D) Statement–1 is false, Statement–2 is true.

11. Statement–1: Micelles are formed by surfactant molecules above the critical micellar concentration (CMC).
because
Statement–2: The conductivity of a solution having surfactant molecules decreases sharply at the CMC.
(A) Statement–1 is true, Statement–2 is true; Statement–2 is a correct explanation for Statement–1
(B) Statement–1 is true, Statement–2 is true; Statement–2 is not a correct explanation for Statement–1

(C) Statement–1 is true, Statement–2 is false
(D) Statement–1 is false, Statement–2 is true.

12. Statement–1: Boron always forms covalent bond.
because
Statement–2: The small size of B^{3+} favours the formation of covalent bond.
(A) Statement–1 is true, Statement–2 is true; Statement–2 is a correct explanation for Statement–1
(B) Statement–1 is true, Statement–2 is true; Statement–2 is not a correct explanation for Statement–1
(C) Statement–1 is true, Statement–2 is false
(D) Statement–1 is false, Statement–2 is true.

13. Statement–1: In water, orthoboric acid behaves as a weak monobasic acid.
because
Statement–2: In water, orthoboric acid acts as a proton donor.
(A) Statement–1 is true, Statement–2 is true; Statement–2 is a correct explanation for Statement–1
(B) Statement–1 is true, Statement–2 is true; Statement–2 is not a correct explanation for Statement–1
(C) Statement–1 is true, Statement–2 is false
(D) Statement–1 is false, Statement–2 is true.

Section–III: Linked Comprehensive Type

This section contains 2 paragraphs C_{14-16} and C_{17-19}. Based upon each paragraph, 3 multiple choice questions have to be answered. Each question has 4 choices (A), (B), (C) and (D), out of which ONLY ONE is correct.

C_{14-16}: Paragraph for Question Nos. 14 to 16

Chemical reactions involve interaction of atoms and molecules. A large number of atoms/molecules (approximately 6.023×10^{23}) are present in a few grams of any chemical compound varying with their atomic/molecular masses. To handle such large numbers conveniently, the mole concept was introduced. This concept has implications in diverse areas such as analytical chemistry, biochemistry, electrochemistry and radiochemistry. The following examples illustrate a typical case, involving chemical/electrochemical reaction, which requires a clear understanding of the mole concept.

A 4.0 molar aqueous solution of NaCl is prepared and 500 mL of this solution is electrolysed. This leads to the evolution of chlorine gas at one of the electrodes (atomic mass: Na = 23, Hg = 200; 1 Faraday = 96500 Coulombs).

14. The total number of moles of chlorine gas evolved is
(A) 0.5 (B) 1.0 (C) 2.0 (D) 3.0

15. If the cathode is a Hg electrode, the maximum weight (g) of amalgam formed from the solution is
(A) 200 (B) 225 (C) 400 (D) 446

16. The total charge (coubombs) required for complete electrolysis is

(A) 24125 (B) 48250 (C) 96500 (D) 193000

C_{17-19}: Paragraph for Question Nos. 17 to 19

The noble gases have closed-shell electronic configuration and are monoatomic gases under normal conditions. The low boiling points of the lighter noble gases are due to weak dispersion forces between the atoms and the absence of other interatomic interactions.

The direct reactions of xenon with fluorine lead to a series of compounds with oxidation numbers + 2, + 4 and + 6. XeF_4 reacts violently with water to give XeO_3. The compounds of xenon exhibit rich stereochemistry and their geometries can be deduced considering the total number of electron pairs in the valence shell.

17. Argon is used in arc welding because of its

(A) low reactivity with metal (B) ability to lower the melting point of metal

(C) flammability (D) high calorific value.

18. The structure of XeO_3 is

(A) linear (B) planar (C) pyramidal (D) T-shaped.

19. XeF_4 and XeF_6 are expected to be

(A) oxidising (B) reducing (C) unreactive (D) strongly basic.

Section–IV: Matrix-Match Type

This section contains 3 questions. Each question contains statements given in two columns which have to be matched. Statements (A, B, C, D) in Column I have to be matched with statements (p, q, r, s) in Column II. The answers to these questions have to be appropriately bubbled as illustrated in the following example.

If the correct matches are A-p, A-s, B-q, B-r, C-p, C-q and D-s, then the correctly bubbled 4×4 matrix should be as follows:

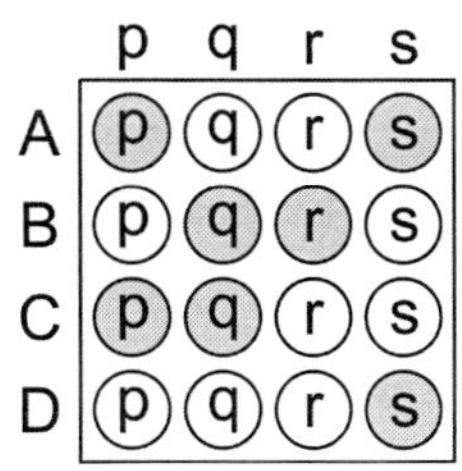

20. Match the complexes in Column I with their properties listed in Column II. Indicate your answer by darkening the appropriate bubbles of the 4×4 matrix given in the ORS.

Column I	*Column II*
(A) $[Co(NH_3)_4(H_2O)_2]Cl_2$	(p) geometrical isomers
(B) $[Pt(NH_3)_2Cl_2]$	(q) paramagnetic
(C) $[Co(H_2O)_5Cl]Cl$	(r) diamagnetic
(D) $[Ni(H_2O)_6]Cl_2$	(s) metal ion with + 2 oxidation state

21. Match gases under specified conditions listed in Column I with their properties/laws in Column II. Indicate your answer by darkening the appropriate bubbles of the 4 × 4 matrix given in the ORS.

Column I	*Column II*
(A) hydrogen gas (P = 200 atm, T = 273 K)	(p) compressibility factor = 1
(B) hydrogen gas (P = 0, T = 273 K)	(q) attractive forces are dominant
(C) CO_2 (P = 1 atm, T = 273 K)	(r) $PV = nRT$
(D) real gas with very large molar volume	(s) $P(V - ab) = nRT$

22. Match the chemical substances in Column I with type of polymers/type of bonds in Column II. Indicate your answer by darkening appropriate bubbles of the 4 × 4 matrix given in the ORS.

Column I	*Column II*
(A) Cellulose	(p) natural polymer
(B) nylon-66	(q) synthetic polymer
(C) protein	(r) amide linkage
(D) sucrose	(s) glycoside linkage

ANSWER KEYS

1. C	2. B	3. A	4. D	5. B	6. A
7. D	8. B	9. B	10. D	11. B	12. A
13. C	14. B	15. D	16. D	17. A	18. C

19. A

20. A-p, q, s; B-p, r, s; C-q, s; D-q, s

21. A-p, s; B-r; C-p, q; D-p, s

22. A-p, s; B-q, r; C-p, r; D-s

SOLUTIONS

Paper-1
Section–I: Straight Objective Type

1. (A)

In 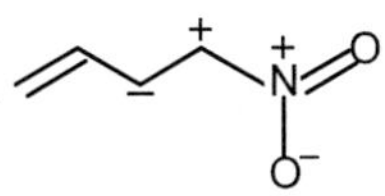 same type of charge (+) is present at the nearest position. Hence it is less stable.

2. (A)

$H_2O(l) \rightleftharpoons H_2O(g)$

(1 bar, 373 K) (1 bar, 373 K)

At 100°C, $H_2O(l)$ is in equilibrium with $H_2O(g)$. Hence $\Delta G = 0$. Since the liquid molecules get converted into gaseous molecules, ΔG = +ve. Hence (A) is correct.

3. (A)

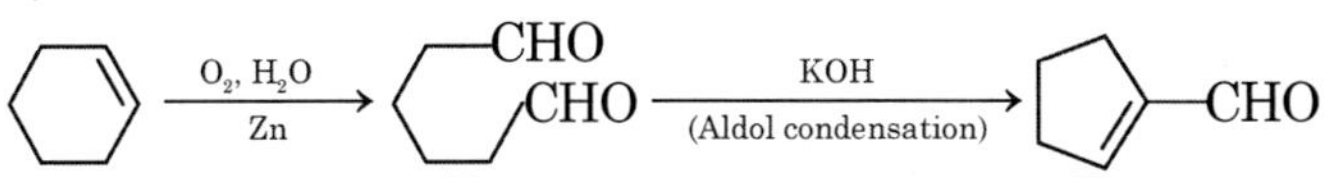

E F

4. (D)

$aG + bH \longrightarrow$ Product

Rate of reaction $\propto [G]^a[H]^b$

$a = 1, \ b = 2$

Overall order of reaction = $a + b = 1 + 2 = 3$

5. (B)

(A) Mn^+: $3d^5 4s^1$

In the presence of strong field ligand CO, the effective configuration of Mn^+ is $3d^6 4s^0$. There are three lone pairs of electrons (d^6) for back bonding with the vacant orbital of carbon in CO.

(B) Fe: $3d^6 4s^2$

In the presence of CO, the effective configuration of Fe(0) is $3d^8$. There are four lone pairs of electrons (d^8) for back bonding with the vacant orbital of carbon in CO.

(C) Cr: $3d^5 4s^1$

In the presence of CO, the effective configuration of Cr(0) is $3d^6$. There are three lone pairs of electrons (d^6) for back bonding with the vacant orbital of carbon in CO.

(D) V: $3d^4 4s^2$

In the presence of CO, the effective configuration of V(0) is $3d^6$. There are three lone pairs of electrons for back bonding with the vacant orbital of carbon of CO.

Hence the C—O back bonding is the lowest in $Fe(CO)_5$.

6. (C)

When a positron is emitted from a nucleus, a proton gets converted into a neutron. Hence the atomic number decreases by one but the atomic mass remains constant.

$$\text{Thus,}\ \frac{\text{Atomic mass}}{\text{Atomic number}} = \frac{23}{10}$$

7. (D)

$Cr_2O_7^{2-} + Fe^{2+} \longrightarrow 2Cr^{3+} + Fe^{3+}$

n factor of $Cr_2O_7^{2-} = 6$; n factor of $Fe^{2+} = 1$

Hence to reduce one mole of potassium dichromate six moles of Fe^{2+} are required. Hence (D) is correct.

8. (A)

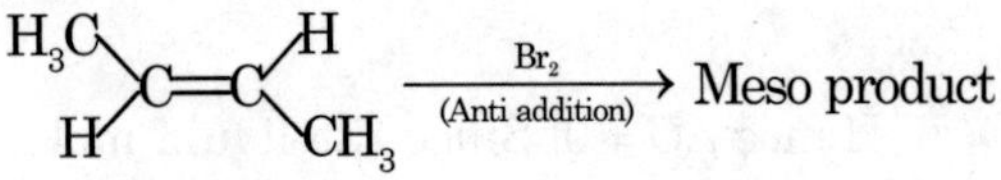

trans-but-2-ene

Hence the bromination of *trans*-but-2-ene gives only one stereoisomer.

9. (B)

$Hg^{2+} + 2KI \longrightarrow \underset{\text{Red}}{HgI_2\downarrow} + 2K^+$

$HgI_2 + \underset{\text{(excess)}}{2KI} \longrightarrow \underset{\text{Soluble}}{K_2HgI_4}$

$HgI_4^{2+} + Co(SCN)_2 \longrightarrow \underset{\text{Blue crystalline precipitate}}{Hg[Co(NCS)_4]}$

Section–II: Assertion-Reason Type

10. (C)

Statement–1 is true and Statement–2 is false.

11. (B)

Statement–1 in true, Statement–2 is true and Statement–2 is not the correct explanation for Statement–1.

Blue colour of the solution is due to the solvated electrons.

12. (C)

Statement–1 is true and Statement–2 is false.

13. (C)

Statement–1 is true and Statement–2 is false.

$C_6H_{12}O_6 + \text{Fehling's solution} \longrightarrow (C_6H_{11}O_7)^- + \underset{\text{Red}}{Cu_2O\downarrow}$

Section–III: Linked Comprehensive Type

14. (C)

In the given reaction, the reagent used is aqueous NaOH and $CHCl_3$.

15. (C)

The electrophile in the given reaction is $:CCl_2$.

$OH^-(aq) + CHCl_3 \rightleftharpoons \bar{C}Cl_3 + H_2O$

$\bar{C}Cl_3 \longrightarrow Cl^- + :CCl_2$

Dichlorocarbene intermediate

16. (B)

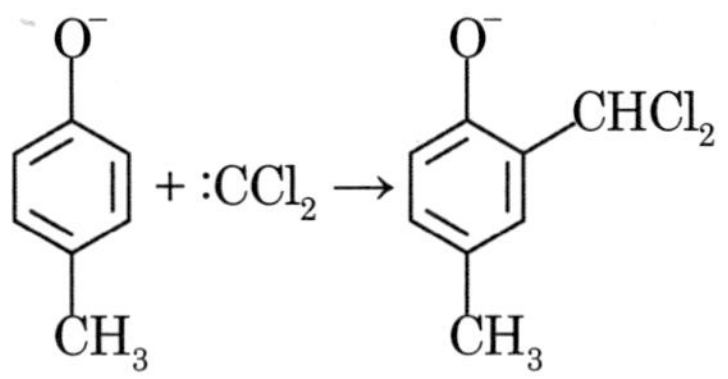

17. (C)

Iodide ion is oxidised by chlorine since the electrode potential of I_2 is less than that of Cl_2.

18. (D)

Mn^{3+} ion is not stable in acid medium since the reaction of Mn^{3+} ion with H_2O is spontaneous.

19. (A)

The blue colour is due to the formation of $Fe_4[Fe(CN)_6]_3$.

Section–IV: Matrix-Match Type

20. (A) The reaction, $2O_2^- \longrightarrow O_2 + O_2^{2-}$, is a redox reaction as well as a disproportionation reaction.

(B) $2CrO_4^{2-} + 2H^+ \longrightarrow Cr_2O_7^{2-} + H_2O$

$Cr_2O_7^{2-}$ ion has a dimeric bridged structure containing tetrahedral chromium ion.

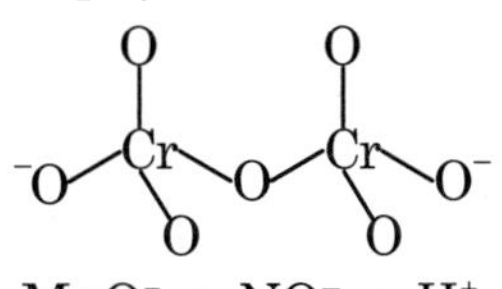

(C) $MnO_4^- + NO_2^- + H^+ \longrightarrow MnO_2 + NO_3^- + OH^-$

It is a redox reaction and one of the products (NO_3^-) has trigonal planar structure:

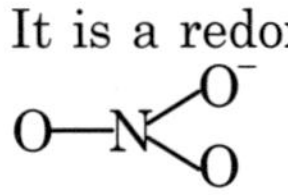

(D) $2HNO_3^- + 3H_2SO_4 + 6FeSO_4 \longrightarrow 2NO + 3Fe_2(SO_4)_3 + 4H_2O$

It is a redox reaction.

21. (A) C_6H_5CHO gives precipitates with 2,4-dinitrophenylhydrazine or ammoniacal $AgNO_3$, and is involved in cyanohydrin formation.

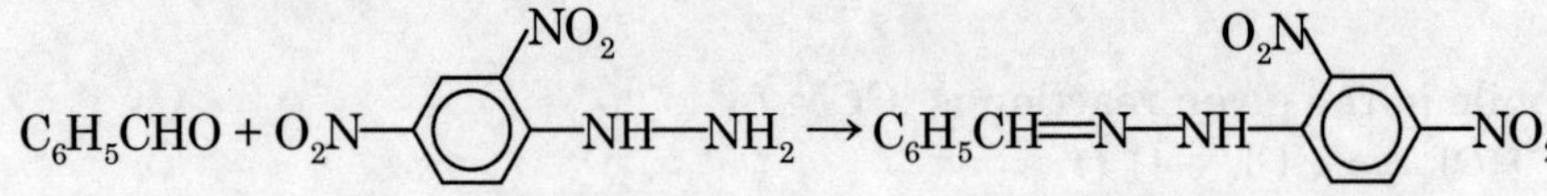

Orange-red precipitate

$$C_6H_5CHO + Ag_2O \xrightarrow{NH_4OH} C_6H_5COO^- + Ag\downarrow$$

Grey-white precipitate

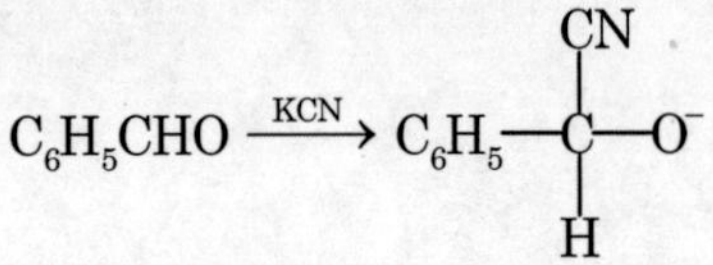

(B) $CH_3C{\equiv}CH$ gives precipitate with ammoniacal $AgNO_3$ solution.

$$CH_3C{\equiv}CH \xrightarrow{\text{ammoniacal } AgNO_3} CH_3C{\equiv}C^-Ag^+\downarrow$$

White precipitate

(C) CN^- ion gives precipitate with ammoniacal $AgNO_3$ solution, is a nucleophile and is involved in cyanohydrin formation.

$$AgNO_3 + CN^- \longrightarrow AgCN\downarrow$$

White precipitate

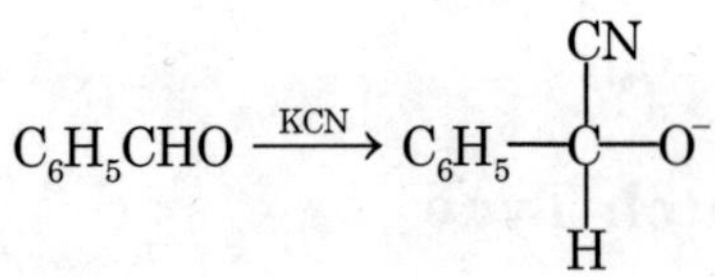

(D) I^- ion gives precipitate with ammoniacal $AgNO_3$ solution and is a nucleophile.

$$AgNO_3 + I^- \longrightarrow AgI\downarrow$$

Yellow precipitate

22. (A) Simple cubic and face centred cubic unit cells have the cell parameters: $a = b = c$ and $\alpha = \beta = \gamma$, and they belong to the same crystal systems.

(B) Cubic and rhombohedral unit cells have the cell parameters: $a = b = c$ and $\alpha = \beta = \gamma$, and they are two different crystal systems.

(C) Cubic and tetragonal unit cells are two different crystal systems.

(D) Hexagonal and monoclinic unit cells are two different crystal systems and have only two crystallography angles of 90° ($\alpha = \beta = 90°$).

PAPER—II

Section–I: Straight Objective Type

1. (C)

The structural isomers for C_6H_{14} are

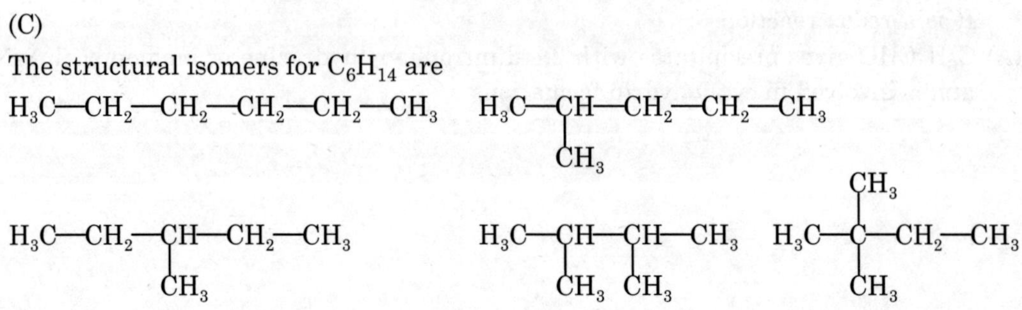

2. (B)

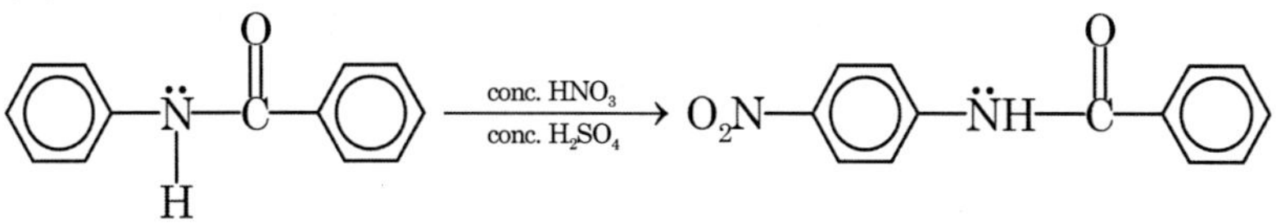

Due to the presence of a lone pair of electrons on the nitrogen atom, it will activate the ring and it will stabilise the intermediate cation at the *o*- and *p*-positions.

3. (A)

$\Delta T_f = i \times K_f \times m$

or $2 = i \times 1.72 \times \dfrac{20}{172} \times \dfrac{1000}{50}$ or $i = \dfrac{2 \times 172 \times 50}{1.72 \times 20 \times 1000} = 0.5$

4. (D)

$O_2^{2-} : (\sigma_{1s})^2 (\sigma_{1s}^*)^2 (\sigma_{2s})^2 (\sigma_{2s}^*)^2 (\sigma_{2p_z})^2 (\pi_{2p_x})^2 (\pi_{2p_y})^2 (\pi_{2p_x}^*)^2 (\pi_{2p_y}^*)^2$; No. of unpaired electrons in $O_2^{2-} = 0$

N═N⟶O; No. of unpaired electrons in $N_2O = 0$

$O{=}\overset{+}{O}{-}\overset{-}{O} \leftrightarrow O{=}O{\rightarrow}O$; No. of unpaired electron in $O_3 = 0$

$O_2^- : (\sigma_{1s})^2(\sigma_{1s}^*)^2(\sigma_{2s})^2 (\sigma_{2s}^*)^2(\sigma_{2p_z})^2(\pi_{2p_x})^2(\pi_{2p_y})^2(\pi_{2p_x}^*)^2(\pi_{2p_y}^*)^1$; No. of unpaired electrons in $O_2^- = 1$

Hence O_2^- is paramagnetic.

5. (B)

$\Delta G = \Delta H - T\Delta S = -54.07 \times 1000 - 298 \times 10 = -57050 \text{ J mol}^{-1}$

$\Delta G = -2.303\ RT \log K$

$= -2.303 \times 8.314 \times 298 \log K = -5705 \log K$

Hence $-57050 = -5705 \log K$

or $\log K = 10$

6. (A)

NO^+, CN^- and N_2 are isoelectronic with CO (fourteen electron system) and hence they all have the same bond order (= 3).

N═O^- is sixteen electron system and its bond order is 2. Hence NO^- has bond order different from that in CO.

7. (D)

The hybridisation of P in P_4 is sp^3.

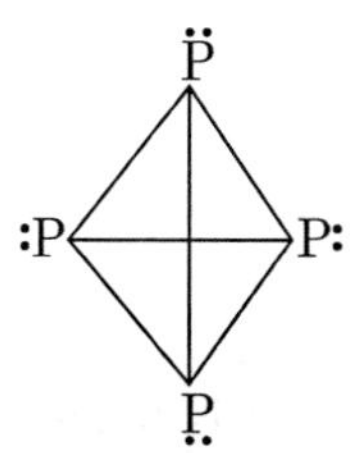

Hybridisation: sp^3; structure: tetrahedral

8. (B)

Extraction of zinc from zinc blende is achieved by roasting followed by reduction with carbon.

9. (B)

$$Br—CH_2—CH_2—Br \xrightarrow{\text{alc. KOH}} CH_2{=}CH—Br \xrightarrow{NaNH_2} HC{\equiv}CH$$

The compound $CH_2{=}CH—Br$ has partial C—Br double bond and it requires a stronger base than alcoholic KOH to remove HBr. Hence the reagent for the conversion is alcoholic KOH followed by $NaNH_2$.

10. (D)

Statement–1 is false and Statement–2 is true.

o-Hydroxybenzoic acid has a lower boiling point than *p*-hydroxybenzoic acid. This is because *o*-hydroxybenzoic acid is involved in intramolecular hydrogen bonding and is a monomer, while *p*-hydroxybenzoic acid is involved in intramolecular hydrogen bonding and is a dimer.

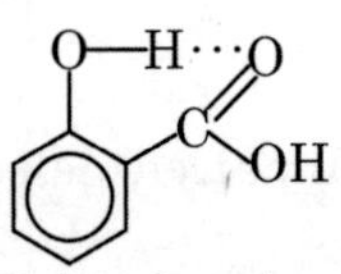

o-Hydroxybenzoic acid
(monomer)

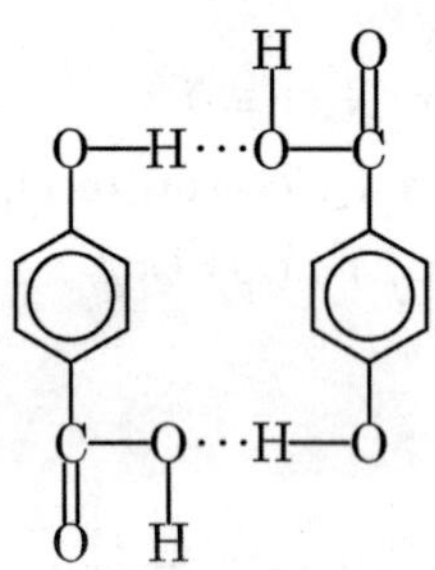

p-Hydroxybenzoic acid
(dimer)
More stronger intermolecular forces increase the boiling point.

Section–II: Assertion-Reason Type

11. (B)

Statement–1 is true, Statement–2 is true and Statement–2 is not a correct explanation for Statement–1. The formation of micelles occurs only above a particular temperature called Kraft temperature (T_k) and above a particular concentration called critical micelle concentration(C_M). Each micelle contains at least one hundred molecules. Hence the conductivity of the solution decreases sharply at the critical micelle concentration.

12. (A)

Statement–1 is true, Statement–2 is true and Statement–2 is a correct explanation for Statement–1.

According to the Fajan's rule, the small cations having high charge density have tendency to form always covalent bonds.

13. (C)

Statement–1 is true and Statement–2 is false. Orthoboric acid (H_3BO_3) is a weak Lewis acid: $H_3BO_3 + H_2O \rightleftharpoons B(OH)_4 + H^+$

Orthoboric acid accepts OH^- ion from water and it does not donate proton.

Section–III: Linked Comprehensive Type

14. (B)

$NaCl \longrightarrow Na^+ + Cl^-$

At anode: $2Cl^- \longrightarrow Cl_2$

Moles of Cl^- ion = 2 in 500 mL

Hence 1 mole of Cl_2 gas is evolved.

15. (D)

Na/Hg (amalgam) formed at the cathode = 2 moles

Weight of 2 moles of Na/Hg = 2 × 223 = 446 g

16. (D)

$2Cl^- - 2e^- \longrightarrow Cl_2$

2 moles of electrons (2 Faraday) are required.

1 F = 96500 C

2 F = 193000 C

Hence total charge required for complete electrolysis of NaCl is 193000 C.

17. (A)

In high temperature metallergical extraction by arc welding, argon is mainly used to provide an inert atmosphere so that oxygen cannot react with metals to form metal oxide. The use of argon in arc welding is due to its low reactivity with metals even at high temperature.

18. (C)

The hybridisation of Xe in XeO_3 is sp^3 and it has a pyramidal structure.

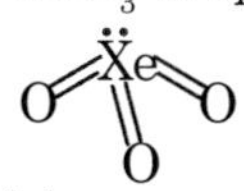

Pyramidal structure of XeO_3

19. (A)

XeF_4 and XeF_6 are expected to undergo oxidation.

$6XeF_4 + 12H_2O \longrightarrow 2XeO_3 + 4Xe + 24HF + 3O_2$

$XeF_6 + 3H_2O \longrightarrow XeO_3 + 6HF$

Section–IV: Matrix-Match Type

20. (A) $[Co(NH_3)_4(H_2O)_2]Cl_2$ exhibits geometrical isomerism, is paramagnetic and has metal ion with + 2 oxidation state. Co^{2+} is a $3d^7$ system having three unpaired electrons.

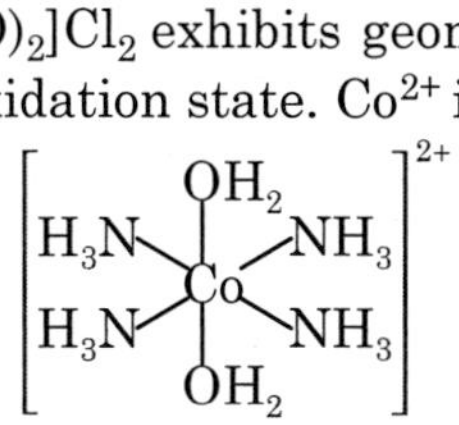

trans-$[Co(NH_3)_4(H_2O_2)]^{2+}$

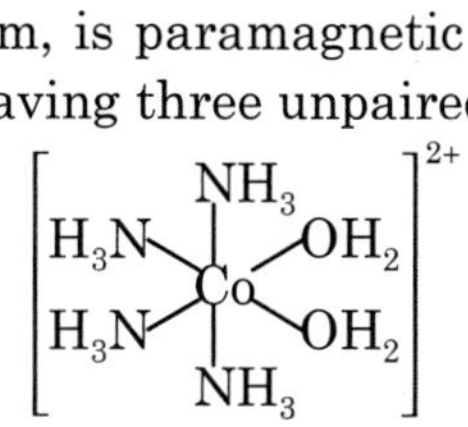

cis-$[Co(NH_3)_4(H_2O)_2]^{2+}$

(B) $[Pt(NH_3)_2Cl_2]$ exhibits geometrical isomerism, is diamagnetic and has metal ion with + 2 oxidation state. Pt^{2+} is a $5d^8$ system having no unpaired electron (dsp^2 hybridisation).

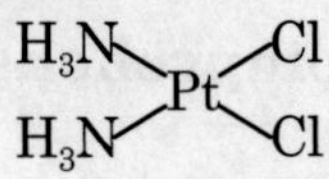

cis-$[Pt(NH_3)_2Cl_2]$

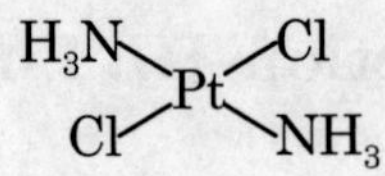

trans-$[Pt(NH_3)_2Cl_2]$

(C) $[Co(H_2O)_5Cl]Cl$ has metal ion with + 2 oxidation state and is paramagnetic. Co^{2+} is a $3d^7$ system having three unpaired electrons.

(D) $[Ni(H_2O)_6]Cl_2$ has metal ion with + 2 oxidation state and is paramagnetic. Ni^{2+} is a $3d^8$ system having two unpaired electrons. H_2O being a weak field ligand, the hybridisation of Ni in $[Ni(H_2O)_6]Cl_2$ is sp^3d^2.

21. (A) The compressibility factor of H_2 gas ($P = 200$ atm, $T = 273$ K) $\neq 1$. $Z = PV_m/RT$ at high pressure and low temperature. It follows the real gas equation $P(V - nb) = RT$. van der Waals equation $\left(P + \frac{an^2}{V^2}\right)(V - nb) = RT$ reduces to $P(V - nb) = RT$ at high pressure and low temperature since $a = 0$ at high pressure and low temperature.

(B) At $P = 0$, Z for H_2 gas is 1 and it increases continuously with increasing pressure.

(C) The molecules of CO_2 have larger intermolecular attractive forces under normal conditions of temperature and pressure (1 atm, 273 K).

(D) At very larger molecular volume, $Z \neq 1$. $Z = PV_m/RT$.

22. (A) Cellulose is a natural polymer and has glycoside linkage.

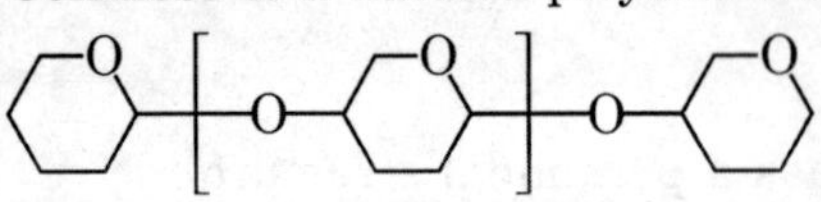

Glycoside linkage

(B) Nylon-6,6 is a synthetic polymer and has amide linkage.

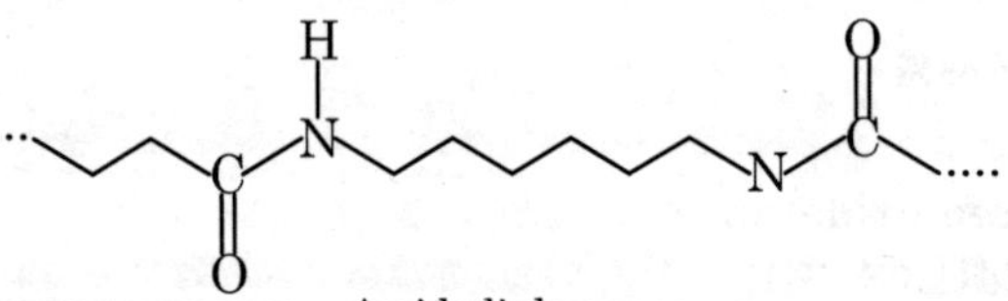

Amide linkage

(C) Protein is a natural polymer and has amide linkage.

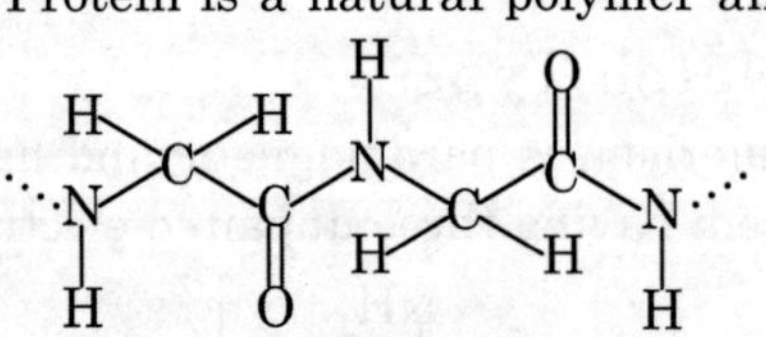

Amide linkage

(D) Sucrose has glycoside linkage.

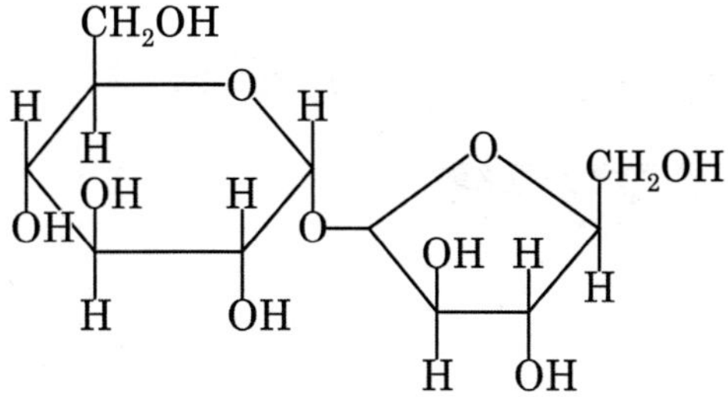

Glycoside linkage

I.I.T. JEE Objective Question Paper—2008

Section–I: Straight Objective Type

This section contains 6 multiple choice questions. Each question has 4 choices (A), (B), (C) and (D), out of which ONLY ONE is correct.

1. Hyperconjugation involves overlap of the following orbitals

(A) $\sigma-\sigma$ (B) $\sigma-p$ (C) $p-p$ (D) $\pi-\pi$

2. The major product of the following reaction is

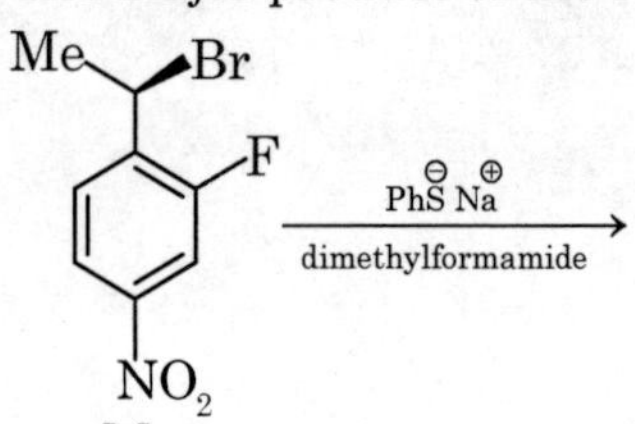

(A) 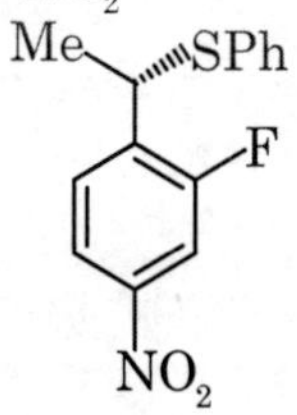

(B) 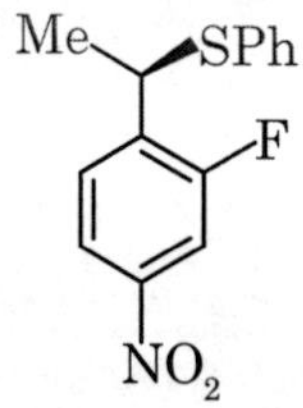

(C) 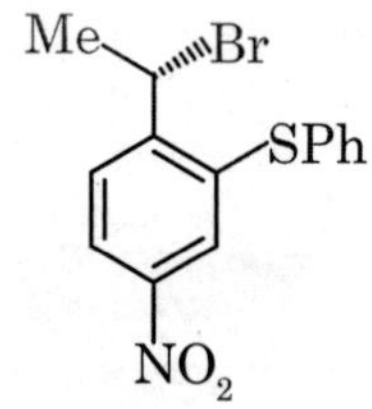

(D) 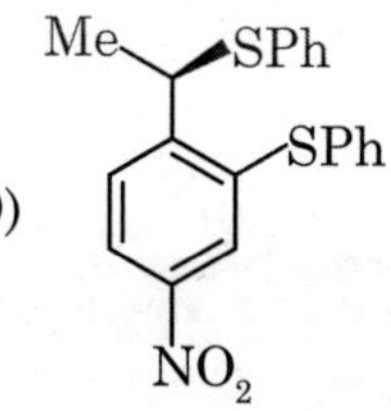

3. Aqueous solution of $Na_2S_2O_3$ on reaction with Cl_2 gives

(A) $Na_2S_4O_6$ (B) $NaHSO_4$ (C) NaCl (D) NaOH

4. Native silver metal forms a water soluble complex with a dilute aqueous solution of NaCN in the presence of

(A) nitrogen (B) oxygen (C) carbon dioxide (D) argon

5. Under the same reaction conditions, initial concentration of 1.386 mol dm^{-1} of a substance becomes half in 40 seconds and 20 seconds through first order and zero order kinetics, respectively. Ratio (k_1/k_0) of the rate constants for first order (k_1) and zero order (k_0) of the reactions is

(A) 0.5 $mol^{-1}\ dm^3$ (B) 1.0 mol dm^{-3} (C) 1.5 mol dm^{-3} (D) 2.0 $mol^{-1}\ dm^3$

6. 2.5 mL of $(2/5)M$ weak monoacidic base ($K_b = 1 \times 10^{-12}$ at 25°C) is titrated with $(2/15)M$ HCl in water at 25°C. The concentration of H^+ at equivalence point is ($K_w = 1 \times 10^{-14}$ at 25°C)

(A) $3.7 \times 10^{-13}\ M$ (B) $3.2 \times 10^{-7}\ M$ (C) $3.2 \times 10^{-2}\ M$ (D) $2.7 \times 10^{-2}\ M$

Section–II: Multiple Correct Answers Type

This section contains 4 multiple correct answer(s) type questions. Each question has 4 choices (A), (B), (C) and (D), out of which ONE or MORE is/are correct.

7. The correct statement(s) about the compound given below is(are)

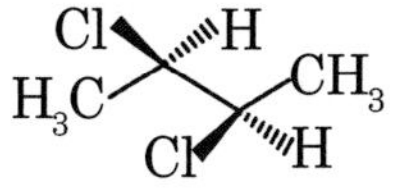

(A) The compound is optically active

(B) The compound possesses centre of symmetry

(C) The compound possesses plane of symmetry

(D) The compound possesses axis of symmetry.

8. The correct statement(s) concerning the structures E, F and G is(are)

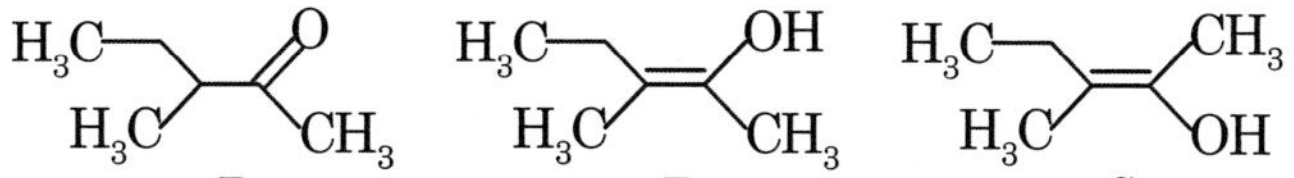

E F G

(A) E, F and G are resonance structures

(B) E, F and G are tautomers

(C) F and G are geometrical isomers

(D) F and G are diastereomers.

9. A solution of colourless salt H on boiling with excess NaOH produces a non-flammable gas. The gas evolution ceases after some time. Upon addition of Zn dust to the same solution, the gas evolution restarts. The colourless salt(s) H is(are)

(A) NH_4NO_3 (B) NH_4NO_2 (C) NH_4Cl (D) $(NH_4)_2SO_4$

10. A gas described by van der Waals equation

(A) behaves similar to an ideal gas in the limit of large molar volumes

(B) behaves similar to an ideal gas in the limit of large pressures

(C) is characterized by van der Waals coefficients that are dependent on the identity of the gas but are independent of the temperature

(D) has the pressure that is lower than the pressure exerted by the same gas behaving ideally.

Section–III: Reasoning Type

This section contains 4 reasoning type questions. Each question has 4 choice (A), (B), (C) and (D), out of which ONLY ONE is correct.

11. Statement–1: Bromobenzene upon reaction with Br_2/Fe gives 1,4-dibromobenzene as the major product.

Statement–2: In bromobenzene, the inductive effect of the bromo group is more dominant than the mesomeric effect in directing the incoming electrophile.

(A) Statement–1 is true, Statement–2 is true; Statement–2 is a correct explanation for Statement–1

(B) Statement–1 is true, Statement–2 is true; Statement–2 is not a correct explanation for Statement–1
(C) Statement–1 is true, Statement–2 is false
(D) Statement–1 is false, Statement–2 is true.

12. Statement–1: Pb^{4+} compounds are stronger oxidizing agents than Sn^{4+} compounds.
Statement–2: The higher oxidation states for the group 14 elements are more stable for the heavier members of the group due to 'inert pair effect'.
(A) Statement–1 is true, Statement–2 is true; Statement–2 is a correct explanation for Statement–1
(B) Statement–1 is true, Statement–2 is true; Statement–2 is not a correct explanation for Statement–1
(C) Statement–1 is true, Statement–2 is false
(D) Statement–1 is false, Statement–2 is true.

13. Statement–1: The plot of atomic number (y-axis) versus number of neutrons (x-axis) for stable nuclei shows a curvature towards x-axis from the line of 45° slope as the atomic number is increased.
Statement–2: Proton-proton electrostatic repulsions begin to overcome attractive forces involving protons and neutrons in heavier nuclides.
(A) Statement–1 is true, Statement–2 is true; Statement–2 is a correct explanation for Statement–1
(B) Statement–1 is true, Statement–2 is true; Statement–2 is not a correct explanation for Statement–1
(C) Statement–1 is true, Statement–2 is false
(D) Statement–1 is false, Statement–2 is true.

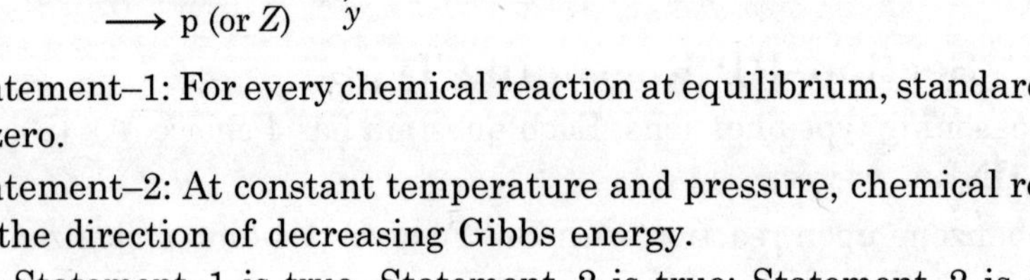

14. Statement–1: For every chemical reaction at equilibrium, standard Gibbs energy of reaction is zero.
Statement–2: At constant temperature and pressure, chemical reactions are spontaneous in the direction of decreasing Gibbs energy.
(A) Statement–1 is true, Statement–2 is true; Statement–2 is a correct explanation for Statement–1
(B) Statement–1 is true, Statement–2 is true; Statement–2 is not a correct explanation for Statement–1
(C) Statement–1 is true, Statement–2 is false
(D) Statement–1 is false, Statement–2 is true.

Section–IV: Linked Comprehension Type

This section contains 3 paragraphs. Based upon each paragraph, 3 multiple choice questions have to be answered. Each question has 4 choices (A), (B), (C) and (D), out of which ONLY ONE is correct.

Paragraph for Question Nos. 15 to 17

In the following reaction sequence, products I, J and L are formed. K represents a reagent.

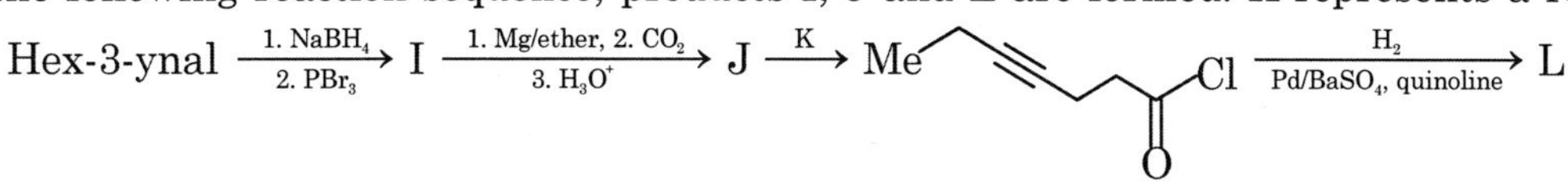

15. The structure of the product I is

(A) 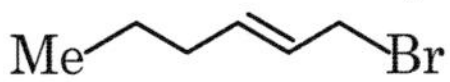

(B) 

(C) 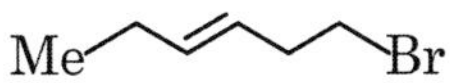

(D) 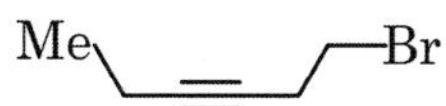

16. The structures of compounds J and K, respectively, are

(A) 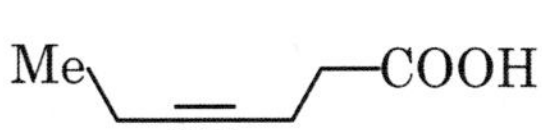 and $SOCl_2$

(B) 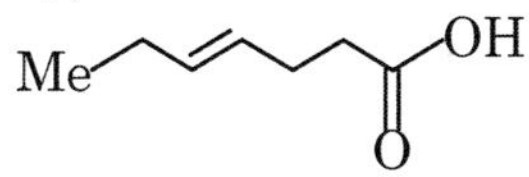 and SO_2Cl_2

(C) 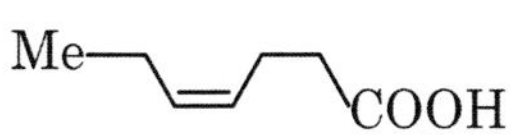 and $SOCl_2$

(D) 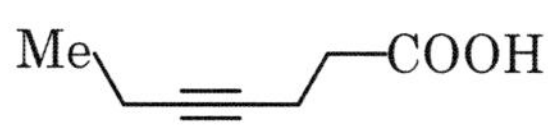 and CH_3SO_2Cl

17. The structure of product L is

(A) 

(B) 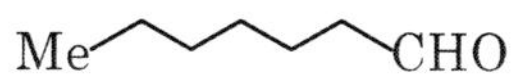

(C) 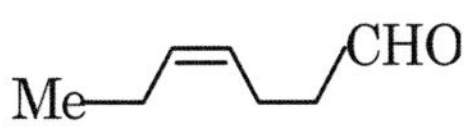

(D) 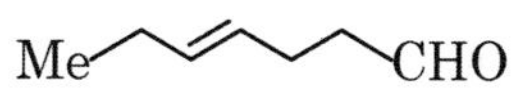

Paragraph for Question Nos. 18 to 20

There are some deposits of nitrates and phosphates in earth's crust. Nitrates are more soluble in water. Nitrates are difficult to reduce under the laboratory conditions but microbes do it easily. Ammonia forms large number of complexes with transition metal ions. Hybridization easily explains the ease of sigma donation capability of NH_3 and PH_3. Phosphine is a flammable gas and is prepared from white phosphorus.

18. Among the following, the correct statement is

(A) Phosphates have no biological significance in humans

(B) Between nitrates and phosphates, phosphates are less abundant in earth's crust

(C) Between nitrates and phosphates, nitrates are less abundant in earth's crust

(D) Oxidation of nitrates is possible in soil.

19. Among the following, the correct statement is

(A) Between NH_3 and PH_3, NH_3 is a better electron donor because the lone pair of electrons occupies spherical *s* orbital and is less directional

(B) Between NH_3 and PH_3, PH_3 is a better electron donor because the lone pair of electrons occupies sp^3 orbital and is more directional

(C) Between NH_3 and PH_3, NH_3 is a better electron donor because the lone pair of electrons occupies sp^3 orbital and is more directional

(D) Between NH_3 and PH_3, PH_3 is a better electron donor because the lone pair of electrons occupies spherical *s* orbital and is less directional.

20. White phosphorus on reaction with NaOH gives PH_3 as one of the products. This is a

(A) dimerization reaction (B) disproportionation reaction

(C) condensation reaction (D) precipitation reaction.

Paragraph for Question Nos. 21 to 23

Properties such as boiling point, freezing point and vapour pressure of a pure solvent change when solute molecules are added to get homogeneous solution. These are called colligative properties. Applications of colligative properties are very useful in day-to-day life. One of its examples is the use of ethylene glycol and water mixture as anti-freezing liquid in the radiator of automobiles. A solution *M* is prepared by mixing ethanol and water. The mole fraction of ethanol in the mixture is 0.9.

Given: Freezing point depression constant of water (K_f^{water}) = 1.86 K kg mol^{-1}; Freezing point depression constant of ethanol (K_f^{ethanol}) = 2.0 K kg mol^{-1}; Boiling point elevation constant of water (K_b^{water}) = 0.52 K kg mol^{-1}; Boiling point elevation constant of ethanol (K_b^{ethanol}) = 1.2 K kg mol^{-1}; Standard freezing point of water = 273 K; Standard freezing point of ethanol = 155.7 K; Standard boiling point of water = 373 K; Standard boiling point of ethanol = 351.5 K; Vapour pressure of pure water = 32.8 mmHg; Vapour pressure of pure ethanol = 40 mmHg; Molecular weight of water = 18 g mol^{-1}; Molecular weight of ethanol = 46 g mol^{-1}.

In answering the following questions, consider the solutions to be ideal dilute solutions and solutes to be non-volatile and non-dissociative.

21. The freezing point of the solution *M* is

(A) 268.7 K (B) 268.5 K (C) 234.2 K (D) 150.9 K

22. The vapour pressure of the solution *M* is

(A) 39.3 mmHg (B) 36.0 mmHg (C) 29.5 mmHg (D) 28.8 mmHg

23. Water is added to the solution *M* such that the mole fraction of water in the solution becomes 0.9. The boiling point of this solution is

(A) 380.4 K (B) 376.2 K (C) 375.5 K (D) 354.7 K

ANSWER KEYS

1. A	2. B	3. B	4. B	5. A	6. D
7. A, D	8. B, C, D	9. A, B	10. A, C, D	11. B	12. C
13. A	14. D	15. D	16. A	17. C	18. C
19. C	20. B	21. D	22. B	23. B	

SOLUTIONS

1. (A)
Hyperconjugation involves the overlap of σ C—H bonding electrons with the *p*-orbital of unsaturated system, carbocation or carbon free radical.

2. (B)
The given organic compound is a haloalkane as well as haloarene. Since PhS^- is a weak base, it cannot cause nucleophilic substitution by benzyne mechanism. Since the NO_2 group is present at the *meta* position with respect to the fluorine atom, S_NAr mechanism is also ruled out. Hence, the given compound reacts with PhS^- by S_{N^2} mechanism.

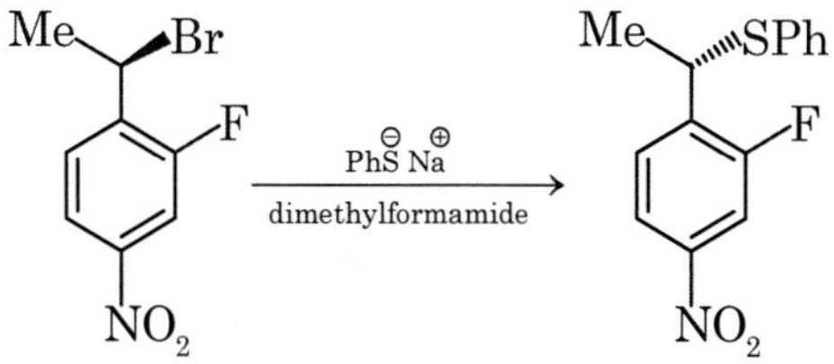

3. (B)
Chlorine is a stronger oxidising agent than iodine. Hence $Na_2S_2O_3$ on oxidation forms $NaHSO_4$ rather than sodium tetrathionate.
$Na_2S_2O_3 + 4Cl_2 + 5H_2O \longrightarrow 2NaHSO_4 + 8HCl$

4. (B)
Silver is oxidised to Ag^+ by oxygen in the presence of cyanide ion. The reaction of Ag^+ with CN^- gives a complex ion, $[Ag(CN)_2]^-$.

5. (A)
For first order reaction, $k_1 = \dfrac{0.693}{t_{1/2}} = \dfrac{0.693}{40}$ sec^{-1}. For zero order reaction,

$$k_0 = \frac{[A_0]}{2t_{1/2}} = \frac{1.386}{2 \times 20} \text{ mol dm}^{-3} \text{ sec}^{-1}$$

$$\frac{k_1}{k_0} = \frac{0.693 \times 2 \times 20}{40 \times 1.386} = \frac{1}{2} = 0.5 \text{ mol}^{-1} \text{ dm}^3$$

6. (D)
$BOH + HCl \longrightarrow BCl + H_2O$

$$2.5 \times 10^{-3} \times \frac{2}{5} = x \times 10^{-3} \times \frac{2}{15}$$

or $$x = 7.5 \text{ mL}$$

$$[BCl] = \frac{2.5 \times \frac{2}{5}}{10} = 0.1M$$

$$\begin{array}{llll} B^+ & + H_2O \rightleftharpoons & BOH + & H^+ \\ 0.1(1-h) & & 0.1\,h & 0.1\,h \end{array}$$

$$K_h = \frac{K_w}{K_b} = \frac{[\text{BOH}][\text{H}^+]}{[\text{B}^+]} = \frac{0.1\,h^2}{(1-h)}$$

$$\frac{1\times10^{-14}}{1\times10^{-12}} = 1\times10^{-2} = \frac{0.1\,h^2}{(1-h)}$$

or h = 0.27 (We cannot ignore the value of h with respect to 1 because it is more than 0.1.)

Hence $[H^+] = 0.1 \times 0.27 = 2.7 \times 10^{-2} M$

7. (A, D)

The given compound is represented in saw-horse projection as

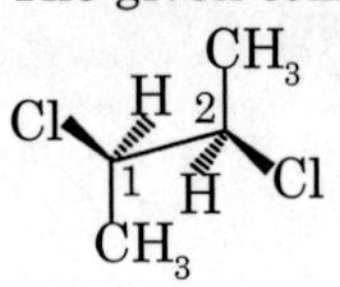

It does not have a plane of symmetry or a centre of symmetry. Hence it is optically active. The compound on rotation about C_1—C_2, keeping C_2 fixed and C_1 rotated by 180°, gives the following compound:

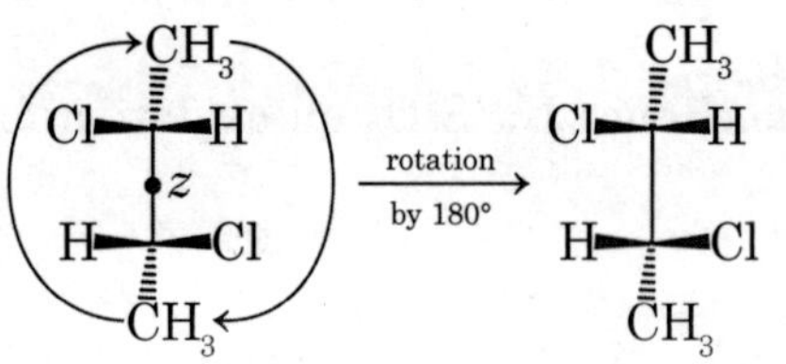

The compound possesses 180° axis of symmetry (two fold axis of symmetry) passing through the point z perpendicular to the plane of paper.

8. (B, C, D)

E is a keto form and F and G are its enol forms. F and G are also geometrical isomers and diastereoisomers.

9. (A, B)

Since the colourless salt produces NH_3 gas (non-inflammable) on boiling with excess of NaOH, it is an ammonium salt. On addition of zinc dust (a reducing agent) to this solution, only the nitrites and nitrates will liberate NH_3 gas again.

$$3Zn + NO_2^- + 5OH^- \longrightarrow 3ZnO_2^{2-} + H_2O + NH_3$$

$$4Zn + NO_3^- + 7OH^- \longrightarrow 4ZnO_2^{2-} + 2H_2O + NH_3$$

10. (A, C, D)

A real gas following the van der Waals equation behaves similar to an ideal gas at low pressures (or large molar volumes). The van der Waals coefficients a and b are independent of temperature and characteristic of a gas. The pressure exerted by a real gas is lower than the pressure exerted by the same gas under ideal conditions since some pressure is lost by the real gas due to the presence of intermolecular force of attraction.

11. (B)

The Statement–2 is correct because the inductive effect is a permanent effect and the inductive effect of halogen atom always dominates its mesomeric effect. But the Statement–2 cannot explain the formation of 1,4-dibromobenzene as the major product. The reason for the formation of 1,4-dibromobenzene is that Br is an *ortho* and *para* directing group due to the resonance stabilization of arenium ion. The prominent attack of Br^+ occurs at the *para* position due to the steric effect.

12. (C)
Statement–1 is true and Statement–2 is false. Since Pb^{4+} ion is less stable than Sn^{4+} ion, Pb^{4+} ion is a stronger oxidising agent than Sn^{4+} ion. Due to the inert pair effect, the stability of higher oxidation state of group 14 elements decreases on moving down the group.

13. (A)
Proton-proton repulsions begin to dominate over proton-neutron attractions with the increase of value of Z. Hence the Statement–2 is correct. In order to stabilize the heavier nuclides, more number of neutrons are to be added so that the proton-neutron attractions overcome the proton-proton repulsions. Hence the Statement–2 is the correct explanation of Statement–1.

14. (D)
For a chemical reaction at equilibrium, Gibbs free energy change of reaction (ΔG) is zero but the standard Gibbs free energy change of reaction (ΔG^0) may not be zero. Hence the Statement–1 is false but the Statement–2 is true.

15. (D)

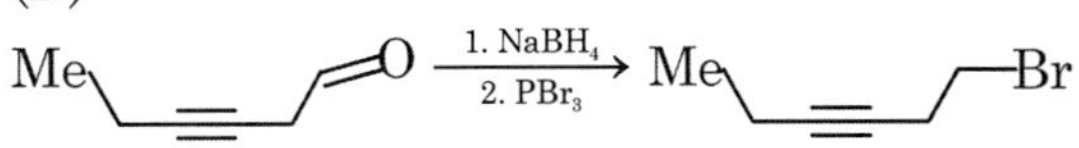

I

16. (A)

J

17. (C)

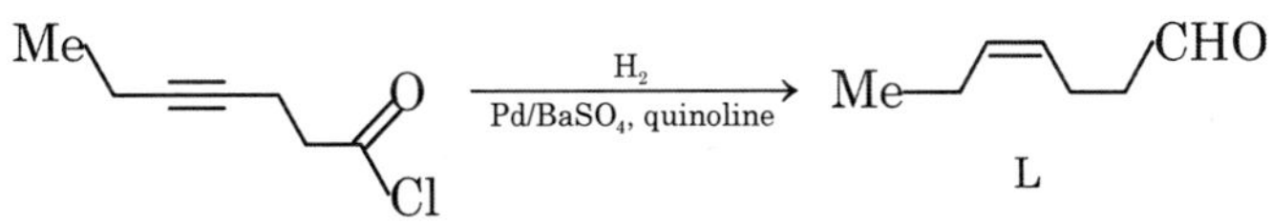

L

18. (C)
Nitrates are less abundant than phosphates in the earth's crust because they are soluble in water. Moreover, they can also be reduced by the numerous microbes present in the earth's crust.

19. (C)
NH_3 is a better electron donor than PH_3 because the lone pair of electrons on the nitrogen atom occupies an sp^3 hybridised orbital and is more directional due to the smaller size of nitrogen atom.

20. (B)

$$\overset{0}{P}_4 + 3NaOH + 3H_2O \longrightarrow 3NaH_2\overset{+1}{P}O_2 + \overset{-1}{P}H_3$$

This is a disproportionation reaction.

21. (D)

Since the mole fraction of ethanol in the mixture is 0.9, ethanol is the solvent and water is the solute. Assuming water to be a non-volatile solute, the freezing point of the solution should be less than the freezing point of the pure solvent (ethanol). Thus, the freezing point of the solution is less than 155.7 K.

22. (B)

It is given that the solute is to be treated as non-volatile and non-dissociative. Thus, vapour pressure of solution = $P^0_{\text{solvent}}\ x_{\text{solvent}} = 40 \times 0.9 = 36.0$ mmHg.

23. (B)

Since the mole fraction of water in solution is 0.9, water is the solvent and ethanol is the solute and has to be considered as non-volatile and non-dissociative. Mole fraction of ethanol in the solution = 0.1. Molality of ethanol in solution = 0.1/(0.9 × 18) = 6.17. $\Delta T_b = K_b m = 0.52 \times 6.17 = 3.2$. Hence the boiling point of solution = 373 + 3.2 = 376.2 K.

I.I.T. JEE Objective Question Paper—2009

Section–I: Single Correct Choice Type

This section contains 8 multiple choice questions. Each question has 4 choices (A), (B), (C) and (D), for its answer out of which ONLY ONE is correct.

1. Among the electrolytes Na_2SO_4, $CaCl_2$, $Al_2(SO_4)_3$ and NH_4Cl, the most effective coagulating agent for Sb_2S_3 sol is
(A) Na_2SO_4 (B) $CaCl_2$ (C) $Al_2(SO_4)_3$ (D) NH_4Cl

2. Given that the abundances of isotopes ^{54}Fe, ^{56}Fe and ^{57}Fe are 5%, 90% and 5%, respectively, the atomic mass of Fe is
(A) 55.85 (B) 55.95 (C) 55.75 (D) 56.06

3. The correct acidity order of the following is

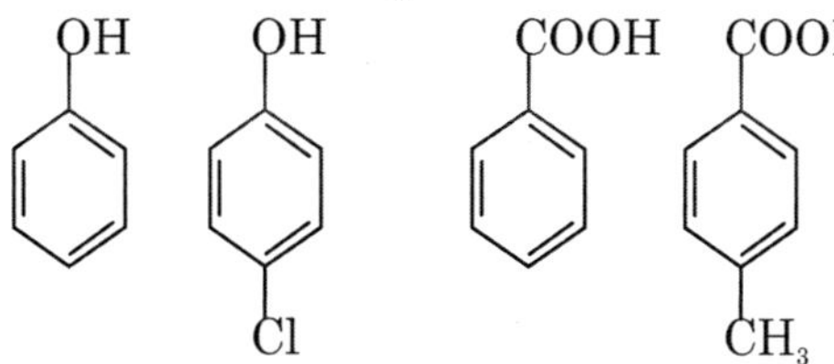

(I) (II) (III) (IV)

(A) (III) > (IV) > (II) > (I) (B) (IV) > (III) > (I) > (II)
(C) (III) > (II) > (I) > (IV) (D) (II) > (III) > (IV) > (I)

4. The IUPAC name of the following compound is

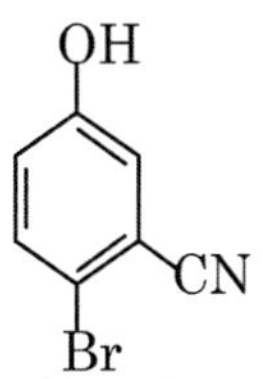

(A) 4-Bromo-3-cyanophenol (B) 2-Bromo-5-hydroxybenzonitrile
(C) 2-Cyano-4-hydroxybromobenzene (D) 6-Bromo-3-hydroxybenzonitrile

5. The term that corrects for the attractive forces present in a real gas in the van der Waals equation is
(A) nb (B) $\frac{an^2}{V^2}$ (C) $-\frac{an^2}{V^2}$ (D) $-nb$

6. Among cellulose, poly(vinyl chloride), nylon and natural rubber, the polymer in which the intermolecular force of attraction is weakest is
(A) nylon (B) poly(vinyl chloride) (C) cellulose (D) natural rubber.

7. The reaction of P_4 with X leads selectively to P_4O_6. The X is
(A) dry O_2 (B) a mixture of O_2 and N_2
(C) moist O_2 (D) O_2 in the presence of aqueous NaOH

8. The Henry's law constant for the solubility of N_2 gas in water at 298 K is 1.0×10^5 atm. The mole fraction of N_2 in air is 0.8. The number of moles of N_2 from air dissolved in 10 moles of water at 298 K and 5 atm pressure is
(A) 4.0×10^{-4} (B) 4.0×10^{-5} (C) 5.0×10^{-4} (D) 4.0×10^{-6}

Section–II: Multiple Correct Choice Type

This section contains 4 multiple choice questions. Each question has 4 choices (A), (B), (C) and (D) for its answer, out of which ONE OR MORE is/are correct.

9. The compound(s) that exhibit(s) geometrical isomerism is(are)
(A) $[Pt(en)Cl_2]$ (B) $[Pt(en)_2]Cl_2$ (C) $[Pt(en)_2Cl_2]Cl_2$ (D) $[Pt(NH_3)_2Cl_2]$

10. The compound(s) formed upon combustion of sodium metal in excess air is(are)
(A) Na_2O_2 (B) Na_2O (C) NaO_2 (D) NaOH

11. The correct statement(s) regarding defects in solids is(are)
(A) Frenkel defect is usually favoured by a very small difference in the sizes of cation and anion.
(B) Frenkel defect is a dislocation defect.
(C) Trapping of an electron in the lattice leads to the formation of F-center.
(D) Schottky defects have no effect on the physical properties of solids.

12. The correct statement(s) about the compound $H_3C(HO)HC—CH{=}CH—CH(OH)CH_3$ (X) is(are)
(A) The total number of stereoisomers possible for X is 6.
(B) The total number of diastereomers possible for X is 3.
(C) If the stereochemistry about the double bond in X is *trans*, the number of enantiomers possible for X is 4.
(D) If the stereochemistry about the double bond in X is *cis*, the number of enantiomers possible for X is 2.

Section–III: Comprehension Type

This section contains 2 groups of questions. Each group has 3 multiple choice questions based on a paragraph. Each question has 4 choices (A), (B), (C) and (D) for its answer, out of which ONLY ONE is correct.

Paragraph for Question Nos. 13 to 15

p-Amino-*N*,*N*-dimethylaniline is added to a strongly acidic solution of X. The resulting solution is treated with a few drops of aqueous solution of Y to yield blue coloration due to the formation of methylene blue. Treatement of the aqueous solution of Y with the reagent potassium hexacyanoferrate(II) leads to the formation of an intense blue precipitate. The precipitate dissolves on excess addition of the reagant. Similarly, treatment of the solution of Y with the solution of potassium hexacyanoferrate(III) leads to a brown coloration due to the formation of Z.

13. The compound X is

(A) $NaNO_3$ (B) NaCl (C) Na_2SO_4 (D) Na_2S

14. The compound Y is

(A) $MgCl_2$ (B) $FeCl_2$ (C) $FeCl_3$ (D) $ZnCl_2$

15. The compound Z is

(A) $Mg_2[Fe(CN)_6]$ (B) $Fe[Fe(CN)_6]$ (C) $Fe_4[Fe(CN)_6]_3$ (D) $K_2Zn_3[Fe(CN)_6]_2$

Paragraph for Question Nos. 16 to 18

A carbonyl compound P, which gives positive iodoform test, undergoes reaction with MeMgBr followed by dehydration to give an olefin Q. Ozonolysis of Q leads to a dicarbonyl compound R, which undergoes intramolecular aldol reaction to give predominantly S.

$$P \xrightarrow[\substack{2.\ H^+,\ H_2O \\ 3.\ H_2SO_4,\ \Delta}]{1.\ MeMgBr} Q \xrightarrow[2.\ Zn,\ H_2O]{1.\ O_3} R \xrightarrow[2.\ \Delta]{1.\ OH^-} S$$

16. The structure of the carbonyl compound P is

(A) 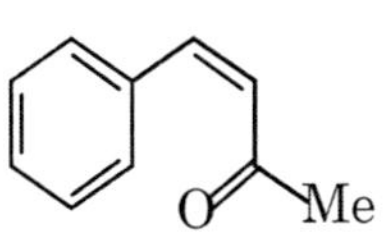

(B) 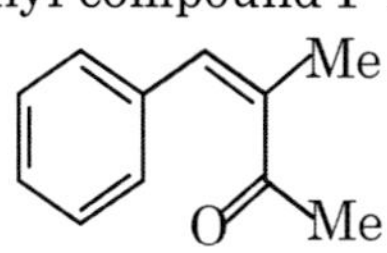

(C) 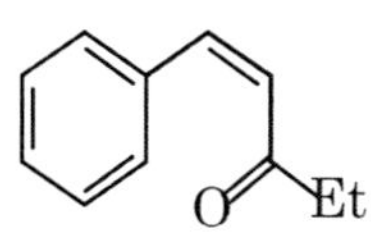

(D) 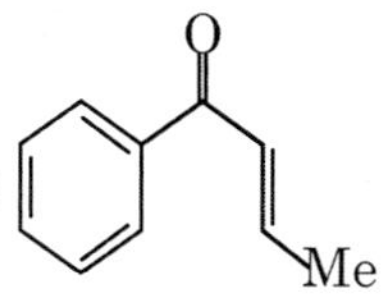

17. The structures of the products Q and R, respectively, are

(A) 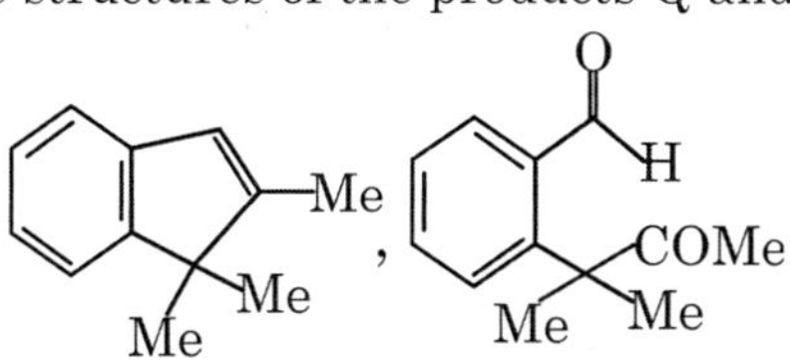

(B) 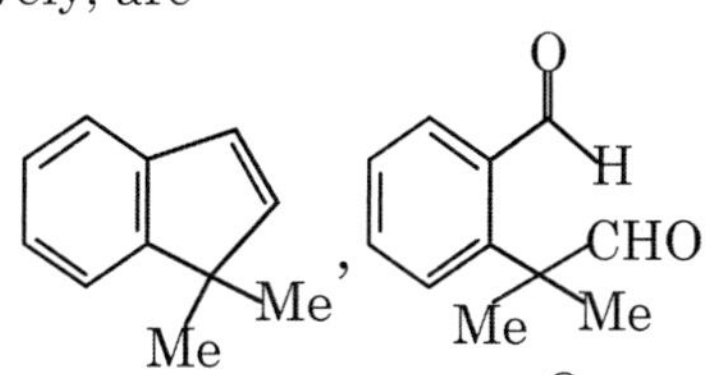

(C) 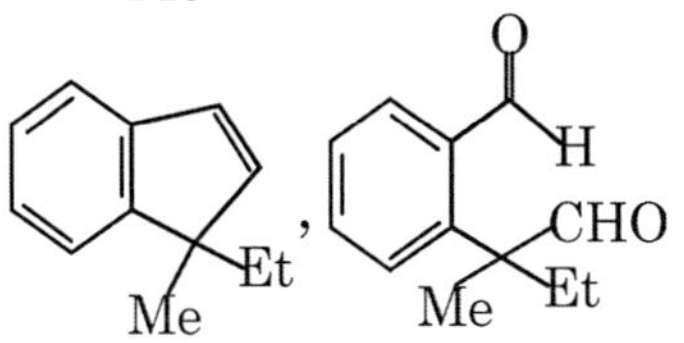

(D) 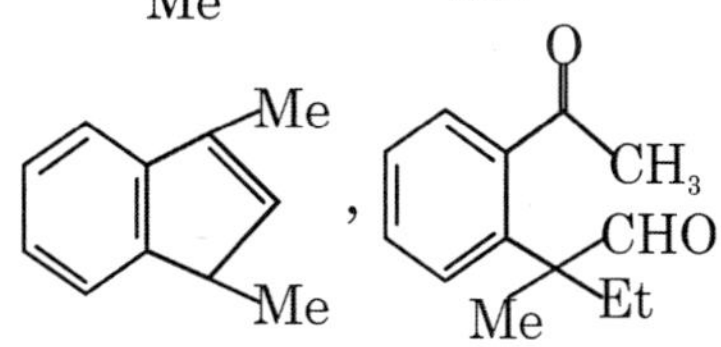

18. The structure of the product S is

(A) 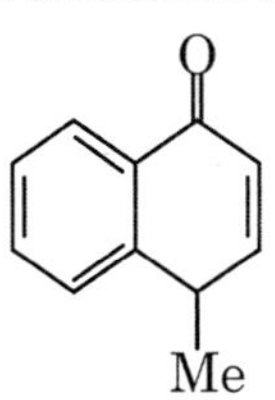

(B) 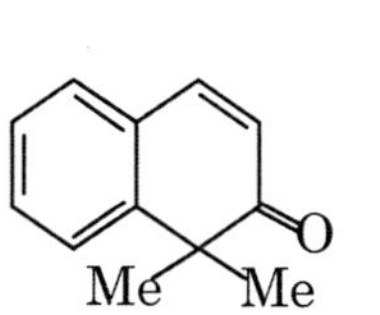

(C) 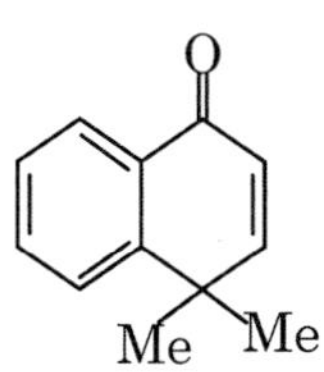

(D) 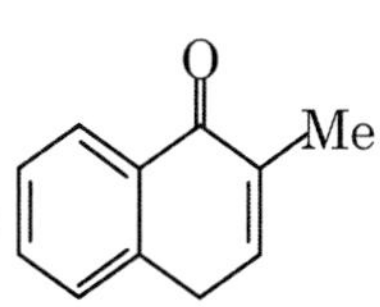

Section–IV: Matrix-Match Type

This section contains 2 questions. Each question contains statements given in two columns, which have to be matched. The statements in Column I are labelled A, B, C and D, while the statements in Column II are labelled p, q, r, s and t. Any given statement in Column I can have correct matching with ONE OR MORE statement(s) in Column II. The appropriate bubbles corresponding to the answers to these questions have to be darkened as illustrated in the following example:

If the correct matches are A – p, s and t; B – q and r; C – p and q; and D – s and t; then the correct darkening of bubbles will look like the following:

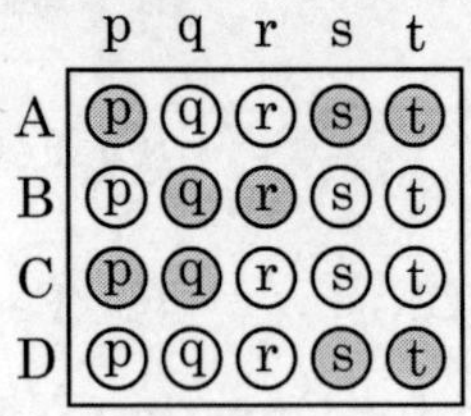

19. Match each of the compounds in Column I with its characteristic reaction(s) in Column II.

Column I	Column II
(A) $CH_3CH_2CH_2CN$	(p) Reduction with Pd–C/H_2
(B) $CH_3CH_2OCOCH_3$	(q) Reduction with $SnCl_2$/HCl
(C) $CH_3—CH=CH—CH_2OH$	(r) Development of foul smell on treatment with chloroform and alcohalic KOH
(D) $CH_3CH_2CH_2CH_2NH_2$	(s) Reduction with diisobutylaluminium hydride (DIBAL-H)
	(t) Alkaline hydrolysis

20. Match each of the diatomic molecules in Column I with its property/properties in Column II.

Column I	Column II
(A) B_2	(p) Paramagnetic
(B) N_2	(q) Undergoes oxidation
(C) O_2^-	(r) Undergoes reduction
(D) O_2	(s) Bond order $\geq$ 2
	(t) Mixing of *s* and *p* orbitals

ANSWER KEYS

1. C	2. B	3. A	4. B	5. B	6. D
7. B	8. A	9. C, D	10. A, B	11. B, C	12. A, D
13. D	14. C	15. B	16. B	17. A	18. B

19. A – p, q, s, t; B – s, t; C – p; D – r

20. A – p, r, t; B – s, t; C – p, q; D – p, q, s

SOLUTIONS

1. (C)

Since Sb_2S_3 is a negative sol, $Al_2(SO_4)_3$ will be the most effective coagulant due to higher charge density on Al^{3+} in accordance with Hardy–Schulze rule. Order of effectiveness of cations: $Al^{3+} > Ca^{++} > Na^+ > NH_4^+$.

2. (B)

$$\overline{A} = \frac{\sum A_i X_i}{\sum X_i}$$

$\overline{A} = 54 \times 0.05 + 56 \times 0.90 + 57 \times 0.05$ (where $\overline{A}$ is atomic mass of Fe); $\sum x_i = 1$

$\overline{A} = 55.95$

3. (A)

Carboxylic acid is a stronger acid than phenol. The presence of electron donating methyl group decreases acidic strength while presence of electron withdrawing halogen increases acidic strength. Decreasing order of acid strength: III > IV > II.

4. (B)

The priority of CN is the highest.

5. (B)

The measure of force of attraction for n moles of real gas $= \dfrac{an^2}{V^2}$

$$\left(P + \frac{an^2}{V^2}\right)(V - nb) = nRT$$

6. (D)

Both nylon and cellulose have intermolecular hydrogen bonding whereas polyvinyl chloride has dipole-dipole interaction. Natural rubber has London forces which are the weakest.

7. (B)

$P_4 + 3O_2 \xrightarrow{N_2} P_4O_6$ (exclusively)

(N_2 is used to retard the further oxidation)

8. (A)

According to Henry's law

$P_{N_2} = K_{N_2} X_{N_2}$ where K is the Henry's constant (in atm) and X_{N_2} is mole fraction of N_2.

$P_{N_2} = X_{N_2} P = 0.8 \times 5 \text{ atm} = 4.0 \text{ atm}$

Hence 4 atm = 1.0×10^5 atm $\times X_{N_2}$

$$4 \times 10^{-5} = \frac{n_{N_2}}{n_{N_2} + n_{H_2O}} = \frac{n_{N_2}}{n_{N_2} + 10}$$

$n_{N_2} = 4 \times 10^{-4}$ moles

9. (C, D)

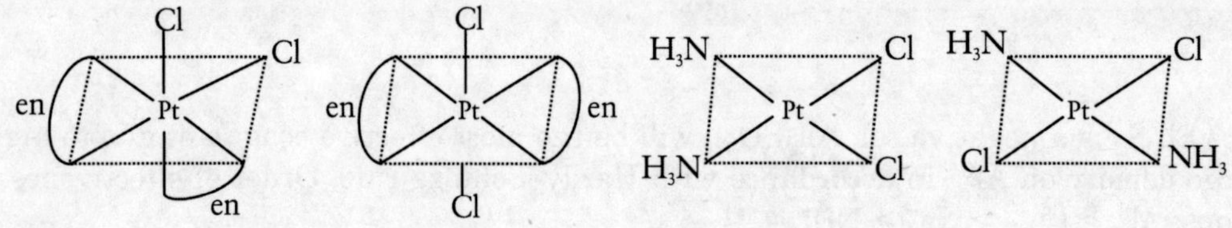

cis *trans* *cis* *trans*

10. (A, B)

Combustion of sodium metal in excess of air yields Na_2O and Na_2O_2.

$4Na + O_2 \xrightarrow{\Delta} 2Na_2O$

$2Na + O_2 \xrightarrow{\Delta} Na_2O_2$

11. (B, C)

Frenkel defect is a dislocation defect, observed when the size of cation and anion differ significantly. F-center is formed when an anion is lost from the lattice and vacancy is filled by trapping an electron. Schottky defect changes the density of a crytalline solid.

12. (A, D)

$$CH_3-\overset{*}{C}H(OH)-CH{=}CH-\overset{*}{C}H(OH)-CH_3$$

(I) *d* *cis* *d*
(II) *d* *trans* *d*
(III) *l* *cis* *l*
(IV) *l* *trans* *l*
(V) *d* *cis* *l*
(VI) *d* *trans* *l*

I and II are enantiomers, II and IV are enatiomers, V is meso due to the presence of plane of symmetry while VI is meso due to the presence of center of symmetry. A total of six stereoisomers are possible while number of diastereomers are four (I or III, II or IV, V and VI). If double bond in X is *trans*, number of enantiomers is two (II and IV). If double bond in X is *cis*, number of enantiomers is two (I and III).

13. (D)

$Na_2S + 2H^+ \longrightarrow H_2S + 2Na^+$

$FeCl_3 + H_2S \longrightarrow FeCl_2 + 2HCl + S$

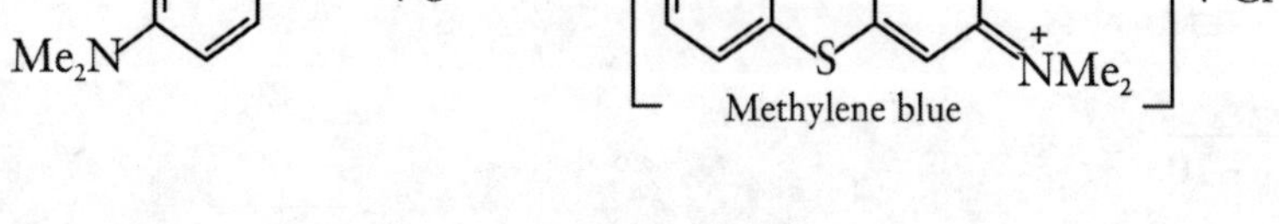

Methylene blue

14. (C)

The compound Y is $FeCl_3$ because when it is treated with $K_4[Fe(CN)_6]$ it gives intense blue precipitate of $Fe_4[Fe(CN)_6]_3$, which dissolves in excess $K_4[Fe(CN)_6]$.

$$4Fe^{3+} + 3\left[Fe(CN)_6\right]^{4-} \longrightarrow Fe_4\left[Fe(CN)_6\right]_3$$

15. (B)

The compound Z is $Fe[Fe(CN)_6]$ (brown colour) and is formed due to reaction of $K_3[Fe(CN)_6]$ and $FeCl_3$ (Y).

$$Fe^{3+} + \left[Fe(CN)_6\right]^{3-} \longrightarrow Fe\left[Fe(CN)_6\right]$$

Solution for the question Nos. 16 to 18

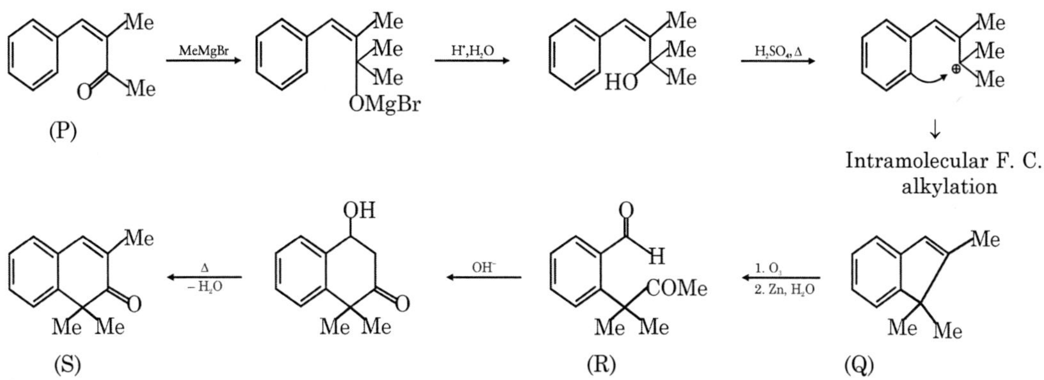

19. (A – p, q, s, t; B – s, t; C – p; D – r)

20. (A – p, r, t; B – s, t; C – p, q; D – p, q, s) (According to MOT)

I.I.T. JEE Objective Question Paper—2010

PAPER—1

Useful Data

Atomic Numbers: Be 4; C 6; N 7; O 8; Al 13; Si 14; Cr 24; Fe 26; Zn 30; Br 35

1 amu	=	1.66×10^{-27} kg	R	=	0.082 L-atm K^{-1} mol^{-1}
h	=	6.626×10^{-34} J s	N_A	=	6.022×10^{23}
m_e	=	9.1×10^{-31} kg	e	=	1.6×10^{-19} C
c	=	3.0×10^{8} m s^{-1}	F	=	96500 C mol^{-1}
R_H	=	2.18×10^{-18} J	$4\pi\epsilon_0$	=	1.11×10^{-1} J^{-1} C^2 m^{-1}

SECTION – I (Single Correct Choice Type)

This section contains 8 multiple choice questions. Each question has four choices (A), (B), (C) and (D) out of which only one is correct.

1. The synthesis of 3-octyne is achieved by adding a bromoalkane into a mixture of sodium amide and alkyne. The bromoalkane and alkyne respectively are

(A) $BrCH_2CH_2CH_2CH_2CH_3$ and $CH_3CH_2C{\equiv}CH$

(B) $BrCH_2CH_2CH_3$ and $CH_3CH_2CH_2C{\equiv}CH$

(C) $BrCH_2CH_2CH_2CH_2CH_3$ and $CH_3C{\equiv}CH$

(D) $BrCH_2CH_2CH_2CH_3$ and $CH_3CH_2C{\equiv}CH$

2. The correct statement about the following disaccharide is

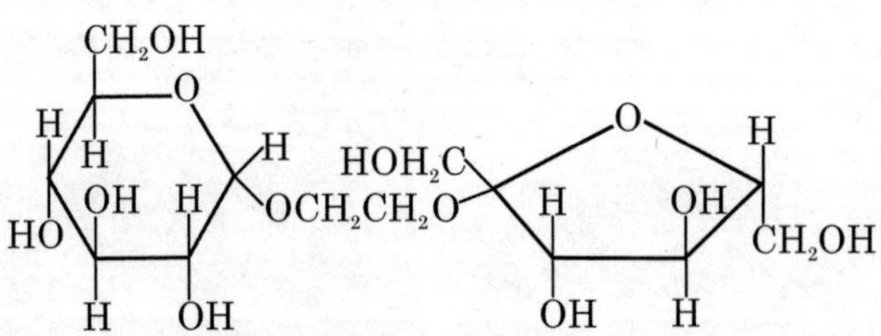

(A) Ring (a) is pyranonse with α-glycosidic link

(B) Ring (a) is furanonse with α-glycosidic link

(C) Ring (b) is furanonse with α-glycosidic link

(D) Ring (b) is pyranonse with β-glycosidic link.

3. In the reaction —OCH_3 $\xrightarrow{HBr}$, the products are

(A) Br——OCH_3 and H_2

(B) —Br and CH_3Br

(C) —Br and CH_3OH

(D) —OH and CH_3Br

4. Plots showing the variation of the rate constant (k) with temperature (T) are given below. The plot that follows Arrhenius equation is

(A)
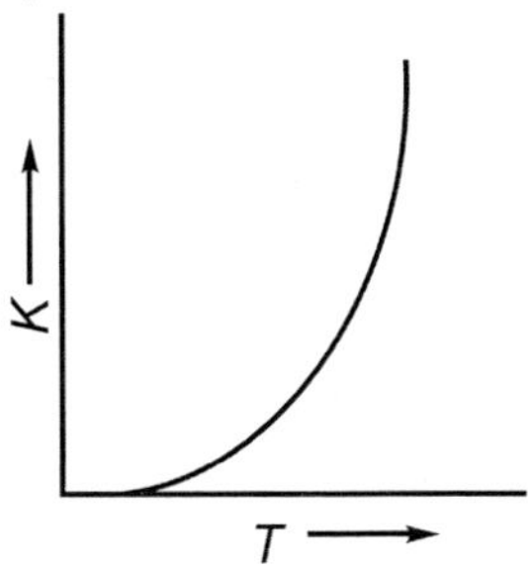

(B)
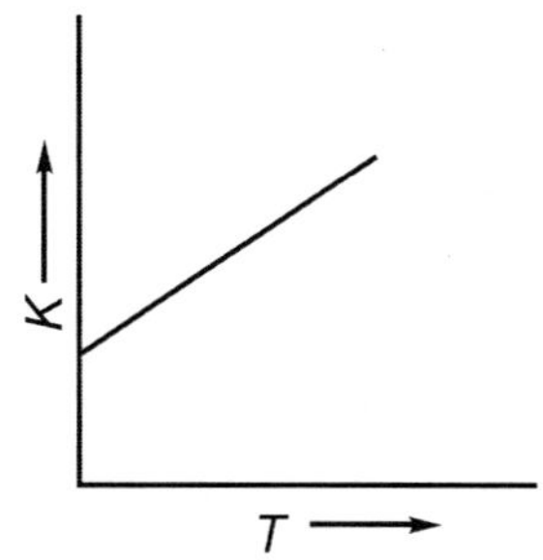

(C)
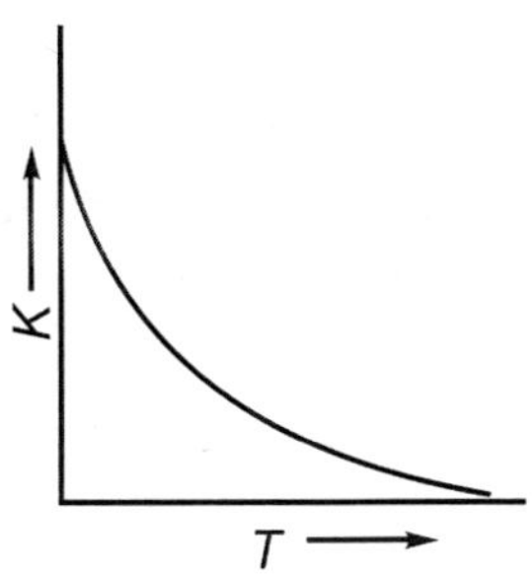

(D)
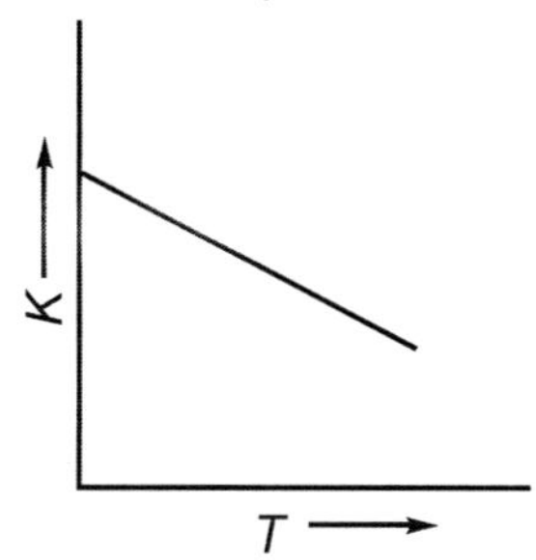

5. The species which by definition has ZERO standard molar enthalpy of formation at 298 K is

(A) $Br_2(g)$ (B) $Cl_2(g)$ (C) $H_2O(g)$ (D) $CH_4(g)$

6. The bond energy (in kcal mol^{-1}) of a C—C single bond is approximately

(A) 1 (B) 10 (C) 100 (D) 1000

7. The correct structure of ethylenediaminetetraacetic acid (EDTA) is

(A) $(HOOC—CH_2)_2N—CH{=}CH—N(CH_2—COOH)_2$

(B) $(HOOC)_2N—CH_2—CH_2—N(COOH)_2$

(C) $(HOOC—CH_2)_2N—CH_2—CH_2—N(CH_2—COOH)_2$

(D) 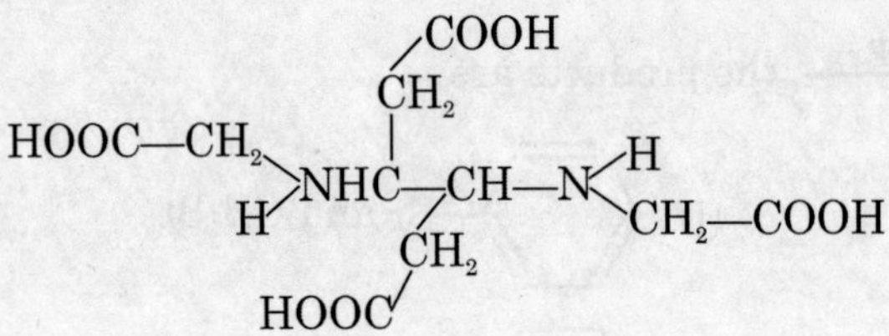

8. The ionization isomer of $[Cr(H_2O)_4Cl(NO_2)]Cl$ is

(A) $[Cr(H_2O)_4(NO_2)_2]Cl_2$ (B) $[Cr(H_2O)_4Cl_2](NO_2)$

(C) $[Cr(H_2O)_4Cl(ONO)]Cl$ (D) $[Cr(H_2O)_4Cl(NO_2)]H_2O$

SECTION – II (Multiple Correct Choice Type)

This section contains 5 multiple choice questions. Each question has four choices (A), (B), (C) and (D) out of which one or more may be correct.

9. In the Newman projection for 2,2-dimethylbutane,

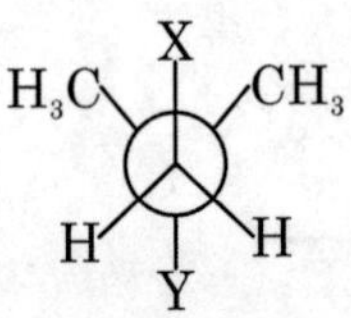

X and Y can respectively be

(A) H and H (B) H and C_2H_5 (C) C_2H_5 and H (D) CH_3 and CH_3

10. Aqueous solutions of HNO_3, KOH, CH_3COOH and CH_3COONa of identical concentrations are provided. The pair(s) of solutions which form a buffer upon mixing is(are)

(A) HNO_3 and CH_3COOH (B) KOH and CH_3COONa

(C) HNO_3 and CH_3COONa (D) CH_3COOH and CH_3COONa

11. In the reaction 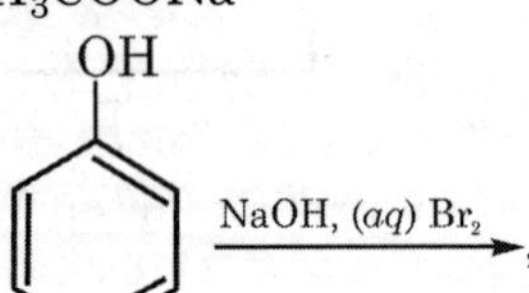

, the intermediate(s) is(are)

(A) 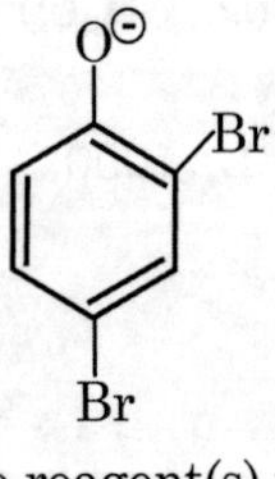 (B) 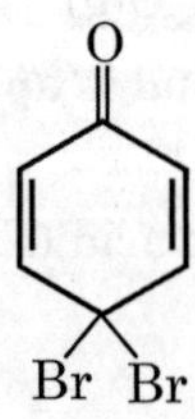 (C)  (D) 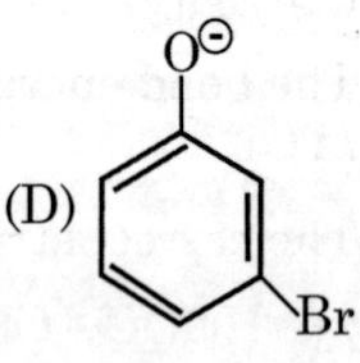

12. The reagent(s) used for softening the temporary hardness of water is(are)

(A) $Ca_3(PO_4)_2$ (B) $Ca(OH)_2$ (C) Na_2CO_3 (D) NaOCl

13. Among the following, the intensive property is (properties are)

(A) molar conductivity (B) electromotive force

(C) resistance (D) heat capacity

SECTION – III (Paragraph Type)

This section contains 2 paragraphs. Based upon the first paragraph 2 multiple choice questions and based upon the second paragraph 3 multiple choice questions have to be answered. Each of these questions has four choices (A), (B), (C) and (D) out of which only one is correct.

Paragraph for Question Nos. 14 to 15

The concentration of potassium ions inside a biological cell is at least twenty times higher than the outside. The resulting potential difference across the cell is important in several processes such as transmission of nerve impulses and maintaining the ion balance. A simple model for such a concentration cell involving a metal M is:

$M(s) | M^+(aq; 0.05 \text{ molar}) || M^+(aq; 1 \text{ molar}) | M(s)$

For the above electrolytic cell, the magnitude of the cell potential $|E_{cell}| = 70$ mV.

14. For the above cell

(A) $E_{cell} < 0; \Delta G > 0$ (B) $E_{cell} > 0; \Delta G < 0$ (C) $E_{cell} < 0; \Delta G^0 > 0$ (D) $E_{cell} > 0; \Delta G^0 > 0$

15. If the 0.05 molar solution of M^+ is replaced by 0.0025 molar M^+ solution, then the magnitude of the cell potential would be

(A) 35 mV (B) 70 mV (C) 140 mV (D) 700 mV

Paragraph for Question Nos. 16 to 18

Copper is the most noble of the first row transition metals and occurs in small deposits in several countries. Ores of copper include chalcanthite ($CuSO_4 \cdot 5H_2O$), atacamite ($Cu_2Cl(OH)_3$), cuprite (Cu_2O), copper glance (Cu_2S) and malachite ($Cu_2(OH)_2CO_3$). However, 80% of the world copper production comes from the ore of chalcopyrite ($CuFeS_2$). The extraction of copper from chalcopyrite involves partial roasting, removal of iron and self-reduction.

16. Partial roasting of chalcopyrite produces

(A) Cu_2S and FeO (B) Cu_2O and FeO (C) CuS and Fe_2O_3 (D) Cu_2O and Fe_2O_3

17. Iron is removed from chalcopyrite as

(A) FeO (B) FeS (C) Fe_2O_3 (D) $FeSiO_3$

18. In self-reduction, the reducing species is

(A) S (B) O^{2-} (C) S^{2-} (D) SO_2

SECTION - IV (Integer Type)

This section contains TEN questions. The answer to each question is a single digit integer ranging from 0 to 9. The correct digit below the question number in the ORS is to be bubbled.

19. A student performs a titration with different burettes and finds titre values of 25.2 mL, 25.25 mL and 25.0 mL. The number of significant figures in the average titre value is

20. The concentration of R in the reaction R $\rightarrow$ P was measured as a function of time and the following data is obtained:

[R] (molar)	1.0	0.75	0.40	0.10
t (min.)	0.0	0.05	0.12	0.18

The order of the reaction is

21. The number of neutrons emitted when $^{235}_{92}U$ undergoes controlled nuclear fission to $^{142}_{54}Xe$ and $^{90}_{38}Sr$ is

22. The total number of basic groups in the following form of lysine is

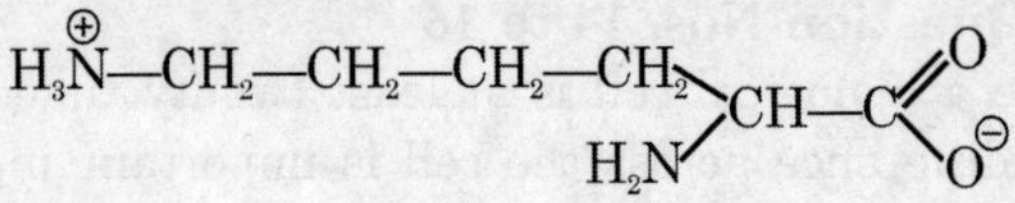

23. The total number of cyclic isomers possible for a hydrocarbon with the molecular formula C_4H_6 is

24. In the scheme given below, the total number of intramolecular aldol condensation products formed from Y is

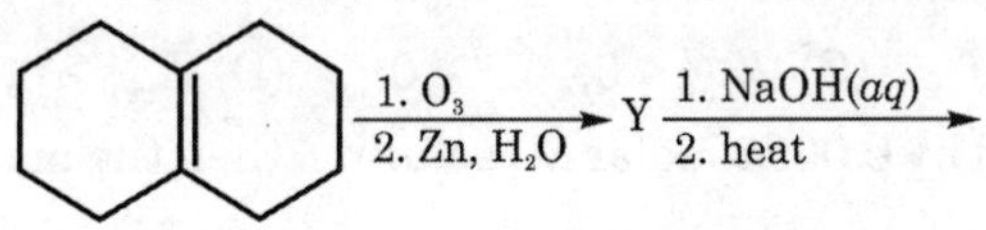

25. Amongst the following, the total number of compounds soluble in aqueous NaOH is

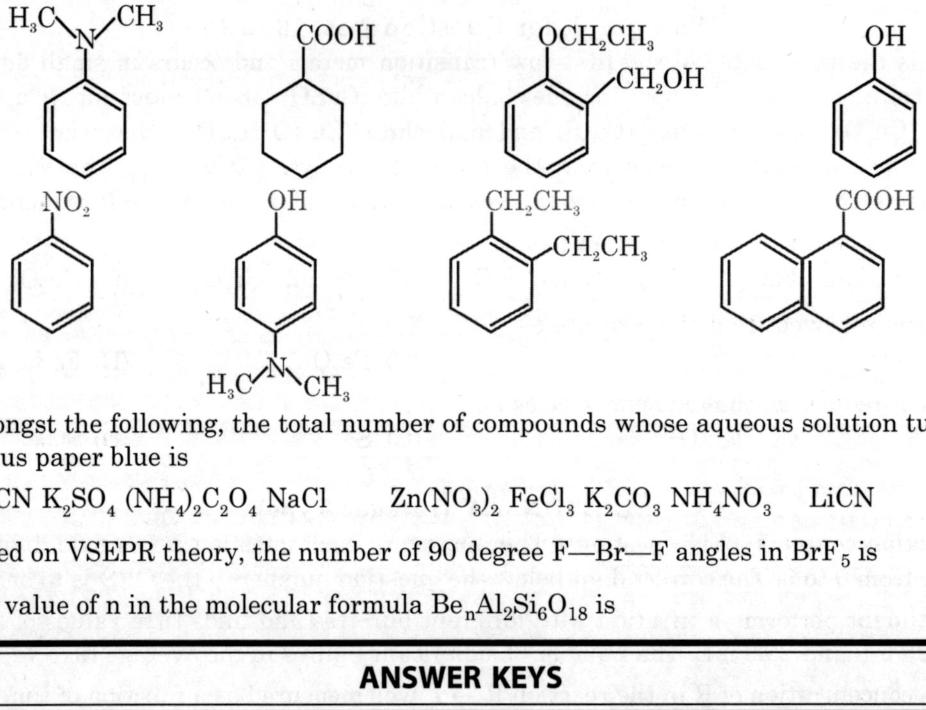

26. Amongst the following, the total number of compounds whose aqueous solution turns red litmus paper blue is

KCN K_2SO_4 $(NH_4)_2C_2O_4$ NaCl $Zn(NO_3)_2$ $FeCl_3$ K_2CO_3 NH_4NO_3 LiCN

27. Based on VSEPR theory, the number of 90 degree F—Br—F angles in BrF_5 is

28. The value of n in the molecular formula $Be_nAl_2Si_6O_{18}$ is

ANSWER KEYS

1. D	2. A	3. D	4. A	5. B	6. C
7. C	8. B	9. B, D	10. C, D	11. A, B, C	12. B, C, D
13. A, B	14. B	15. C	16. B	17. D	18. C
19. 3	20. 0	21. 3	22. 2	23. 5	24. 1
25. 4	26. 3	27. 0	28. 3		

SOLUTIONS

1. (D)

$$CH_3—CH_2—C≡CH \xrightarrow{NaNH_2} CH_3—CH_2—C≡C^-Na^+$$

$$CH_3—CH_2—CH_2—CH_2\boxed{Br + Na^+}{}^-C≡C—CH_2—CH_3 \longrightarrow$$

$$CH_3—CH_2—CH_2—CH_2—C≡C—CH_2—CH_3$$

3-octyne

2. (A)

Ring (a) is pyranose with α-glycosidic link.

3. (D)

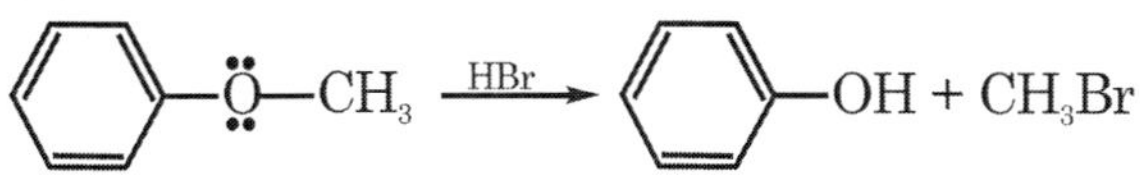

4. (A)

$k = Ae^{-E_a/RT}$

As $T\uparrow$; $k\uparrow$ exponentially

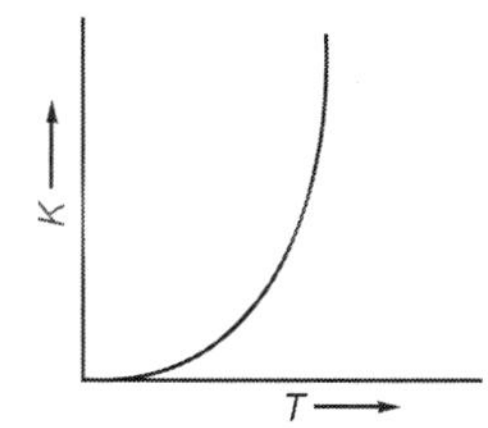

5. (B)

Chlorine is a gas at 298 K and bromine is a liquid at 298 K.

6. (C)

The bond energy of a C—C single bond is approximately 100 kcal mol^{-1}.

7. (C)

The correct structure of ethylenediaminetetraacetic acid (EDTA) is (C).

8. (B)

Cl^- ion is replaced by NO_2^- ion in the coordination sphere.

9. (B, D)

$$\begin{array}{c} \quad\quad\quad\quad CH_3 \\ \quad\quad\quad\quad | \\ H_3\underset{4}{C}—\underset{3}{CH_2}—C_2—\underset{1}{CH_3} \\ \quad\quad\quad\quad | \\ \quad\quad\quad\quad CH_3 \end{array}$$

On $C_2—C_3$ bond axis, X = CH_3, Y = CH_3 and on $C_1—C_2$ bond axis, X = H, Y = C_2H_5.

10. (C, D)

In the option (C), if HNO_3 is present in limiting amount, then the mixture will behave as a buffer. In the option (D), the mixture contains a weak acid (CH_3COOH) and its salt CH_3COONa and this mixture behaves as a buffer.

11. (A, B, C)

The phenoxide ion is para* and ortho directing (*preferably),

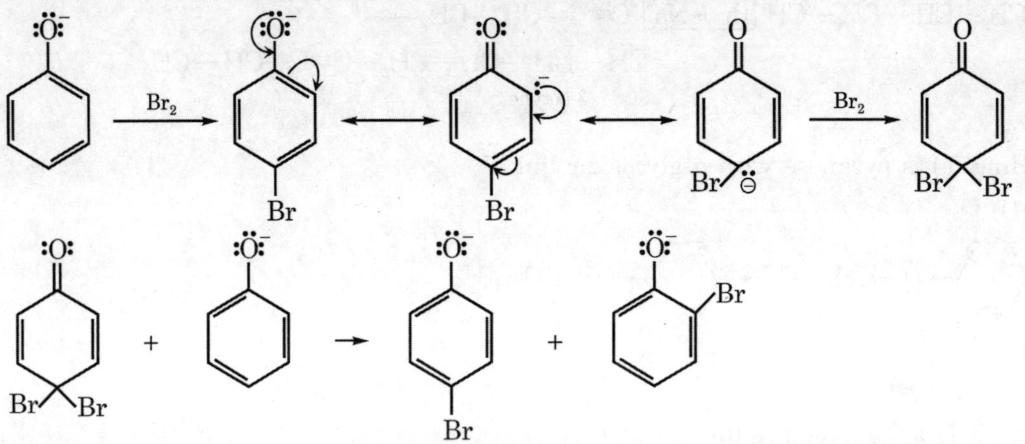

12. (B, C, D)

$$Ca(HCO_3)_2 + Ca(OH)_2 \rightarrow 2CaCO_3 \downarrow + 2H_2O$$

(Clarke's method)

$$NaOCl + H_2O \rightleftharpoons HOCl + NaOH$$

$$OH^- + HCO_3^- \rightarrow CO_3^{2-} + H_2O$$

$$Ca(HCO_3)_2 + Na_2CO_3 \rightarrow CaCO_3 \downarrow + 2NaHCO_3$$

13. (A, B)

Resistance and heat capacity are mass dependent properties and hence are extensive property.

14. (B)

$$M(s) + M^+_{(aq)1M} \longrightarrow M^+_{(aq).05M} + M(s)$$

According to Nernst equation, $E_{cell} = 0 - \dfrac{2.303\,RT}{F}\log\dfrac{M^+_{.05M}}{M^+_{1M}}$

$$= 0 - \frac{2.303\,RT}{F}\log(5 \times 10^{-2}) = +ve$$

Hence $|E_{cell}| = E_{cell} = 0.70$ V and $\Delta G < 0$ for the feasibility of the reaction.

15. (C)

From the above equation $\dfrac{2.303\,RT}{F} = 0.0538$. Hence $E_{cell} = E^0_{cell} - \dfrac{0.0538}{1}\log 0.0025$

$$= 0 - \frac{0.0538}{1}\log 0.0025 \approx 0.13988 \text{ V} \approx 140 \text{ mV}$$

16. (B)

$2CuFeS_2 + O_2 \rightarrow Cu_2S + 2FeS + SO_2 \uparrow$

$2Cu_2S + 3O_2 \rightarrow 2Cu_2O + 2SO_2 \uparrow$

$2FeS + 3O_2 \rightarrow 2FeO + 2SO_2 \uparrow$

17. (D)

$FeO + SiO_2 \rightarrow FeSiO_3$ (slag)

18. (C)

$Cu_2S + 2Cu_2O \rightarrow \underset{\text{(Blister copper)}}{6Cu} + SO_2 \uparrow$

$S^{2-} \rightarrow S^{4+}$ is oxidation, hence S^{2-} is the reducing agent.

19. (3)

20. (0)

From the given data, (for zero order reaction)

$$k_I = \frac{x}{t} = \frac{0.25}{0.05} = 5 \quad \text{and} \quad k_{II} = \frac{x}{t} = \frac{0.60}{0.12} = 5$$

21. (3)

$${}^{235}_{92}U \longrightarrow {}^{142}_{54}Xe + {}^{90}_{38}Sr + 3\,{}^{1}_{0}n$$

Hence 3 neutrons are emitted.

22. (2)

$-\overset{\overset{\displaystyle O}{\|}}{C}-O^{\ominus}$ and $-NH_2$ are the basic groups in lysine.

23. (5)

In C_4H_6, the possible cyclic isomers are

24. (1)

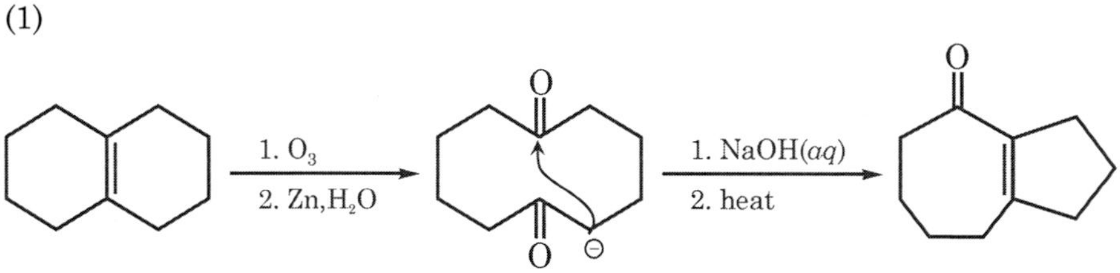

25. (4)

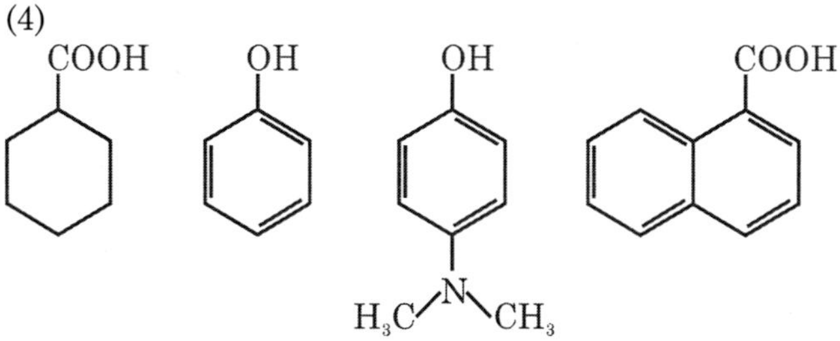

These four compounds are soluble in aqueous NaOH.

26. (3)

KCN, K_2CO_3 and LiCN are basic in nature and their aqueous solution turns red litmus paper blue.

27. (0)

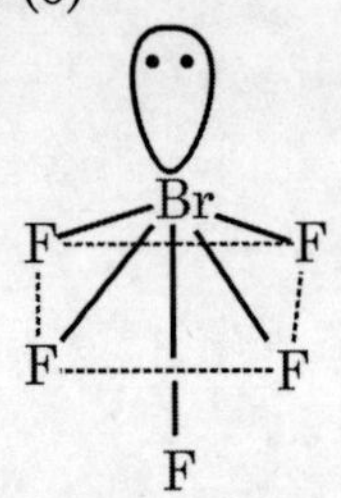

All the four planar bonds (F—Br—F) will reduce from 90° to 84.8° after lp–bp repulsion.

28. (3)

$Be_3Al_2Si_6O_{18}$ (beryl)

According to charge balance in a molecule, $2n + 6 + 24 - 36 = 0$ or $n = 3$.

I.I.T. JEE Objective Question Paper—2010

PAPER—2

Useful Data

Atomic Numbers: B 5; C 6; N 7; O 8; F 9; Na 11; Si 14; P 15; S 16; Cl 17; Ti 22; V 23; Cr 24; Ni 28; Cu 29; Br 35; Rh 45; Sn 50; Xe 54; Tl 81

1 amu	=	1.66×10^{-27} kg	e	=	1.6×10^{-19} C
R	=	0.082 L-atm K^{-1} mol^{-1}	c	=	3.0×10^{8} m s^{-1}
h	=	6.626×10^{-34} J s	F	=	96500 C mol^{-1}
N_A	=	6.022×10^{23}	R_H	=	2.18×10^{-18} J
m_e	=	9.1×10^{-31} kg	$4\pi\epsilon_0$	=	1.11×10^{-10} J^{-1} C^2 m^{-1}

SECTION – I (Single Correct Choice Type)

This section contains 6 multiple choice questions. Each question has four choices (A), (B), (C) and (D) out of which only one is correct.

1. The compounds P, Q and S were separately subjected to nitration using HNO_3/H_2SO_4 mixture. The major product formed in each case respectively, is

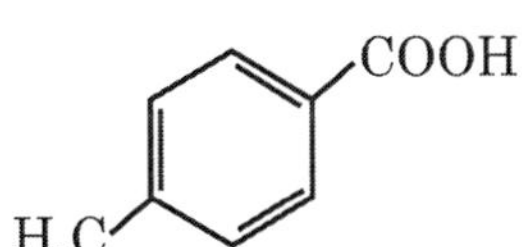
P

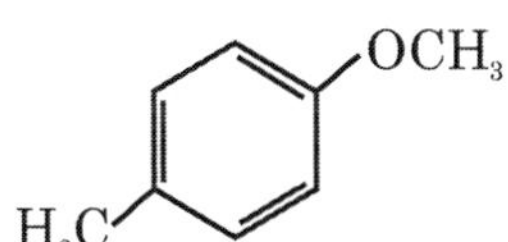
Q

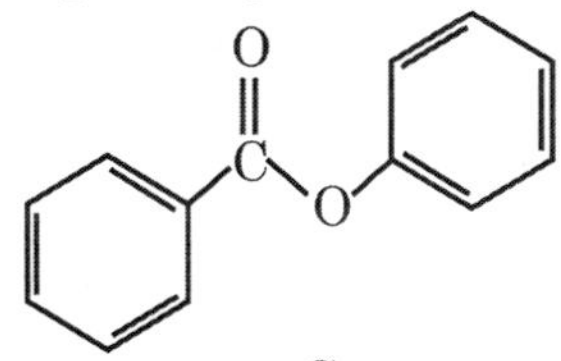
S

(A) 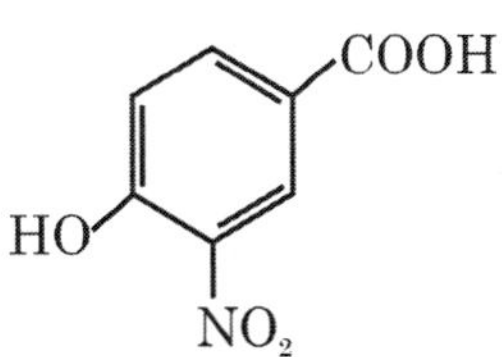 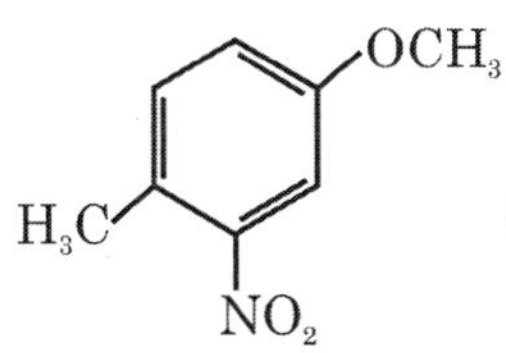 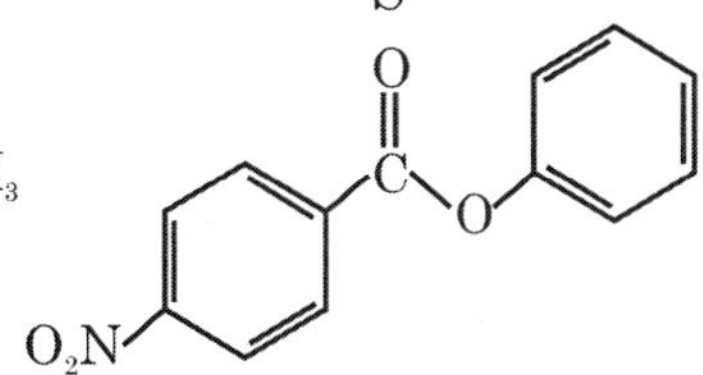

(B) 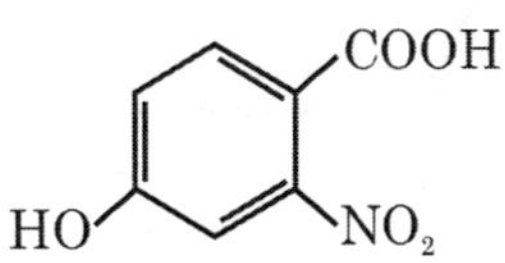 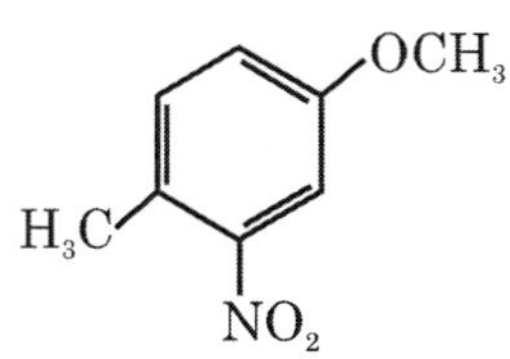 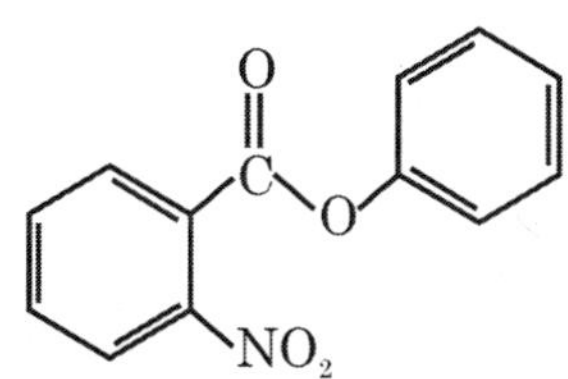

(C) 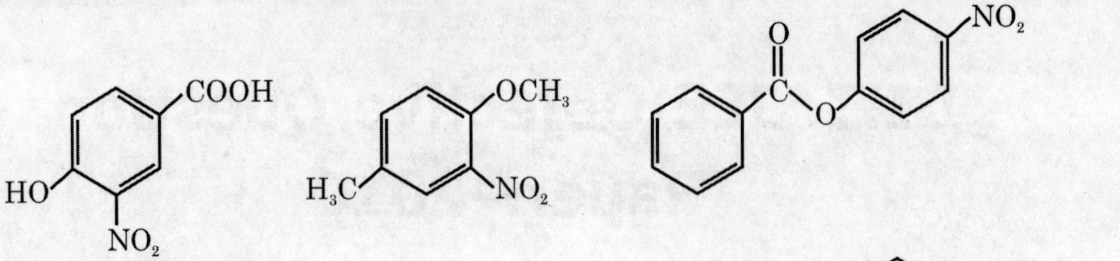

(D) 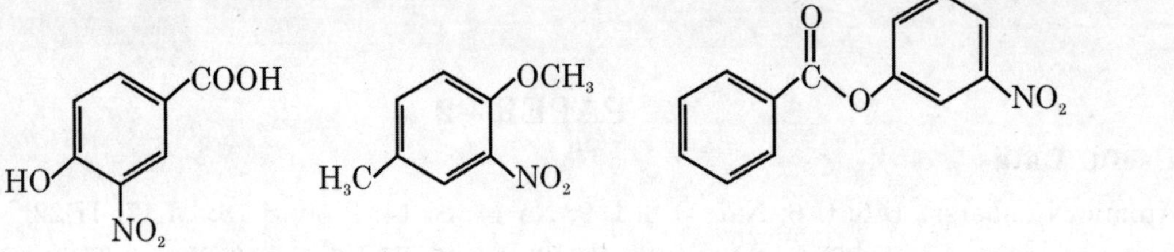

2. Assuming that Hund's rule is violated, the bond order and magnetic nature of the diatomic molecule B_2 is
(A) 1 and diamagnetic
(B) 0 and diamagnetic
(C) 1 and paramagnetic
(D) 0 and paramagnetic

3. The packing efficiency of the two-dimensional square unit cell shown below is

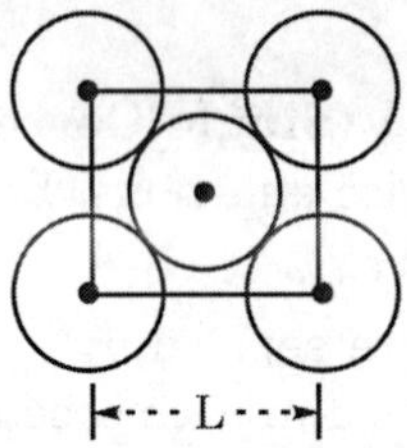

(A) 39.27% (B) 68.02% (C) 74.05% (D) 78.54%

4. The complex showing a spin-only magnetic moment of 2.82 B.M. is
(A) $Ni(CO)_4$ (B) $[NiCl_4]^{2-}$ (C) $Ni(PPh_3)_4$ (D) $[Ni(CN)_4]^{2-}$

5. In the reaction 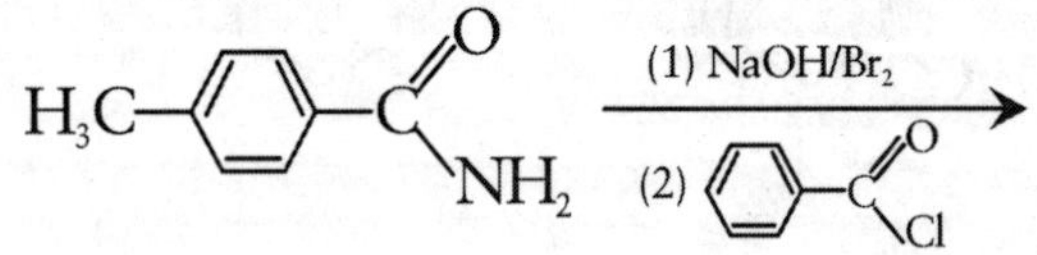

T, the structure of the product T is

(A) 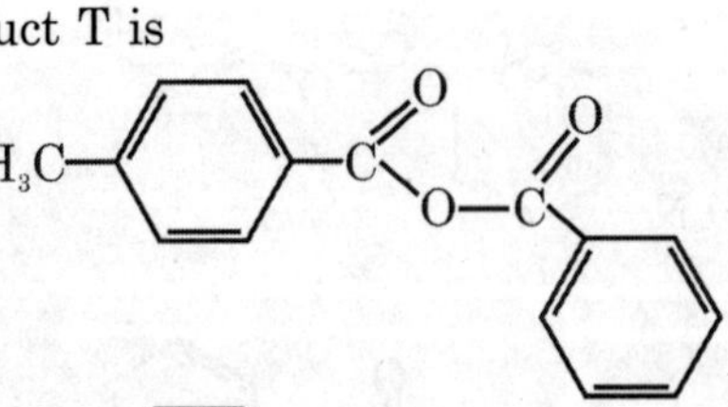

(B) 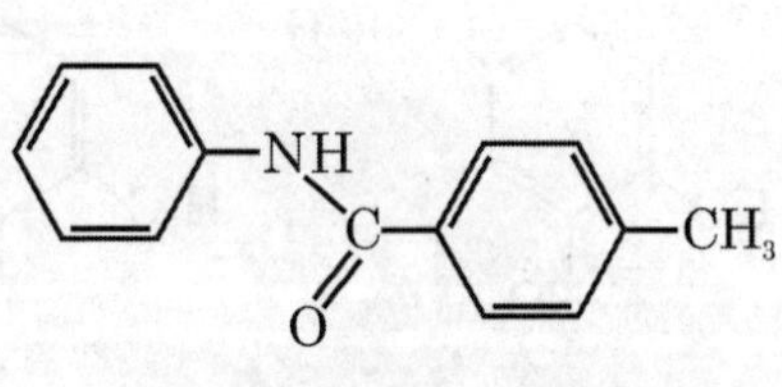

(C) 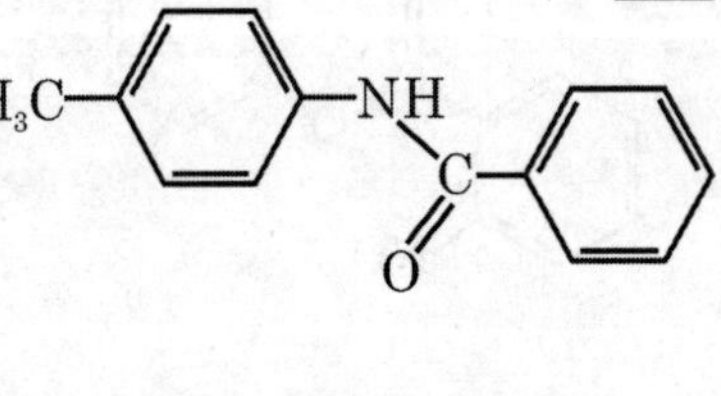

(D) 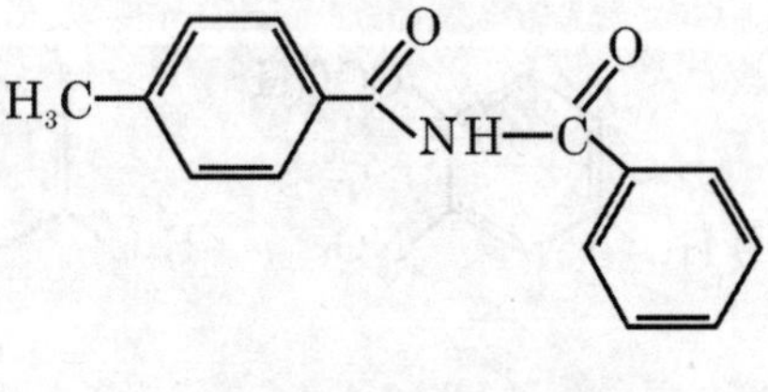

6. The species having pyramidal shape is
(A) SO_3 (B) BrF_3 (C) SiO_3^{2-} (D) OSF_2

SECTION – II (Integer Type)

This section contains 5 questions. The answer to each question is a single-digit integer, ranging from 0 to 9. The correct digit below the question no. in the ORS is to be bubbled.

7. Silver (atomic weight = 108 g mol^{-1}) has a density of 10.5 g cm^{-3}. The number of silver atoms on a surface of area 10^{-12} m^2 can be expressed in scientific notation as $y \times 10^x$. The value of x is

8. Among the following, the number of elements showing only one non-zero oxidation state is
O, Cl, F, N, P, Sn, Tl, Na, Ti

9. One mole of an ideal gas is taken from a to b along two paths denoted by the solid and the dashed lines as shown in the graph below. If the work done along the solid line path is w_s and that along the dotted line path is w_d, then the integer closest to the ratio w_d/w_s is

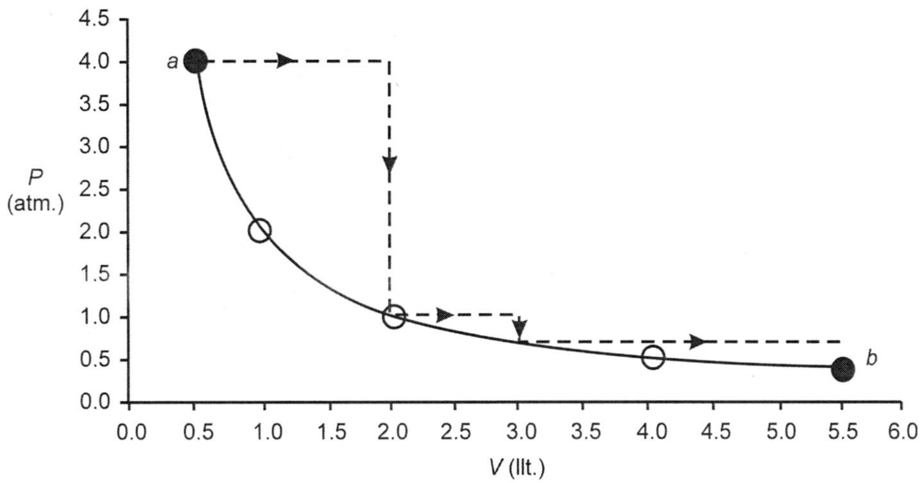

10. The total number of diprotic acids among the following is

H_3PO_4 H_2SO_4 H_3PO_3 H_2CO_3 $H_2S_2O_7$
H_3BO_3 H_3PO_2 H_2CrO_4 H_2SO_3

11. Total number of geometrical isomers for the complex $[RhCl(CO)(PPh_3)(NH_3)]$ is

SECTION – III (Paragraph Type)

This section contains 2 paragraphs. Based upon each of the paragraphs 3 multiple choice questions have to be answered. Each of these questions has four choices (A), (B), (C) and (D) out of which only one is correct.

Paragraph for questions 12 to 14

Two aliphatic aldehydes P and Q react in the presence of aqueous K_2CO_3 to give compound R, which upon treatment with HCN provides compound S. On acidification and heating, S gives the product shown below:

12. The compounds P and Q respectively are

(A) 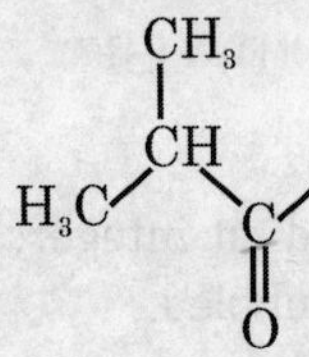 and 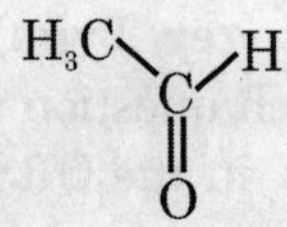

(B) 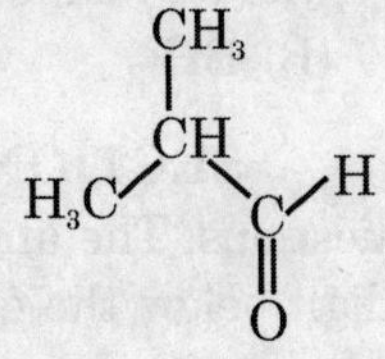 and 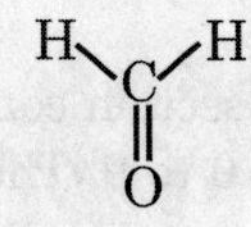

(C) 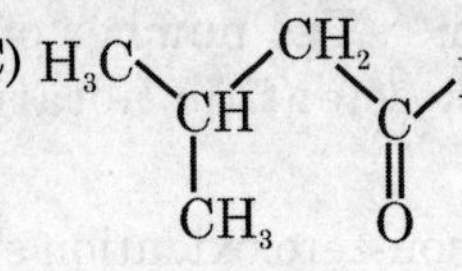 and 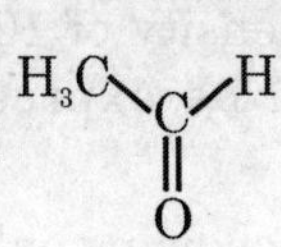

(D) 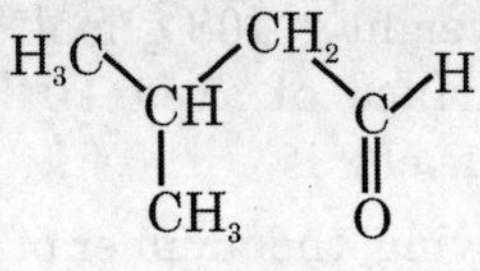 and 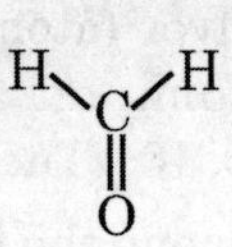

13. The compound R is

(A) 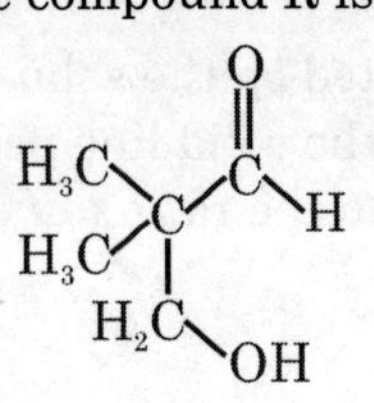

(B) 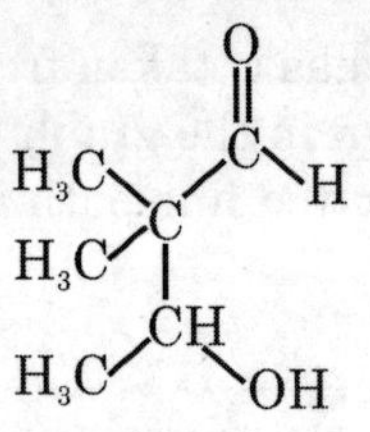

(C) 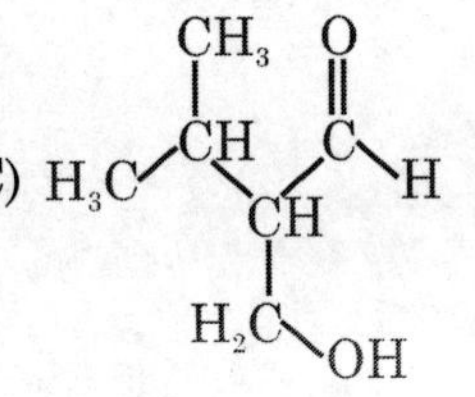

(D) 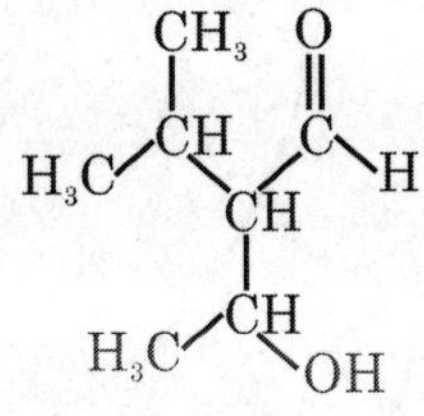

14. The compound S is

(A) 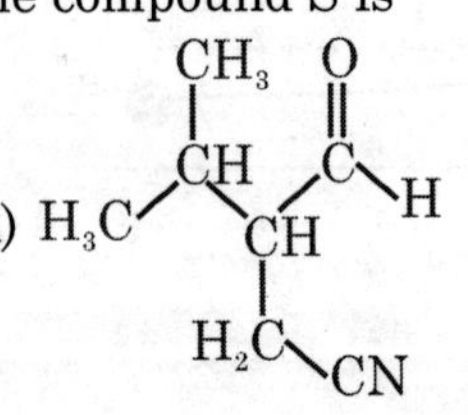

(B) 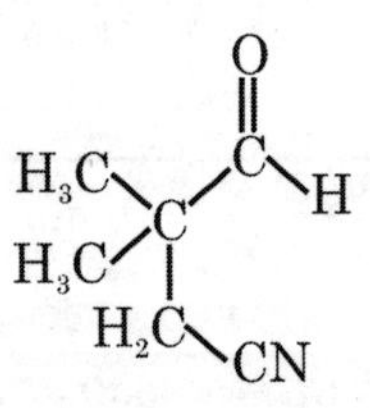

(C) 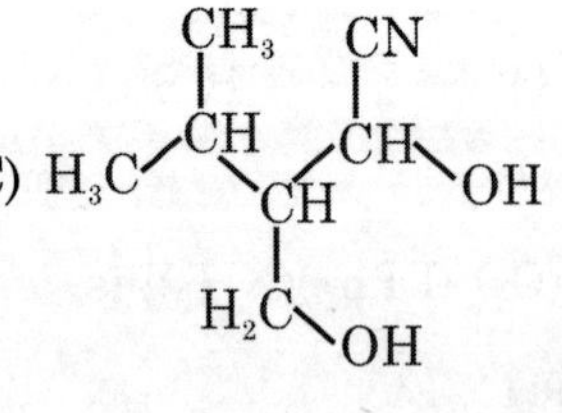

(D) 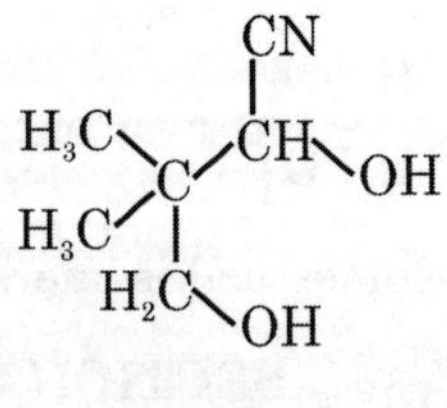

Paragraph for questions 15 to 17

The hydrogen-like species Li^{2+} is in a spherically symmetric state S_1 with one radial node. Upon absorbing light the ion undergoes transition to a state S_2. The state S_2 has one radial node and its energy is equal to the ground state energy of the hydrogen atom.

15. The state S_1 is

(A) 1*s* (B) 2*s* (C) 2*p* (D) 3*s*

16. Energy of the state S_1 in units of the hydrogen atom ground state energy is

(A) 0.75 (B) 1.50 (C) 2.25 (D) 4.50

17. The orbital angular momentum quantum number of the state S_2 is

(A) 0 (B) 1 (C) 2 (D) 3

SECTION – IV (Matrix Type)

This section contains 2 questions. Each question has four statements (A, B, C and D) given in Column I and five statements (p, q, r, s and t) in Column II. Any given statement in Column I can have correct matching with one or more statement(s) given in Column II. For example, if for a given question, statement B matches with the statements given in q and r, then for that particular question, against statement B, darken the bubbles corresponding to q and r in the ORS.

18. Match the reactions in Column I with appropriate options in Column II.

Column I | Column II

(A) 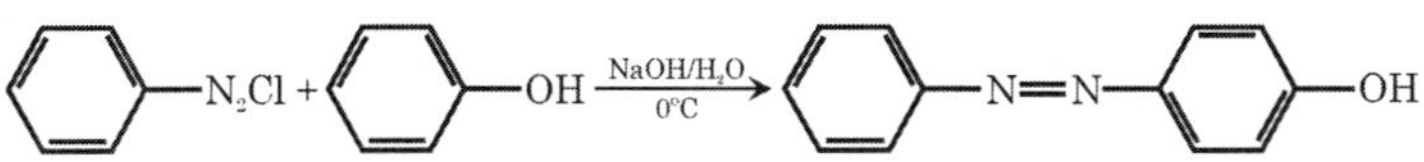

(p) Racemic mixture

(B) 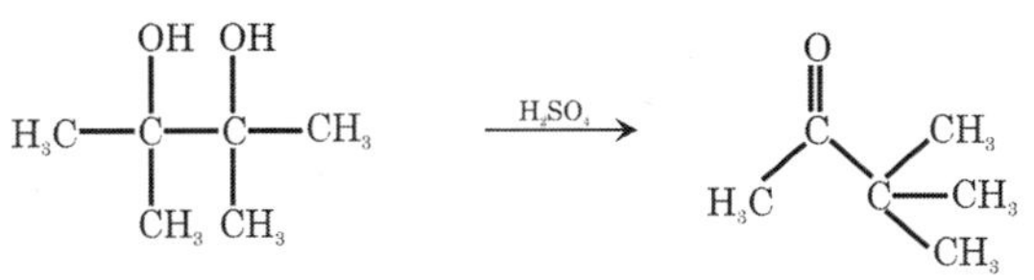

(q) Addition reaction

(C) 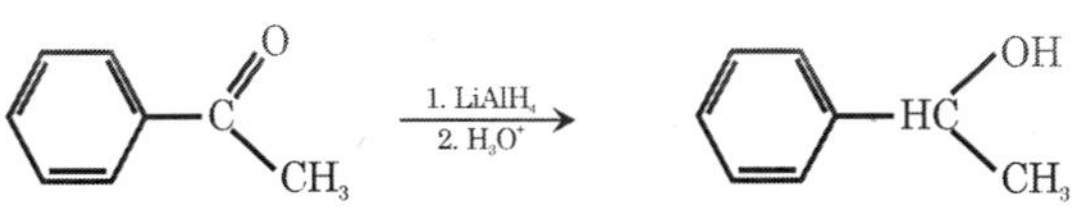

(r) Substitution reaction

(D) 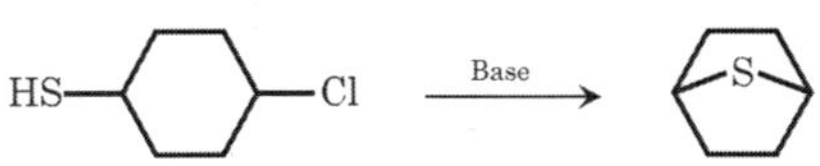

(s) Coupling reaction

(t) Carbocation intermediate

19. All the compounds listed in Column I react with water. Match the result of the respective reactions with the appropriate options listed in Column II.

Column I	Column II
(A) $(CH_3)_2SiCl_2$	(p) Hydrogen halide formation
(B) XeF_4	(q) Redox reaction
(C) Cl_2	(r) Reacts with glass
(D) VCl_5	(s) Polymerization
	(t) O_2 formation

ANSWER KEYS

1. C	2. A	3. D	4. B	5. C	6. D
7. 7	8. 2	9. 2	10. 6	11. 3	12. B
13. A	14. D	15. B	16. C	17. B	

18. A–r, s, t; B–t; C–p, q; D–r
19. A–p, s; B–p, q, r, t; C–p, q, t; D–p

SOLUTIONS

1. (C)

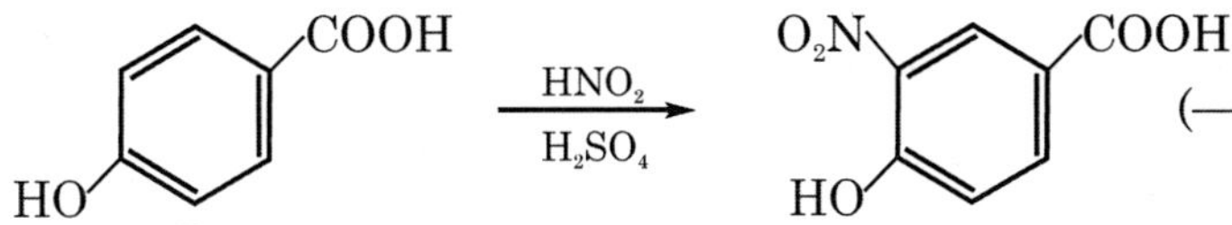

(—OH group is more activating)

P

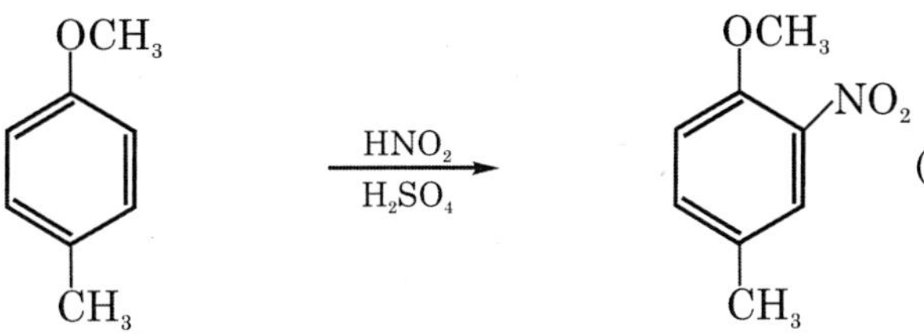

(—OCH_3 group is more activating)

Q

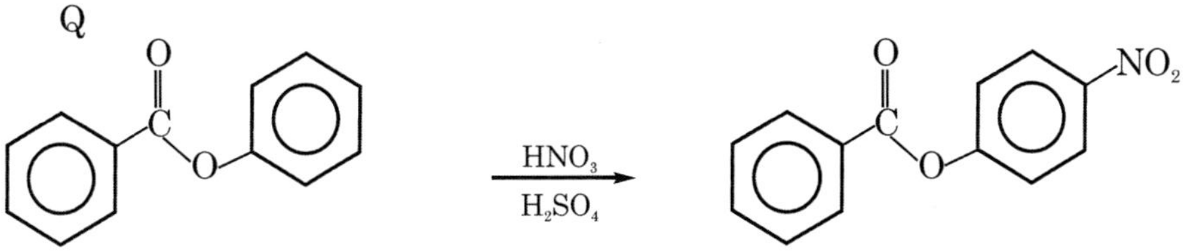

2. (A)

B_2 (10) = $\sigma_{1s}{}^2\sigma^*{}_{1s}{}^2\sigma_{2s}{}^2\sigma^*{}_{2s}{}^2\pi_{2p_x}{}^2$

Bond order = $\dfrac{6-4}{2} = 1$

Nature of B_2 molecule is diamagnetic since it does not contain any unpaired electron.

3. (D)

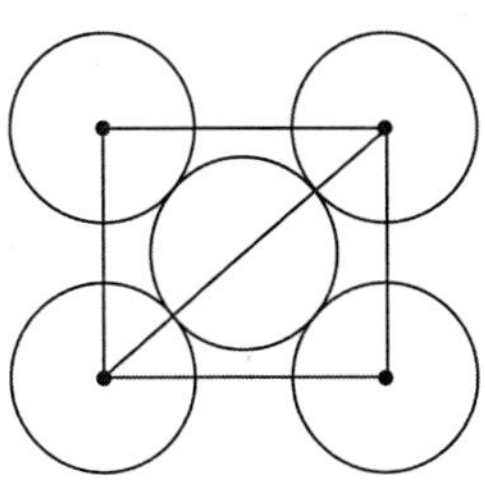

a = $(2\sqrt{2}r)$

P. F. = $\dfrac{2\times\pi r^2}{\left(2\sqrt{2}r\right)^2} = \dfrac{2\pi r^2}{8r^2} = \dfrac{\pi}{4} = \dfrac{3.14159}{4} = 0.7854$. Hence the packing efficiency is 78.54%.

4. (B)

$[NiCl_4]^{2-}$, oxidation state of Ni = + 2

Ni (28) = $3d^8 4s^2$

$Ni^{2+} = 3d^8$

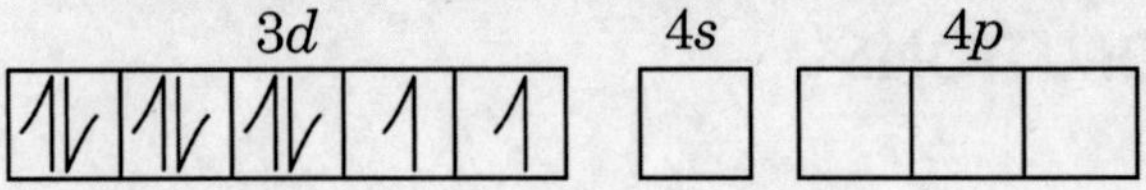

Cl^- being a weak ligand, outer orbitals are involved in the hybridisation.

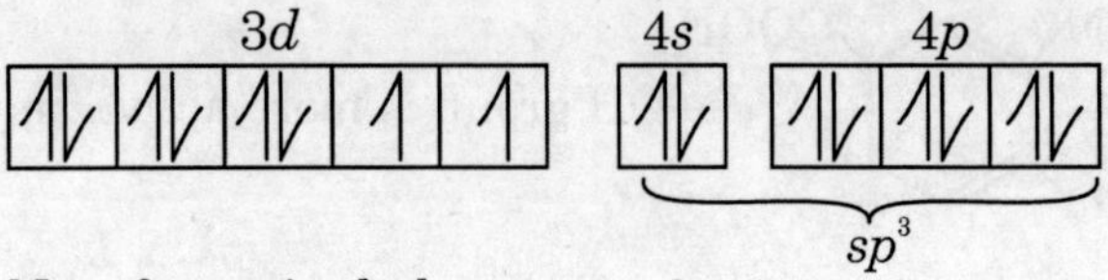

No. of unpaired electrons = 2

Magnetic moment μ = 2.82 B. M.

5. (C)

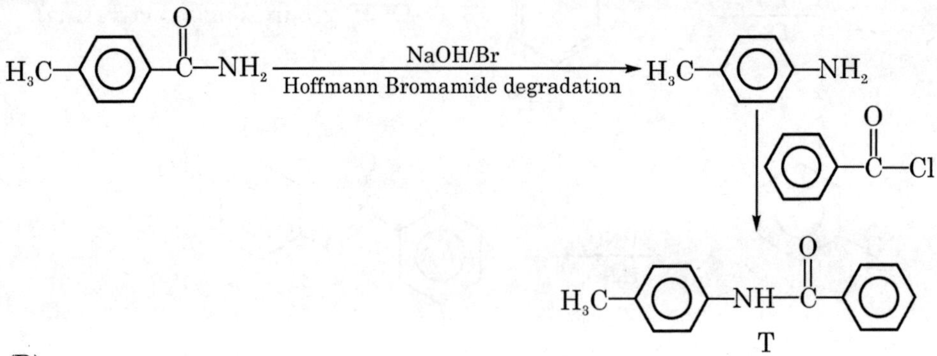

6. (D)

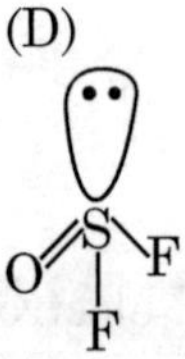

Trignonal pyramid

7. (7)

$d = \dfrac{\text{mass}}{V} \Rightarrow 10.5$ g/cc means in 1 cc $\Rightarrow$ 10.5g of Ag present

Number of atoms of Ag in 1 cc $\Rightarrow \dfrac{10.5}{108} \times N_A$

In 1 cm, number of atoms of Ag $= \sqrt[3]{\dfrac{10.5}{108} N_A}$

In 1 cm^2, number of atoms of Ag $= \left(\dfrac{10.5}{108} N_A\right)^{2/3}$

In 10^{-12} m^2 or 10^{-8} cm^2, number of atoms of Ag $= \left(\dfrac{10.5}{108} N_A\right)^{2/3} \times 10^{-8}$

$$= \left(\frac{1.05 \times 6.022 \times 10^{24}}{108}\right)^{2/3} \times 10^{-8} = 1.5 \times 10^7$$

Hence, $x = 7$

8. (2)
Na and F exhibit only one non-zero oxidation state.

9. (2)
$w_d = 4 \times 1.5 + 1 \times 1 + 2.5 \times 2/3 = 8.65$
Thus the process is isothermal.

$$w_s = 2 \times 2.303 \log \frac{5.5}{0.5} = 2 \times 2.303 \times \log 11 = 2 \times 2.303 \times 1.0414 = 4.79$$

$$\frac{w_d}{w_s} = \frac{8.65}{4.79} = 1.80 \simeq 2$$

10. (6)
6 : H_2SO_4, H_2CO_3, $H_2S_2O_7$, H_2CrO_4, H_3PO_3, H_2SO_3

11. (3)

12. (B)

13. (A)

14. (D)

Solution to Questions 12-14:

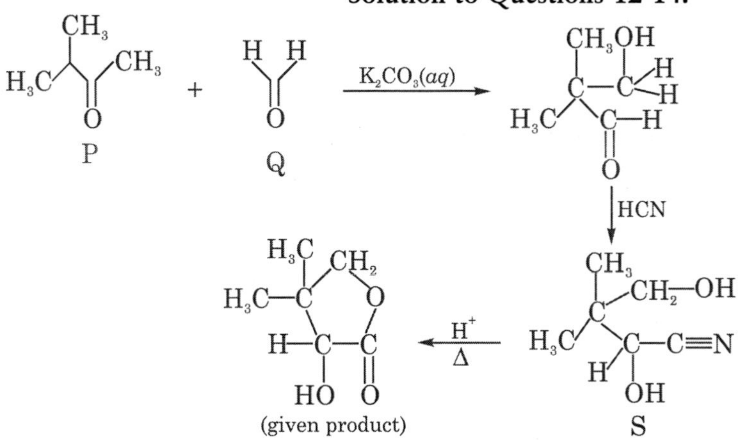

15. (B)
For S_1 (spherically symmetrical), radial node = 1
Hence $n - 1 = 1$
Or $n = 2$
For S_2, radial node = 1

$$E_{S_2} = \frac{-13.6 \times 2^2}{n^2} = E_H \text{ in ground state} = -13.6$$

$$E_{S_2} = \frac{-13.6 \times 9}{n^2} = -13.6 \text{ Or } n = 3$$

Hence state S_1 is $2s$ and state S_2 is $3p$.

16. (C)

$$\frac{E_{S_1}}{E_{H(ground)}} = \frac{-13.6 \times 9}{4 \times (-13.6)} = 2.25$$

17. (B)

The azimuthal quantum number for $S_2 = l = 1$

18. (A–r, s, t; B–t; C–p, q; D–r)

(A) This is an electrophilic substitution reaction and a coupled product is formed. Hence, this is also a coupling reaction.

(B) This is pinacole-pinacolone rearrangement and in this reaction the intermediate is a carbocation.

(C) This is an addition of the carbonyl compound and both the enanciomers are formed. Hence, mixture is formed.

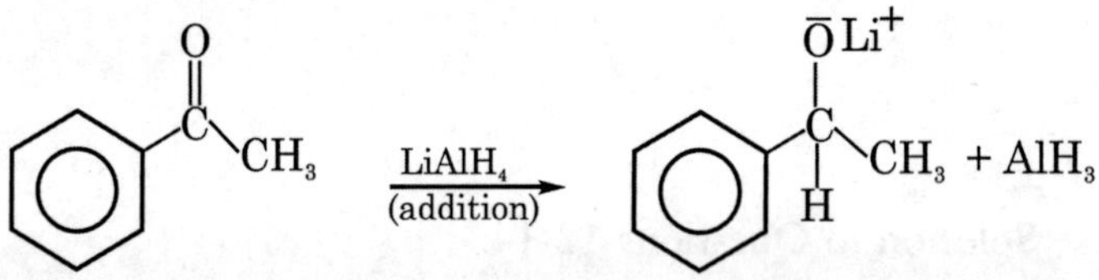

(D) This is a nucleophilic substitution reaction.

19. (A–p, s; B–p, q, r, t; C–p, q, t; D–p)

(A) $(CH_3)_2SiCl_2 + 2H_2O \longrightarrow (CH_3)_2Si(OH)_2 + 2HCl$

$(CH_3)_2Si(OH)_2$ undergoes polymerization to form silicones.

(B) $3XeF_4 + 6H_2O \longrightarrow XeO_3 + 2Xe + 12HF + 1\frac{1}{2}O_2$

(C) $Cl_2 + H_2O \longrightarrow 2HCl + \frac{1}{2}O_2$

(D) $VCl_5 + H_2O \longrightarrow VOCl_3 + 2HCl$

I.I.T. JEE Objective Question Paper—2011

PAPER—1

SECTION - I (Total marks: 21)
(Single Correct Answer Type)

This section contains 7 multiple choice questions. Each question has four choices (A), (B), (C) and (D) out of which only one is correct.

1. Extra pure N_2 can be obtained by heating
   (A) NH_3 with CuO (B) NH_4NO_3 (C) $(NH_4)_2CrO_7$ (D) $Ba(N_3)_4$

2. Dissolving 120 g of urea (mol. wt. 60) in 1000 g of water gave a solution of density 1.15 g/mL. The molarity of the solution is
   (A) 1.78*M* (B) 2.00*M* (C) 2.05*M* (D) 2.22*M*

3. Bombardment of aluminium by α-particle leads to its artificial disintegration in two ways, (i) and (ii) as shown. Products X, Y and Z respectively are,

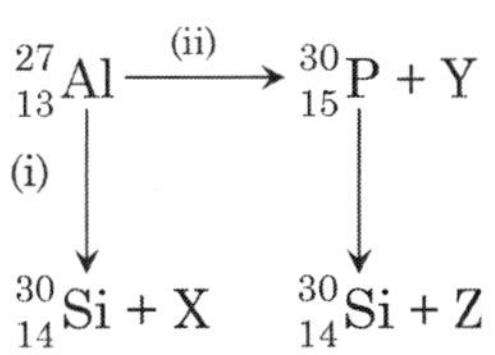

   (A) proton, neutron, positron (B) neutron, positron, proton
   (C) proton, positron, neutron (D) positron, proton, neutron

4. Geometrical shapes of the complexes formed by the reaction of Ni^{2+} with Cl^-, CN^- and H_2O, respectively, are
   (A) octahedral, tetrahedral and square planar
   (B) tetrahedral, square planar and octahedral
   (C) square planar, tetrahedral and octahedral
   (D) octahedral, square planar and tetrahedral.

5. The major product of the following reaction is

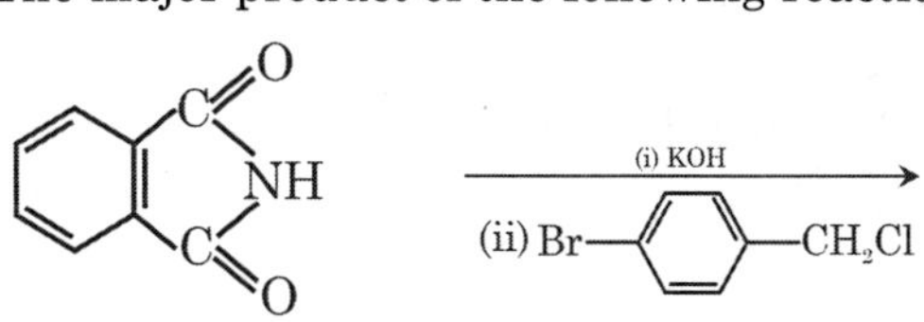

(A) 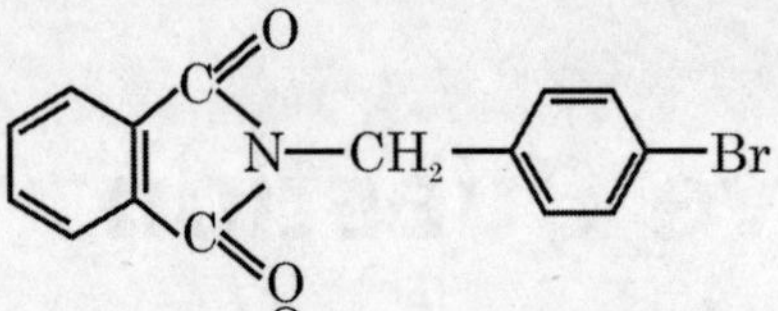 (B) 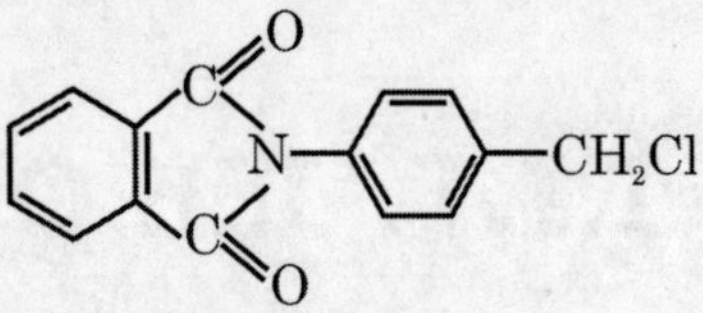

(C) 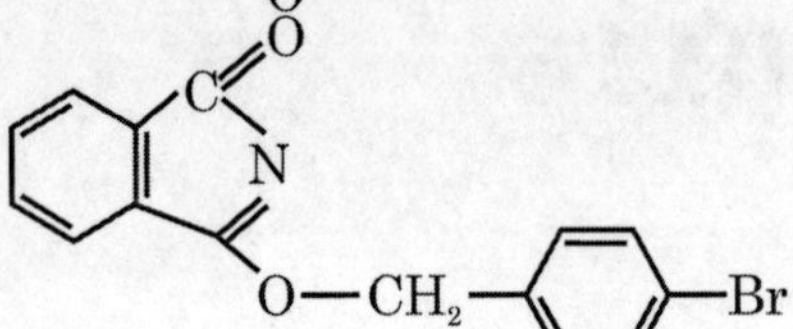 (D) 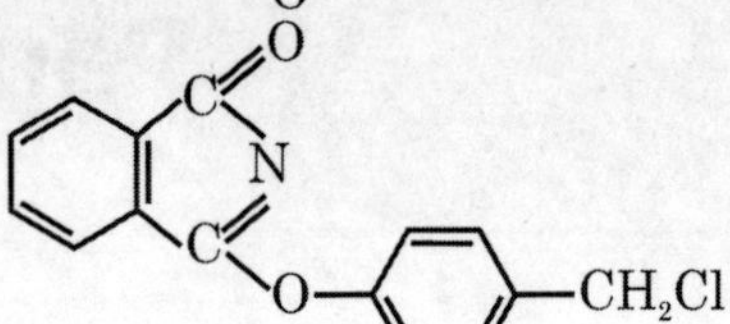

6. Among the following compounds, the most acidic is
   (A) *p*-nitrophenol (B) *p*-hydroxybenzoic acid
   (C) *o*-hydroxybenzoic acid (D) *p*-toluic acid.

7. $AgNO_3(aq)$ was added to an aqueous KCl solution gradually and the conductivity of the solution was measured. The plot of conductance (L) versus the volume of $AgNO_3$ is

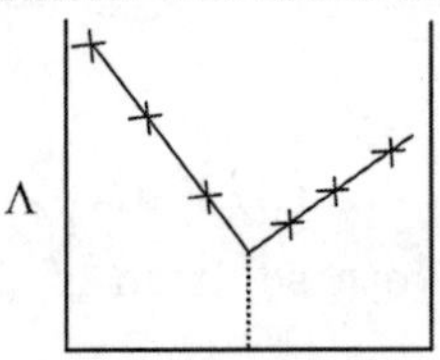

volume
(P)

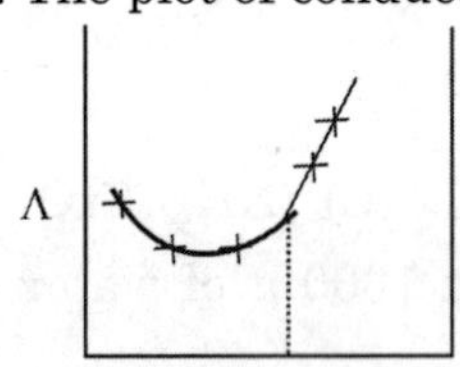

volume
(Q)

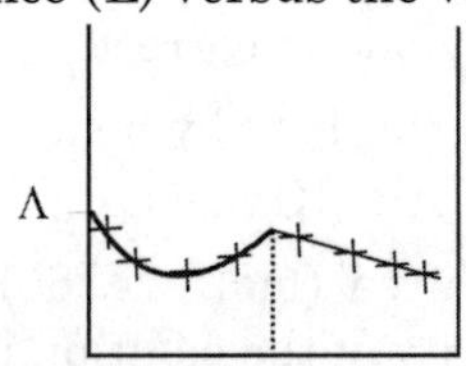

volume
(R)

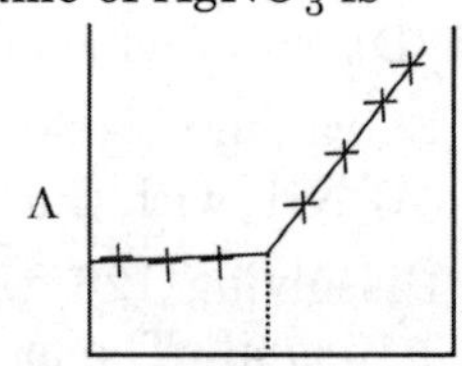

volume
(S)

   (A) (P) (B) (Q) (C) (R) (D) (S)

SECTION – II (Total marks: 16)
(Multiple Correct Answer Type)

This section contains 4 multiple choice questions. Each question has four choices (A), (B), (C) and (D) out of which one or more may be correct.

8. Amongst the given options, the compound(s) in which all the atoms are in one plane in all the possible conformations (if any), is(are)
   (A) 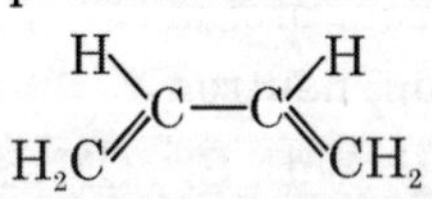 (B) 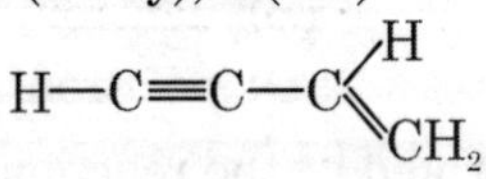 (C) $H_2C{=}C{=}O$ (D) $H_2C{=}C{=}CH_2$

9. Extraction of metal from the ore cassiterite involves
   (A) carbon reduction of an oxide ore (B) self-reduction of a sulphide ore
   (C) removal of copper impurity (D) removal of iron impurity.

10. According to kinetic theory of gases
    (A) collisions are always elastic
    (B) heavier molecules transfer more momentum to the wall of the container
    (C) only a small number of molecules have very high velocity
    (D) between collisions, the molecules move in straight lines with constant velocities.

11. The correct statement(s) pertaining to the adsorption of a gas on a solid surface is(are)

(A) Adsorption is always exothermic

(B) Physisorption may transform into chemisorption at high temperature

(C) Physiosorption increases with increasing temperature but chemisorption decreases with increasing temperature

(D) Chemisorption is more exothermic than physisorption, however it is very slow due to higher energy of activation.

SECTION – III (Total marks: 15)
(Paragraph Type)

This section contains 2 paragraphs. Based upon one of paragraphs 2 multiple choice questions and based on the other paragraph 3 multiple choice questions have to be answered. Each of these questions has four choice (A), (B), (C) and (D) out of which only one is correct.

Paragraph for Question Nos. 12 and 13

An acyclic hydrocarbon P, having molecular formula C_6H_{10}, gave acetone as the only organic product through the following sequence of reactions, in which Q is an intermediate organic compound.

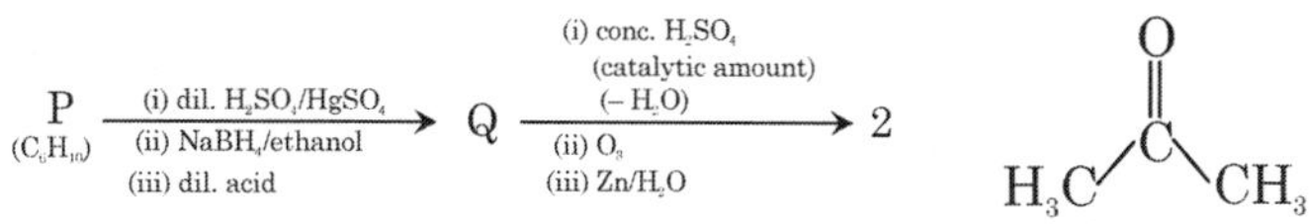

12. The structure of compound P is

(A) $CH_3CH_2CH_2CH_2—C≡C—H$

(B) $H_3CH_2C—C≡C—CH_2CH_3$

(C) 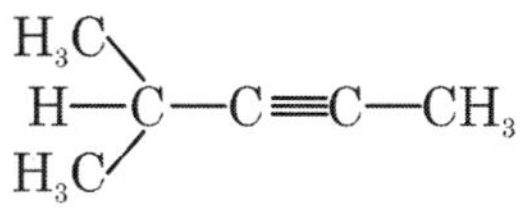

(D) 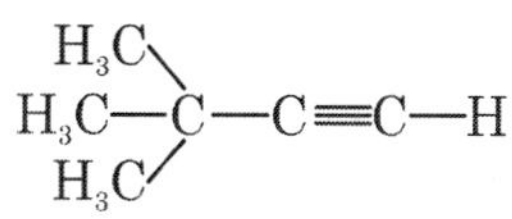

13. The structure of the compound Q is

(A) 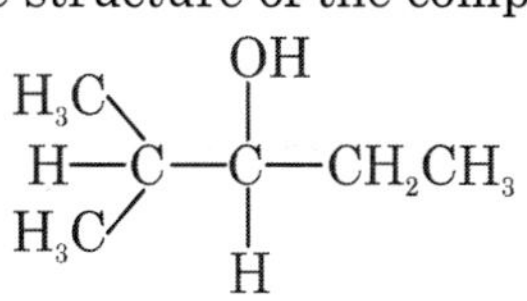

(B) 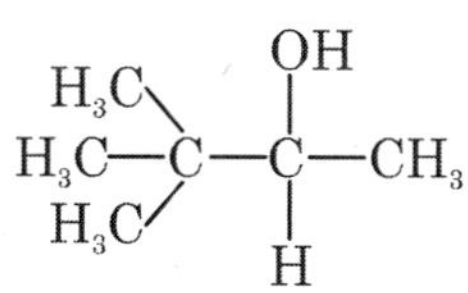

(C) 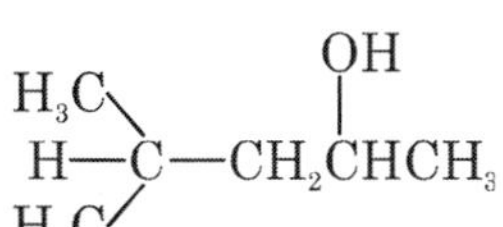

(D) 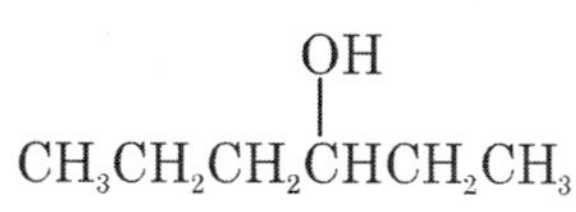

Paragraph for Question Nos. 14 to 16

When a metal rod M is dipped into an aqueous colourless concentrated solution of compound N, the solution turns light blue. Addition of aqueous NaCl to the blue solution gives a white precipitate O. Addition of aqueous NH_3 dissolves O and gives an intense blue solution.

14. The metal rod M is

(A) Fe (B) Cu (C) Ni (D) Co

15. The compound N is

(A) $AgNO_3$ (B) $Zn(NO_3)_2$ (C) $Al(NO_3)_3$ (D) $Pb(NO_3)_2$

16. The final solution contains

(A) $[Pb(NH_3)_4]^{2+}$ and $[CoCl_4]^{2-}$ (B) $[Al(NH_3)_4]^{3+}$ and $[Cu(NH_3)_4]^{2+}$

(C) $[Ag(NH_3)_2]^{+}$ and $[Cu(NH_3)_4]^{2+}$ (D) $[Ag(NH_3)_2]^{+}$ and $[Ni(NH_3)_6]^{2+}$

SECTION – IV (Total marks: 28)
(Integer Answer Type)

This section contains 7 questions. The answer to each of the question is a single digit integer, ranging from 0 to 9. The bubble corresponding to the correct answer is to darkened in the ORS.

17. The difference in the oxidation numbers of the types of sulphur atoms in $Na_2S_4O_6$ is

18. A decapeptide (Mol. Wt. 796) on complete hydrolysis gives glycine (Mol. Wt. 75), alanine and phenylalanine. Glycine contributes 47.0% to the total weight of the hydrolysed products. The number of glycine units present in the decapeptide is

19. The work function (ϕ) of some metals is listed below. The number of metals which will show photoelectric effect when light of 300 nm wavelenght falls on the metal is

Metal	Li	Na	K	Mg	Cu	Ag	Fe	Pt	W
ϕ (eV)	2.4	2.3	2.2	3.7	4.8	4.3	4.7	6.3	4.75

20. The maximum number of electrons that can have principal quantum number, $n = 3$, and spin quantum number, $m_s = -\frac{1}{2}$, is

21. Reaction of Br_2 with Na_2CO_3 in aqueous solution gives sodium bromide and sodium bromate with evolution of CO_2 gas. The number of sodium bromide molecules involved in the balanced chemical equation is

22. To an evacuated vessel with movable piston under external pressure of 1 atm, 0.1 mol of He and 1.0 mol of an unknown compound (vapour pressure 0.68 atm. at 0°C) are introduced. Considering the ideal gas behaviour, the total volume (in L) of the gases at 0°C is close to

23. The total number of alkenes possible by dehydrobromination of 3-bromo-3-cyclopentylhexane using alcoholic KOH is

ANSWER KEYS

1. D	2. C	3. A	4. B	5. A	6. C
7. D	8. B, C	9. A, C, D	10. A, B, C, D	11. A, B, D	12. D
13. B	14. B	15. A	16. C	17. 5	18. 6
19. 4	20. 9	21. 5	22. 7	23. 5	

SOLUTIONS

1. (D)

$$Ba(N_3)_2 \xrightarrow{\Delta} Ba + 3N_2$$

2. (C)

Total mass of solution = 1000 + 120 = 1120 g

$$\text{Total volume of solution in L} = \frac{1120}{1.15} \times 10^3$$

$$M = \frac{W}{M.W.} \times \frac{1}{V(\text{in L})} = \frac{120}{60} \times \frac{1.15 \times 10^3}{1120} = 2.05M$$

3. (A)

$$^{27}_{13}Al + ^{4}_{2}\alpha \longrightarrow ^{30}_{14}Si + ^{1}_{1}p(X)$$

$$^{27}_{13}Al + ^{4}_{2}\alpha \longrightarrow ^{30}_{15}P + ^{1}_{0}n(Y)$$

$$^{30}_{15}P \longrightarrow ^{30}_{14}Si + ^{0}_{+1}\beta(Z)$$

4. (B)

$[NiCl_4]^{2-} \longrightarrow$ Tetrahedral

$[Ni(CN)_4]^{2-} \longrightarrow$ Square planar

$[Ni(H_2O)_6]^{2+} \longrightarrow$ Octahedral

5. (A)

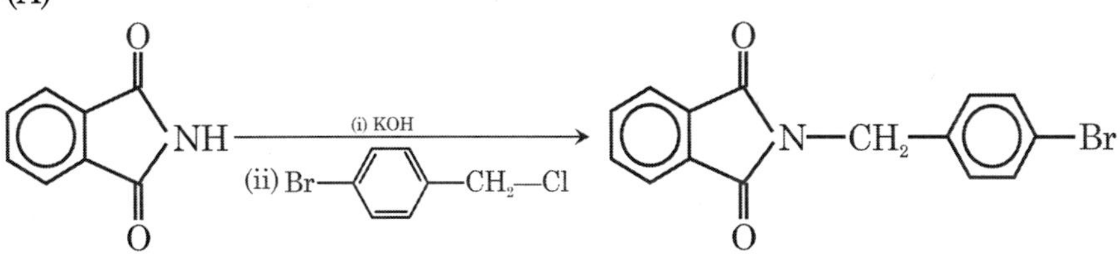

The partial double bond character along C—Br bond prevents the attact of nucleophile at phenylic position.

6. (C)

Due to the ortho effect *o*-hydroxybenzoic acid is the strongest acid and the correct order of decreasing K_a is

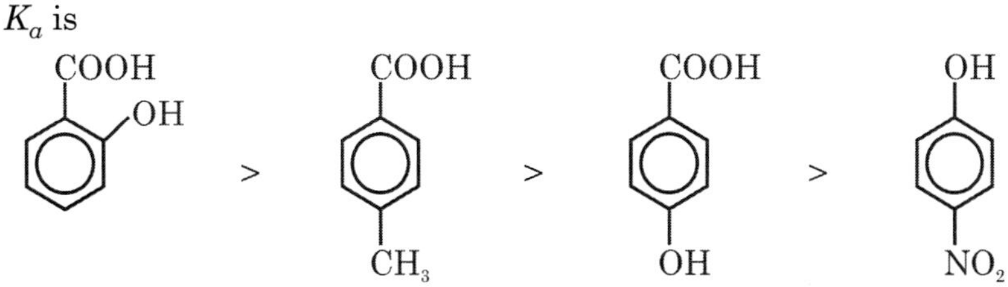

7. (D)

$AgNO_3(aq) + KCl(aq) \longrightarrow AgCl(s) + KNO_3(aq)$

Initially there is aqueous KCl solution. As solution of $AgNO_3$ is added, AgCl(*s*) is formed. Hence the conductivity of solution is almost compensated (or slightly increased) by the formation of KNO_3. After the end point the conductivity increases more repidly because of addition of excess $AgNO_3$ solution.

8. (B, C)

Along the C—C single bond conformations are possible in butadiene in which all the atoms may not lie in the same plane. In $H_2C{=}C{=}CH_2$, the hydrogen atoms are not in the same plane.

9. (A, C, D)

$SnO_2 + 2C \rightarrow 2CO + Sn$

The ore cassiterite contains the impurity of Fe, Mn, W and traces of Cu.

10. (A, B, C, D)

11. (A, B, D)

12. (D)

13. (B)

Solution for the Q. Nos. 12 to 13:

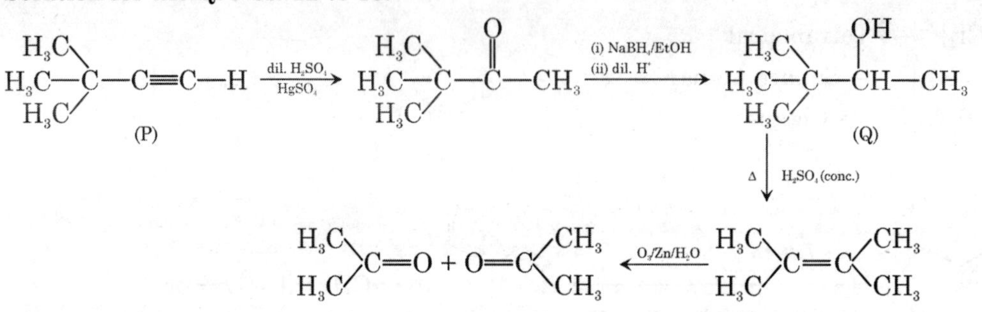

14. (B)

$$\underset{M}{Cu} + \underset{N}{2AgNO_3} \longrightarrow \underset{Blue}{Cu(NO_3)_2} + 2Ag$$

While Cu is partially oxidized to $Cu(NO_3)_2$, the remaining $AgNO_3$ reacts with NaCl.

15. (A)

16. (C)

$$\underset{N}{AgNO_3} + NaCl \longrightarrow \underset{O}{AgCl\downarrow} + NaNO_3$$

$AgCl + 2NH_3 \longrightarrow [Ag(NH_3)_2]^+Cl^-$

$Cu(NO_3)_2 + 4NH_4OH \longrightarrow [Cu(NH_3)_4]^{2+}$

17. (5)

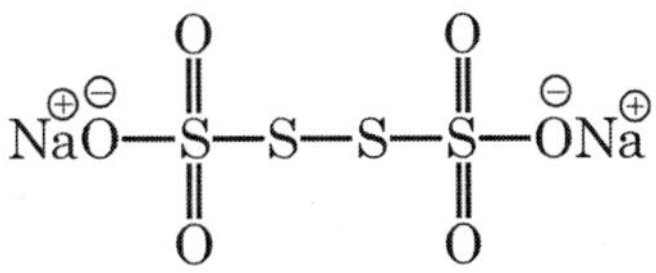

S will have oxidation number = + 5, 0

Difference in oxidation number = 5

18. (6)

For n-units of glycine, $\dfrac{n \times 75}{(796 + 9 \times 18)} \times 100 = 47$

or $n = 6$

19. (4)

The energy associated with incident photon $= \dfrac{hc}{\lambda}$

or $\quad E = \dfrac{6.6 \times 10^{-34} \times 3 \times 10^{8}}{300 \times 10^{-9}}$ J

E in eV $= \dfrac{6.6 \times 10^{-34} \times 3 \times 10^{8}}{300 \times 10^{-9} \times 1.6 \times 10^{-19}} = 4.16$ eV

Hence the number of metals showing photoelectric effects will be 4, i.e. Li, Na, K, Mg.

20. (9)

For principal quantum number ($n = 3$), number of orbitals $= n^2 = 9$.

Hence, the number of electrons with $m_s = -\dfrac{1}{2}$ will be 9.

21. (5)

$3Br_2 + 3Na_2CO_3 \longrightarrow 5NaBr + NaBrO_3 + 3CO_2$

Hence, the number of NaBr molecules is 5.

22. (7)

For any ideal gas, $PV = nRT$

$0.32 \times V = 0.1 \times 0.0821 \times 273$

$V = 7$ L

(unknown compound X will not follow ideal gas equation)

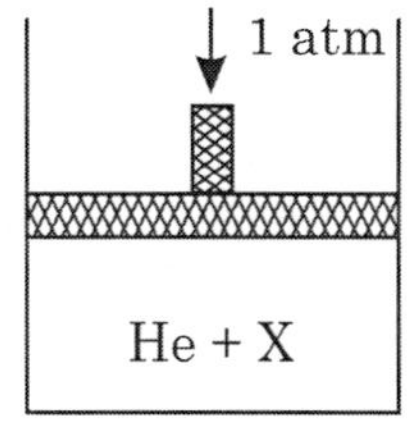

For He, $n = 0.1, P = 0.32$ atm.
$V = ?, T = 273$ K

23. (5)

Total no. of alkenes will be 5.

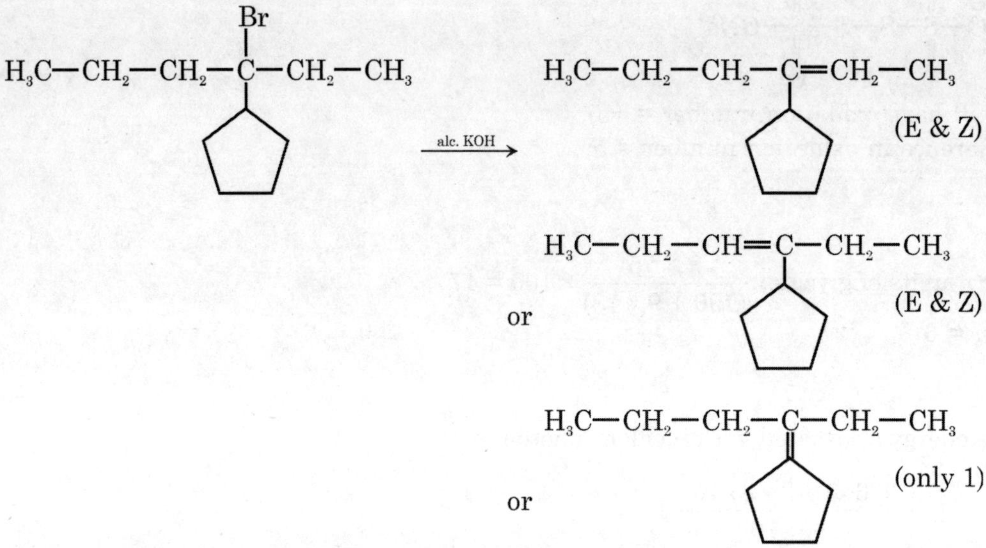

I.I.T. JEE Objective Question Paper—2011

PAPER—2

SECTION – I (Total marks: 24)
(Single Correct Answer Type)

This section contains 8 multiple choice questions. Each question has four choices (A), (B), (C) and (D) out of which only one is correct.

1. The freezing point (in °C) of a solution containing 0.1 g of $K_3[Fe(CN)_6]$ (Mol. Wt. 329) in 100 g of water ($K_f = 1.86$ K kg mol^{-1}) is

(A) -2.3×10^{-2} (B) -5.7×10^{-2} (C) -5.7×10^{-3} (D) -1.2×10^{-2}

2. Amongst the compounds given, the one that would form a brilliant colored dye on treatment with $NaNO_2$ in dil. HCl followed by addition to an alkaline solution of β-naphthol is

(A) 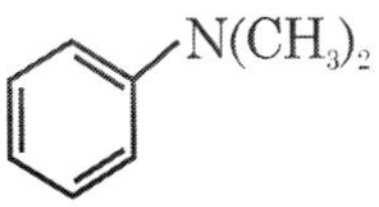

(B) 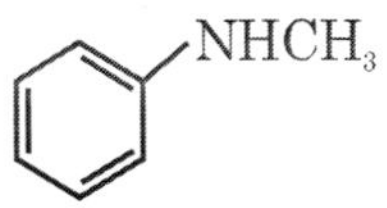

(C) 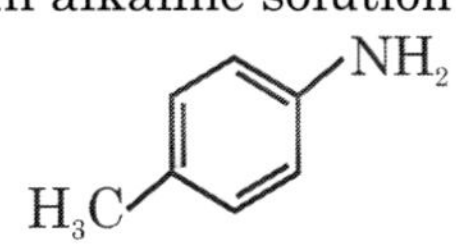

(D) 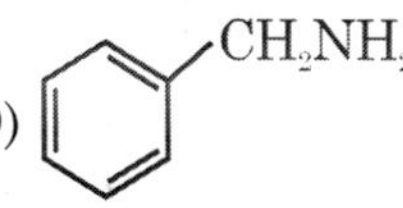

3. The major product of the following reaction is

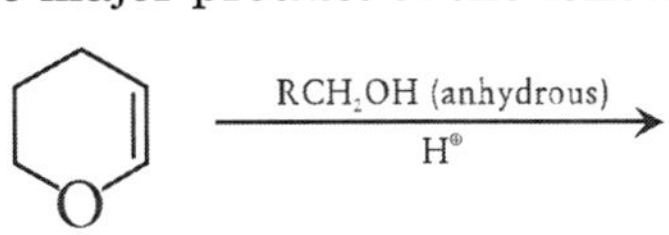

(A) a hemiacetal (B) an acetal (C) an ether (D) an ester.

4. The following carbohydrate is

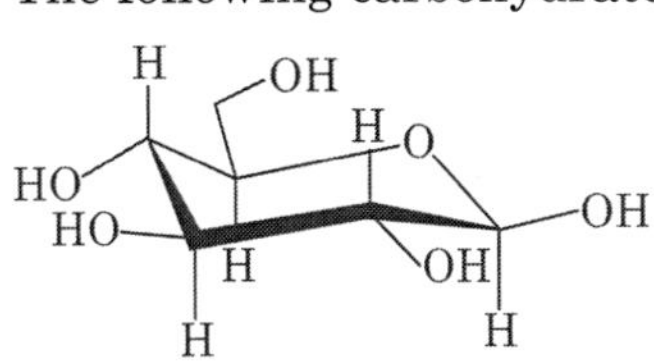

(A) a ketohexose (B) an aldohexose (C) an α-furanose (D) an α-pyranose.

5. Oxidation states of the metal in the minerals haematite and magnetite, respectively, are

(A) II, III in haematite and III in magnetite (B) II, III in haematite and II in magnetite

(C) II in haematite and II, III in magnetite (D) III in haematite and II, III in magnetite

6. Among the following complexes (K–P)

$K_3[Fe(CN)_6]$ (K), $[Co(NH_3)_6]Cl_3$ (L), $Na_3[Co(oxalate)_3]$ (M), $[Ni(H_2O)_6]Cl_2$ (N), $K_2[Pt(CN)_4]$ (O) and $[Zn(H_2O)_6](NO_3)_2$ (P), which are diamagnetic?

(A) K, L, M, N (B) K, M, O, P (C) L, M, O, P (D) L, M, N, O

7. Passing H_2S gas into a mixture of Mn^{2+}, Ni^{2+}, Cu^{2+} and Hg^{2+} ions in an acidified aqueous solution precipitates

(A) CuS and HgS (B) MnS and CuS (C) MnS and NiS (D) NiS and HgS

8. Consider the following cell reaction: $2Fe(s) + O_2(g) + 4H^+(aq) \rightarrow 2Fe^{2+}(aq) + 2H_2O(l)$ $E^0 = 1.67$ V

At $[Fe^{2+}] = 10^{-3} M$, $P(O_2) = 0.1$ atm and $pH = 3$, the cell potential at 25°C is

(A) 1.47 V (B) 1.77 V (C) 1.87 V (D) 1.57 V

SECTION – II (Total marks: 16)
(Multiple Correct Answer(s) Type)

This section contains 4 multiple choice questions. Each question has four choices (A), (B), (C) and (D) out of which one or more may be correct.

9. Reduction of the metal centre in aqueous permanganate ion involves

(A) 3 electrons in neutral medium (B) 5 electrons in neutral medium

(C) 3 electrons in alkaline medium (D) 5 electrons in acidic medium.

10. The correct functional group X and the reagent/reaction conditions Y in the following scheme are

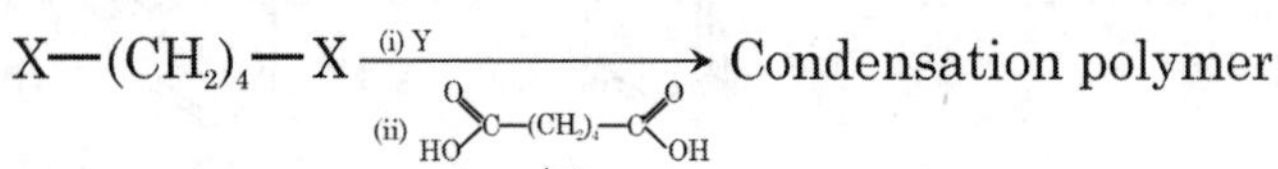

(A) X = $COOCH_3$, Y = H_2/Ni/heat (B) X = $CONH_2$, Y = H_2/Ni/heat

(C) X = $CONH_2$, Y = Br_2/NaOH (D) X = CN, Y = H_2/Ni/heat.

11. For the first order reaction, $2N_2O_5(g) \longrightarrow 4NO_2(g) + O_2(g)$

(A) the concentration of the reactant decreases exponentially with time

(B) the half-life of the reaction decreases with increasing temperature

(C) the half-life of the reaction depends on the initial concentration of the reactant

(D) the reaction proceeds to 99.6% completion in eight half-life duration.

12. The equilibrium $2Cu^{I} \rightleftharpoons Cu^0 + Cu^{II}$, in aqueous medium at 25°C shifts towards the left in the presence of

(A) NO_3^- (B) Cl^- (C) SCN^- (D) CN^-

SECTION – III (Total marks: 24)
(Integer Answer Type)

This section contains 6 questions. The answer to each of the questions is a single-digit integer, ranging from 0 to 9. The bubble corresponding to the correct answer is to be darkened in the ORS.

13. The maximum number of isomers (including stereoisomers) that are possible on monochlorination of the following compound is

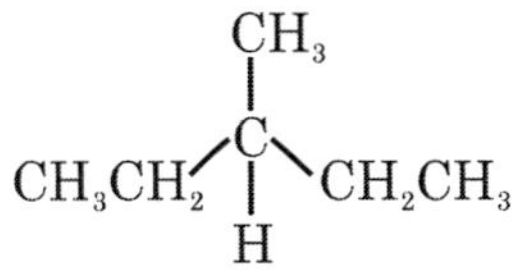

14. The total number of contributing structure showing hyperconjugation (involving C—H bonds) for the following carbocation is

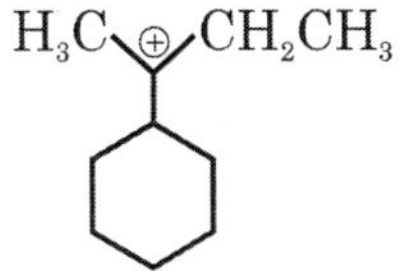

15. Among the following, the number of compounds that can react with PCl_5 to give $POCl_3$ is

O_2, CO_2, SO_2, H_2O, H_2SO_4, P_4O_{10}

16. The volume (in mL) of 0.1*M* $AgNO_3$ required for complete precipitation of chloride ions present in 30 mL of 0.01*M* solution of $[Cr(H_2O)_5Cl]Cl_2$, as silver chloride is close to

17. In 1 L saturated solution of AgCl $[K_{sp}(AgCl) = 1.6 \times 10^{-10}]$, 0.1 mol of CuCl $[K_{sp}(CuCl) = 1.0 \times 10^{-6}]$ is added. The resultant concentration of Ag^+ in the solution is 1.6×10^{-x}. The value of x is

18. The number of hexagonal faces that are present in a truncated octahedron is

SECTION – IV (Total marks: 16)
(Matrix-Match Type)

This section contains 2 questions. Each question has four statements (A, B, C and D) given in Column I and five statements (p, q, r, s and t) in Column II. Any given statement in Column I can have correct matching with one or more statement(s) given in Column II. For example, if for a given question, statement B matches with the statement given q and r, then for the particular question, against statement B, darken the bubbles corresponding to q and r in the ORS.

19. Match the transformations in column I with appropriate options in column II.

Column I	Column II
(A) $CO_2(s) \longrightarrow CO_2(g)$	(p) phase transition
(B) $CaCO_3(s) \longrightarrow CaO(s) + CO_2(g)$	(q) allotropic change
(C) $2H \longrightarrow H_2(g)$	(r) ΔH is positive
(D) $P_{(white,\ solid)} \longrightarrow P_{(red,\ solid)}$	(s) ΔS is positive
	(t) ΔS is negative

20. Match the reactions in column I with appropriate types of steps/reactive intermediate involved in these reactions as given in column II.

Column I	Column II
(A) 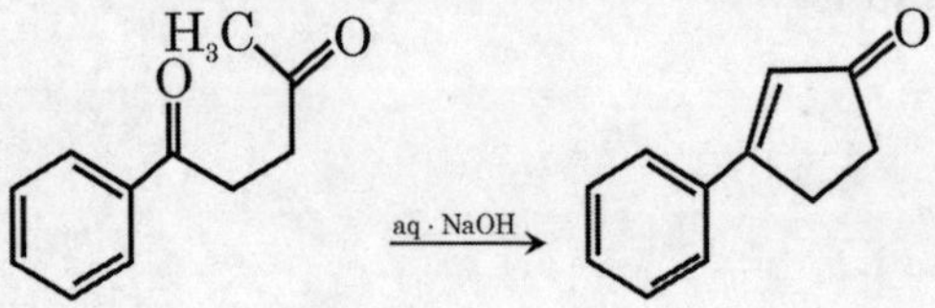	(p) Nucleophilic substitution
(B) 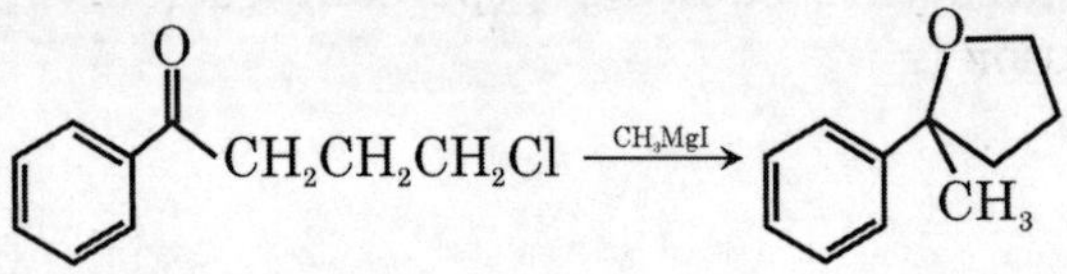	(q) Electrophilic substitution
(C) 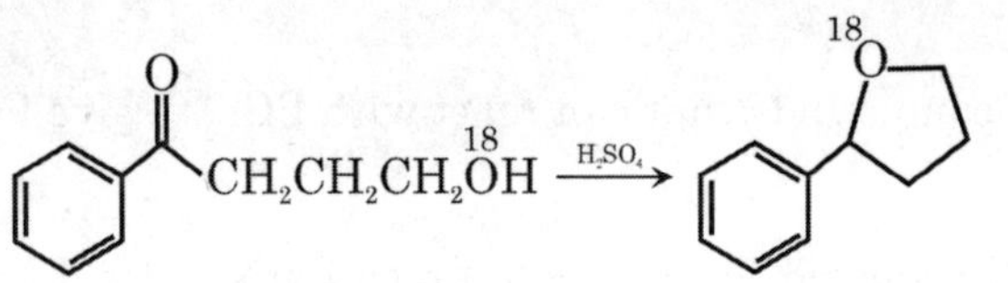	(r) Dehydration
(D) 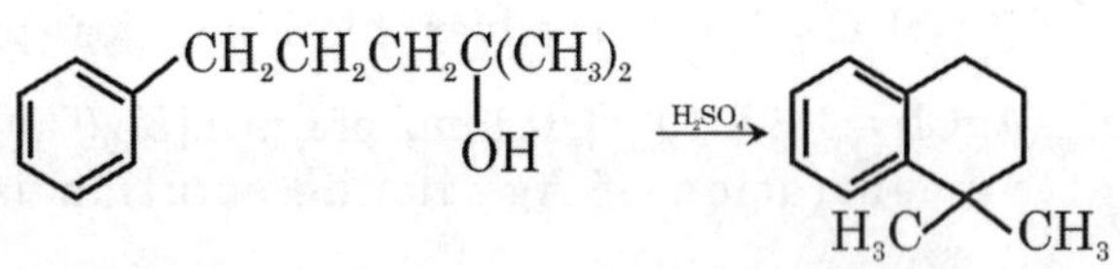	(s) Nucleophilic addition
	(t) Carbanion

ANSWER KEYS

1. A	2. C	3. B	4. B	5. D	6. C
7. A	8. D	9. A, D	10. A, B, C, D	11. A, B, D	12. B, C, D
13. 8	14. 6	15. 5	16. 6	17. 7	18. 8

19. A – p, r, s; B – r, s; C – t; D – p, q, t
20. A – r, s, t; B – p, s; C – r, s; D – q, r

SOLUTIONS

1. (A)

$K_3[Fe(CN)_6] \longrightarrow 3K^+ + [Fe(CN)_6]^{3-}$

$i = 4$

$$\Delta T_f = K_f \times i \times \frac{m}{M} \times \frac{1000}{W} = 1.86 \times 4 \times \frac{0.1}{329} \times \frac{1000}{100} = 2.3 \times 10^{-2}$$

$T_f = -2.3 \times 10^{-2}$

2. (C)

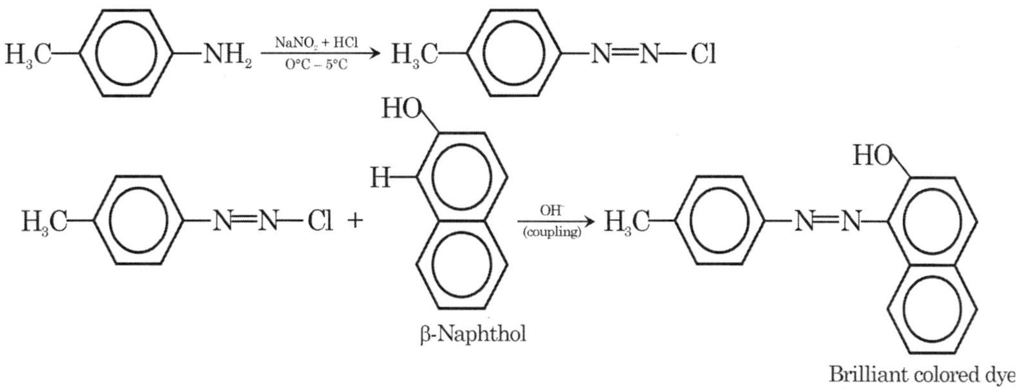

3. (B)

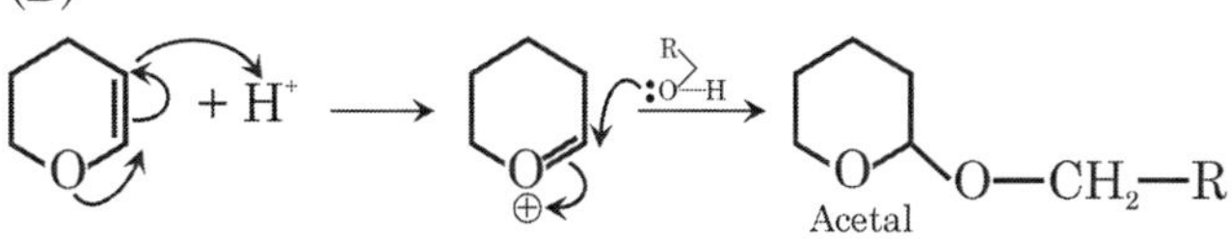

4. (B)

The carbohydrate is an aldohexose.

5. (D)

Haematite: Fe_2O_3: $2x + 3 \times (-2) = 0$

$x = 3$

Magnetite: Fe_3O_4 [an equimolar mixture of FeO and Fe_2O_3]

FeO: $x - 2 = 0$ or $x = 2$

Fe_2O_3: $x = 3$

6. (A)

Following compounds are diamagnetic: L: $[Co(NH_3)_6]Cl_3$; M: $Na_3[Co(ox)_3]$; O: $K_2[Pt(CN)_4]$; P: $[Zn(H_2O)_6](NO_3)_2$.

7. (A)

H_2S in presence of aqueous acidified solution precipitates as sulphide of Cu and Hg other than Pb^{2+}, Bi^{3+}, Cd^{2+}, As^{3+}, Sb^{3+} and Sn^{2+}.

8. (D)

$2Fe(s) + O_2(g) + 4H^+(aq) \longrightarrow 2Fe^{2+}(aq) + 2H_2O(l)$

$n = 4$ (no. of moles of electrons involved)

$$E_{cell} = E^0_{cell} - \frac{0.0591}{n}\log Q = 1.67 - \frac{0.0591}{4}\log\frac{(10^{-3})^2}{0.1\times(10^{-3})^4} \quad \{\because [H^+] = 10^{-pH}\}$$

$$= 1.67 - 0.10 = 1.57 \text{ V}$$

9. (A, D)

In acidic medium,

$$MnO_4^- + 8H^+ + 5e^- \longrightarrow Mn^{2+} + 4H_2O$$

In neutral medium,

$$MnO_4^- + 2H_2O + 3e^- \longrightarrow MnO_2 + 4OH^-$$

Hence the number of electrons loose in acidic and neutral medium is 5 and 3 electrons respectively.

10. (A, B, C, D)

Condensation polymers are formed by condensation of a diol or diamine with dicarboxylic acids.

Hence X may be $-\overset{O}{\overset{\|}{C}}-OR$ or $-\overset{O}{\overset{\|}{C}}-NH_2$ or $-C\equiv N$

$$H_3CO-\overset{O}{\overset{\|}{C}}-(CH_2)_4-\overset{O}{\overset{\|}{C}}-OCH_3 \xrightarrow{H_2/Ni} HO-H_2C(CH_2)_4-CH_2-OH \xrightarrow{HO-\overset{O}{\overset{\|}{C}}-(CH_2)_4-\overset{O}{\overset{\|}{C}}-OH} \text{Polyester}$$

$$H_2N-\overset{O}{\overset{\|}{C}}-(CH_2)_4-\overset{O}{\overset{\|}{C}}-NH_2 \xrightarrow{H_2/Ni} H_2N-CH_2(CH_2)_4-CH_2-NH_2 \xrightarrow{HO-\overset{O}{\overset{\|}{C}}-(CH_2)_4-\overset{O}{\overset{\|}{C}}-OH} \text{Polyamide}$$

$\downarrow$ Br_2/OH^-, Δ; Hoffmann bromamide reaction

$$H_2N-(CH_2)_4-NH_2 \xrightarrow{HO-\overset{O}{\overset{\|}{C}}-(CH_2)_4-\overset{O}{\overset{\|}{C}}-OH} \text{Polyamide}$$

$$N\equiv C-(CH_2)_4-C\equiv N \xrightarrow{H_2/Ni} H_2N-CH_2-(CH_2)_4-CH_2-NH_2 \xrightarrow{HO-\overset{O}{\overset{\|}{C}}-(CH_2)_4-\overset{O}{\overset{\|}{C}}-OH} \text{Polyamide}$$

11. (A, B, D)

For first order reaction, $[A] = [A]_0e^{-kt}$

Hence the concentration of $[NO_2]$ decreases exponentially.

Also, $t_{1/2} = \dfrac{0.693}{K}$, which is independent of concentration and $t_{1/2}$ decreases with the increase of temperature.

$$t_{99.6} = \frac{2.303}{K}\log\left(\frac{100}{0.4}\right) \quad \Rightarrow \quad t_{99.6} = \frac{2.303}{K}(2.4) = 8 \times \frac{0.693}{K} = 8t_{1/2}$$

12. (B, C, D)

Cu^{2+} ions will react with CN^- and SCN^- forming $[Cu(CN)_4]^{3-}$ and $[Cu(SCN)_4]^{3-}$ leading the reaction in the backward direction.

$Cu^{2+} + 2CN^- \longrightarrow Cu(CN)_2 \quad \Rightarrow \quad 2Cu(CN)_2 \longrightarrow 2CuCN + (CN)_2$

$CuCN + 3CN^- \longrightarrow [Cu(CN)_4]^{3-}$

$Cu^{2+} + 4SCN^- \longrightarrow [Cu(SCN)_4]^{3-}$

Cu^{2+} also combines with Cl^- to give $CuCl_2$ which reacts with Cu to produce CuCl pushing the reaction in the backward direction.

$CuCl_2 + Cu \longrightarrow 2CuCl \downarrow$

13. (8)

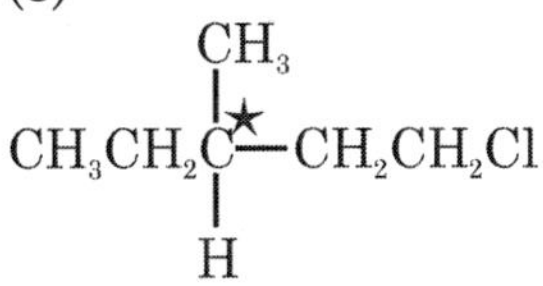
Enantiomeric pair = 2

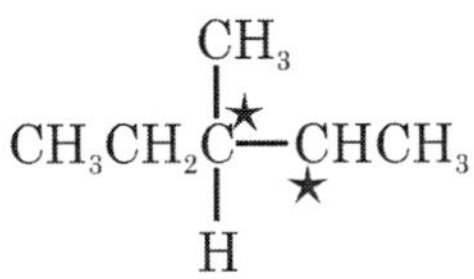
Two enantiomeric pairs = 4

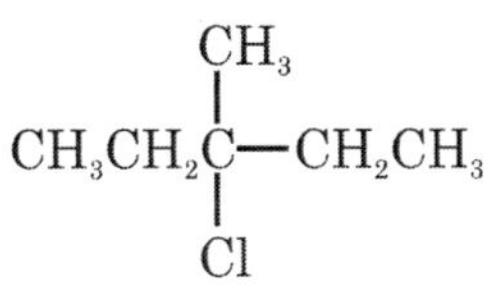
1

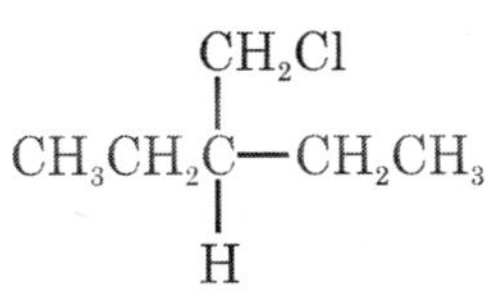
1

Total number of isomers = 2 + 4 + 1 + 1 = 8

14. (6)

Six H atoms are there.

15. (5)

$H_2O, O_2, SO_2, CO_2, H_2SO_4$

$PCl_5 + H_2O \xrightarrow{\text{Insufficient water}} POCl_3 + 2HCl$

$2PCl_5 + O_2 \longrightarrow 2POCl_3 + 2Cl_2$

$PCl_5 + SO_2 \longrightarrow POCl_3 + SOCl_2$

$PCl_5 + CO_2 \longrightarrow POCl_3 + COCl_2$

$2PCl_5 + H_2SO_4 \longrightarrow 2POCl_3 + SO_2Cl_2 + 2HCl$

16. (6)

Number of ionisable Cl^- in $[Cr(H_2O)_5Cl]Cl_2$ is 2

Hence millimoles of Cl^- = 30 × 0.01 × 2 = 0.6

Hence millimoles of Ag^+ required = 0.6

Thus $0.6 = 0.1\ V$

$V = 6$ mL

17. (7)

Let the solubility of AgCl be x mol litre^{-1} and that of CuCl be y mol litre^{-1}.

$$\underset{}{AgCl} \rightleftharpoons \underset{x}{Ag^+} + \underset{x}{Cl^-}$$

$$CuCl \rightleftharpoons \underset{y}{Cu^+} + \underset{y}{Cl^-}$$

Hence K_{sp} of AgCl = $[Ag^+][Cl^-]$

$1.6 \times 10^{-10} = x(x + y)$ ...(i)

Similarly K_{sp} of CuCl = $[Cu^+][Cl^-]$

$1.6 \times 10^{-6} = y(x + y)$ ...(ii)

On solving (i) and (ii), we get $[Ag^+] = 1.6 \times 10^{-7}$

Hence $x = 7$

18. (8)

19. (A–p, r, s); (B–r, s); (C–t); (D–p, q, t)

20. (A–r, s, t); (B–p, s); (C–r, s); (D–q, r)

I.I.T. JEE Objective Question Paper—2012

PAPER—1

SECTION - I
(Single Correct Answer Type)

This section contains 10 multiple choice questions. Each question has four choices (A), (B), (C) and (D) out of which only one is correct.

1. A compound M_pX_q has cubic close packing (ccp) arrangement of X. Its unit cell structure is shown below. The empirical formula of the compound is

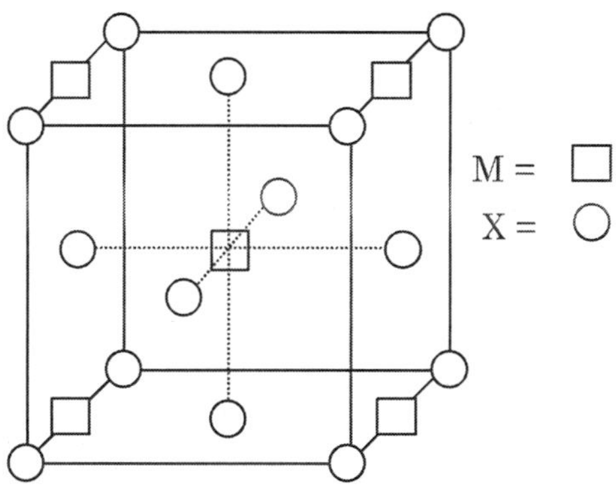

(A) MX (B) MX_2 (C) M_2X (D) M_5X_{14}

2. The carboxyl functional group (—COOH) is present in
(A) picric acid (B) barbituric acid (C) ascorbic acid (D) aspirin.

3. As per IUPAC nomenclature, the name of the complex $[Co(H_2O)_4(NH_3)_2]Cl_3$ is
(A) Tetraaquadiaminecobalt(III) chloride (B) Tetraaquadiamminecobalt(III) chloride
(C) Diaminetetraaquacobalt(III) chloride (D) Diamminetetraaquacobalt(III) chloride.

4. In allene (C_3H_4), the type(s) of hybridization of the carbon atoms is(are)
(A) *sp* and sp^3 (B) *sp* and sp^2 (C) only sp^2 (D) sp^2 and sp^3

5. The kinetic energy of an electron in the second Bohr orbit of a hydrogen atom is [a_0 is Bohr radius]

(A) $\frac{h^2}{4\pi^2 m a_0^2}$ (B) $\frac{h^2}{16\pi^2 m a_0^2}$ (C) $\frac{h^2}{32\pi^2 m a_0^2}$ (D) $\frac{h^2}{64\pi^2 m a_0^2}$

6. Which ordering of compounds is according to the decreasing order of the oxidation state of nitrogen?
(A) HNO_3, NO, NH_4Cl, N_2 (B) HNO_3, NO, N_2, NH_4Cl
(C) HNO_3, NH_4Cl, NO, N_2 (D) NO, HNO_3, NH_4Cl, N_2

7. For one mole of a van der Waals gas when $b = 0$ and $T = 300$ K, the PV *vs* $1/V$ plot is shown below. The value of the van der Waals constant a (atm litre^2 mol^{-2}) is

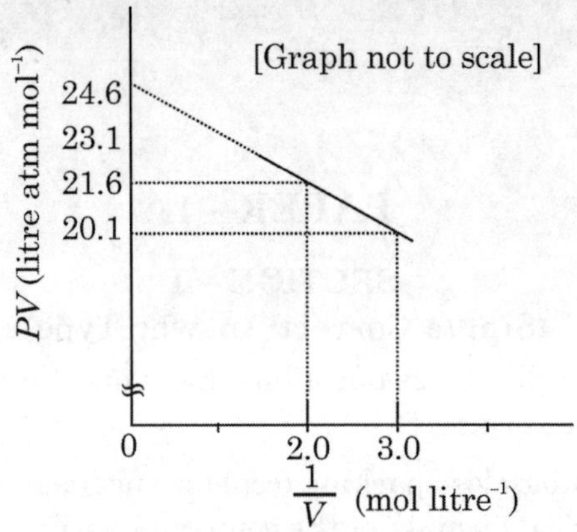

(A) 1.0 (B) 4.5 (C) 1.5 (D) 3.0

8. The number of aldol reaction(s) that occurs in the given transformation is

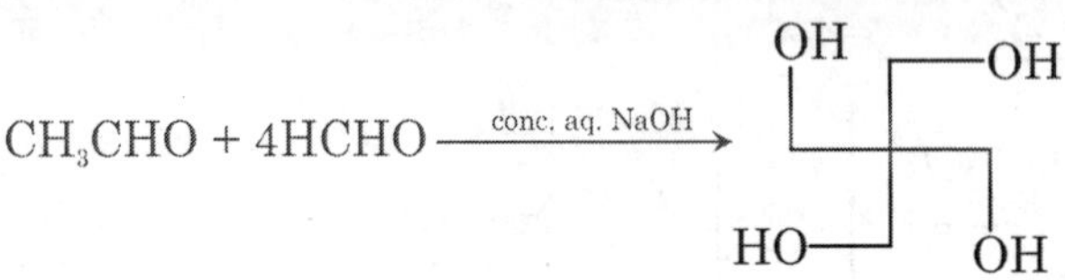

(A) 1 (B) 2 (C) 3 (D) 4

9. The colour of light absorbed by an aqueous solution of $CuSO_4$ is
(A) orange-red (B) blue-green (C) yellow (D) violet.

10. The number of optically active products obtained from the complete ozonolysis of the given compound is

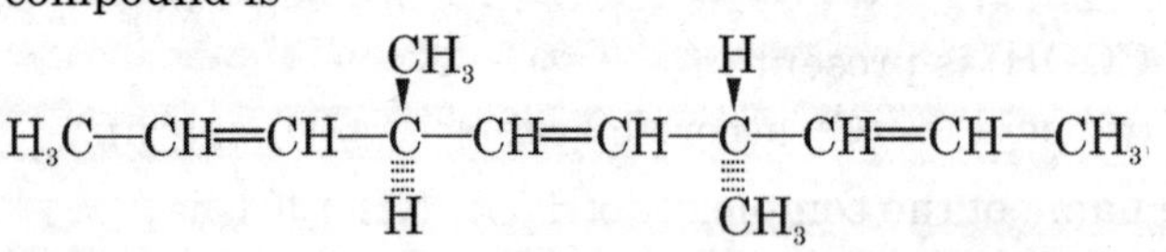

(A) 0 (B) 1 (C) 2 (D) 4

SECTION – II
(Multiple Correct Answer(s) Type)

This section contains 5 multiple choice questions. Each question has four choices (A), (B), (C) and (D) out of which one or more are correct.

11. Which of the following hydrogen halides react(s) with $AgNO_3(aq)$ to give a precipitate that dissolves in $Na_2S_2O_3(aq)$?
(A) HCl (B) HF (C) HBr (D) HI

12. Identify the binary mixture(s) that can be separated into individual compounds, by differential extraction, as shown in the given scheme.

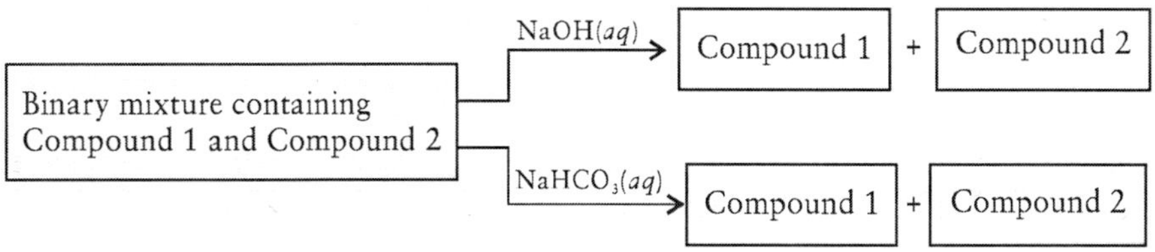

(A) C_6H_5OH and C_6H_5COOH
(B) C_6H_5COOH and $C_6H_5CH_2OH$
(C) $C_6H_5CH_2OH$ and C_6H_5OH
(D) $C_6H_5CH_2OH$ and $C_6H_5CH_2COOH$

13. For an ideal gas, consider only *P*-*V* work in going from an initial state *X* to the final state *Z*. The final state *Z* can be reached by either of the two paths shown in the figure. Which of the following choice(s) is(are) correct? [Take ΔS as change in entropy and w as work done]

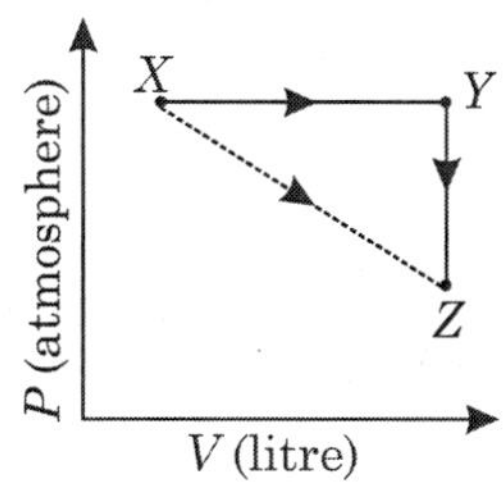

(A) $\Delta S_{x \to z} = \Delta S_{x \to y} + \Delta S_{y \to z}$
(B) $w_{x \to z} = w_{x \to y} + w_{y \to z}$
(C) $w_{x \to y \to z} = w_{x \to y}$
(D) $\Delta S_{x \to y \to z} = \Delta S_{x \to y}$

14. Which of the following molecules, in pure form, is(are) unstable at room temperature?

15. Choose the correct reason(s) for the stability of the lyophobic colloidal particles.
(A) Preferential adsorption of ions on their surface from the solution
(B) Preferential adsorption of solvent on their surface from the solution
(C) Attraction between different particles having opposite charges on their surface
(D) Potential difference between the fixed layer and the diffused layer of opposite charges around the colloidal particles.

SECTION – III
(Integer Answer Type)

This section contains 5 questions. The answer to each question is a single digit integer, ranging from 0 to 9 (both inclusive).

16. 29.2% (*w*/*w*) HCl stock solution has density of 1.25 g mL^{-1}. The molecular mass of HCl is 36.5 g mol^{-1}. The volume (mL) of stock solution required to prepare a 200 mL solution of 0.4*M* HCl is

17. The substituents R_1 and R_2 for nine peptides are listed in the table given below. How many of these peptides are positively charged at $pH = 7.0$?

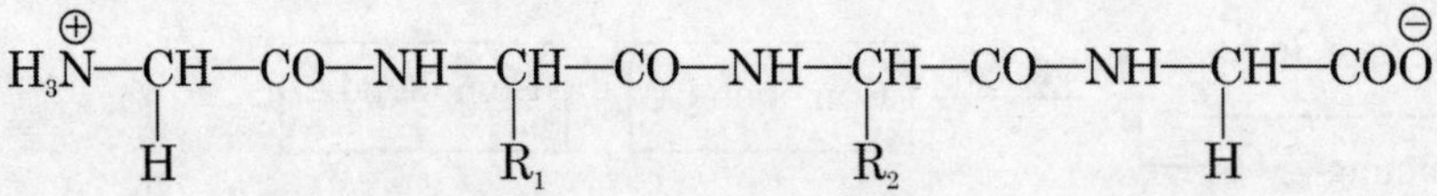

Peptide	R_1	R_2
I	H	H
II	H	CH_3
III	CH_2COOH	H
IV	CH_2CONH_2	$(CH_2)_4NH_2$
V	CH_2CONH_2	CH_2CONH_2
VI	$(CH_2)_4NH_2$	$(CH_2)_4NH_2$
VII	CH_2COOH	CH_2CONH_2
VIII	CH_2OH	$(CH_2)_4NH_2$
IX	$(CH_2)_4NH_2$	CH_3

18. An organic compound undergoes first-order decomposition. The time taken for its decomposition to 1/8 and 1/10 of its initial concentration are $t_{1/8}$ and $t_{1/10}$ respectively. What is the value of $([t_{1/8}]/[t_{1/10}] \times 10)$? (take $\log_{10} 2 = 0.3$)

19. When the following aldohexose exists in its D-configuration, the total number of stereoisomers in its pyranose form is

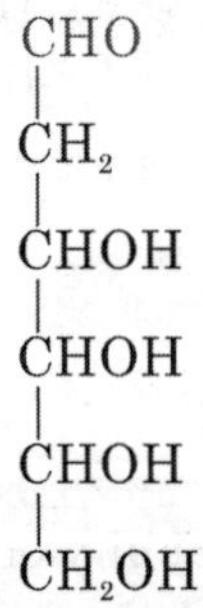

20. The periodic table consists of 18 groups. An isotope of copper, on bombardment with protons, undergoes a nuclear reaction yielding element X as shown below. To which group, element X belongs in the periodic table?

$${}^{63}_{29}Cu + {}^{1}_{1}H \rightarrow 6\,{}^{1}_{0}n + \alpha + 2\,{}^{1}_{1}H + X$$

ANSWER KEYS

1. B	2. D	3. D	4. B	5. C	6. B
7. C	8. C	9. A	10. A	11. A, C, D	12. B, D
13. A, C	14. B, C	15. A, D	16. 8	17. 4	18. 9
19. 8	20. 8				

SOLUTIONS

1. (B)

$$X = 8 \times \frac{1}{8} + 6 \times \frac{1}{2} = 4$$

$$M = 4 \times \frac{1}{4} + 1 = 2$$

Hence, unit cell formula of the compound is M_2X_4 and the empirical formula of the compound is MX_2.

2. (D)

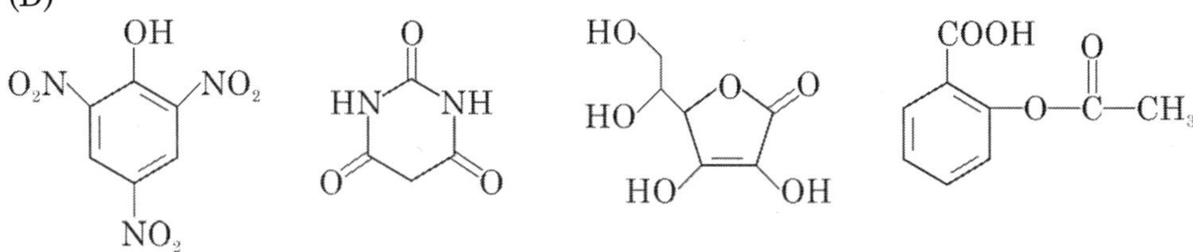

(Picric Acid) (Barbituric Acid) (Ascorbic Acid) (Aspirin)

3. (D)

$[Co(H_2O)_4(NH_3)_2]Cl_3$

Diamminetetraaquacobalt(III) chloride

4. (B)

$$\overset{sp^2}{H_2C}=\overset{sp}{C}=\overset{sp^2}{CH_2}$$

5. (C)

According to Bohr's postulate,

$$mvr = \frac{nh}{2\pi}$$

Hence, $$v = \frac{nh}{2\pi mr}$$

$$KE = \frac{1}{2}mv^2 = \frac{1}{2}m\left(\frac{nh}{2\pi mr}\right)^2$$

But $$r = \frac{a_0 \times n^2}{Z}$$

Hence, for second Bohr orbit

$$r = \frac{a_0 \times 2^2}{1} = 4a_0$$

$$KE = \frac{1}{2}m\left(\frac{2^2h^2}{4\pi^2m^2 \times (4a_0)^2}\right) = \frac{h^2}{32\pi^2 ma_0^2}$$

6. (B)

$$H\overset{+5}{N}O_3,\ \overset{+2}{N}O,\ \overset{0}{N_2},\ \overset{-3}{N}H_4Cl$$

7. (C)

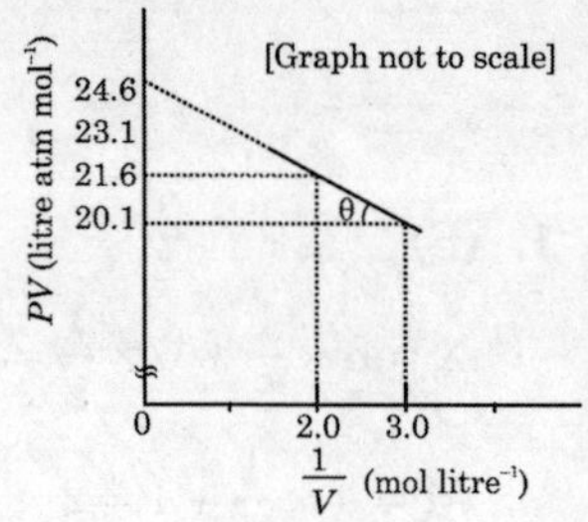

van der Waals equation for 1 mole of real gas is

$$\left(P+\frac{a}{V^2}\right)(V-b)=RT$$

But $b = 0$ (given)

or $$\left(P+\frac{a}{V^2}\right)(V)=RT$$

or $$PV=-a\times\frac{1}{V}+RT \quad \Rightarrow \quad y=mx+c$$

Slope $= \tan(\pi-\theta) = -a$

$$\tan\theta = a = \frac{24.6-20.1}{3-2} = 1.5 \quad \text{or} \quad \tan\theta = \frac{24.6-20.1}{3-2} = 1.5$$

8. (C)

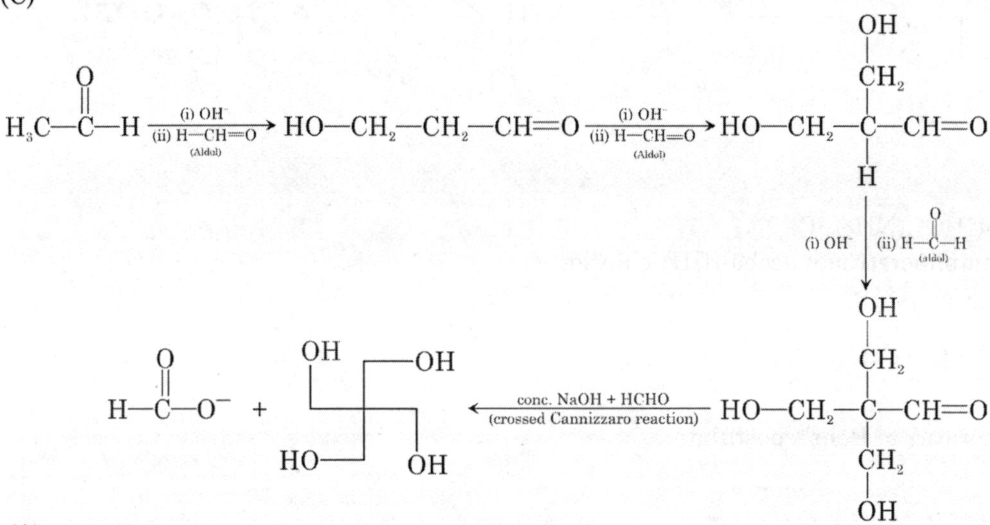

9. (A)

Aqueous solution of copper(II) sulphate absorbs orange red ligts apears lue (complementary colour).

10. (A)

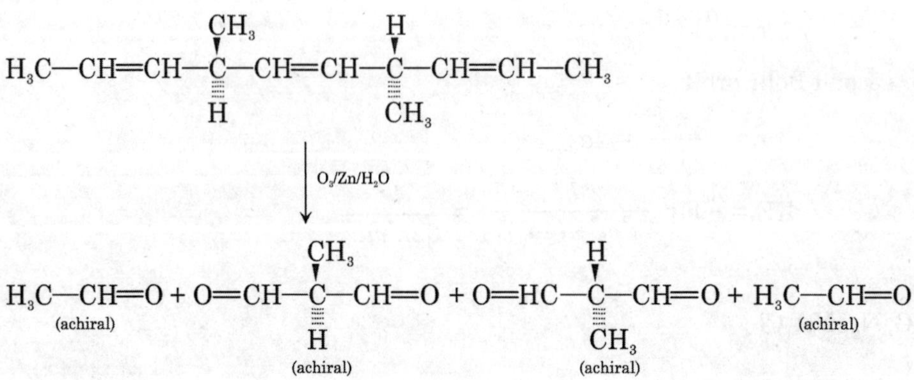

11. (A, C, D)

$HX + AgNO_3 \longrightarrow AgX \downarrow + HNO_3$ (X = Cl, Br, I)

$AgX + 2Na_2S_2O_3 \longrightarrow Na_3[Ag(S_2O_3)_2] + NaX$

12. (B, D)

(A) Both are soluble in NaOH and hence they are inseparable.

(B) Only benzoic acid (C_6H_5COOH) is soluble in NaOH and $NaHCO_3$, while benzyl alcohol ($C_6H_5CH_2OH$) is not. Hence, they are separable.

(C) NaOH can enable separation between benzyl alcohol ($C_6H_5CH_2OH$) and phenol (C_6H_5OH) since only the latter is soluble in NaOH. However, in $NaHCO_3$, both are insoluble. Hence, they are inseparable.

(D) α-Phenyl acetic acid ($C_6H_5CH_2COOH$) is soluble in NaOH and $NaHCO_3$, while benzyl alcohol ($C_6H_5CH_2OH$) is not. Hence, they are separable.

13. (A, C)

$\Delta S_{X\to Z} = \Delta S_{X\to Y} + \Delta S_{Y\to Z}$ [entropy (S) is a state function and hence it is additive]

$w_{X\to Y\to Z} = w_{X\to Y}$ (work done in $Y \to Z$ is zero since it is an isochoric process)

14. (B, C)

The compounds  and 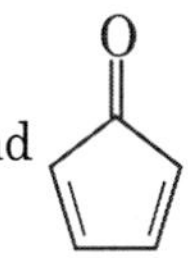 being antiaromatic are unstable at room temperature.

15. (A, D)

The lyophobic colloids are stable due to the preferential adsorption of ions on their surface from solution and potential difference between the fixed layer and the diffused layer of opposite charges around the colloidal particles makes lyophobic sol stable.

16. (8)

Stock solution of HCl = 29.2% (w/w)

Molarity of stock solution of HCl = $\dfrac{29.2 \times 1000 \times 1.25}{36.5 \times 100}$

If volume of stock solution required = V mL, $V \times \dfrac{29.2}{36.5} \times \dfrac{1000}{80} = 200 \times 0.4$ or $V = 8$ mL

17. (4)

Peptides with isoelectric point > 7, would exist as cation in neutral solution (pH = 7). Positively charged peptides = 4 (IV, VI, VIII and IX)

18. (9)

$$t_{1/8} = \frac{2.303 \log 8}{k} = \frac{2.303 \times 3 \log 2}{k}$$

$$t_{1/10} = \frac{2.303}{k} \log 10 = \frac{2.303}{k} \Rightarrow \left[\frac{t_{1/8}}{t_{1/10}}\right] \times 10 = \frac{\left(\dfrac{2.303 \times 3 \log 2}{k}\right)}{\left(\dfrac{2.303}{k}\right)} \times 10 = 9$$

19. (8)

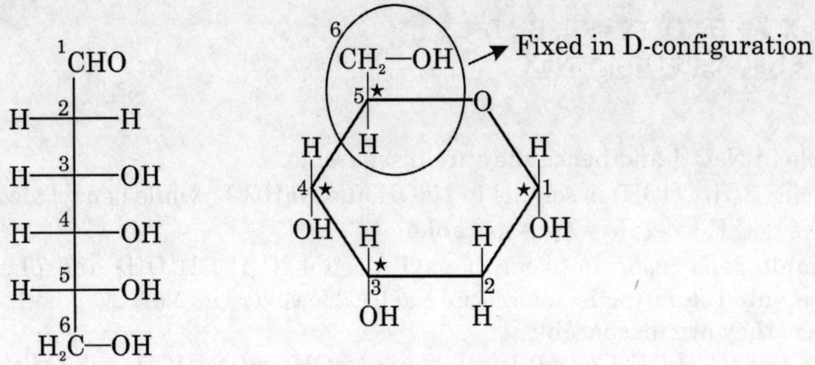

Hence, total number of stereoisomers in pyranose form of D-configuration = $2^3 = 8$.

20. (8)

$${}^{63}_{29}Cu + {}^{1}_{1}H \rightarrow 6\,{}^{1}_{0}n + {}^{4}_{2}He + 2\,{}^{1}_{1}H + {}^{A}_{Z}X$$

Mass number: $63 + 1 = 1 \times 6 + 4 + 1 \times 2 + A$

or $A = 64 - 12 = 52$

Atomic number: $29 + 1 = 6 \times 0 + 2 + 2 \times 1 + Z$

or $Z = 30 - 4 = 26$

Hence, ${}^{A}_{Z}X = {}^{52}_{26}Fe$. Hence, X is in the group 8 in the periodic table.

I.I.T. JEE Objective Question Paper—2012

PAPER—2

SECTION - I
(Single Correct Answer Type)

This section contains 8 multiple choice questions. Each question has four choices (A), (B), (C) and (D) out of which only one is correct.

1. The major product H of the given reaction sequence is

$CH_3-CH_2-CO-CH_3 \xrightarrow{CN^-} G \xrightarrow[\text{Heat}]{95\%\ H_2SO_4} H$

(A) $CH_3-CH=\underset{\underset{CH_3}{|}}{C}-COOH$ (B) $CH_3-CH=\underset{\underset{CH_3}{|}}{C}-CN$

(C) $CH_3-CH_2-\overset{\overset{OH}{|}}{\underset{\underset{CH_3}{|}}{C}}-COOH$ (D) $CH_3-CH=\underset{\underset{CH_3}{|}}{C}-CO-NH_2$

2. $NiCl_2\{P(C_2H_5)_2(C_6H_5)\}_2$ exhibits temperature dependent magnetic behaviour (paramagnetic/diamagnetic). The coordination geometries of Ni^{2+} in the paramagnetic and diamagnetic states are respectively

(A) tetrahedral and tetrahedral (B) square planer and square planar
(C) tetrahedral and square planar (D) square planer and tetrahedral.

3. In the cyanide extraction process of silver from argentite ore, the oxidising and reducing agents used are

(A) O_2 and CO respectively (B) O_2 and Zn dust respectively
(C) HNO_3 and Zn dust respectively (D) HNO_3 and CO respectively.

4. The reaction of white phosphorous with aqueous NaOH gives phosphine along with another phosphorus containing compound. The reaction type, the oxidation states of phosphorus in phosphine and the other product are respectively

(A) redox reaction, – 3, – 5 (B) redox reaction, + 3, + 5
(C) disproportionation reaction, – 3, + 5 (D) disproportionation reaction, – 3, + 3

5. The shape of XeO_2F_2 molecule is

(A) trigonal bipyramidal (B) square planar
(C) tetrahedral (D) see-saw.

6. For a dilute solution containing 2.5 g of a non-volatile non-electrolyte solute in 100 g of water, the elevation in boiling point at 1 atm pressure is 2°C. Assuming concentration of solute is much lower than the concentration of solvent, the vapour pressure (mm of Hg) of the solution is ($K_b = 0.76$ K kg mol^{-1})

(A) 724 (B) 740 (C) 736 (D) 718

7. The compound that undergoes decarboxylation most readily under mild condition is

(A) 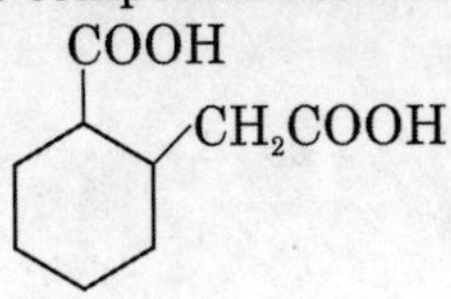

(B) 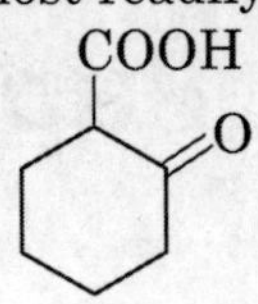

(C) 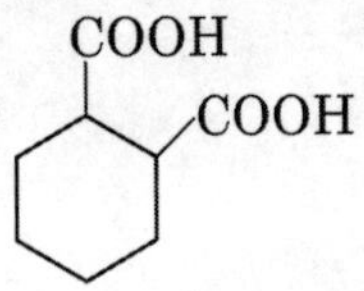

(D) 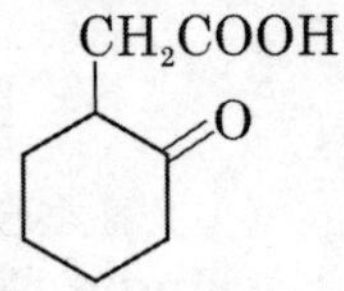

8. Using the data provided, calculate the multiple bond energy (kJ mol^{-1}) of a C≡C bond in C_2H_2. That energy is (bond energy of C—H bond 350 kJ mol^{-1})

$2C(s) \longrightarrow 2C(g)$ $\Delta H = 1410$ kJ mol^{-1}

$2C(s) \longrightarrow 2C(g)$ $\Delta H = 1410$ kJ mol^{-1}

$H_2(g) \longrightarrow 2H(g)$ $\Delta H = 330$ kJ mol^{-1}

(A) 1165 (B) 837 (C) 865 (D) 815

SECTION – II
(Paragraph Type)

This section contains 6 multiple choice questions relating to three paragraphs with two questions on each paragraph. Each question has four choices (A), (B), (C) and (D) out of which only one is correct.

Paragraph for Questions 9 and 10

In the following reaction sequence, the compound J is an intermediate.

$$\text{I} \xrightarrow[\text{CH}_3\text{COONa}]{(\text{CH}_3\text{CO})_2\text{O}} \text{J} \xrightarrow[\text{(ii) SOCl}_2\text{, (iii) anhyd. AlCl}_3]{\text{(i) H}_2\text{, Pd/C}} \text{K}$$

J ($C_9H_8O_2$) gives effervescence on treatment with $NaHCO_3$ and a positive Baeyer's test.

9. The compound I is

(A) 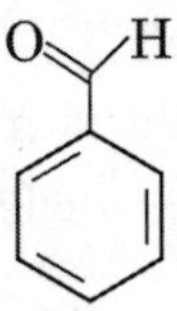

(B) 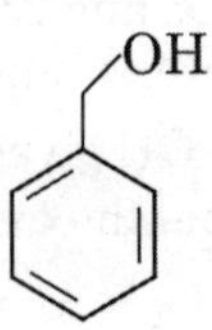

(C) 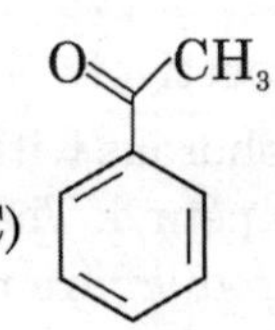

(D) 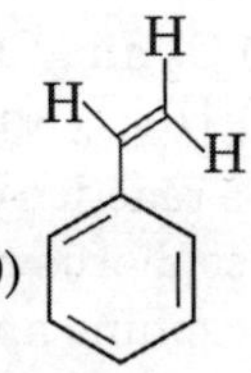

10. The compound K is

(A) 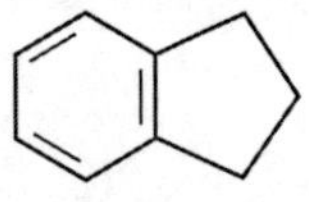

(B) 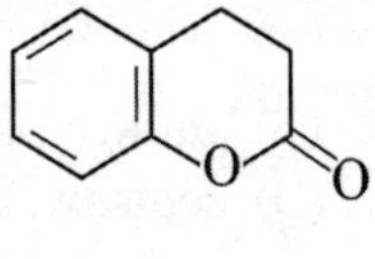

(C) 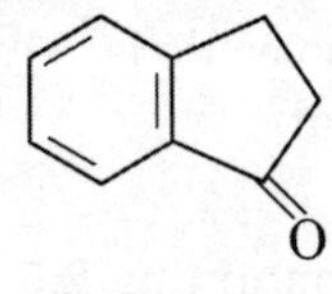

(D) 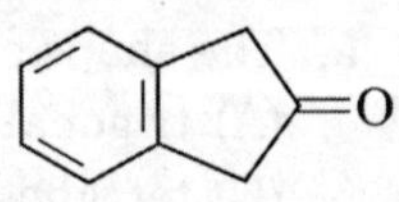

Paragraph for Questions 11 and 12

The electrochemical cell shown below is a concentration cell.
$M | M^{2+}$ (saturated solution of a sparingly soluble salt, MX_2) | | M^{2+} (0.001 mol dm^{-3}) | M
The emf of the cell depends on the difference in concentrations of M^{2+} ions at the two electrodes. The emf of the cell at 298 K is 0.059 *V*.

11. The value of ΔG (kJ mol^{-1}) for the given cell is (1 F = 96500 C mol^{-1})
(A) – 5.7 (B) 5.7 (C) 11.4 (D) – 11.4

12. The solubility product (K_{sp}; mol^3 dm^{-9}) of MX_2 at 298 K based on the information available for the given concentration cell is ($2.303 \times R \times 2.98/F = 0.059$ *V*)
(A) 1×10^{-15} (B) 4×10^{-15} (C) 1×10^{-12} (D) 4×10^{-12}

Paragraph for Questions 13 and 14

Bleaching powder and bleach solution are produced on a large scale and used in several household products. The effectiveness of bleach solution is often measured by iodometry.

13. Bleaching powder contains is salt of an oxoacid as one of its components. The anhydride of that oxoacid is
(A) Cl_2O (B) Cl_2O_7 (C) ClO_2 (D) Cl_2O_6

14. 25 mL of household solution was mixed with 30 mL of 0.50*M* KI and 10 mL of 4*N* acetic acid. In the titration of the liberated iodine, 48 mL of 0.25*N* $Na_2S_2O_3$ was used to reach the end point. The molarity of the household bleach solution is
(A) 0.48*M* (B) 0.96*M* (C) 0.24*M* (D) 0.024*M*

SECTION – III
(Multiple Correct Answer(s) Type)

The section contains 6 multiple choice questions. Each question has four choices (A), (B), (C) and (D) out of which one or more are correct.

15. The reversible expansion of an ideal gas under adiabatic and isothermal conditions is shown in the figure. Which of the following statement(s) is (are) correct?

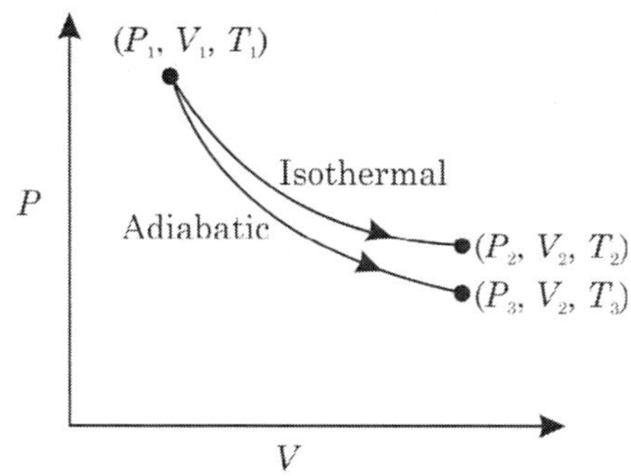

(A) $T_1 = T_2$ (B) $T_3 > T_1$
(C) $w_{isothermal} > w_{adiabatic}$ (D) $\Delta U_{isothermal} > \Delta U_{adiabatic}$

16. For the given aqueous reactions, which of the statement(s) is(are) true?

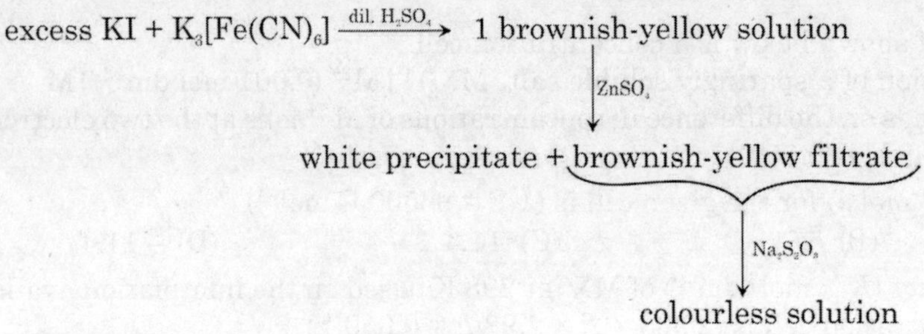

(A) The first reaction is a redox reaction
(B) White precipitate is $Zn_3[Fe(CN)_6]_2$
(C) Addition of filtrate to starch solution gives blue colour
(D) White precipitate is soluble in NaOH solution.

17. With reference to the scheme given, which of the given statement(s) about T, U, V and W is(are) correct?

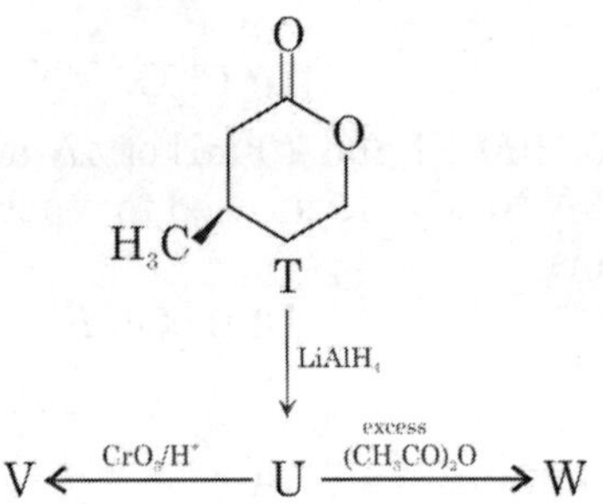

(A) T is soluble in hot aqueous NaOH
(B) U is optically active
(C) Molecular formula of W is $C_{10}H_{18}O_4$
(D) V gives effervescence on treatment with aqueous $NaHCO_3$.

18. Which of the given statement(s) about N, O, P and Q with respect to M is(are) correct?

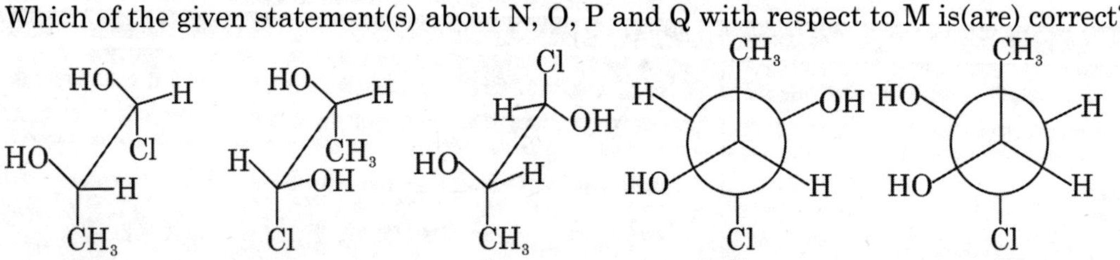

(A) M and N are non-mirror image stereoisomers
(B) M and O are identical
(C) M and P are enantiomers
(D) M and Q are identical.

19. With respect to graphite and diamond, which of the statement(s) given below is(are) correct?
(A) Graphite is harder than diamond.
(B) Graphite has higher electrical conductivity than diamond.

(C) Graphite has higher thermal conductivity than diamond.
(D) Graphite has higher C—C bond order than diamond.

20. The given graphs/data I, II, III and IV represent general trends observed for different physisorption and chemisorption processes under mild conditions of temperature and pressure. Which of the following choice(s) about I, II, III and IV is (are) correct?

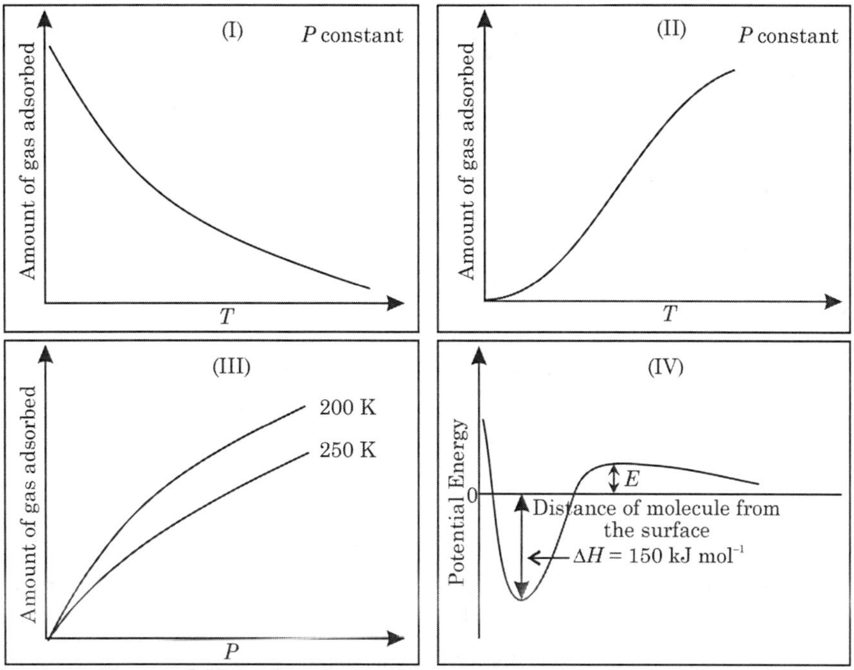

(A) I is physisorption and II is chemisorption
(B) I is physisorption and III is chemisorption
(C) IV is chemisorption and II is chemisorption
(D) IV is chemisorption and III is chemisorption.

ANSWER KEYS

1. A	2. C	3. B	4. C	5. D	6. A
7. B	8. D	9. A	10. C	11. D	12. B
13. A	14. C	15. A, C, D	16. A, C, D	17. A, C, D	18. A, B, C
19. B, D	20. A, C				

SOLUTIONS

1. (A)

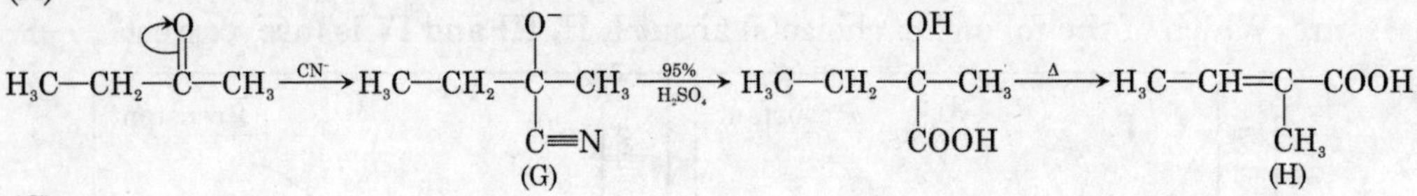

2. (C)

In both the complexes, Ni exists as Ni^{2+}

In sp^3 (tetrahedral geometry)

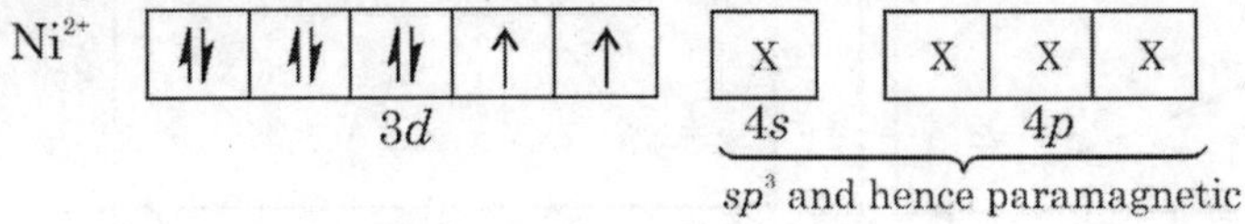

(x represents electron pair donated by the ligand)

In dsp^2 (square planar geometry)

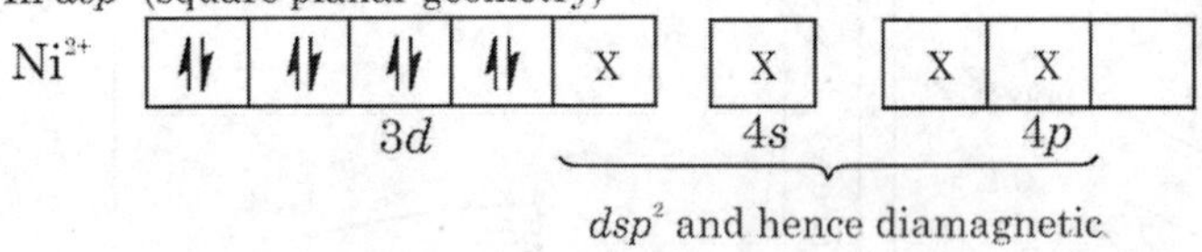

3. (B)

The reactions involved in cyanide extraction process are:

$$\underset{\text{(argentite ore)}}{Ag_2S} + 4NaCN \longrightarrow 2Na[Ag(CN)_2] + Na_2S$$

$$4Na_2S + \underset{\text{(oxidising agent)}}{5[O_2]} + 2H_2O \longrightarrow 2Na_2SO_4 + 4NaOH + 2S$$

$$2Na[Ag(CN)_2] + \underset{\text{(reducing agent)}}{Zn} \longrightarrow Na_2[Zn(CN)_4] + 2Ag\downarrow$$

4. (C)

The balanced disproportionation reaction involving white phosphorus with aqueous NaOH is $\overset{0}{P_4} + 3NaOH + 3H_2O \longrightarrow \overset{-3}{P}H_3 + 3NaH_2\overset{+1}{P}O_2$

Since the option involving + 1 oxidation state is missing, it may be considered that NaH_2PO_2 formed has undergone thermal decomposition as follows:

$$2NaH_2PO_2 \xrightarrow{\Delta} Na_2H\overset{+5}{P}O_4 + PH_3$$

Although heating is not mentioned in the question, the other product as per available options seems to be Na_2HPO_4 (oxidation state of P = + 5).

5. (D)

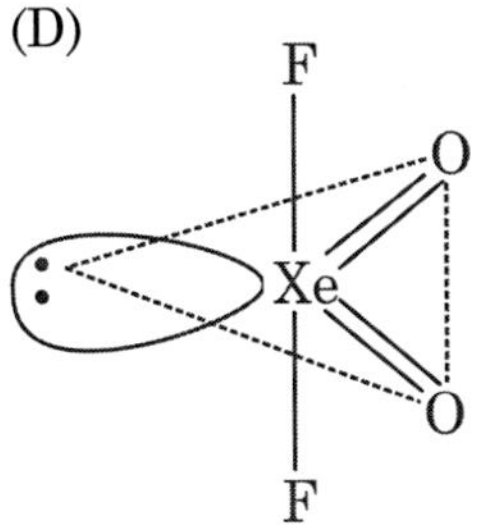

Hybridization: sp^3d

Shape: see-saw

6. (A)

B ⟶ Solute; A ⟶ Solvent

$W_B = 2.5$ g, $W_A = 100$ g

$\Delta T_b = 2^0$

$$\frac{p^0 - p_s}{p^0} = x_B = \frac{n_B}{n_B + n_A} \Rightarrow \frac{p^0 - p_s}{p^0} = \frac{n_B}{n_A} \quad \text{Hence, } n_B \ll n_A$$

$$\frac{p^0 - p_s}{p^0} = \frac{n_B}{n_A}$$

$$\frac{760 - P_{soln}}{760} = \frac{2.5/M}{\frac{100}{18} \times \frac{1000}{1000}} = \frac{m \times 18}{1000} \quad \text{...(i)}$$

From boiling point elevation, we have

$$2 = 0.76 \times m \quad \text{or} \quad m = \frac{2}{0.76} \quad \text{...(ii)}$$

On equating Equns. (i) and (ii), we get $P_{soln} = 724$ mm

7. (B)

β-keto acids undergo decarboxylation easily.

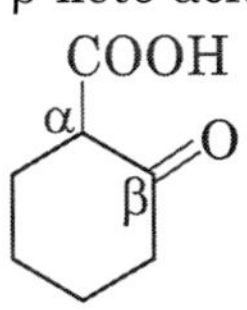

(β-keto acid)

8. (D)

(i) $2C(s) + H_2(g) \longrightarrow H{-}C{\equiv}C{-}H(g)$ $\Delta H = 225$ kJ mol^{-1}

(ii) $2C(s) \longrightarrow 2C(g)$ $\Delta H = 1410$ kJ mol^{-1}

(iii) $H_2(g) \longrightarrow 2H(g)$ $\Delta H = 330$ kJ mol^{-1}

From Equn. (i), we have

$$225 = [2 \times \Delta H_{C(s) \to C(g)} + 1 \times BE_{H-H}] - [2 \times BE_{C-H} + 1 \times BE_{C\equiv C}]$$

or $$225 = [1410 + 1 \times 330] - [2 \times 350 + 1 \times BE_{C\equiv C}]$$

or $$225 = [1410 + 330] - [700 + BE_{C\equiv C}]$$

or $\quad 225 = 1740 - 700 - BE_{C=C}$

or $\quad 225 = 1040 - BE_{C=C}$

or $\quad BE_{C=C} = 1040 - 225 = 815 \text{ kJ mol}^{-1}$

9. (A)

10. (C)

Sol. 9–10

$$\underset{(I)}{Ph{-}CHO} \xrightarrow[\substack{CH_3COONa \\ \text{(Perkin condensation)}}]{(CH_3CO)_2O} \underset{\substack{(J) \\ \text{(Cinnamic acid)}}}{Ph{-}CH{=}CH{-}COOH} \xrightarrow{NaHCO_3} \text{(Effervescence)}$$

$$\text{(J)} \xrightarrow[\text{(Bayer's reagent)}]{1\%\text{ cold, alk. }KMnO_4} \text{(Positive test)}$$

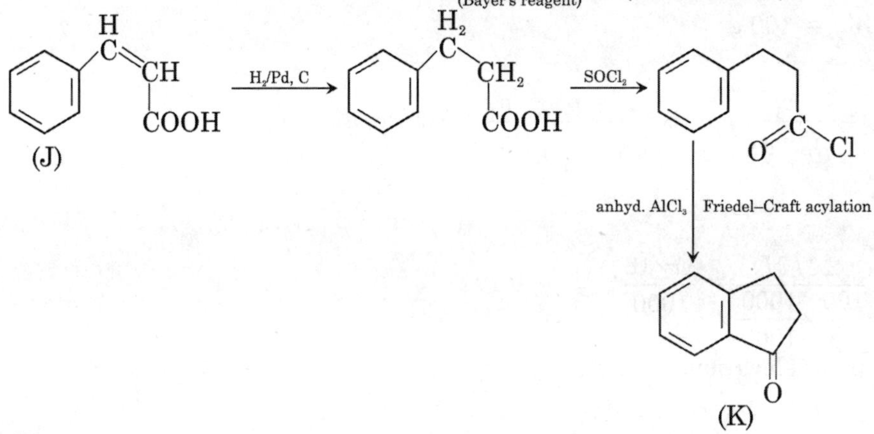

11. (D)

At anode: $M(s) + 2X^-(aq) \longrightarrow MX_2(aq) + 2e^-$

At cathode: $M^{+2}(aq) + 2e^- \longrightarrow M(s)$

n factor of the cell reaction is 2.

$\Delta G = -nFE_{cell} = -2 \times 96500 \times 0.059 = -11387$ J/mole $= -11.387$ KJ/mole $= -11.4$ KJ/mole

12. (B)

$M|M^+(\text{sat.})\|M^{2+}(0.001M)$

$$E_{cell} = \frac{-0.059}{n}\log\frac{[M^{2+}]_a}{[M^{2+}]_c}$$

$$0.059 = \frac{0.059}{2}\log\frac{[0.001]}{[M^{2+}]_a}$$

$[M^{2+}]_a = 10^{-5} = S$ (solubility of salt in saturated solution)

$$\underset{(S)}{MX_2} = \underset{(S)}{M^{2+}} + \underset{(2S)}{2X^-(aq)}$$

$K_{sp} = 4S^3 = 4 \times (10^{-5})^3 = 4 \times 10^{-15}$

13. (A)

$$\underset{\text{(Bleaching powder)}}{Ca(OCl)Cl} \longrightarrow Ca^{2+} + OCl^- + Cl^-$$

$$\underset{\text{(oxoacid)}}{HOCl} \longrightarrow H^+ + OCl^-$$

$$2HOCl \xrightarrow{\Delta} H_2O + Cl_2O$$

14. (C)

$$N_1V_1 = N_2V_2$$

$$N_1 \times 25 = 0.25 \times 48$$

Or $\quad N_1 = \dfrac{0.25 \times 48}{25}$

Or $\quad N_1 = 0.48M$

Hence, molarity $= 0.48/2 = 0.24$

15. (A, C, D)

$T_1 = T_2$ since the process is isothermal.

The work done in adiabatic process is less than in isothermal process because area covered by isothermal curve is more than the area covered by the adiabatic curve. In an adiabatic process expansion occurs by using internal energy and hence it decreases while in an isothermal process temperature remains constant and as a result there is no change in internal energy.

16. (A, C, D)

$$K_3\left[\overset{+3}{Fe}(CN)_6\right] + KI(\text{excess}) \longrightarrow K_4\left[\overset{+2}{Fe}(CN)_6\right] + \underset{\text{Brownish yellow solution}}{KI_3} \quad \text{(redox reaction)}$$

$$\underset{\text{(Brownish yellow filtrate)}}{I_3^-} + 2Na_2S_2O_3 \longrightarrow \underset{\text{Clear solution}}{Na_2S_4O_6} + 2NaI + I^-$$

$$K_4[Fe(CN)_6] + ZnSO_4 \longrightarrow \underset{\text{(White ppt.)}}{K_2Zn_3[Fe(CN)_6]_2} \xrightarrow{NaOH} \underset{\text{Soluble}}{Na_2[Zn(OH)_4]}$$

17. (A, C, D)

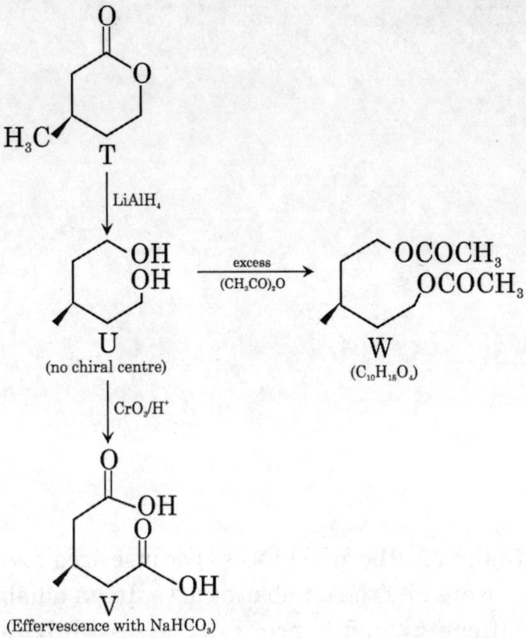

18. (A, B, C)

Converting all the structures in the Fischer projection, we have

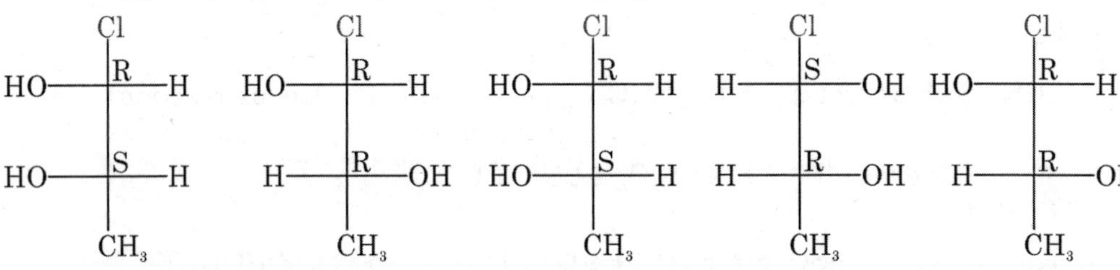

M N O P Q

M and N are diastereoisomers. M and O are identical. M and P are enantiomers. M and Q are diastereoisomers.

19. (B, D)

Diamond is harder than graphite. Graphite is good conductor of electricity than diamond. Diamond is better thermal conductor than graphite. In graphite, the C—C bond acquires double bond character and hence higher bond order than in diamond.

20. (A, C)

Under mild conditions of temperature, the extent of physisorption decreases with increase in temperature and hence the graph (I) is correct. Since the initial temperature provided meets the necessary activation energy needed for chemical bond formation for chemisorption, the graph (II) is correct. The graph (IV) is correct for chemisorption since the plot is now of potential energy *vs* distance of molecule from surface.

I.I.T. JEE Objective Question Paper—2013 (Main)

1. Which of the following complex species is not expected to exhibit optical isomerism?

(A) $[Co(en)_3]^{3+}$ (B) $[Co(en)_2Cl_2]^+$

(C) $[Co(NH_3)_3Cl_3]$ (D) $[Co(en)(NH_3)_2Cl_2]^+$

2. Which one of the following molecules is expected to exhibit diamagnetic behaviour?

(A) C_2 (B) N_2 (C) O_2 (D) S_2

3. A solution of (–)-1-chloro-1-phenylethane in toluene recemises slowly in the presence of a small amount of $SbCl_5$, due to the formation of

(A) carbanion (B) carbene (C) carbocation (D) free radical.

4. Given: $E^0_{Cr^{3+}/Cr} = -0.74$ V ; $E^0_{MnO_4^-/Mn^{2+}} = 1.51$ V ; $E^0_{Cr_2O_7^{2-}/Cr^{3+}} = 1.33$ V ; $E^0_{Cl/Cl^-} = 1.36$ V

Based on the data given above, strongest oxidising agent will be

(A) Cl (B) Cr^{3+} (C) Mn^{2+} (D) MnO_4^-

5. A piston filled with 0.04 mol of an ideal gas expands reversibly from 50.0 mL to 375 mL at a constant temperature of 37.0°C. As it does so, it absorbs 208 J of heat. The values of q and w for the process will be (R = 8.314 J/mol K) (ln 7.5 = 2.01)

(A) $q = +208$ J, $w = -208$ J (B) $q = -208$ J, $w = -208$ J

(C) $q = -208$ J, $w = +208$ J (D) $q = +208$ J, $w = +208$ J

6. The molarity of a solution obtained by mixing 750 mL of 0.5M HCl with 250 mL of 2M HCl will be

(A) 0.875M (B) 1.00M (C) 1.75M (D) 0.975M

7. Arrange the following compounds in order of decreasing acidity.

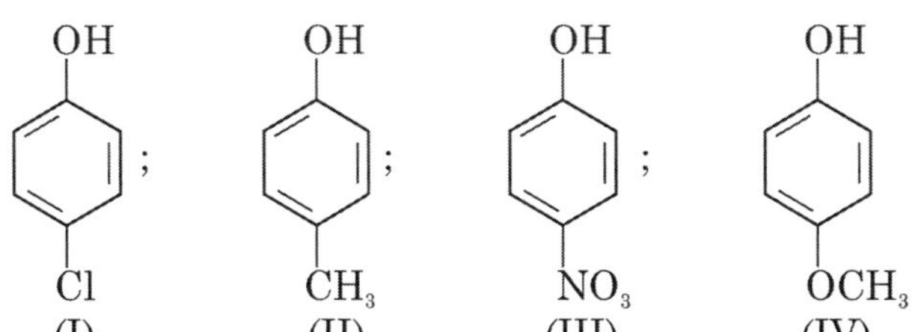

(I) (II) (III) (IV)

(A) II > IV > I > III (B) I > II > III > IV (C) III > I > II > IV (D) IV > III > I > II

8. For gaseous state, if most probable speed is denoted by C^*, average speed by $\bar{C}$ and root mean square speed by C, then for a large number of molecules the ratios of these speed are

(A) $C^* : \bar{C} : C = 1.225 : 1.128 : 1$
(B) $C^* : \bar{C} : C = 1.128 : 1.225 : 1$
(C) $C^* : \bar{C} : C = 1 : 1.128 : 1.225$
(D) $C^* : \bar{C} : C = 1 : 1.225 : 1.228$

9. The rate of a reaction doubles when its temperature changes from 300 K to 310 K. Activation energy of such a reaction will be ($R = 8.314$ J K^{-1} mol^{-1} and log 2 = 0.301)

(A) 53.6 kJ mol^{-1}
(B) 48.6 kJ mol^{-1}
(C) 58.5 kJ mol^{-1}
(D) 60.5 kJ mol^{-1}

10. A compound with molecular mass 180 is acylated with CH_3COCl to get a compound with molecular mass 390. The number of amino groups present per molecule of the former compound is

(A) 2
(B) 5
(C) 4
(D) 6

11. Which of the following arrangements does not represent the correct order of the property stated against it?

(A) $V^{2+} < Cr^{2+} < Mn^{2+} < Fe^{2+}$: paramagnetic behaviour
(B) $Ni^{2+} < Co^{2+} < Fe^{2+} < Mn^{2+}$: ionic size
(C) $Co^{3+} < Fe^{3+} < Cr^{3+} < Sc^{3+}$: stability in aqueous solution
(D) Sc < Ti < Cr < Mn : number of oxidation states.

12. The order of stability of the following carbocations is

$H_2C{=}CH{-}\overset{\oplus}{C}H_2$; $H_3C{-}CH_2{-}\overset{\oplus}{C}H_2$;

I II

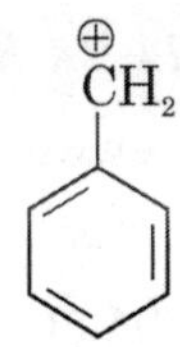

III

(A) II > III > I
(B) I > II > III
(C) III > I > II
(D) III > II > I

13. Consider the following reaction

$$xMnO_4^- + yC_2O_4^{2-} + zH^+ \longrightarrow xMn^{2+} + 2yCO_2 + \frac{z}{2}H_2O$$

The values of x, y and z in the reaction are, respectively

(A) 5, 2 and 16
(B) 2, 5 and 8
(C) 2, 5 and 16
(D) 5, 2 and 8

14. Which of the following is the wrong statement?

(A) ONCl and ONO^- are not isoelectronic
(B) O_3 molecule is bent
(C) Ozone is violet-black in solid state
(D) Ozone is a diamagnetic gas

15. A gaseous hydrocarbon gives upon combustion 0.72 g of water and 3.08 g of CO_2. The empirical formula of the hydrocarbon is

(A) C_2H_4
(B) C_3H_4
(C) C_6H_5
(D) C_7H_8

16. In which of the following pairs of molecules/ions, both the species are not likely to exist?

(A) H_2^+, He_2^{2-}
(B) H_2^-, He_2^{2-}
(C) H_2^{2+}, He_2
(D) H_2^-, He_2^{2+}

17. Which of the following exists as covalent crystals in the solid state?

(A) Iodine
(B) Silicon
(C) Sulphur
(D) Phosphorus

18. Synthesis of each molecule of glucose in photosynthesis involves

(A) 18 molecules of ATP (B) 10 molecules of ATP
(C) 8 molecules of ATP (D) 6 molecules of ATP

19. The coagulating power of electrolytes having ions Na^+, Al^{3+} and Ba^{2+} for arsenic sulphide sol increases in the order

(A) $Al^{3+} < Ba^{2+} < Na^+$ (B) $Na^+ < Ba^{2+} < Al^{3+}$ (C) $Ba^{2+} < Na^+ < Al^{3+}$ (D) $Al^{3+} < Na^+ < Ba^{2+}$

20. Which of the following represents the correct order of increasing first ionization enthalpy for Ca, Ba, S, Se and Ar?

(A) Ca < S < Ba < Se < Ar (B) S < Se < Ca < Ba < Ar
(C) Ba < Ca < Se < S < Ar (D) Ca < Ba < S < Se < Ar

21. Energy of an electron is given by $E = -2.178 \times 10^{-18}\,\text{J}\left(\frac{Z^2}{n^2}\right)$. Wavelength of light required to excite an electron in a hydrogen atom from level $n = 1$ to $n = 2$ will be
($h = 6.62 \times 10^{-34}$ J s and $c = 3.0 \times 10^8$ m s^{-1})

(A) 1.214×10^{-7} m (B) 2.816×10^{-7} m (C) 6.500×10^{-7} m (D) 8.500×10^{-7} m

22. Compound (A), C_8H_9Br, gives a white precipitate when warmed with alcoholic $AgNO_3$. Oxidation of (A) gives an acid (B), $C_8H_6O_4$. (B) easily forms anhydride on heating. Identify the compound (A).

(A) 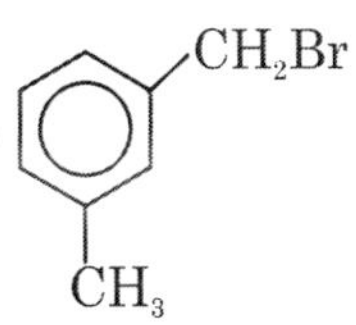
(B) 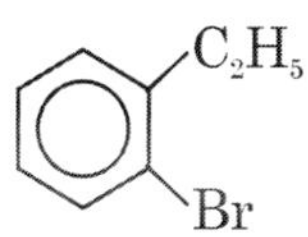
(C) 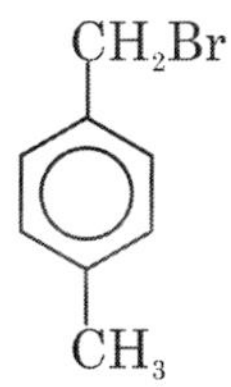
(D) 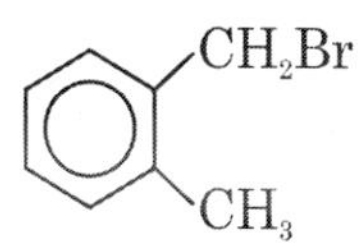

23. Four successive members of the first row transition elements are listed below with atomic numbers. Which one of them is expected to have the highest $E^0_{M^{3+}/M^{2+}}$ value?

(A) Cr(Z = 24) (B) Mn(Z = 25) (C) Fe(Z = 26) (D) Co(Z = 27)

24. How many litres of water must be added to 1 litre of an aqueous solution of HCl with a pH of 1 to create an aqueous solution with pH of 2?

(A) 0.1 L (B) 0.9 L (C) 2.0 L (D) 9.0 L

25. The first ionisation enthalpy of Na is 5.1 eV. The value of electron gain enthalpy of Na^+ will be

(A) – 2.55 eV (B) – 5.1 eV (C) – 10.2 eV (D) + 2.55 eV

26. An organic compound A upon reacting with NH_3 gives B. On heating B gives C. C in presence of KOH reacts with Br_2 to give $CH_3CH_2NH_2$. A is

(A) CH_3COOH (B) $CH_3CH_2CH_2COOH$
(C) $CH_3—CH(CH_3)—COOH$ (D) CH_3CH_2COOH

27. Stability of the species Li_2, Li_2^-, and Li_2^+ increases in the order of

(A) $Li_2 < Li_2^+ < Li_2^-$ (B) $Li_2^- < Li_2^+ < Li_2$ (C) $Li_2 < Li_2^- < Li_2^+$ (D) $Li_2^- < Li_2 < Li_2^+$

28. An unknown alochol is treated with the "Lucas reagent" to determine whether the alcohol is primary, secondary or tertiary. Which alcohol reacts fastest and by what mechanism?

(A) secondary alcohol by S_{N^1} (B) tertiary alcohol by S_{N^1}

(C) secondary alcohol by S_{N^2} (D) tertiary alcohol by S_{N^2}

29. The gas leaked from a storage tank of the Union Carbide plant in Bhopal gas tragedy was

(A) Methylisocyanate (B) Methylamine (C) Ammonia (D) Phosgene

30. Experimentally it was found that a metal oxide has formula $M_{0.98}O$. Metal M is present as M^{2+} and M^{3+} in its oxide. Fraction of the metal which exists as M^{3+} would be

(A) 7.01% (B) 4.08% (C) 6.05% (D) 5.08%

ANSWER KEYS

1. B	2. A, B	3. C	4. D	5. A	6. A
7. C	8. C	9. A	10. B	11. A	12. D
13. C	14. All statements are correct and there is no answer.				15. D
16. C	17. B	18. A	19. B	20. C	21. A
22. D	23. D	24. D	25. B	26. D	27. B
28. B	29. A	30. B			

SOLUTIONS

1. (B)

$[Co(NH_3)_3Cl_3]$ exists in two forms (facial and meridomial).

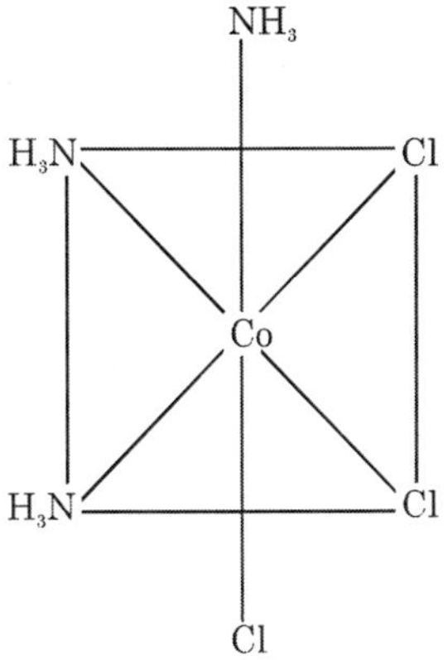

facial

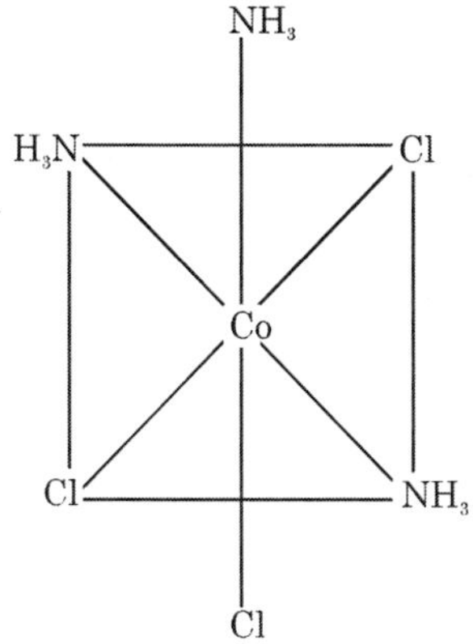

meridonial

Both of these forms contain plane of symmetry and are achiral. Hence, $[Co(NH_3)_3Cl_3]$ does not show optical isomerism.

2. (A) and (B)

C_2—Diamagnetic; N_2—Diamagnetic; O_2—Paramagnetic; S_2—Paramagnetic

3. (C)

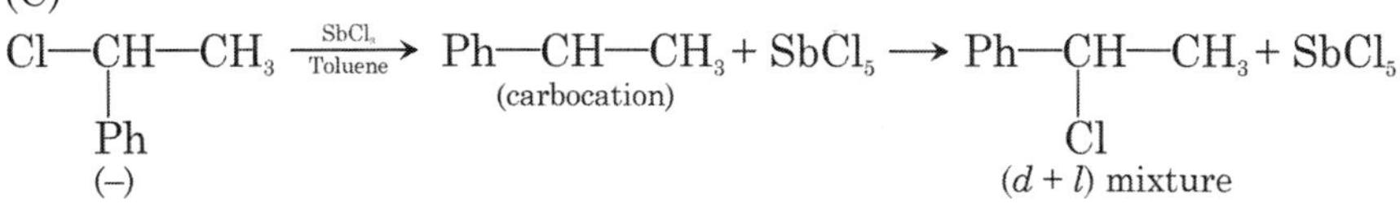

4. (D)

Higher the E^0 value better is the oxidising agent. Hence, MnO_4^- is the strongest oxidising agent.

5. (A)

The process is isothermal expansion. Hence, $q = -w$, $\Delta u = 0$.

$q = +208$ J, $w = -208$ J (expansion work)

6. (A)

$$M_f = \frac{M_1V_1 + M_2V_2}{V_1 + V_2} = \frac{0.5 \times \frac{3}{4} + 2 \times \frac{1}{4}}{1} = 0.875M$$

7. (C)

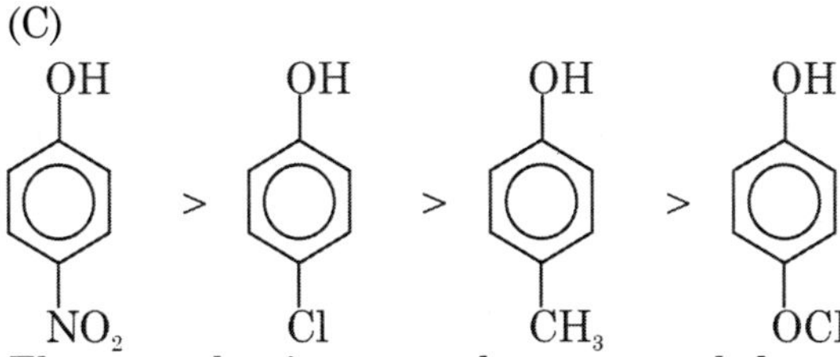

Electron releasing group decreases and electron withdrawing group increases acidic strength.

8. (C)

$C^* = \text{most probable speed} = \sqrt{\dfrac{2RT}{M}}$

$\bar{C} = \text{root mean square speed} = \sqrt{\dfrac{8RT}{\pi M}}$

$C^* < \bar{C} < C$

$C^* : \bar{C} : C = 1 : \sqrt{\dfrac{4}{\pi}} : \sqrt{\dfrac{3}{2}} = 1 : 1.128 : 1.225$

9. (A)

$$\log\frac{K_2}{K_1} = \frac{-E_a}{2.030R}\left(\frac{1}{T_2} - \frac{1}{T_1}\right)$$

$$\frac{K_2}{K_1} = 2; \quad T_2 = 310\text{ K}, \quad T_1 = 300\text{ K}$$

$$\log 2 = \frac{-E_a}{2.303 \times 8.134}\left(\frac{1}{310} - \frac{1}{300}\right)$$

$$E_a = 53598.6\text{ J/mol} = 53.6\text{ kJ/mol}$$

10. (B)

$$R\text{—}NH_2 + CH_3\text{—}\overset{\overset{\Large O}{\|}}{C}\text{—}Cl \rightarrow R\text{—}NH\text{—}\overset{\overset{\Large O}{\|}}{C}\text{—}CH_3 + HCl$$

Each $CH_3\text{—}\overset{\overset{O}{\|}}{C}$ addition increases the molecular weight by 42.
Total increase in molecular weight = 390 – 180 = 210

Hence, the number of NH_2 groups $= \dfrac{210}{42} = 5$

11. (A)

(A) V^{2+} = 3 unpaired electrons
Cr^{2+} = 4 unpaired electrons
Mn^{2+} = 5 unpaired electrons
Fe^{2+} = 4 unpaired electrons
Hence, the order of paramagnetic behaviour is: $V^{2+} < Cr^{2+} = Fe^{2+} < Mn^{2+}$

(B) Ionic size decreases from left to right in a period.

(C) Co^{3+}/Co^{2+} = 1.97 V; Fe^{3+}/Fe^{2+} = 0.77 V; Cr^{3+}/Cr^{2+} = – 0.41 V
Sc^{3+} is highly stable (It does not exhibit + 2 oxidation state).

(D) The oxidation state increases as we move from group 3 to group 7 in the same period.

12. (D)

The order of stability of carbocations is

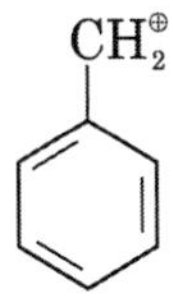 $> CH_2{=}CH{-}\overset{\oplus}{C}H_2 > CH_3{-}CH_2{-}\overset{\oplus}{C}H_2$

Benzyl Allyl Propyl

13. (C)

$$2MnO_4^- + 5C_2O_4^{2-} + 16H^+ \longrightarrow 2Mn^{2+} + 10CO_2 + 8H_2O$$

$x = 2$, $y = 5$ and $z = 16$

14. (All statements are correct and there is no answer)

(A) $ONCl = 8 + 7 + 17 = 32$ electrons

$ONO^- = 8 + 7 + 8 + 1 = 24$ electrons

(B) 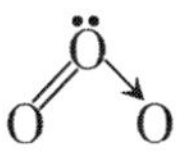 Central atom O is sp^2 hybridised with 1 lone pair, hence bent shape.

(C) Ozone is violet-black in solid state.

(D) O_3 has no unpaired electrons, hence is diamagnetic.

15. (D)

18 g H_2O contains 2 g H

$\therefore$ 0.72 g H_2O contains 0.08 g H

44 g CO_2 contains 12 g C

$\therefore$ 3.08 g CO_2 contains 0.84 g C

$$\therefore \quad C : H = \frac{0.84}{12} : \frac{0.08}{1} = 0.07 : 0.08 = 7 : 8$$

$\therefore$ Empirical formula $= C_7H_8$

16. (C)

H_2^{2+} : Bond order $= 0$

$$He_2 : \text{Bond order} = \frac{2-2}{2} = 0$$

Hence, both H_2^{2+} and He_2 do not exist.

17. (B)

Silicon exists as covalent crystal in solid state (Network like structure, like diamond).

18. (A)

$$6CO_2 + 12NADPH + 18ATP \longrightarrow C_6H_{12}O_6 + 12NADP + 18\ ADP$$

19. (B)

According to Hardy Schulze rule, greater the charge on cation, greater is its coagulating power for negetively charged solution. Hence, order of coagulating power: $Na^+ < Ba^{2+} < Al^{3+}$.

20. (C)

Order of increasing ΔH_{IE_1} : Ba < Ca < Se < S < Ar

Ba < Ca ; Se < S. On moving top to bottom in a group, size increases. So ionisation enthalpy decreases.

Ar: Maximum value of ionisation enthalpy since it is a noble gas.

21. (A)

$$\Delta E = 2.178 \times 10^{-18}\left(\frac{1}{1^2} - \frac{1}{2^2}\right) = \frac{hc}{\lambda}$$

$$2.178 \times 10^{-18}\left(\frac{1}{1^2} - \frac{1}{2^2}\right) = \frac{6.62 \times 10^{-34} \times 3.0 \times 10^8}{\lambda}$$

Hence, $\lambda \approx 1.214 \times 10^{-7}$ m

22. (D)

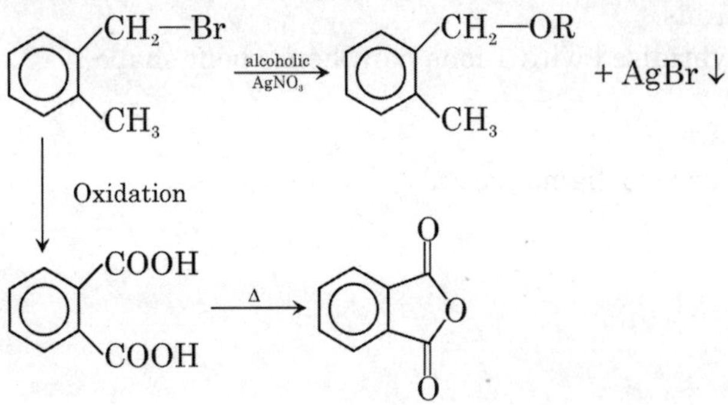

Phthalic anhydride

23. (D)

$E^0_{Cr^{3+}/Cr^{2+}} = -\ 0.41$ V; $E^0_{Mn^{3+}/Mn^{2+}} = +\ 1.57$ V; $E^0_{Fe^{3+}/Fe^{2+}} = +\ 0.77$ V; $E^0_{Co^{3+}/Co^{2+}} = +\ 1.97$ V

E^0 value normaly increases from left to right in the period of *d*-block elements. Some E^0 values are exceptionally higher due to stability of product ion. For example, $E^0_{Mn^{3+}/Mn^{2+}} =$ + 1.57 V; $E^0_{Co^{3+}/Co^{2+}} = +\ 1.97$ V.

24. (D)

pH = 1, $[H^+] = 10^{-1} = 0.1M$

pH = 2, $[H^+] = 10^{-2} = 0.01M$

$$M_1V_1 = M_2V_2$$

$$0.1 \times 1 = 0.01 \times V_2$$

$$V_2 = 10 \text{ L}$$

Volume of water added = 10 – 1 = 9 L

25. (B)

$Na \longrightarrow Na^+ + e^-$ 1stI.E

$Na^+ + e^- \longrightarrow Na$ Electron gain enthalpy of Na^+, $\Delta H_{eg} = -\ 5.1$ eV (because the reaction is reversed)

26. (D)

$$\underset{(A)}{CH_3CH_2-\overset{\overset{O}{\|}}{C}-OH} \xrightarrow{NH_3} \underset{(B)}{CH_3CH_2COONH_4} \xrightarrow{\Delta} \underset{(C)}{CH_3-CH_2-\overset{\overset{O}{\|}}{C}-NH_2} \xrightarrow[\text{Hofmann bromamide reaction}]{Br_2, KOH} CH_3-CH_2-NH_2$$

27. (B)

$$Li_2\,(6) = \sigma 1s^2 \sigma^* 1s^2 \sigma 2s^2;\ \text{B.O.} = \frac{4-2}{2} = 1$$

$$Li_2^+\,(5) = \sigma 1s^2 \sigma^* 1s^2 \sigma 2s^1;\ \text{B.O.} = \frac{3-2}{2} = 0.5$$

$$Li_2^-\,(7) = \sigma 1s^2 \sigma^* 1s^2 \sigma 2s^2\, \sigma 2s^1;\ \text{B.O.} = \frac{4-3}{2} = 0.5$$

Li_2^+ is more stable than Li_2^- because Li_2^- has more number of antibonding electrons.

28. (B)

The reaction of alcohol with Lucas reagent is mostly an S_N1 reaction and the rate of reaction is directly proportional to the carbocation stability formed in the reaction. Since 3° R—OH forms 3° carbocation, it will react fastest.

29. (A)

Methyl isocyanate, $CH_3-N=C=O$ (gas)

30. (B)

Consider one mole of the oxide.

Moles of M = 0.98, Moles of O^{2-} = 1

Let moles of $M^{3+} = x$

Moles of $M^{2+} = 0.98 - x$

Doing charge balance, we have

$$(0.98 - x) \times 2 + 3x - 2 = 0$$

or $$1.96 - 2x + 3x - 2 = 0$$

or $$x = 0.04$$

Hence, percentage of $M^{3+} = \dfrac{0.04}{0.98} \times 100 = 4.08\%$.

I.I.T. JEE Objective Question Paper—2013 (Advanced)

PAPER—1

SECTION - I

(Only One Option Correct Type)

This section contains 10 multiple choice questions. Each question has four choices (A), (B), (C) and (D) out of which only one is correct.

1. In the reaction, $P + Q \longrightarrow R + S$ the time taken for 75% reaction of P is twice the time taken for 50% reaction of P. The concentration of Q varies with reaction time as shown in the figure. The overall order of the reaction is

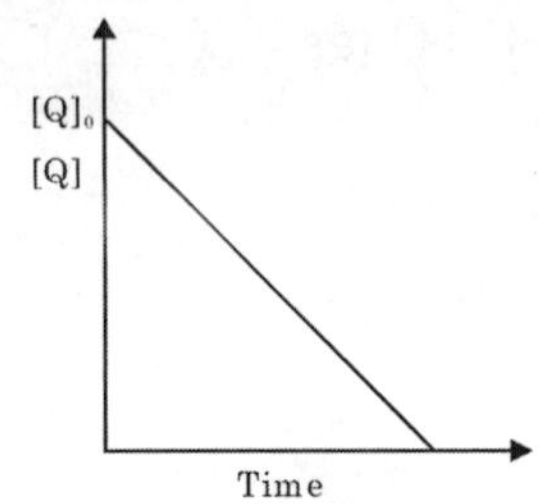

(A) 2 (B) 3 (C) 0 (D) 1

2. Consider the following complex ions, P, Q and R

$P = [FeF_6]^{3-}$, $Q = [V(H_2O)_6]^{2+}$ and $R = [Fe(H_2O)_6]^{2+}$

The correct order of the complex ions, according to their spin-only magnetic moment values (in B.M.) is

(A) $R < Q < P$ (B) $Q < R < P$ (C) $R < P < Q$ (D) $Q < P < R$

3. The arrangement of X^- ions around A^+ ion in solid AX is given in the figure (not drawn to scale). If the radius of X^- is 250 pm, the radius of A^+ is

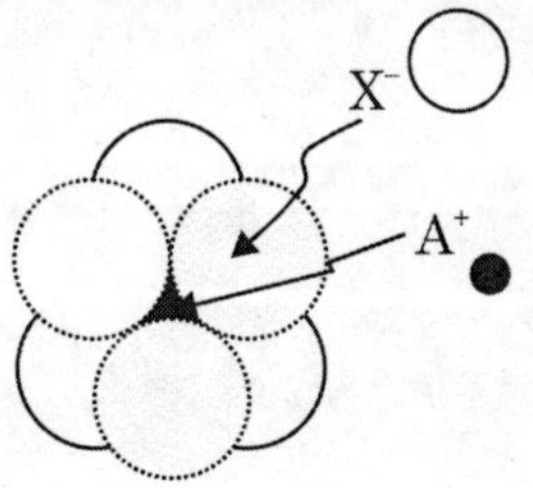

(A) 104 pm (B) 125 pm (C) 183 pm (D) 57 pm

4. Concentrated nitric acid, upon long standing, turns yellow-brown due to the formation of
(A) NO (B) NO_2 (C) N_2O (D) N_2O_4

5. The compound that does not liberate CO_2, on treatment with aqueous sodium bicarbonate solution, is
(A) Benzoic acid (B) Benzenesulphonic acid
(C) Salicylic acid (D) Carbolic acid (phenol)

6. Sulfide ores are common for the metals
(A) Ag, Cu and Pb (B) Ag, Cu and Sn (C) Ag, Mg and Pb (D) Al, Cu and Pb

7. Methylene blue, from its aqueous solution, is adsorbed on activated charcoal at 25°C. For this process, the correct statement is
(A) The adsorption requires activation at 25°C.
(B) The adsorption is accompanied by a decrease in enthalpy.
(C) The adsorption increases with increase of temperature.
(D) The adsorption is irreversible.

8. KI in acetone, undergoes S_{N^2} reaction with each of P, Q, R and S. The rates of the reaction vary as

$H_3C—Cl$

P

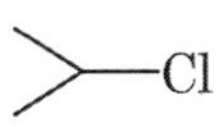

Q

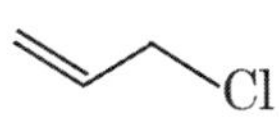

R

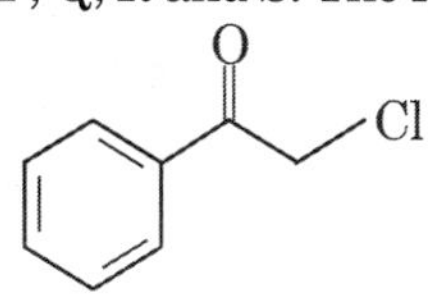

S

(A) P > Q > R > S (B) S > P > R > Q (C) P > R > Q > S (D) R > P > S > Q

9. The standard enthalpies of formation of $CO_2(g)$, $H_2O(l)$ and glucose(s) at 25°C are – 400 kJ/mol, – 300 kJ/mol, respectively. The standard enthalpy of combusion per gram of glucose at 25°C is
(A) + 2900 kJ (B) – 2900 kJ (C) – 16.11 kJ (D) + 16.11 kJ

10. Upon treatment with ammoniacal H_2S, the metal ion that precipitates as a sulfide is
(A) Fe(III) (B) Al(III) (C) Mg(II) (D) Zn(II)

SECTION – II
(One or More Options Correct Type)

This section contains 5 multiple choice questions. Each question has four choices (A), (B), (C) and (D) out of which one or more are correct.

11. The initial rate of hydrolysis of methyl acetate (1*M*) by a weak acid (HA, 1*M*) is 1/100th of that of a strong acid (HX, 1*M*), at 25°C. The K_a of HA is
(A) 1×10^{-4} (B) 1×10^{-5} (C) 1×10^{-6} (D) 1×10^{-3}

12. The hyperconjugative stabilities of *tert*-butyl cation and 2-butene, respectively, are due to
(A) $\sigma \to p$ (empty) and $\sigma \to \pi^*$ electron delocalisations.
(B) $\sigma \to \sigma^*$ and $\sigma \to \pi$ electron delocalisations.
(C) $\sigma \to p$ (filled) and $\sigma \to \pi$ electron delocalisations.
(D) p (filled) $\to \sigma^*$ and $\sigma \to \pi^*$ electron delocalisations.

13. The pair(s) of coordination complexes/ions exhibiting the same kind of isomerism is(are)

(A) $[Cr(NH_3)_5Cl]Cl_2$ and $[Cr(NH_3)_4Cl_2]Cl$ (B) $[Co(NH_3)_4Cl_2]^+$ and $[Pt(NH_3)_2(H_2O)Cl]^+$

(C) $[CoBr_2Cl_2]^{2-}$ and $[PtBr_2Cl_2]^{2-}$ (D) $[Pt(NH_3)_3(NO_3)]Cl$ and $[Pt(NH_3)_3Cl]Br$

14. Among P, Q, R and S, the aromatic compound(s) is/are

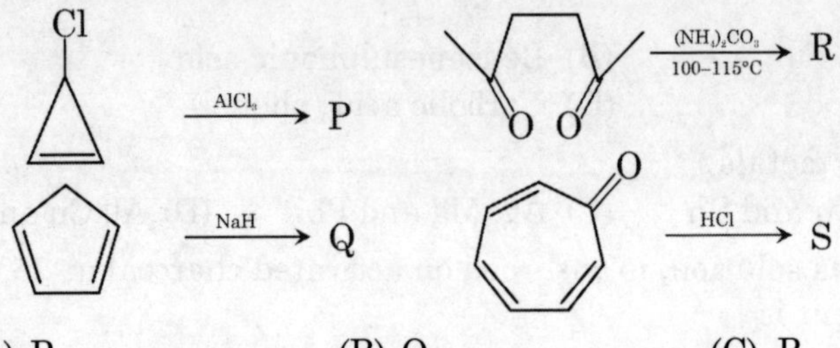

(A) P (B) Q (C) R (D) S

15. Benzene and naphthalene form an ideal solution at room temperature. For this process, the true statement(s) is(are)

(A) ΔG is positive (B) ΔS_{system} is positive (C) $\Delta S_{surroundings} = 0$ (D) $\Delta H = 0$

SECTION – III
(Integer Value Correct Type)

This section contains 5 questions. The answer to each of the questions is a single digit integer, ranging from 0 to 9 (both inclusive).

16. The atomic masses of He and Ne are 4 and 20 amu, respectively. The value of the de Broglie wavelength of He gas at – 73°C is "M" times that of the de Broglie wavelength of Ne at 727°C. M is

17. $EDTA^{4-}$ is ethylenediaminetetraacetate ion. The total number of N—Co—O bond angles in $[Co(EDTA)]^{1-}$ complex ion is

18. The total number of carboxylic acid groups in the product P is

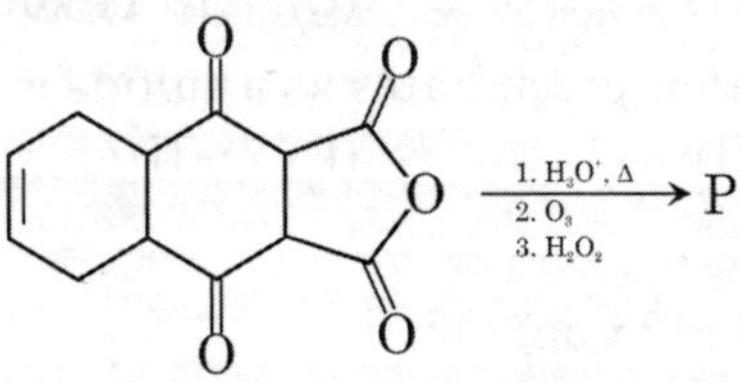

19. A tetrapeptide has—COOH group on alanine. This produces glycine (Gly), valine (Val), phenyl alanine (Phe) and alanine (Ala), on complete hydrolysis. For this tetrapeptide, the number of possible sequences (primary structures) with—NH_2 group attached to a chiral center is

20. The total number of lone-pairs of electrons in melamine is

ANSWER KEYS

1. D	2. B	3. A	4. B	5. D	6. A
7. B	8. B	9. C	10. D	11. A	12. A
13. B, D	14. A, B, C, D	15. B, C, D	16. 5	17. 8	18. 2
19. 4	20. 6				

SOLUTIONS

1. (D)
Overall order of reaction can be decided by the data given: $t_{75\%} = 2t_{50\%}$
Hence it is a first order reaction with respect to P.
From graph [Q] is linearly decreasing with time, i.e. the order of reaction with respect to Q is zero and the rate expression is: $r = k[P]^1[Q]^0$.

2. (B)
$P = Fe^{3+}$ (no. of unpaired electrons = 5)
$Q = V^{2+}$ (no. of unpaired electrons = 3)
$R = Fe^{2+}$ (no. of unpaired electrons = 4)
Since all ligands are weak field, the number of unpaired electrons remains same in the complex ion. Magnetic moment is calculated using the equation: $\mu = \sqrt{n(n+2)}$ B.M.

3. (A)
According to the given figure, A^+ is present in the octahedral void of X^-. The limiting radius in octahedral void is related to the radius of sphere as $r_{void} = 0.414 r_{sphere}$
$r_{A^+} = 0.414 r_{X^-} = 0.414 \times 250$ pm $= 103.5 \approx 104$ pm

4. (B)
$4HNO_3 \longrightarrow 2H_2O + 4NO_2 + O_2$
NO_2 remains dissolved in nitric acid colouring it yellow or even red at higher temperature.

5. (D)
pK_a of PhOH (carbolic acid) is 9.98 and that of carbonic acid (H_2CO_3) is 6.63. Hence, phenol does not produce effervescence with HCO_3^- ion.

6. (A)
Sulfide ore of Ag → Argentite (Ag_2S), Pb → Galena (PbS), Cu → Copper glance (Cu_2S)

7. (B)
Adsorption of methylene blue on activated charcoal is physical adsorption. Hence, it is characterised by decrease in enthalpy.

8. (B)
Relative reactivity for S_N2 reaction in the given structures is

Substrate	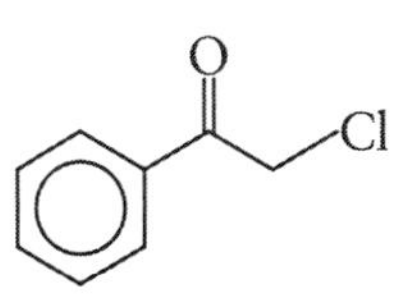 (S)	CH_3Cl (P)	$H_2C{=}CH{-}CH_2Cl$ (R)	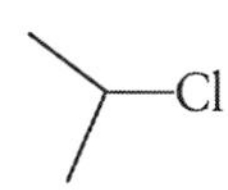 (Q)
Relative rates towards S_N2	100000	200	79	0.02

9. (C)

Combustion of glucose

$C_6H_{12}O_6 + 6O_2 \longrightarrow 6CO_2 + 6H_2O$

$\Delta H_{combustion} = (6 \times \Delta H_f CO_2 + 6 \times \Delta H_f H_2O) - \Delta H_f C_6H_{12}O_6$

$= (6 \times -400 + 6 \times -300) - (-1300) = -2900$ kJ/mol $= -2900/180$ kJ/g $= -16.11$ kJ/g

10. (D)

Among Fe^{3+}, Al^{3+}, Mg^{2+}, Zn^{2+} ions only Zn^{2+} ion is precipitated with ammonical H_2S as ZnS.

11. (A)

Rate in weak acid $= \dfrac{1}{100}$ (rate in strong acid)

Hence $[H^+]_{weak\ acid} = \dfrac{1}{100}[H^+]_{strong\ acid}$ or $[H^+]_{weak\ acid} = \dfrac{1}{100}M = 10^{-2}M$

Hence, $C\alpha = 10^{-2}$ and $K_a = 10^{-4}$

12. (A)

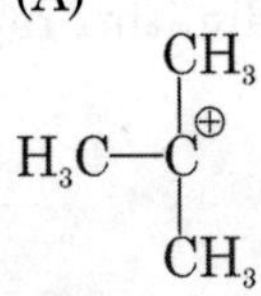

σ – *p*(empty)

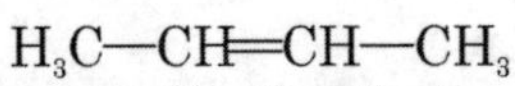

σ – π*(antibonding)

13. (B, D)

$[Co(NH_3)_4Cl_2]^+$ (an octahedral complex) and $[Pt(NH_3)_2(H_2O)Cl]^+$ (a square planar complex) will exhibit geometrical isomerism.

$[Pt(NH_3)_3(NO)_3]Cl$ and $[Pt(NH_3)_3Cl]Br$ will show ionization isomerism.

14. (A, B, C, D)

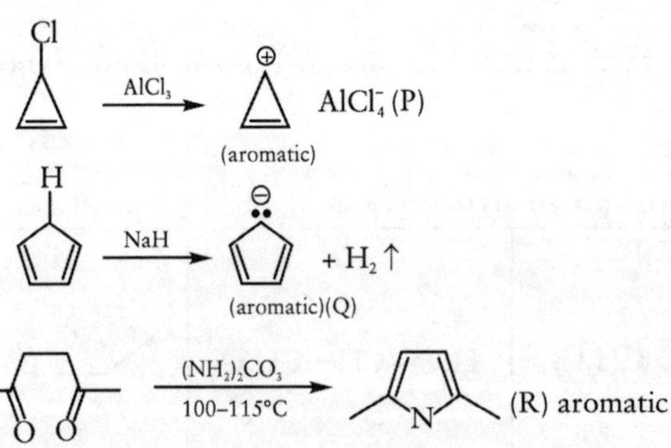

Mechanism

$(NH_4)_2CO_3 \xrightarrow{\Delta} 2NH_3 + CO_2 + H_2O$

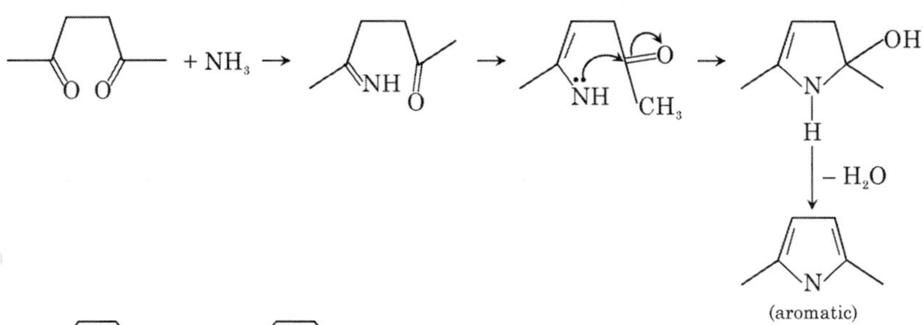

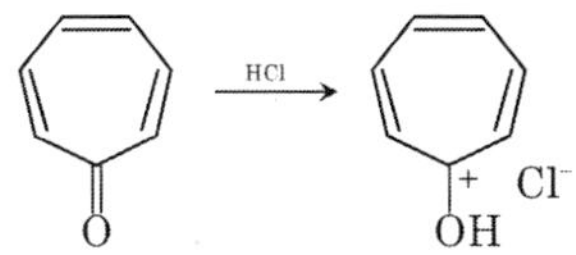
(S) aromatic

15. (B, C, D)

For ideal solution, $\Delta S_{\text{system}} > 0$, $\Delta S_{\text{surrounding}} = 0$, $\Delta H_{\text{mixing}} = 0$

16. (5)

$$\lambda = \frac{h}{mV} = \frac{h}{\sqrt{2\text{M K.E.}}} \qquad \text{(since K.E.} \propto T)$$

$$\text{or } \lambda \propto \frac{1}{\sqrt{\text{M}T}}$$

For two gases, $\dfrac{\lambda_{\text{He}}}{\lambda_{\text{Ne}}} = \sqrt{\dfrac{\text{M}_{\text{Ne}}T_{\text{Ne}}}{\text{M}_{\text{He}}T_{\text{He}}}} = \sqrt{\dfrac{20}{4} \times \dfrac{1000}{200}} = 5$

17. (8)

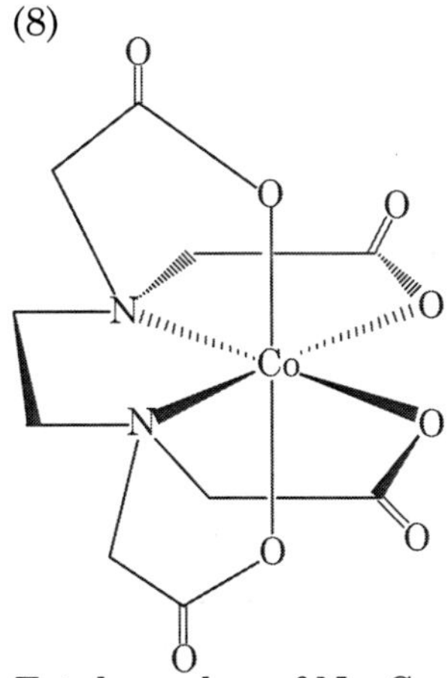

Total number of N—Co—O bond angles is 8.

18. (2)

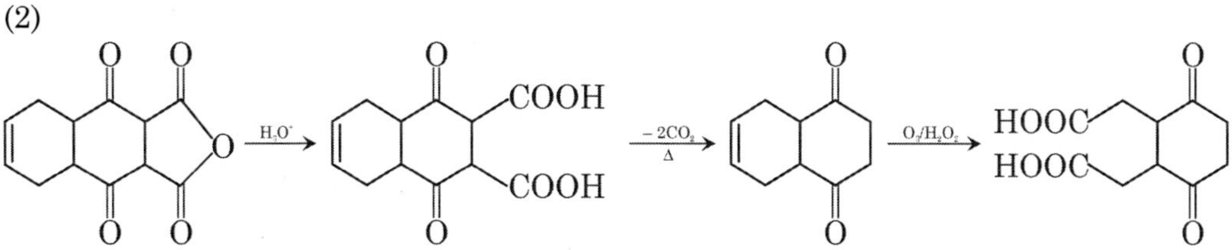

19. (4)

Because —COOH group of tetrapeptide is intact on alanine, its NH_2 must be participating in condensation.

Hence alanine is at one terminus, – – – A.

To fill the 3 blanks, the possible options are

(i) When NH_2 group attached to non-chiral carbon

G	V	P
G	P	V

(ii) When NH_2 group attached to chiral carbon

V	G	P	P	V	G
V	P	G	P	G	V

where, Glycine (G)
Valine (V)
Phenyl alanine (P)
Alanine (A)

Hence the number of possible sequence is 4.

20. (6 lone pairs)

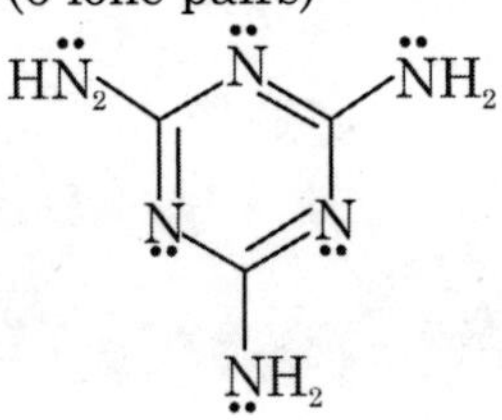

Melamine

I.I.T. JEE Objective Question Paper—2013 (Advanced)

PAPER—2

SECTION - I
(One or More Options Correct Type)

This section contains 8 multiple choice questions. Each question has four choices (A), (B), (C) and (D) out of which one or more are correct.

1. The K_{sp} of Ag_2CrO_4 is 1.1×10^{-12} at 298 K. The solubility (in mol/L) of Ag_2CrO_4 in a $0.1M$ $AgNO_3$ solution is
   (A) 1.1×10^{-11} (B) 1.1×10^{-10} (C) 1.1×10^{-12} (D) 1.1×10^{-9}
2. In the following reaction, the product(s) formed is(are)

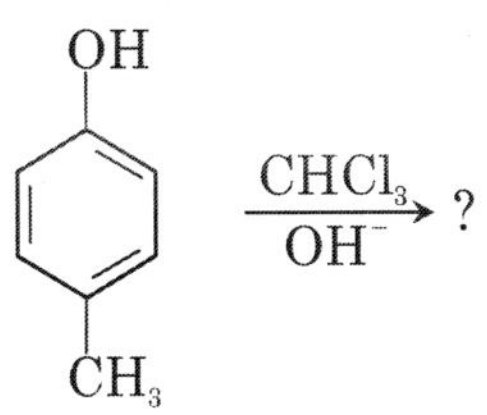

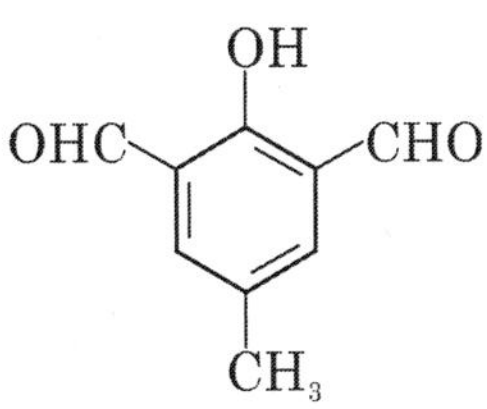

P

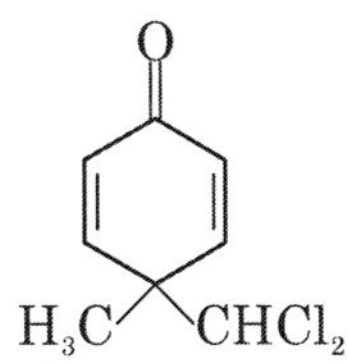

Q

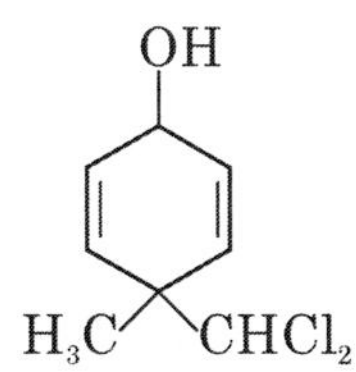

R

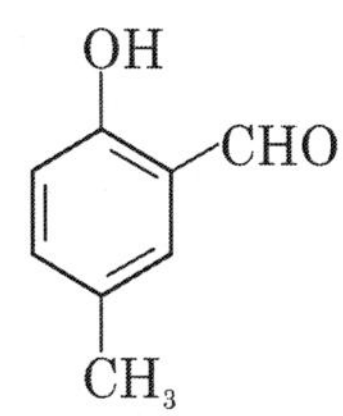

S

   (A) P (major) (B) Q (minor) (C) R (minor) (D) S (major)
3. The major product(s) of the following reaction is(are)

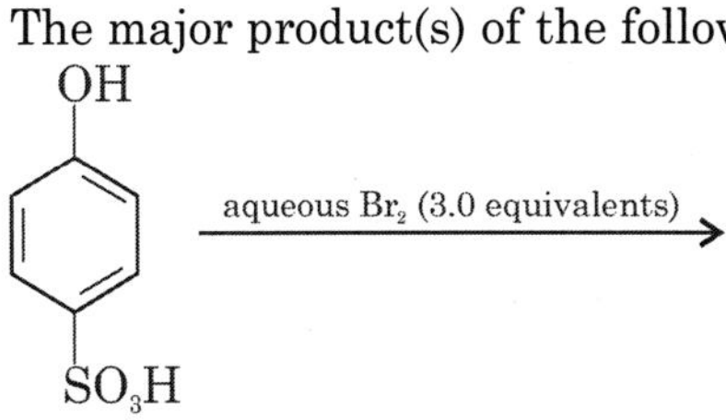

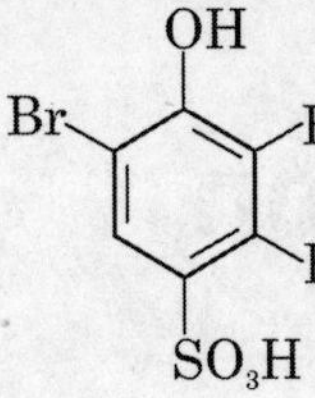
P

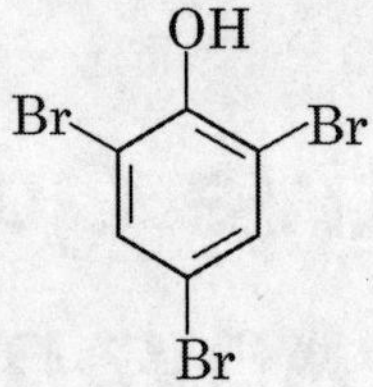
Q

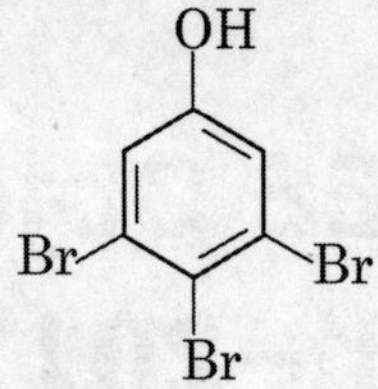
R

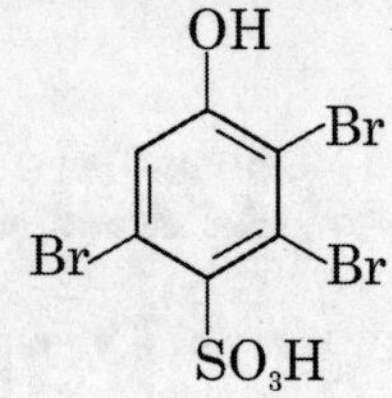
S

(A) P (B) Q (C) R (D) S

4. After completion of the reactions (I and II), the organic compound(s) in the reaction mixtures is(are)

Reaction I :

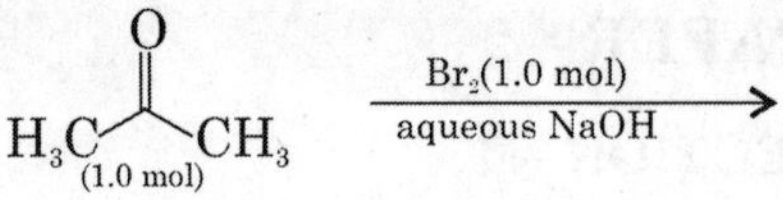

Reaction II :

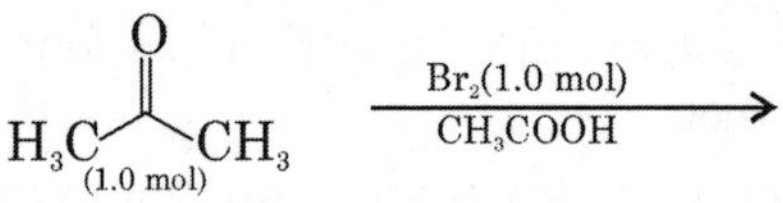

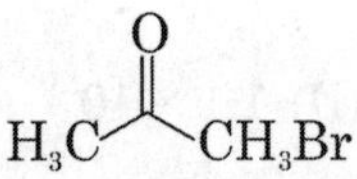
P

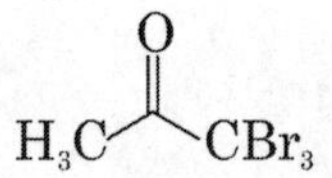
Q

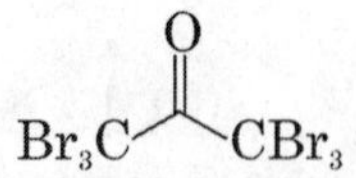
R

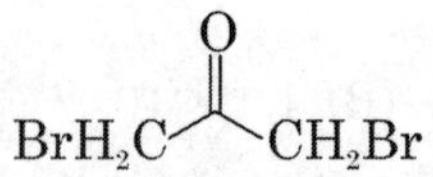
S

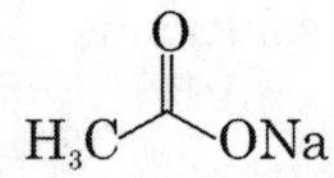
T

$CHBr_3$
U

(A) Reaction I: P and Reaction II: P
(B) Reaction I: U, acetone and Reaction II: Q, acetone
(C) Reaction I: T, U, acetone and Reaction II: P
(D) Reaction I: R, acetone and Reaction II: S

5. The correct statement(s) about O_3 is(are)
(A) O—O bond lengths are equal.
(B) Thermal decomposition of O_3 is endothermic.
(C) O_3 is diamagnetic in nature.
(D) O_3 has a bent structure.

6. In the nuclear transmutation, ${}^{9}_{4}Be + X \longrightarrow {}^{8}_{4}Be + Y$
(X, Y) is(are)
(A) (γ, n) (B) (p, D) (C) (n, D) (D) (γ, p)

7. The carbon-based reduction method is not used for the extraction of
(A) tin from SnO_2 (B) iron from Fe_2O_3
(C) aluminium from Al_2O_3 (D) magnesium from $MgCO_3 \cdot CaCO_3$

8. The themal dissociation equilibrium of $CaCO_3(s)$ is studied under different conditions.
$CaCO_3(s) \rightleftharpoons CaO(s) + CO_2(g)$
For this equilibrium, the correct statement(s) is(are)
(A) ΔH is dependent on T
(B) K is independent of the initial amount of $CaCO_3$

(C) K is dependent on the pressure of CO_2 at a given T

(D) ΔH is independent of the catelyst, if any.

SECTION – II (Paragraph Type)

This section contains 4 paragraphs each describing theory, experiment, data, etc. Eight questions relate to four paragraphs with two questions on each paragraph. Each question of a paragraph has only one correct answer among the four choices (A), (B), (C) and (D).

Paragraph for Questions 9 and 10

An aqueous solution of a mixture of two inorganic salts, when treated with dilute HCl, gave a precipitate (P) and a filtrate (Q). The precipitate P was found to dissolve in hot water. The filtrate (Q) remained unchanged, when treated with H_2S in a dilute mineral acid medium. However, it gave a precipitate (R) with H_2S in an ammoniacal medium. The precipitate R gave a coloured solution (S), when treated with H_2O_2 in an aqueous NaOH medium.

9. The precipitate P contains

(A) Pb^{2+} (B) Hg_2^{2+} (C) Ag^+ (D) Hg^{2+}

10. The coloured solution S contains

(A) $Fe_2(SO_4)_3$ (B) $CuSO_4$ (C) $ZnSO_4$ (D) Na_2CrO_4

Paragraph for Questions 11 and 12

P and Q are isomers of dicarboxylic acid $C_4H_4O_4$. Both decolorize Br_2/H_2O. On heating, P forms the cyclic anhydride.

Upon treatment with dilute alkaline $KMnO_4$, P as well as Q could produce one or more than one from S, T and U.

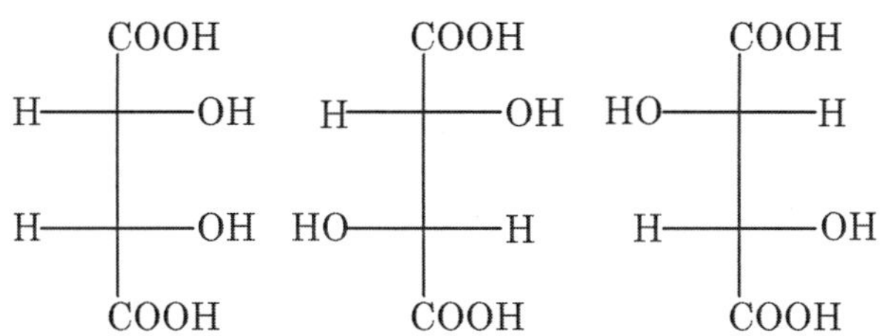

S T U

11. Compounds formed form P and Q are, respectively

(A) Optically active S and optically active pair (T, U)

(B) Optically inactive S and optically inactive pair (T, U)

(C) Optically active pair (T, U) and optically active S

(D) Optically inactive pair (T, U) and optically inactive S.

12. In the following reaction sequences V and W are respectively:

$$Q \xrightarrow[\Delta]{H_2/Ni} V$$

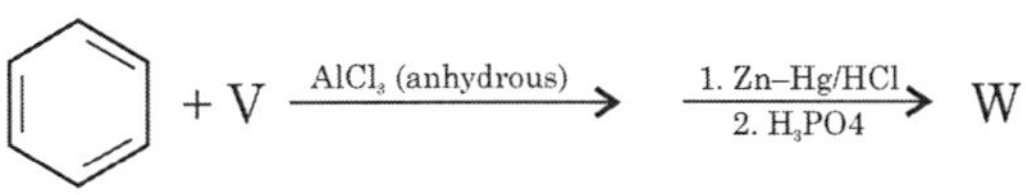

(A) 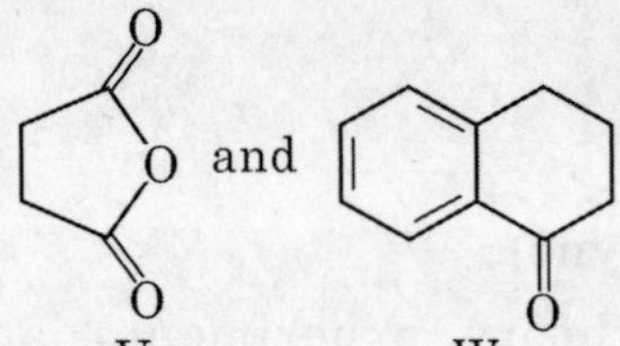

V and W

(B) 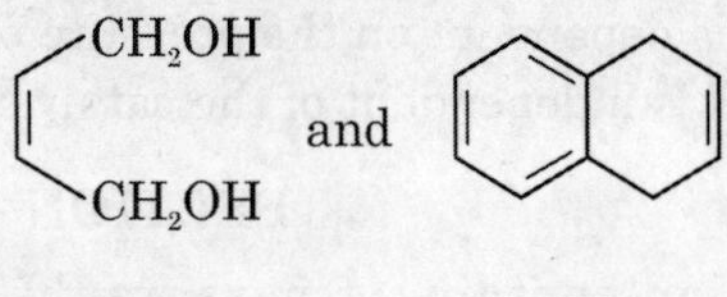

V and W

(C) 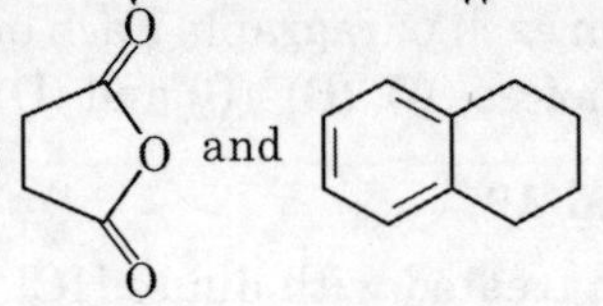

V and W

(D) 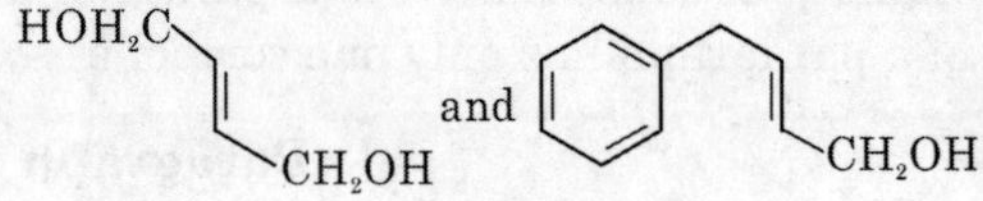

V and W

Paragraph for Questions 13 and 14

A fixed mass m of a gas is subjected to transformation of states from K to L to M to N and back to K as shown

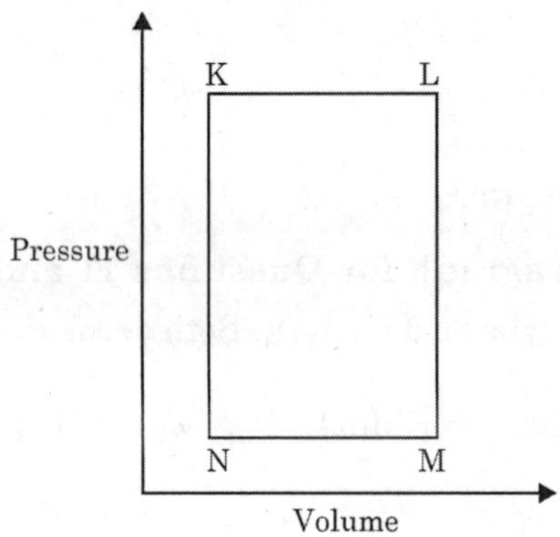

13. The succeeding operations that enable this transformation of states are

(A) Heating, cooling, heating, cooling
(B) Cooling, heating, cooling, heating
(C) Heating, cooling, cooling, heating
(D) Cooling, heating, heating, cooling.

14. The pair of isochoric processes among the transormation of states is

(A) K to L and L to M
(B) L to M and N to K
(C) L to M and M to N
(D) M to N and N to K

Paragraph for Questions 15 and 16

The reactions of Cl_2 gas with cold-dilute and hot-concentrated NaOH in water give sodium salts of two (different) oxoacids of chlorine, P and Q, respectively. The Cl_2 gas reacts with SO_2 gas, in presence of charcoal, to give a product R. R reacts with white phosphorus to give a compound S. On hydrolysis, S gives an oxoacid of phosphorus T.

15. P and Q, respectively, are the sodium salts of

(A) hypochlorous and chloric acids
(B) hypochlorous and chlorous acids
(C) chloric and perchloric acids
(D) chloric and hypochlorous acids

16. R, S and T, respectively, are

(A) SO_2Cl_2, PCl_5 and H_3PO_4
(B) SO_2Cl_2, PCl_3 and H_3PO_3
(C) $SOCl_2$, PCl_3 and H_3PO_2
(D) $SOCl_2$, PCl_5 and H_3PO_4

SECTION – III (Matching List Type)

This section contains 4 multiple choice questions. Each question has matching lists. The codes for the lists have choces (A), (B), (C) and (D) out of which only one is correct.

17. The unbalanced chemical reactions given in List I show missing reagent or condition (?) which are provided in List II. Match List I with List II and select the correct answer using the code given below the lists:

List I	List II
P. $PbO_2 + H_2SO_4 \xrightarrow{?} PbSO_4 + O_2$ + other product	1. NO
Q. $Na_2S_2O_3 + H_2O \xrightarrow{?} NaHSO_4$ + other product	2. I_2
R. $N_2H_4 \xrightarrow{?} N_2$ + other product	3. Warm
S. $XeF_2 \xrightarrow{?} Xe$ + other product	4. Cl_2

Codes:	P	Q	R	S
(A)	4	2	3	1
(B)	3	2	1	4
(C)	1	4	2	3
(D)	3	4	2	1

18. Match the chemical conversions in List I with the appropriate reagent in List II and select the correct answer using the code given below the lists:

List I **List II**

P. 

1. (i) $Hg(OAc)_2$; (ii) $NaBH_4$

Q. 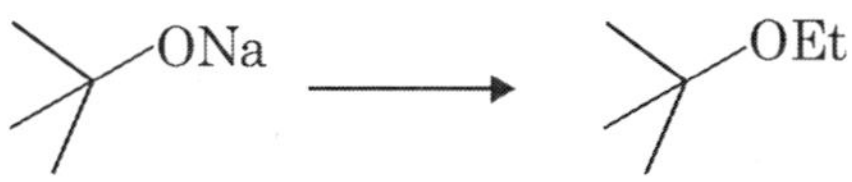

2. NaOEt

R. 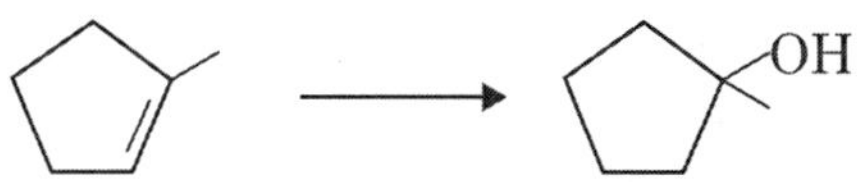

3. EtBr

S. 

4. (i) BH_3; (ii) $H_2O_2/NaOH$

Codes:	P	Q	R	S
(A)	2	3	1	4
(B)	3	2	1	4
(C)	2	3	4	1
(D)	3	2	4	1

19. An aqueous solution of X is added slowly to an aqueous solution of Y as shown in List I. The variation in conductivity of these reactions is given in List II. Match List I with List II and select the correct answer using the code given below the lists:

List I	List II
P. $(C_2H_5)_3N + CH_3COOH$ (X, Y)	1. Conductivity decreases and then increases
Q. KI (0.1*M*) + $AgNO_3$(0.01*M*) (X, Y)	2. Conductivity decreases and then does not change much
R. CH_3COOH + KOH (X, Y)	3. Conductivity increases and then does not change much
S. NaOH + HI (X, Y)	4. Conductivity does not change much and then increases

Codes:	P	Q	R	S
(A)	3	4	2	1
(B)	4	3	2	1
(C)	2	3	4	1
(D)	1	4	3	2

20. The standard reduction potential data at 25°C is given below.

E^0 (Fe^{3+}, Fe^{2+}) = + 0.77 V;
E^0 (Fe^{2+}, Fe) = – 0.44 V;
E^0 (Cu^{2+}, Cu) = + 0.34 V;
E^0 (Cu^{+}, Cu) = + 0.52 V;
E^0 $(O_2(g) + 4H^+ + 4e^- \longrightarrow 2H_2O)$ = + 1.23 V;
E^0 $(O_2(g) + 2H_2O + 4e^- \longrightarrow 4OH^-)$ = + 0.40 V;
E^0 (Cr^{3+}, Cr) = – 0.74 V;
E^0 (Cr^{2+}, Cr) = – 0.91 V

Match E^0 of the rebox pair in List I with the values given in List II and select the correct answer using the code given below the lists:

List I	List II
P. E^0 (Fe^{3+}, Fe)	1. – 0.18 V
Q. E^0 $(4H_2O \rightleftharpoons 4H^+ + 4OH^-)$	2. – 0.4 V
R. E^0 $(Cu^{2+} + Cu \longrightarrow 2Cu^+)$	3. – 0.04 V
S. E^0 (Cr^{3+}, Cr^{2+})	4. – 0.83 V

Codes:	P	Q	R	S
(A)	4	1	2	3
(B)	2	3	4	1
(C)	1	2	3	4
(D)	3	4	1	2

ANSWER KEYS

1. B	2. B, D	3. B	4. C	5. A, C, D	6. A, B
7. C, D	8. A, B, D	9. A	10. D	11. B	12. A
13. C	14. B	15. A	16. A	17. D	18. A
19. A	20. D				

SOLUTIONS

1. (B)

$$Ag_2CrO_4 \rightleftharpoons 2Ag^+ + CrO_4^{2}$$

$$s \qquad 0.1 + 2s \qquad s$$

$$\approx 0.1$$

$$1.1 \times 10^{-12} = (0.1)^2 s$$

$$s = 1.1 \times 10^{-10}$$

2. (B, D)

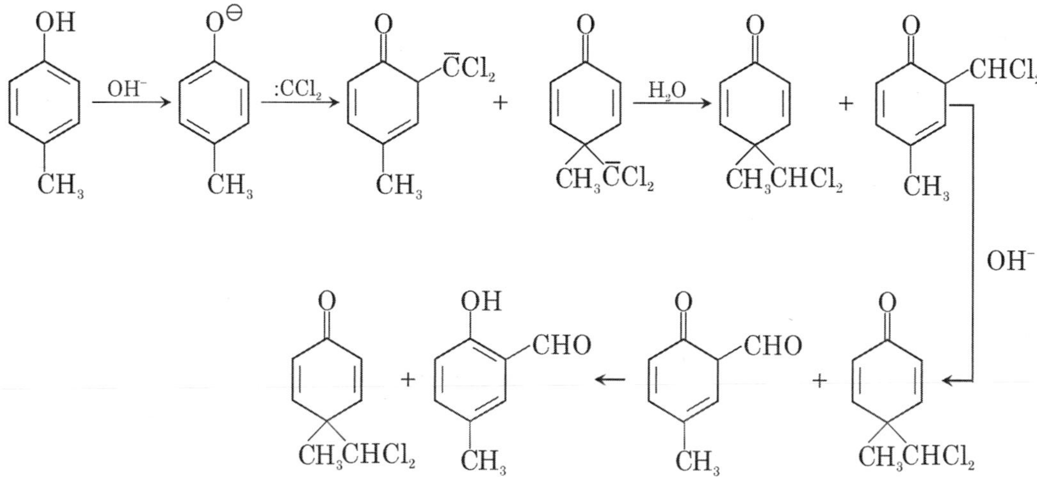

3. (B)

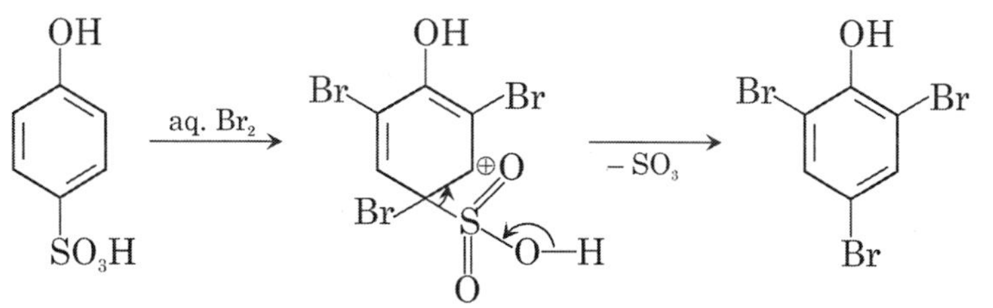

4. (C)

Reaction I: 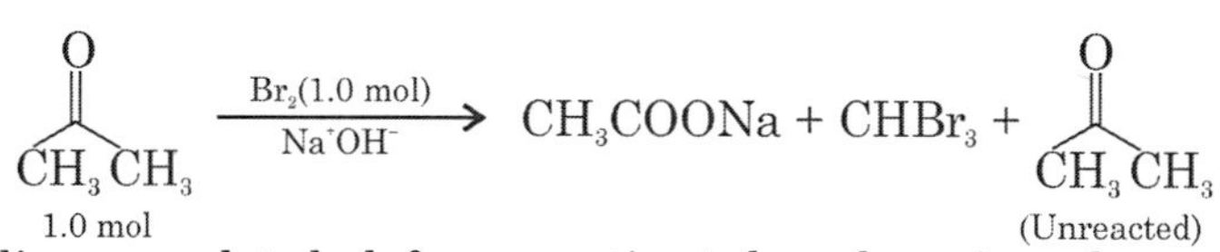

In basic medium complete haloform reaction takes place since the rate of reaction increases with each α-halogenation.

Reaction II: 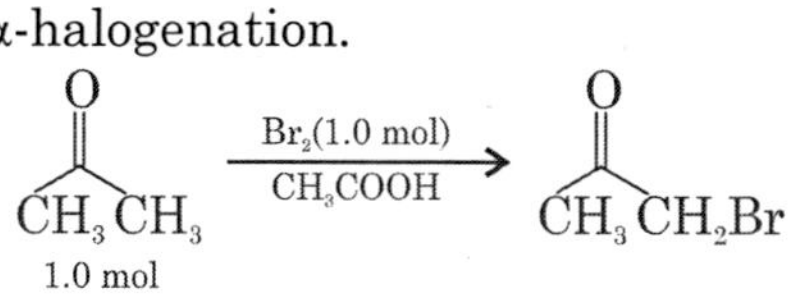

In acidic medium, monohalogenation takes place with 1 mol of halogen.

5. (A, C, D)

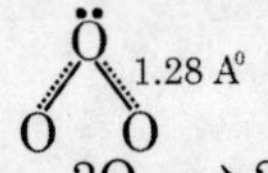

All electrons are paired and hence diamagnetic.

$3O_2 \longrightarrow 2O_3$ $\Delta H = + 142$ kJ/mol

6. (A, B)

${}^{9}_{4}Be + X \longrightarrow {}^{8}_{4}Be + Y$

If X is ${}^{0}_{0}\gamma$ then Y is ${}^{1}_{0}n$

If X is ${}^{1}_{1}p$ then Y is ${}^{2}D$

7. (C, D)

SnO_2 and Fe_2O_3 are reduced by carbon reduction method.

Al_2O_3, $MgCO_3 \cdot CaCO_3$ are reduced by electrolytic reduction method.

8. (A, B, D)

(A) $\Delta H_2 - \Delta H_1 = C_{P(\gamma \times n)}(T_2 - T_1)$

and C_p depends on temperature. Hence enthalpy also depends on temperature.

(B) $CaCO_3(s) \rightleftharpoons CaO(s) + CO_2(g)$ $k_p = (p_{CO_2})_{\text{at equlibrium}}$

For a given reaction.

K_{eq} depends only on temperature.

(C) K_{eq} depends only on temperature.

(D) Enthalpy of reaction is independent of the catalyst. Catalyst generaly changes activation energy.

9. (A)

10. (D)

Sol. (9 and 10)

$Pb^{2+} + 2Cl^- \longrightarrow PbCl_2$ (white ppt.)
soluble in hot water

$Pb^{2+}/Cr^{3+} \xrightarrow{HCl(aq)} PbCl_2 + CrCl_3$

(P) (Q) White ppt. Filtrate

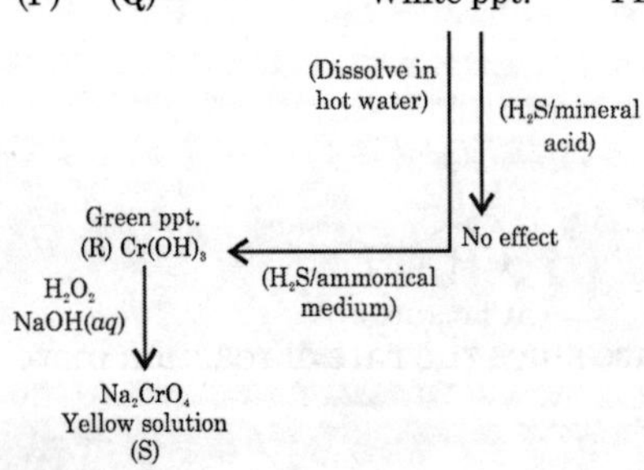

11. (B)

12. (A)

Sol. (11 and 12)

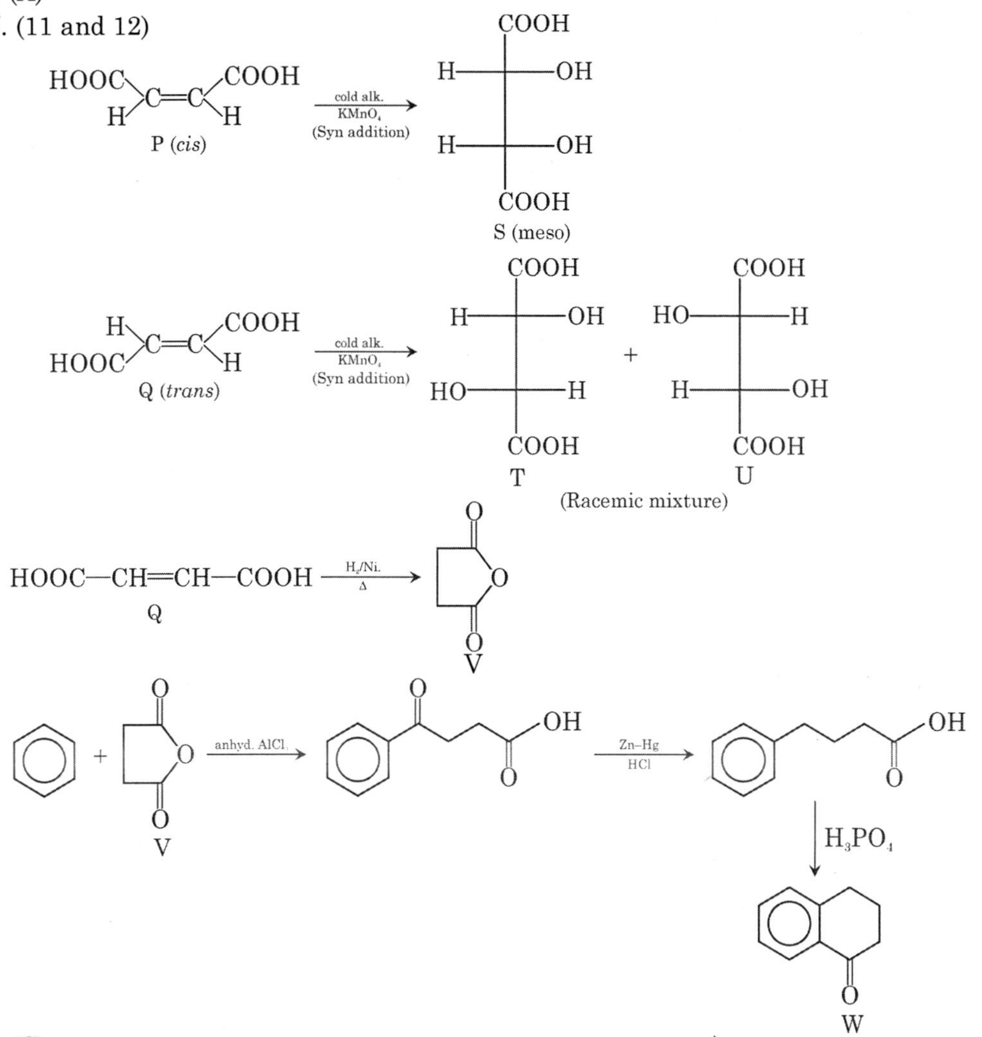

13. (C)

K → L ; V ↑ at constant P

Hence T↑ (Heating)

L → M ; P ↓ at constant V

Hence T↓ (Cooling)

M → N ; V ↓ at constant P

Hence T↓ (Cooling)

N → K ; P ↑ at constant V

Hence T↑ (Heating)

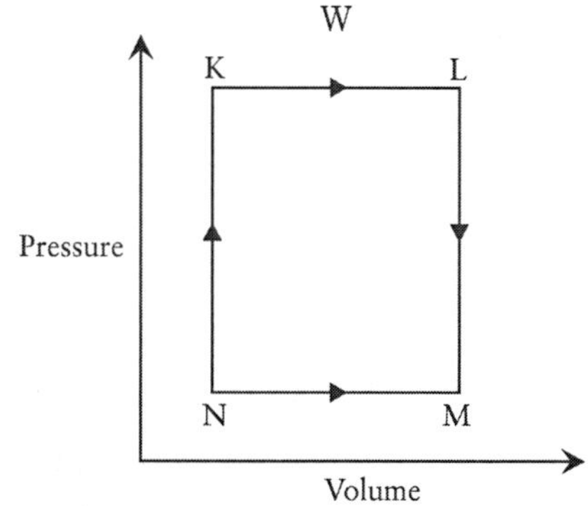

14. (B)

L → M

M → K

Both are having constant volume, therefore these processes are isochoric.

15. (A)

16. (A)

Sol. (15 and 16)

Cl_2 + cold dil. NaOH $\longrightarrow$ NaOCl + NaCl

Cl_2 + hot conc. NaOH $\longrightarrow$ $NaClO_3$ + NaCl

NaOCl is salt of hypochlorous acid = P

$NaClO_3$ is salt of chloric acid = Q

$Cl_2 + SO_2 \xrightarrow{\text{charcoal}} SO_2Cl_2$ (R)

$SO_2Cl_2 + P_4 \longrightarrow PCl_5$ (S) + SO_2

$PCl_5 + H_2O \longrightarrow H_3PO_4$ (T) + HCl

17. (D)

(P) $2PbO_2 + H_2SO_4 \xrightarrow{\text{warm}} 2PbSO_4 + O_2 + 3H_2O$

(Q) $Na_2S_2O_3 + H_2O \xrightarrow{+\,Cl_2} NaHSO_4 + HCl$

(R) $N_2H_4 \xrightarrow{+\,I_2} NaHSO_4 + HCl$

(S) $XeF_2 \xrightarrow{+\,NO} Xe + NOF$

18. (A)

P. 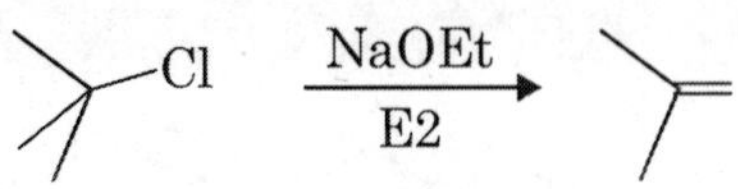

Q. 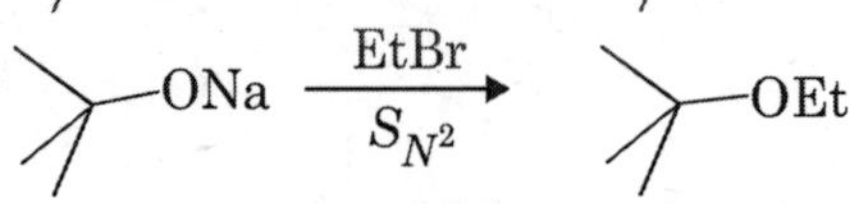

R. 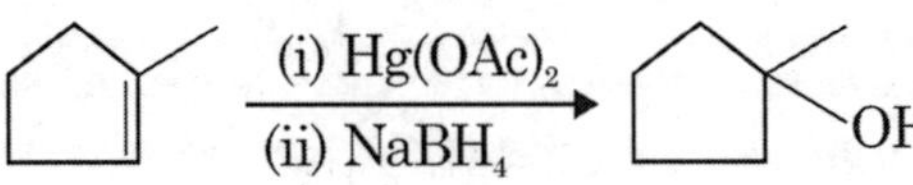

(Markownikov addition)

S. 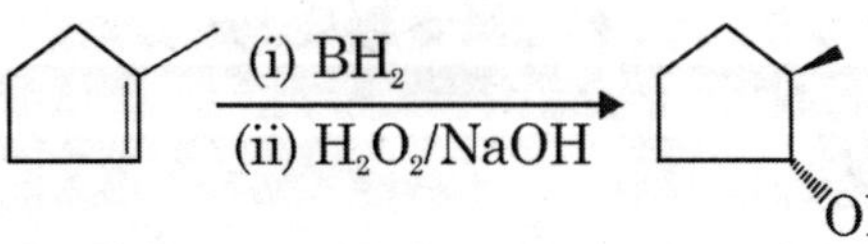

(Antimarkownikov addition)

19. (A)

$$\overset{X}{(C_2H_5)_3N} + \overset{Y}{CH_3COOH} \longrightarrow CH_3COO^-(aq) + (C_2H_5)_3NH^+(aq)$$

(P) Since CH_3COOH is a weak acid, its conductivity is already less. On addition of weak base, acid-base reaction takes place and new ions are created. Hence, conductivity increases.

(Q) KI (0.1*M*) + $AgNO_3$ (0.01*M*) $\longrightarrow$ AgI $\downarrow$ + $KNO_3(aq)$

Since the only reaction taking place is precipitation of AgI and in place of Ag^+, K^+ is coming in the solution, conductivity remains nearly constant and then increases.

(R) $CH_3COOH + KOH \longrightarrow CH_3COOH(aq) + H_2O$

$OH^-(aq)$ is getting replaced by CH_3COO^-, which has poorer conductivity. Hence, conductivity dereases and then after the end point, due to common ion effect, no further creation of ions takes place. Hence, conductivity remains nearly same.

(S) $NaOH + HI \longrightarrow NaI(aq) + H_2O$

Since H^+ is getting replaced by Na^+ conductivity dereases and after the end point due to OH^-, conductivity increases.

Hence, answer is: (P) – (3) ; (Q) – (4) ; (R) – (2) ; (S) – (1)

20. (D)

(P) $E^0_{Fe^{3+},Fe}$ $Fe^{3+} \xrightarrow[n=1]{+0.77\text{ V}} Fe^{2+} \xrightarrow[n=2]{-0.44\text{ V}} Fe$

$Fe^{3+} \longrightarrow Fe$: x V, $n = 3$

Hence $1 \times 0.77 + 2 \times (-0.44) = 3 \times x$

or $x = -\dfrac{0.11}{3}\text{ V} \simeq -0.04\text{ V}$

(Q) $4H_2O \rightleftharpoons 4H^+ + 4OH^-$

$$\begin{array}{ll} 2H_2O \longrightarrow O_2 + 4H^+ + 4e^- & -1.23\text{ V} \\ + & \\ O_2 + 2H_2O + 4e^- \longrightarrow 4OH^- & +0.4\text{ V} \\ \hline 4H_2O \rightleftharpoons 4H^+ + 4OH^- & -0.83\text{ V} \end{array}$$

(R) $E^0_{(Cu^{2+}+Cu \to 2Cu^+)}$ $Cu^{2+} \xrightarrow[n=1]{x\text{ V}} Cu^{+} \xrightarrow[n=1]{-0.52\text{ V}} Cu$

$Cu^{2+} \longrightarrow Cu$: $+0.34$ V, $n = 2$

$x \times 1 + 0.52 \times 1 = 0.34 \times 2$

$x = 0.16\text{ V}$

$$\begin{array}{ll} Cu^{2+} + e^- \longrightarrow Cu^+ & +0.16\text{ V} \\ Cu \longrightarrow Cu^+ + e^- & -0.52\text{ V} \\ \hline Cu^{2+} + Cu \longrightarrow 2Cu^+ & -0.36\text{ V} \end{array}$$

However, in the given option, – 0.18 V is printed.

(S) $E^0_{(Cr^{3+},\, Cu^{2+})}$ $Cr^{3+} \xrightarrow[n=1]{x} Cr^{2+} \xrightarrow[n=2]{-0.91\text{ V}} Cr$

$Cr^{3+} \longrightarrow Cr$: -0.74 V, $n = 3$

$x \times 1 + 2 \times (-0.91) = 3 \times (-0.74)$

or $x - 1.82 = -2.22$

or $x = -0.4\text{ V}$

Hence, (P) – (3) ; (Q) – (4) ; (R) – (1) ; (S) – (2)

I.I.T. JEE Objective Question Paper—2014 (Main)

1. Which one of the following properties is not shown by NO?
   (A) It combines with oxygen to form nitrogen dioxide
   (B) Its bond order is 2.5
   (C) It is diamagnetic in gaseous state
   (D) It is a neutral oxide.

2. If Z is a compressibility factor, van der Waals equation at low pressure can be written as
   (A) $Z = 1 - \frac{Pb}{RT}$ (B) $Z = 1 + \frac{Pb}{RT}$ (C) $Z = 1 + \frac{RT}{Pb}$ (D) $Z = 1 - \frac{a}{VRT}$

3. The metal that cannot be obtained by electrolysis of an aqueous solution of its salts is
   (A) Cu (B) Cr (C) Ag (D) Ca

4. Resistance of $0.2M$ solution of an electrolyte is 50 Ω. The specific conductance of the solution is 1.4 S m^{-1}. The resistance of $0.5M$ solution of the same electrolyte is 280 Ω. The molar conductivity of $0.5M$ solution of the electrolyte in S m^2 mol^{-1} is
   (A) 5×10^3 (B) 5×10^2 (C) 5×10^{-4} (D) 5×10^{-3}

5. CsCl crystallises in body centred cubic lattice. If a is its edge length then which of the following expressions is correct?
   (A) $r_{Cs^+} + r_{Cl^-} = \frac{\sqrt{3}}{2}a$ (B) $r_{Cs^+} + r_{Cl^-} = \sqrt{3}a$ (C) $r_{Cs^+} + r_{Cl^-} = 3a$ (D) $r_{Cs^+} + r_{Cl^-} = \frac{3a}{2}$

6. Consider separate solutions of $0.500M$ $C_2H_5OH(aq)$, $0.100M$ $Mg_3(PO_4)_2(aq)$, $0.250M$ $KBr(aq)$ and $0.125M$ $Na_3OP_4(aq)$ at 25°C. Which statement is true about these solutions, assuming all salts to be strong electrolytes?
   (A) $0.125M$ $Na_3PO_4(aq)$ has the highest osmotic pressure.
   (B) $0.500M$ $C_2H_5OH(aq)$ has the highest osmotic pressure.
   (C) They all have the same osmotic pressure.
   (D) $0.100M$ $Mg_3(PO_4)_2(aq)$ has the highest osmotic pressure.

7. In which of the following reactions H_2O_2 acts as a reducing agent?
   (a) $H_2O_2 + 2H^+ + 2e^- \longrightarrow 2H_2O$
   (b) $H_2O_2 - 2e^- \longrightarrow O_2 + 2H^+$
   (c) $H_2O_2 + 2e^- \longrightarrow 2OH^-$
   (d) $H_2O_2 + 2OH^- - 2e^- \longrightarrow O_2 + 2H_2O$
   (A) (a), (c) (B) (b), (d) (C) (a), (b) (D) (c), (d)

8. In S_N2 reactions, the correct order of reactivity for the following compounds CH_3Cl, CH_3CH_2Cl, $(CH_3)_2CHCl$ and $(CH_3)_3CCl$ is

(A) $CH_3CH_2Cl > CH_3Cl > (CH_3)_2CHCl > (CH_3)_3CCl$

(B) $(CH_3)_2CHCl > CH_3CH_2Cl > CH_3Cl > (CH_3)_3CCl$

(C) $CH_3Cl > (CH_3)_2CHCl > CH_3CH_2Cl > (CH_3)_3CCl$

(D) $CH_3Cl > CH_3CH_2Cl > (CH_3)_2CHCl > (CH_3)_3CCl$

9. The octahedral complex of a metal ion M^{3+} with four monodentate ligands L_1, L_2, L_3 and L_4 absorbs wavelenghts in the region of red, green, yellow and blue, respectively. The increasing order of ligand strength of the four ligands is

(A) $L_3 < L_2 < L_4 < L_1$ (B) $L_1 < L_2 < L_4 < L_3$ (C) $L_4 < L_3 < L_2 < L_1$ (D) $L_1 < L_3 < L_2 < L_4$

10. For the estimation of nitrogen, 1.4 g of organic compound was digested by Kjeldahl method and the evolved ammonia was absorbed in 60 mL of $\frac{M}{10}$ sulphuric acid. The unreacted acid required 20 mL of $\frac{M}{10}$ sodium hydroxide for complete neutralization. The percentage of nitrogen in the compound is

(A) 3% (B) 5% (C) 6% (D) 10%

11. The equivalent conductances of NaCl at concentration C and at infinite dilution are λ_C and λ_∞, respectively. The correct relationship between λ_C and λ_∞ is given as (where the constant B is positive)

(A) $\lambda_C = \lambda_\infty - (B)\sqrt{C}$ (B) $\lambda_C = \lambda_\infty + (B)\sqrt{C}$ (C) $\lambda_C = \lambda_\infty + (B)C$ (D) $\lambda_C = \lambda_\infty - (B)C$

12. For the reaction, $SO_2(g) + \frac{1}{2}O_2(g) \rightleftharpoons SO_3(g)$, if $K_P = K_C(RT)^x$ where the symbols have usual meaning then the value of x is (assuming ideality)

(A) $\frac{1}{2}$ (B) 1 (C) -1 (D) $-\frac{1}{2}$

13. In the reaction, $CH_3COOH \xrightarrow{LiAlH_4} A \xrightarrow{PCl_5} B \xrightarrow{Alc.\ KOH} C$, the product C is

(A) Ethylene (B) Acetyl chloride (C) Acetaldehyde (D) Acetylene

14. Sodium phenoxide when heated with CO_2 under pressure at 125°C yields a product which on acetylation produces C.

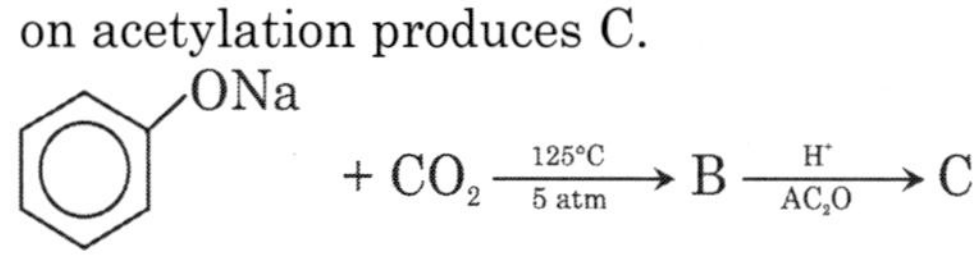

The major product C would be

(A) 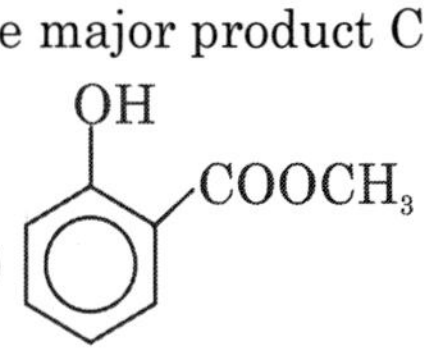

(B) 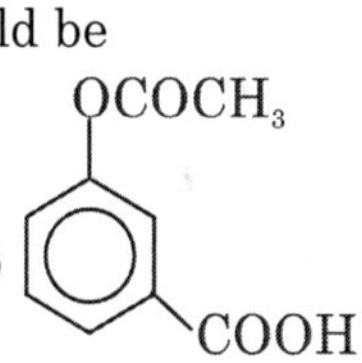

(C) 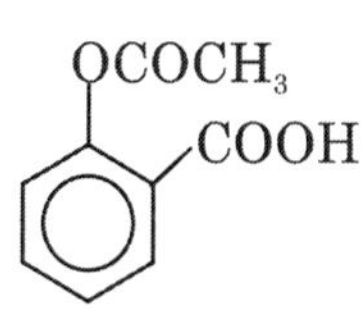

(D) 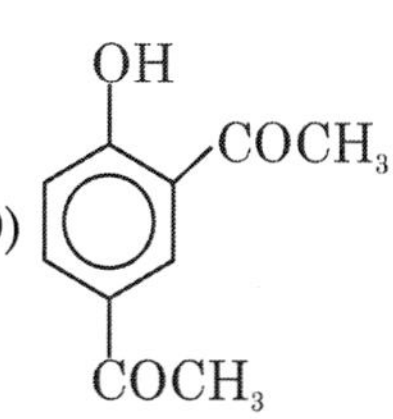

15. On heating an aliphatic primary amine with chloroform and ethanolic potassium hydroxide, the organic compound formed is

(A) an alkyl cyanide (B) an alkyl isocyanide (C) an alkanol (D) an alkanediol

16. The correct statement for the molecule, CsI_3 is

(A) it contains Cs^{3+} and I^- ions.

(B) it contains Cs^{3+}, I^- and lattice I_2 molecule.

(C) it is a covalent molecule.

(D) it contains Cs^+ and I_3^- ions.

17. The equation which is balanced and represents the correct product(s) is

(A) $[Mg(H_2O)_6]^{2+} + (EDTA)^{4-} \xrightarrow{\text{excess NaOH}} [Mg(EDTA)]^{2+} + 6H_2O$

(B) $CuSO_4 + 4KCN \longrightarrow K_2[Cu(CN)_4] + K_2SO_4$

(C) $Li_2O + 2KCl \longrightarrow 2LiCl + K_2O$

(D) $[CoCl(NH_3)_5]^+ + 5H^+ \longrightarrow Co^{2+} + 5NH_4^+ + Cl^-$

18. For which of the following molecule significant $\mu \neq 0$?

(a) 
(b) 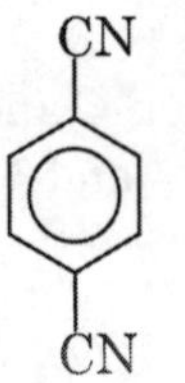
(c) 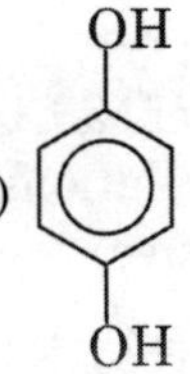
(d) 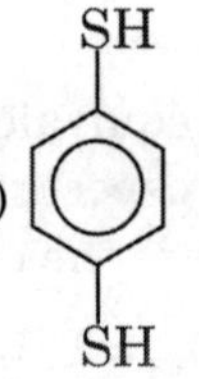

(A) Only (c) (B) (c) and (d) (C) Only (a) (D) (a) and (b)

19. For the non-stoichiometric reaction, 2A + B $\longrightarrow$ C + D, the following kinetic data were obtained in three separate experiments, all at 298 K.

Initial concentration (A)	Initial concentration (B)	Initial rate of formation of C(mol L^{-1} S^{-1})
0.1*M*	0.1*M*	1.2×10^{-3}
0.1*M*	0.2*M*	1.2×10^{-3}
0.2*M*	0.1*M*	2.4×10^{-3}

The rate law for the formation of C is

(A) $\frac{dc}{dt} = k[A][B]^2$ (B) $\frac{dc}{dt} = k[A]$ (C) $\frac{dc}{dt} = k[A][B]$ (D) $\frac{dc}{dt} = k[A]^2[B]$

20. Which series of reactions correctly represents chemical reactions related to iron and its compound?

(A) $Fe \xrightarrow{Cl_2,\text{heat}} FeCl_3 \xrightarrow{\text{heat, air}} FeCl_2 \xrightarrow{Zn} Fe$

(B) $Fe \xrightarrow{O_2,\text{heat}} Fe_3O_4 \xrightarrow{\text{CO, 600°C}} FeO \xrightarrow{\text{CO, 700°C}} Fe$

(C) $Fe \xrightarrow{\text{dil. } H_2SO_4} FeSO_4 \xrightarrow{H_2SO_4,\ O_2} Fe_2(SO_4)_3 \xrightarrow{\text{heat}} Fe$

(D) $Fe \xrightarrow{O_2,\text{heat}} FeO \xrightarrow{\text{dil. } H_2SO_4} FeSO_4 \xrightarrow{\text{heat}} Fe$

21. Considering the basic strength of amines in aqueous solution, which one has the smallest pK_b value?

(A) $(CH_3)_3N$ (B) $C_6H_5NH_2$ (C) $(CH_3)_2NH$ (D) CH_3NH_2

22. Which one of the following bases is not present in DNA?

(A) Cytosine (B) Thymine (C) Quinoline (D) Adenine

23. The correct set of four quantum numbers for the valence elections of rubidium atom ($Z = 37$) is

(A) $5, 1, 1, +\frac{1}{2}$ (B) $5, 0, 1, +\frac{1}{2}$ (C) $5, 0, 0, +\frac{1}{2}$ (D) $5, 1, 0, +\frac{1}{2}$

24. The major organic compound formed by the reaction of 1,1,1-trichloroethane with silver powder is

(A) 2-Butyne (B) 2-Butene (C) Acetylene (D) Ethene

25. Given below are the half-cell reactions:

$Mn^{2+} + 2e^- \longrightarrow Mn;\ E^0 = -1.18$ V

$2(Mn^{3+} + e^- \longrightarrow Mn^{2+});\ E^0 = +1.51$ V

The E^0 for $3Mn^{2+} \longrightarrow Mn + 2Mn^{3+}$ will be

(A) – 0.33 V; the reaction will not occur (B) – 0.33 V; the reaction will occur

(C) – 2.69 V; the reaction will not occur (D) – 2.69 V; the reaction will occur

26. The ratio of masses of oxygen and nitrogen in a particular gaseous mixture is 1 : 4. The ratio of number of their molecule is

(A) 1 : 8 (B) 3 : 16 (C) 1 : 4 (D) 7 : 32

27. Which one is classified as a condensation polymer?

(A) Teflon (B) Acrylonitrile (C) Dacron (D) Neoprene

28. Among the following oxoacids, the correct decreasing order of acid strength is

(A) $HClO_4 > HClO_3 > HClO_2 > HOCl$ (B) $HClO_2 > HClO_4 > HClO_3 > HOCl$

(C) $HOCl > HClO_2 > HClO_3 > HClO_4$ (D) $HClO_4 > HOCl > HClO_2 > HClO_3$

29. For complete combustion of ethanol, $C_2H_5OH(\ell) + 3O_2(g) \longrightarrow 2CO_2(g) + 3H_2O(\ell)$, the amount of heat produced as measured in bomb calorimeter is 1364.47 kJ mol^{-1} at 25°C. Assuming ideality the enthalpy of combustion, $\Delta_C H$, for the reaction will be: ($R = 8.314$ kJ mol^{-1})

(A) – 1460.50 kJ mol^{-1} (B) – 1350.50 kJ mol^{-1}

(C) – 1366.95 kJ mol^{-1} (D) – 1361.95 kJ mol^{-1}

30. The most suitable reagent for the conversion of $R—CH_2—OH \longrightarrow R—CHO$ is

(A) CrO_3 (B) PCC (Pyridinium chlorochromate)

(C) $KMnO_4$ (D) $K_2Cr_2O_7$

ANSWER KEYS

1. C	2. D	3. D	4. C	5. A	6. C
7. B	8. D	9. D	10. D	11. A	12. D
13. A	14. C	15. B	16. D	17. D	18. B
19. B	20. B	21. C	22. C	23. C	24. A
25. C	26. D	27. C	28. A	29. C	30. B

SOLUTIONS

1. (C)

NO is paramagnetic in gaseous state due to the presence of unpaired electron in its structure.

2. (D)

$\left(P + \frac{n^2a}{V^2}\right)(V - nb) = nRT$. For 1 mole, $\left(P + \frac{a}{V^2}\right)(V - b) = RT$

$$PV = RT + Pb - \frac{a}{V} + \frac{ab}{V^2}$$

At low pressure, terms Pb and $\frac{ab}{V^2}$ will be negligible as compared to RT.

Hence, $PV = RT - \frac{a}{V}$

$$Z = 1 - \frac{a}{RTV}$$

3. (D)

During the electrolysis of aqueous solution of s-block elements, H_2 gas is obtained at cathode.

4. (C)

$50 = \frac{1}{K} \times \frac{\ell}{A} \quad \Rightarrow \quad 50 = \frac{1}{1.4} \times \frac{\ell}{A}$

$\frac{\ell}{A} = 70\ \text{m}^{-1} \quad \Rightarrow \quad 280 = \frac{1}{K} \times 70$

$K = \frac{1}{4}\ \text{S m}^{-1}$

$\Lambda_m = \frac{1}{4} \times \left(\frac{1000}{M}\right)(10^{-2}\ \text{m})^3$

$= \frac{1}{4} \times \frac{1000}{0.5} \times 10^{-6} = 500 \times 10^{-6} = 5 \times 10^{-4}\ \text{S m}^2\ \text{mol}^{-1}$

5. (A)

In CsCl structure, Cs^+ ion is in contact with Cl^- ion at the nearest distance which is equal to $\sqrt{3}\frac{a}{2}$.

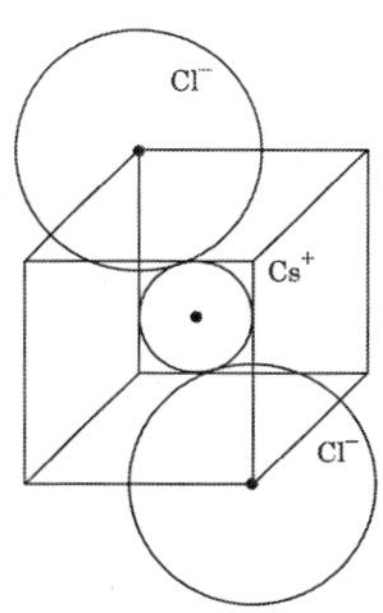

6. (C)

For C_2H_5OH, $\pi = 1 \times 0.5 \times RT$

For KBr, $\pi = 2 \times 0.25 \times RT$

For $Mg_3(PO_4)_2$, $\pi = 5 \times 0.1 \times RT$

For Na_3PO_4, $\pi = 4 \times 0.125 \times RT$

Hence, all are isotonic solutions.

7. (B)

A reducing agent loses electrons during a redox reaction.

Hence, (b, d) is correct.

8. (D)

$$\text{Rate of } S_N2 \text{ reaction} \propto \frac{1}{\text{steric over– crowding in transition state}}$$

9. (D)

Strong field ligands cause higher magnitude of crystal field splitting which is accompanied by the absorption of higher energy radiation.

$\xrightarrow{\text{VIBGYOR}}$ decreasing energy

10. (D)

$$\% \text{ of N} = \frac{1.4 \times \text{milliequivalents of acid consumed}}{\text{mass of organic compound}}$$

$$\text{Meq of acid consumed} = \left(60 \times \frac{1}{10} \times 2\right) - \left(20 \times \frac{1}{10} \times 1\right) = 10$$

$$\text{Hence, } \% \text{ of N} = \frac{1.4 \times 10}{1.4} = 10\%$$

11. (A)

According to Debye Huckel's theory for a strong electrolyte, $\lambda_C = \lambda_\infty - B\sqrt{C}$

12. (D)

For the reaction,

$$SO_2(g) + \frac{1}{2} O_2(g) \rightleftharpoons SO_3(g)$$

$$\Delta n_g = -\frac{1}{2} = x$$

13. (A)

$$CH_3COOH \xrightarrow{LiAlH_4} CH_3CH_2OH \xrightarrow{PCl_5} CH_3CH_2Cl$$

$$CH_3CH_2Cl \xrightarrow{\text{alc. KOH}} H_2C{=}CH_2$$

(Ethylene)

14. (C)

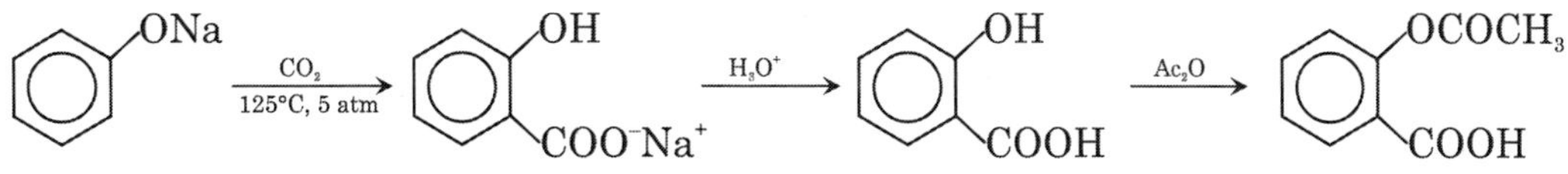

(Aspirin)

15. (B)

$$RNH_2 + CHCl_3 + \underset{\text{(alc.)}}{3KOH} \longrightarrow \underset{\text{(Alkyl isocyanide)}}{RNC\uparrow} + 3KCl + 3H_2O$$

16. (D)

$CsI_3 \longrightarrow Cs^+ + I_3^-$

Cs cannot show +3 oxidation state.

I_2 molecules are too large to be accommodated in lattice.

17. (D)

Equation–1 is not balanced w.r.t. charge.

Equation–2 gives $K_3[Cu(CN)_4]$ as product.

Equation–3 reaction is unfavourable in the forward direction (K_2O is unstable, while Li_2O is stable).

Equation–4 is correct and balanced.

18. (B)

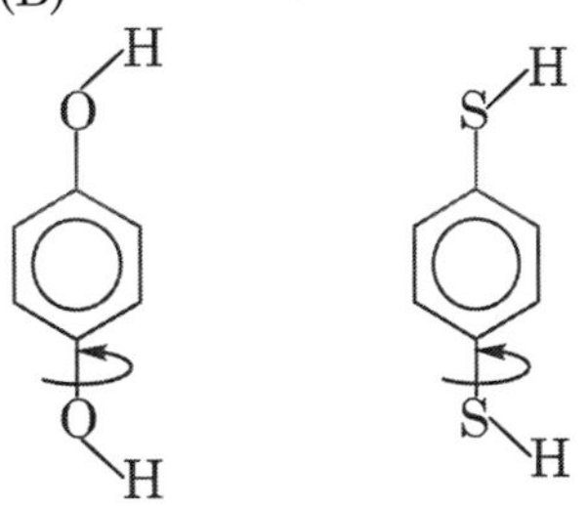

Due to infinite possible conformations in the above cases (of which only one has zero μ); a weighted μ will finally exist.

19. (B)

$R = k[A]^x [B]^y$

$1.2 \times 10^{-3} = k[0.1]^x [0.1]^y \quad \Rightarrow \quad 1.2 \times 10^{-3} = k[0.1]^x [0.2]^y$

$2.4 \times 10^{-3} = k[0.2]^x [0.1]^y$

Solving gives $x = 1$, $y = 0$. $R = k[A]$

20. (B)

In Eq. (A) $FeCl_3$ cannot be reduced when heated in air.

In Eq. (C) $Fe_2(SO_4)_3$ cannot convert to Fe on heating; instead oxide(s) will be formed.

In Eq. (D) $FeSO_4$ cannot be converted to Fe on heating; instead oxide(s) will be formed.

Hence, Eq. (B) is correct.

21. (C)

Aliphatic amines are more basic than aromatic amines.

$(CH_3)_2NH > CH_3NH_2 > (CH_3)_3N$ (among aliphatic amines in water).

22. (C)

Adenine, thymine, cytosine, guanine are bases present in DNA.

Quinoline being an aromatic compound is not present in DNA.

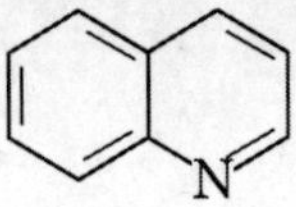

(Quinoline)

23. (C)

${}_{37}Rb = [Kr]\ 5s^1$

$n = 5,\ l = 0,\ m_l = 0,\ s = +\frac{1}{2}$

24. (A)

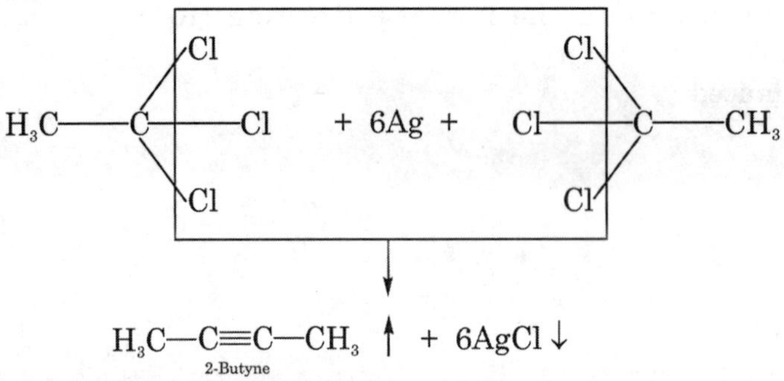

25. (C)

$$Mn^{2+} + 2e^- \longrightarrow Mn \quad E^0 = -1.18\ V$$
$$2Mn^{2+} \longrightarrow 2Mn^{3+} + 2e^- \quad E^0 = -1.51\ V$$
$$3Mn^{2+} \longrightarrow Mn + 2Mn^{3+} \quad E^0 = SOP + SRP$$
$$= -1.18 + (-1.51) = -2.69\ V$$

Negative EMF reflects non-spontaneous cell reaction.

26. (D)

Moles of $O_2 = \frac{w}{32}$ and moles of $N_2 = \frac{4w}{28}$

$$\frac{n_{O_2}}{n_{N_2}} = \frac{w}{32} \times \frac{28}{4w} = \frac{7}{32}$$

27. (C)

Teflon, acrylonitrile and neoprene are addition polymers while dacron is a condensation polymer.

28. (A)

$HClO < HClO_2 < HClO_3 < HClO_4$

Increasing acid strength is due to increase in oxidation state of central atom.

29. (C)

$$C_2H_5OH(\ell) + 3O_2(g) \longrightarrow 2CO_2(g) + 3H_2O(\ell)$$

$$\Delta E = -1364.47 \text{ kJ mol}^{-1}$$

$$\Delta H = ?$$

$$T = 298 \text{ K}$$

$$\Delta n_g = -1$$

$$\Delta H = \Delta E + \Delta n_g RT$$

Hence, $\Delta H = -1366.95 \text{ kJ mol}^{-1}$

30. (B)

$$RCH_2OH \xrightarrow{\text{PCC}} RCHO$$

I.I.T. JEE Objective Question Paper—2014 (Advanced)

PAPER—1

SECTION–I

(One or More Than One Options Correct Type)

This section contains 10 multiple choice questions. Each question has four choices (A), (B), (C) and (D) out of which one or more than one are correct.

1. The correct combination of names for isomeric alcohols with molecular formula $C_4H_{10}O$ is/are

(A) *tert*-butanol and 2-methylpropan-2-ol
(B) *tert*-butanol and 1,1-dimethylethan-1-ol
(C) *n*-butanol and butan-1-ol
(D) isobutyl alcohol and 2-methylpropan-1-ol

2. An ideal gas in a thermally insulated vessel at internal pressure = P_1, volume = V_1 and absolute temperature = T_1 expands irreversibly against zero external pressure, as shown in the diagram. The final internal pressure, volume and absolute temperature of the gas are P_2, V_2 and T_2, respectively. For this expansion,

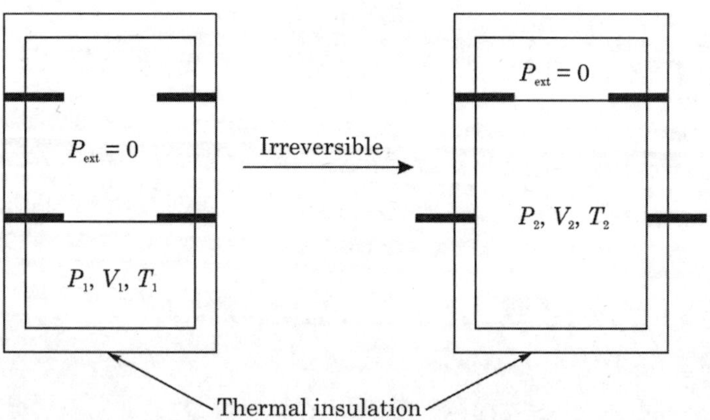

(A) $q = 0$
(B) $T_2 = T_1$
(C) $P_2V_2 = P_1V_1$
(D) $P_2V_2^{\gamma} = P_1V_1^{\gamma}$

3. Hydrogen bonding plays a central role in the following phenomena

(A) Ice floats in water.
(B) Higher Lewis basicity of primary amines than tertiary amines in aqueous solutions.
(C) Formic acid is more acidic than acetic acid.
(D) Dimerisation of acetic acid in benzene.

4. In a galvanic cell, the salt bridge
   (A) does not participate chemically in the cell reaction.
   (B) stops the diffusion of ions from one electrode to another.
   (C) is necessary for the occurrence of the cell reaction.
   (D) ensures mixing of the two electrolytic solutions.

5. For the reaction, $I^- + ClO_3^- + H_2SO_4 \longrightarrow Cl^- + HSO_4^- + I_2$
   The correct statement(s) in the balanced equation is/are
   (A) Stoichiometric coefficient of HSO_4^- is 6. (B) Iodide is oxidized.
   (C) Sulphur is reduced. (D) H_2O is one of the products.

6. The reactivity of compound Z with different halogens under appropriate conditions is given below:

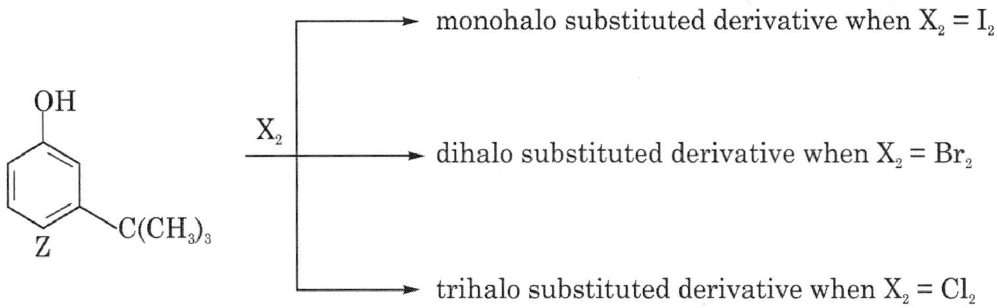

   The observed pattern of electrophilic substitution can be explained by
   (A) the steric effect of the halogen
   (B) the steric effect of the *tert*-butyl group
   (C) the electronic effect of the phenolic group
   (D) the electronic effect of the *tert*-butyl group

7. The correct statement(s) for orthoboric acid is/are
   (A) It behaves as a weak acid in water due to self-ionization.
   (B) Acidity of its aqueous solution increases upon addition of ethylene glycol.
   (C) It has a three dimensional structure due to hydrogen bonding.
   (D) It is a weak electrolyte in water.

8. Upon heating with Cu_2S, the reagent(s) that give copper metal is/are
   (A) $CuFeS_2$ (B) CuO (C) Cu_2O (D) $CuSO_4$

9. The pair(s) of reagents that yield paramagnetic species is/are
   (A) Na and excess of NH_3 (B) K and excess of O_2
   (C) Cu and dilute HNO_3 (D) O_2 and 2-ethylanthraquinol

10. In the reaction shown below, the major product(s) formed is/are

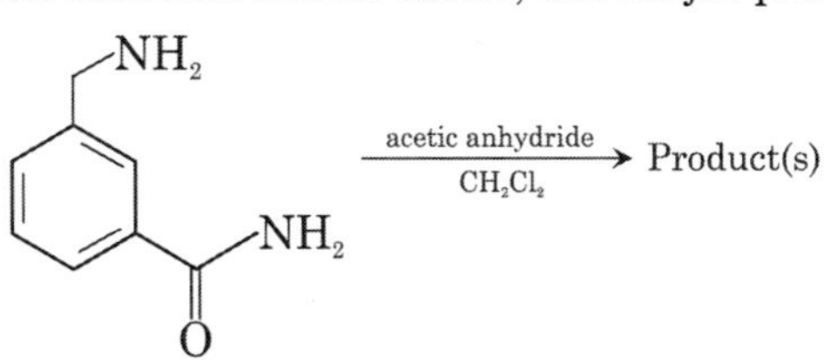

(A) 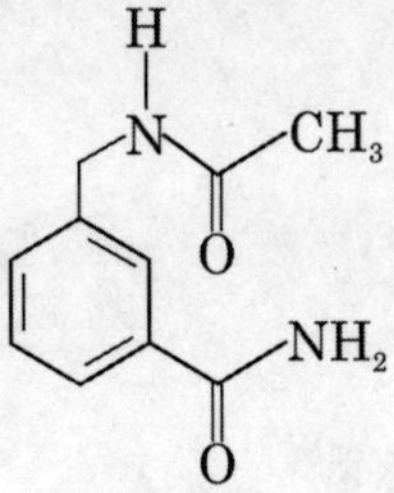 + CH_3COOH

(B) 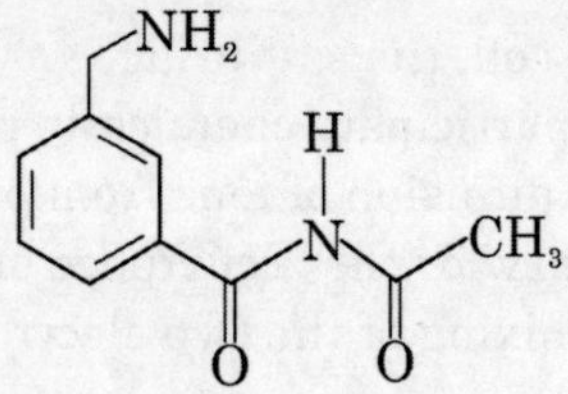 + CH_3COOH

(C) 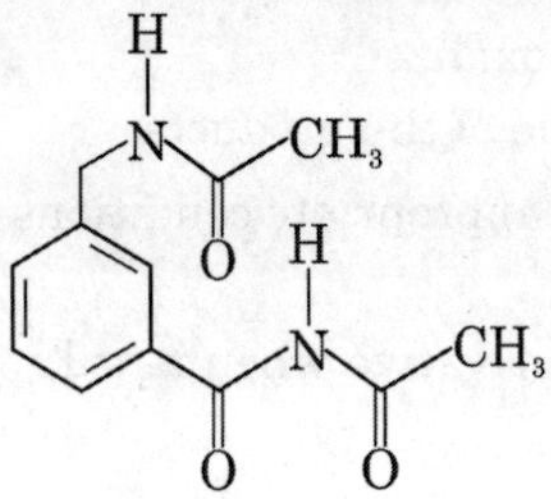 + H_2O

(D) 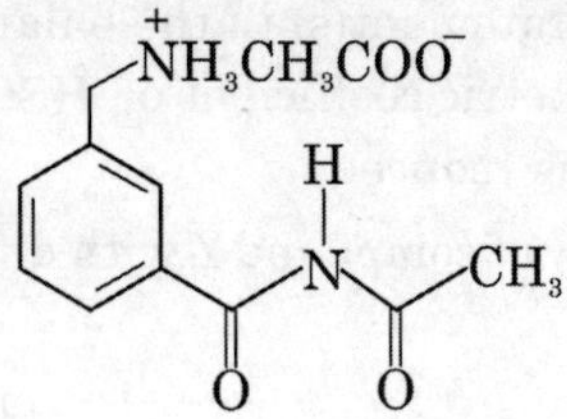

SECTION–II
(Only Integer Value Correct Type)

This section contains 10 questions. Each question, when worked out will result in one integer from 0 to 9 (both inclusive).

11. Among PbS, CuS, HgS, MnS, Ag_2S, NiS, CoS, Bi_2S_3 and SnS_2, the total number of black coloured sulphides is

12. The total number(s) of stable conformers with non-zero dipole moment for the following compound is(are)

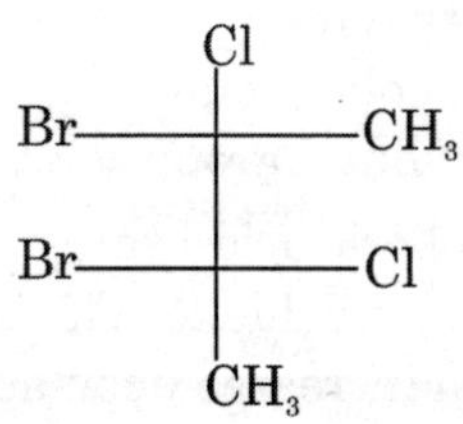

13. Consider the following list of reagents:
Acidified $K_2Cr_2O_7$, alkaline $KMnO_4$, $CuSO_4$, H_2O_2, Cl_2, O_3, $FeCl_3$, HNO_3 and $Na_2S_2O_3$. The total number of reagents that can oxidise aqueous iodide to iodine is

14. A list of species having the formula XZ_4 is given below.

XeF_4, SF_4, SiF_4, BF_4^-, BrF_4^-, $[Cu(NH_3)_4]^{2+}$, $[FeCl_4]^{2-}$, $[CoCl_4]^{2-}$ and $[PtCl_4]^{2-}$.

Defining shape on the basis of the location of X and Z atoms, the total number of species having a square planar shape is

15. Consider all possible isomeric ketones, including stereoisomers of MW = 100. All these isomers are independently reacted with $NaBH_4$ (Note: stereoisomers are also reacted separately). The total number of ketones that give a racemic product(s) is/are

16. In an atom, the total number of electrons having quantum numbers $n = 4$, $|m_l| = 1$ and $m_s = -1/2$ is

17. If the value of Avogadro number is 6.023×10^{23} mol^{-1} and the value of Boltzmann constant is 1.380×10^{-23} J K^{-1}, then the number of significant digits in the calculated value of the universal gas constant is

18. MX_2 dissociates to M^{2+} and X^- ions in an aqueous solution, with a degree of dissociation (α) of 0.5. The ratio of the observed depression of freezing point of the aqueous solution to the value of the depression of freezing point in the absence of ionic dissociation is

19. The total number of *distinct naturally occurring amino acids* obtained by complete acidic hydrolysis of the peptide shown below is

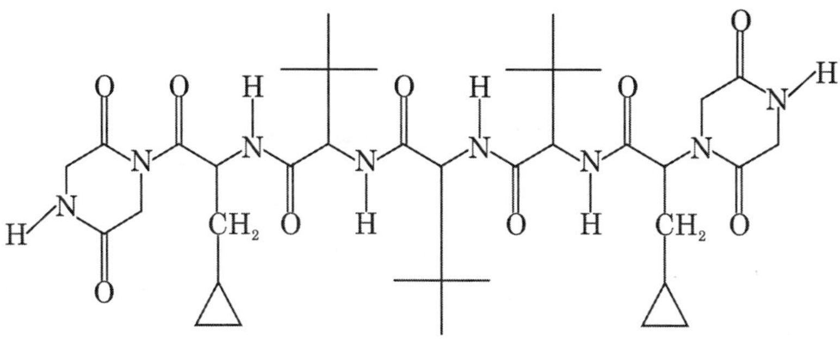

20. A compound H_2X with molar weight of 80 g is dissolved in a solvent having density of 0.4 g mL^{-1}. Assuming no change in volume upon dissolution, the molality of a 3.2 molar solution is

ANSWER KEYS

1. A, C, D	2. A, B, C	3. A, B, D	4. A	5. A, B, D	6. A, B, C
7. B, D	8. B, C, D	9. A, B, C	10. A	11. 6	12. 3
13. 7	14. 4	15. 5	16. 6	17. 4	18. 2
19. 1	20. 8				

SOLUTIONS

1. (A, C, D)

Isomeric alcohols of $C_4H_{10}O$ are

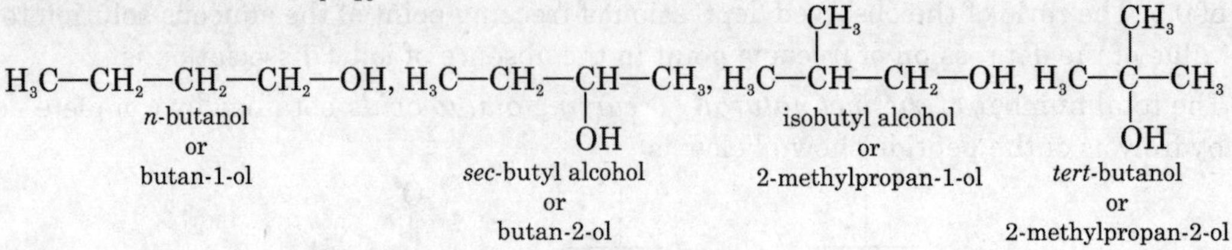

2. (A, B, C)

Since container is thermally insulated, $q = 0$, and it is a case of free expansion. Therefore, $W = 0$ and $\Delta E = 0$. Hence, $T_1 = T_2$. Also, $P_1V_1 = P_2V_2$

3. (A, B, D)

(A) Ice has cage-like structure in which each water molecule is surrounded by four other water molecules tetrahedrally through hydrogen bonding. Due to this density of ice is less than water and it floats on water.

(B) $R{-}NH_2 + H{-}OH \rightleftharpoons R{-}\overset{+}{N}H_3 + OH^-$ (I)

$(R)_3N + H{-}OH \rightleftharpoons (R)_3{-}\overset{+}{N}H + OH^-$ (II)

The cation (I) is more stabilized through hydrogen bonding than the cation (II). Hence, $R{-}NH_2$ is a better base than $(R)_3N$ in aqueous solution.

(C) HCOOH is a stronger acid than CH_3COOH due to the inductive effect and not due to hydrogen bonding.

(D) Acetic acid dimerizes in benzene through intermolecular hydrogen bonding.

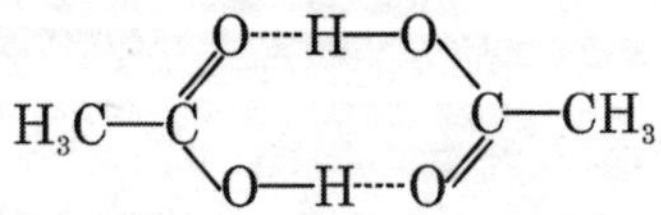

4. (A)

5. (A, B, D)

The balanced equation is: $ClO_3^- + 6I^- + 6H_2SO_4 \rightarrow 3I_2 + Cl^- + 6HSO_4^- + 3H_2O$

6. (A, B, C)

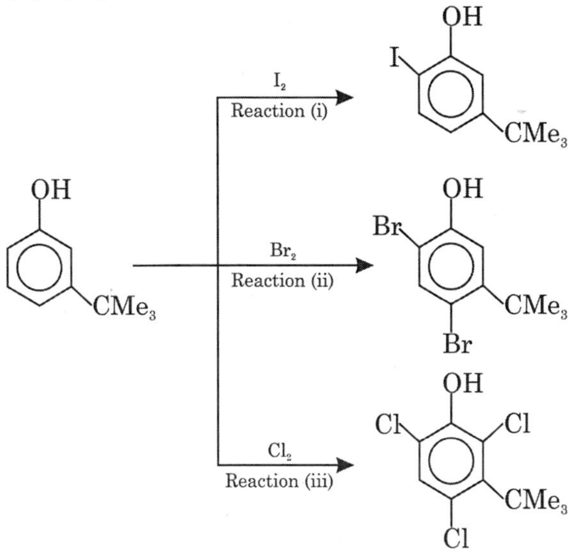

7. (B, D)

(A) H_3BO_3 is a weak monobasic Lewis acid.

$$H_3BO_3 + H—OH \rightleftharpoons B(OH)_4^- + H^+ \quad ...(i)$$

(B) Equilibrium (i) shifts in the forward direction by the addition of syn-diols like ethylene glycol which forms a stable complex with $B(OH)_4^-$.

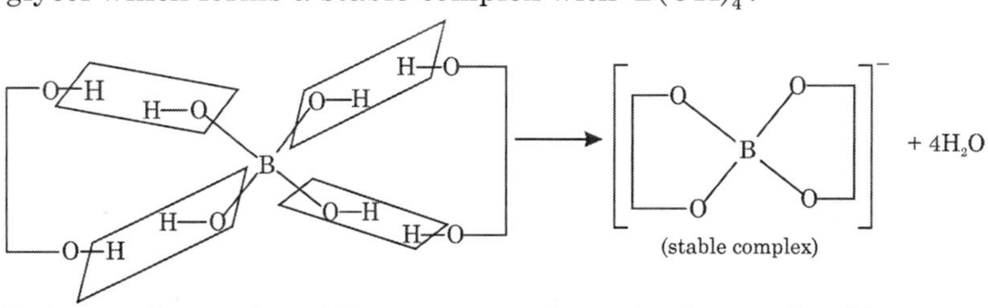

(C) It has a planar sheet like structure due to hydrogen bonding.

(D) H_3BO_3 is a weak electrolyte in water.

8. (B, C, D)

(A) $2CuFeS_2 + O_2 \xrightarrow{\Delta} Cu_2S + 2FeS + SO_2$

(B) $4CuO \xrightarrow{1100°C} 2Cu_2O + O_2$

$2Cu_2O + Cu_2S \xrightarrow{\Delta} 6Cu + SO_2$

(C) $Cu_2S + 2Cu_2O \xrightarrow{\Delta} 6Cu + SO_2$

(D) $CuSO_4 \xrightarrow{720°C} CuO + SO_2 + \frac{1}{2}O_2$

$4CuO \xrightarrow{1100°C} 2Cu_2O + O_2$

$2Cu_2O + Cu_2S \xrightarrow{\Delta} 6Cu + SO_2$

9. (A, B, C)

(A) Sodium (Na) when dissolved in excess liquid ammonia, forms a blue coloured paramagnetic solution.

(B) $K + O_2 \longrightarrow \underset{\text{(Potassium superoxide)}}{KO_2}$

KO_2 is paramagnetic.

(C) $3Cu + \underset{\text{(dilute)}}{8HNO_3} \longrightarrow 3Cu(NO_3)_2 + 2NO + 4H_2O$

NO is paramagnetic.

(D) 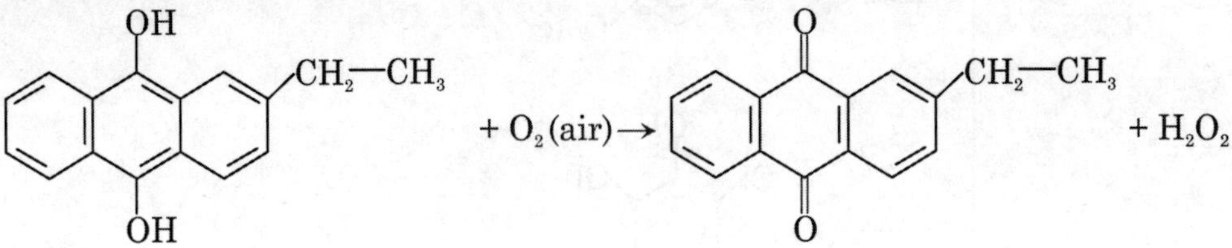

(2-Ethylanthraquinol)

H_2O_2 is diamagnetic

10. (A)

Only amines undergo acetylation and not acid amides.

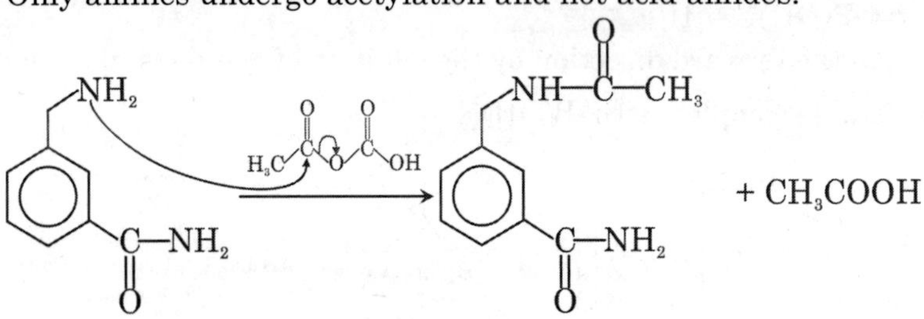

11. (6)

Black coloured sulphides are PbS, CuS, HgS, Ag_2S, NiS, CoS.

Bi_2S_3 in its crystalline form is dark brown but Bi_2S_3 precipitate obtained is black in colour.

12. (3)

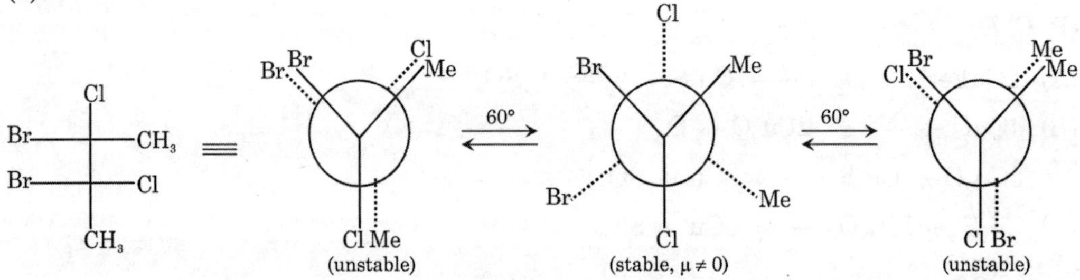

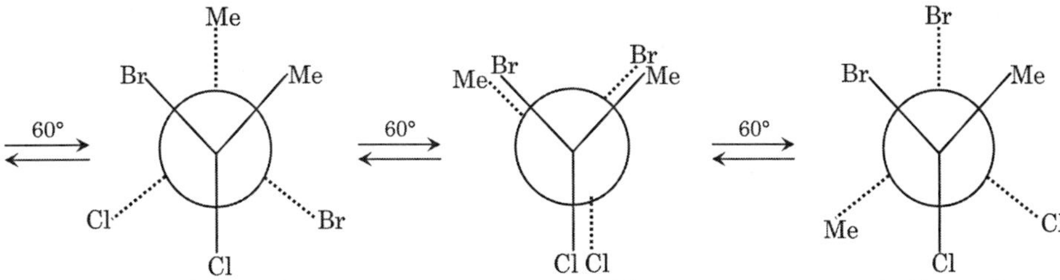

(stable, $\mu \neq 0$) (unstable) (stable, $\mu \neq 0$)

13. (7)

$K_2Cr_2O_7$, $CuSO_4$, H_2O_2, Cl_2, O_3, $FeCl_3$, HNO_3

$K_2Cr_2O_7, 7H_2SO_4 + 6KI \longrightarrow 4K_2SO_4 + Cr_2(SO_4)_3 + 3I_2 + 7H_2O$

$2CuSO_4 + 4KI \longrightarrow Cu_2I_2 + I_2 + 2K_2SO_4$

$H_2O_2 + 2KI \longrightarrow I_2 + 2KOH$

$Cl_2 + 2KI \longrightarrow 2KCl + I_2$

$O_3 + 2KI + H_2O \longrightarrow 2KOH + I_2 + O_2$

$2FeCl_3 + 2KI \longrightarrow 2FeCl_2 + I_2 + 2KCl$

$8HNO_3 + 6KI \longrightarrow 6KNO_3 + 2NO + 4H_2O + 3I_2$

$2KMnO_4 + KI + H_2O \longrightarrow KIO_3 + 2MnO_2 + 2KOH$

14. (4)

$XeF_4 \longrightarrow$ Square planar

$[Cu(NH_3)_4]^{2+} \longrightarrow$ Square planar

$SF_4 \longrightarrow$ See-saw

$BF_4^- \longrightarrow$ Tetrahedral

$[CoCl_4]^{2-} \longrightarrow$ Tetrahedral

$BrF_4^- \longrightarrow$ Square planar

$[PtCl_4]^{2-} \longrightarrow$ Square planar

$SiF_4 \longrightarrow$ Tetrahedral

$[FeCl_4]^{2-} \longrightarrow$ Tetrahedral

15. (5)

(1) 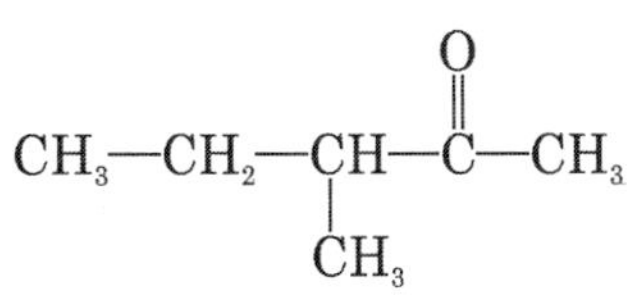 Will not give a racemic mixture on reduction with $NaBH_4$.

(2) 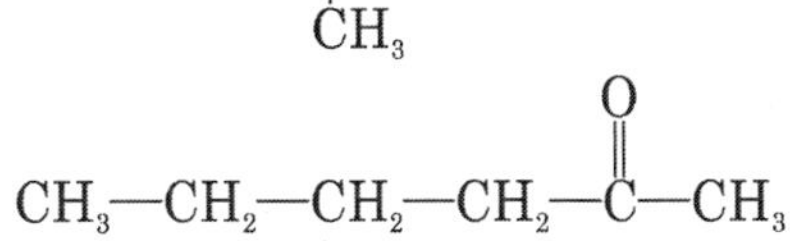 Will give a racemic mixture on reduction with $NaBH_4$.

(3) 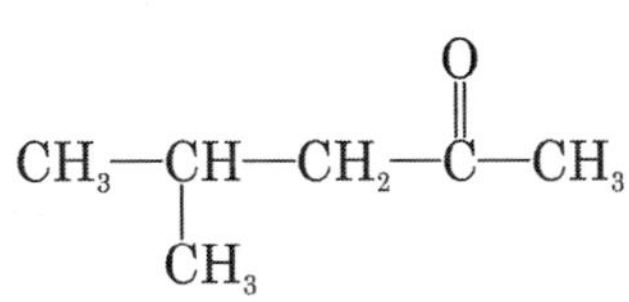 Will give a racemic mixture on reduction with $NaBH_4$.

(4) 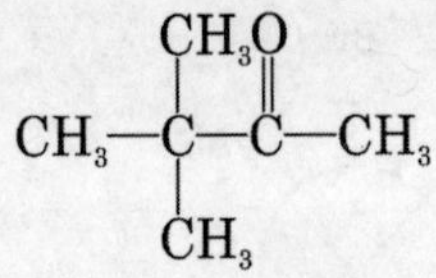

Will give a racemic mixture on reduction with $NaBH_4$.

(5) 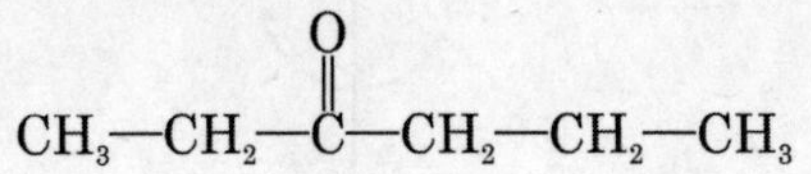

Will give a racemic mixture on reduction with $NaBH_4$.

(6) 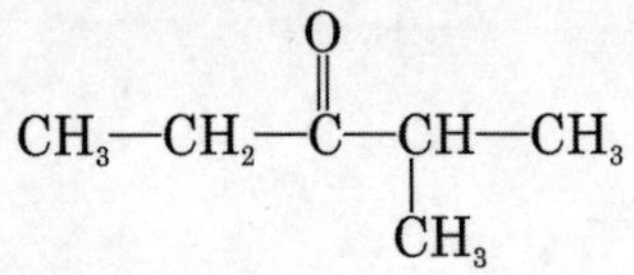

Will give a racemic mixture on reduction with $NaBH_4$.

16. (6)

$n = 4$

$\ell = 0, 1, 2, 3 \quad \Rightarrow \quad |m_\ell| = 1 \Rightarrow \pm 1$

$m_s = -\dfrac{1}{2}$

For $\ell = 0,\ m_\ell = 0$

$\ell = 1,\ m_\ell = -1, 0, +1$

$\ell = 2,\ m_\ell = -2, -1, 0, +1, +2$

$\ell = 3,\ m_\ell = -3, -2, -1, 0, +1, +2, +3$

Hence, six electrons can have $|m_\ell| = 1$ and $m_s = -\dfrac{1}{2}$.

17. (4)

$k = \dfrac{R}{N_A}$

$R = kN_A = 1.380 \times 10^{-23} \times 6.023 \times 10^{23} = 8.31174 \approx 8.312$

18. (2)

$MX_2 \rightleftharpoons M^{2+} + 2X^-$

$1 - \alpha \quad\quad \alpha \quad\quad 2\alpha$

$i = 1 + 2\alpha \quad\quad (\alpha = 0.5)$

$i = 2$

19. (1)

This peptide on complete hydrolysis produced 4 distinct amino acids which are given below:

(1) 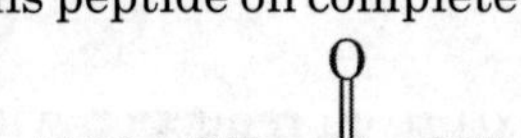

Glycine
(natural)

(2) 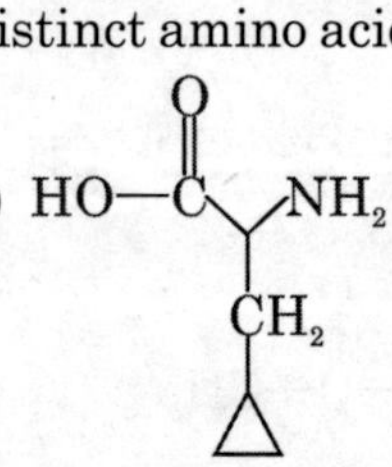

(3) 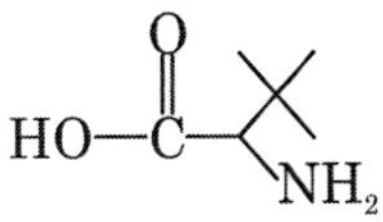

(4) 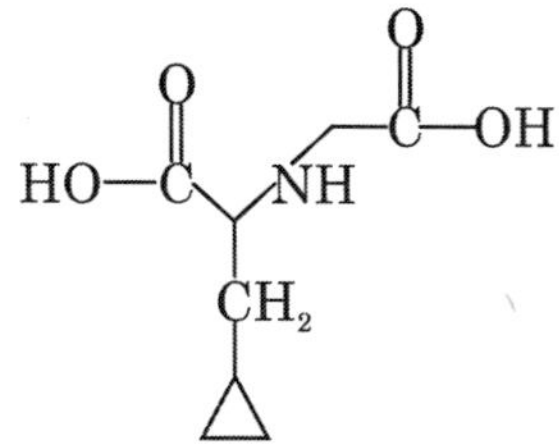

Only glycine is naturally occurring amino acid.

20. (8)

$V_{solution} \approx V_{solvent}$

Since in 1 L solution, 3.2 moles of solute are present, 1 L solution ≈ solvent (d = 0.4 g/mL) ≈ 0.4 kg

$$\text{Hence, molality } (m) = \frac{\text{moles of solute}}{\text{mass of solvent (kg)}} = \frac{3.2}{0.4} = 8$$

I.I.T. JEE Objective Question Paper—2014 (Advanced)

PAPER—2

SECTION - I

(Only One Option Correct Type)

This section contains 10 multiple choice questions. Each question has four choices (A), (B), (C) and (D) out of which only one is correct.

1. Assuming 2*s*-2*p* mixing is not operative, the paramagnetic species among the following is

(A) Be_2 (B) B_2 (C) C_2 (D) N_2

2. For the process, $H_2O(\ell) \longrightarrow H_2O(g)$

at $T = 100°C$ and 1 atmosphere pressure, the correct choice is

(A) $\Delta S_{system} > 0$ and $\Delta S_{surrounding} > 0$ (B) $\Delta S_{system} > 0$ and $\Delta S_{surrounding} < 0$

(C) $\Delta S_{system} < 0$ and $\Delta S_{surrounding} > 0$ (D) $\Delta S_{system} < 0$ and $\Delta S_{surrounding} < 0$

3. For the elementary reaction $M \longrightarrow N$, the rate of disappearance of M increases by a factor of 8 upon doubling the concentration of M. The order of the reaction with respect of M is

(A) 4 (B) 3 (C) 2 (D) 1

4. For the identification of β-naphthol using dye test, it is necessary to use

(A) dichloromethane solution of β-naphthol. (B) acidic solution of β-naphthol.

(C) neutral solution of β-naphthol. (D) alkaline solution of β-naphthol.

5. Isomers of hexane, based on their branching, can be divided into three distinct classes as shown in the figure.

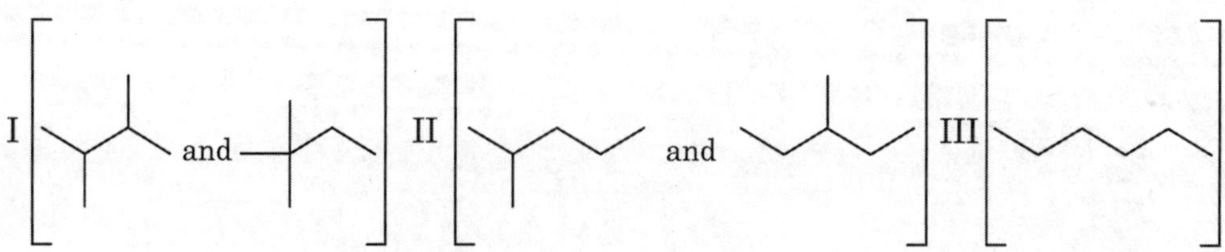

The correct order of their boiling point is

(A) I > II > III (B) III > II > I (C) II > III > I (D) III > I > II

6. The major product in the following reaction is

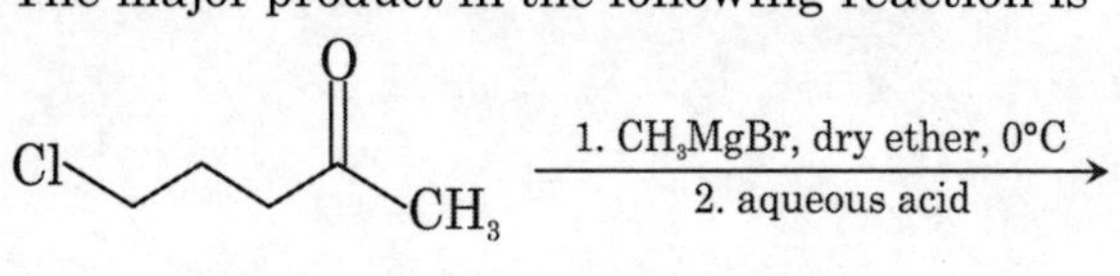

(A) 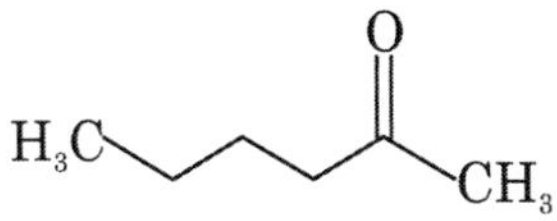

(B) 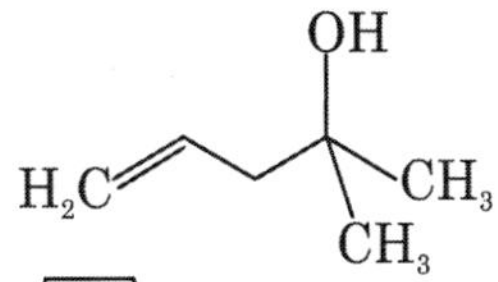

(C) 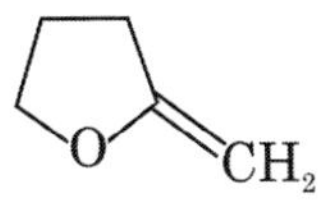

(D) 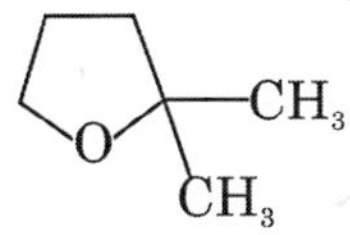

7. Under ambient conditions, the total number of gases released as products in the final step of the reaction scheme shown below is

$$XeF_6 \xrightarrow[\text{hydrolysis}]{\text{Complete}} P + \text{other product}$$

$$P \xrightarrow{OH^-/H_2O} Q \xrightarrow{\text{slow disproportionation in } OH^-/H_2O} \text{Products}$$

(A) 0 (B) 1 (C) 2 (D) 3

8. The product formed in the reaction of $SOCl_2$ with white phosphorus is

(A) PCl_3 (B) SO_2Cl_2 (C) SCl_2 (D) $POCl_3$

9. Hydrogen peroxide in its reaction with KIO_4 and NH_2OH, respectively, is acting as a

(A) reducing agent, oxidising agent
(B) reducing agent, reducing agent
(C) oxidising agent, oxidising agent
(D) oxidising agent, reducing agent.

10. The acidic hydrolysis of ether (X) shown below is fastest when

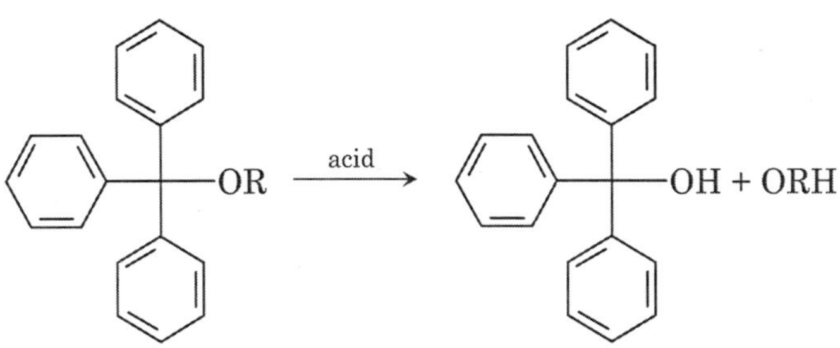

(A) one phenyl group is replaced by a methyl group.
(B) one phenyl group is replaced by a para-methoxyphenyl group.
(C) two phenyl group are replaced by two para-methoxyphenyl groups.
(D) no structural change is made to X.

SECTION – II Comprehension Type (Only One Option Correct)

This section contains three paragraphs, each describing theory, experiments, data, etc. Six questions relate to the three paragraphs with two questions on each paragraph. Each question has only one correct answer among the four given options (A), (B), (C) and (D).

Paragraph for Questions 11 and 12

X and Y are two volatile liquids with molar weights of 10 g mol^{-1} and 40 g mol^{-1}, respectively. Two cotton plugs, one soaked in X and the other soaked in Y, are simultaneously placed at the ends of a tube of length L = 24 cm, as shown in the figure. The tube is filled with an inert gas at 1 atmosphere pressure and a temperature of 300 K. Vapours to X and Y react to form a product which is first observed at a distance d cm from the plug soaked in X. Take X and Y to have equal molecular diameters and assume ideal behaviour for the inert gas and the two vapours.

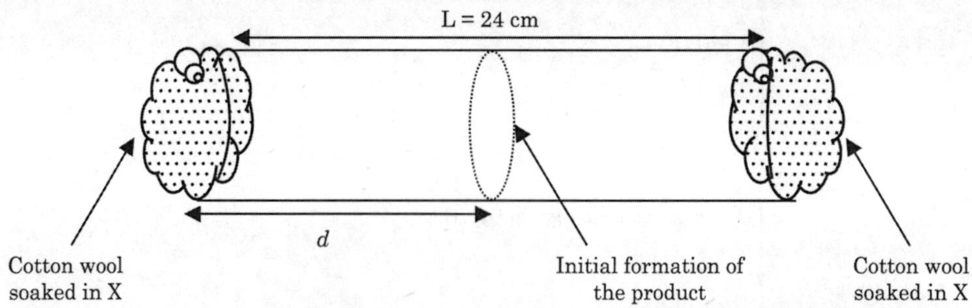

11. The value of d in cm (shown in the figure), as estimated from Graham's law, is
(A) 8 (B) 12 (C) 16 (D) 20

12. The experimental value of d is found to be smaller than the estimate obtained using Graham's law. This is due to
(A) larger mean free path for X as compared to that of Y.
(B) larger mean free path for Y as compared to that of X.
(C) increased collision frequency of Y with the inert gas as compared to that of X with the inert gas.
(D) increased collision frequency of X with the inert gas as compared to that of Y with the inert gas.

Paragraph for Questions 13 and 14

Schemes 1 and 2 describe sequential transformation of alkynes M and N. Consider only the major products formed in each step for both schemes.

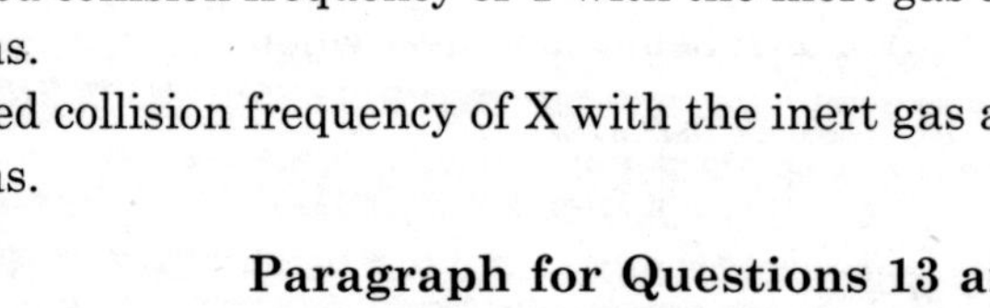

Scheme–1

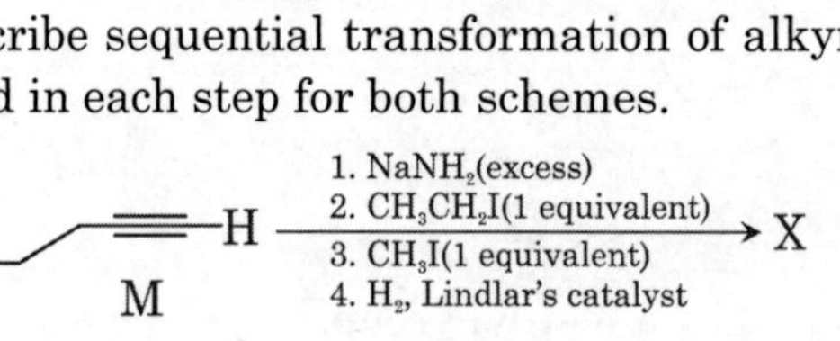

Scheme–2

13. The product X is

(A) 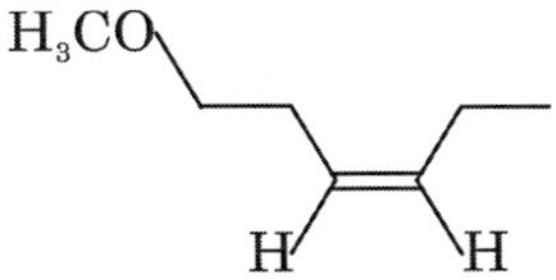

(B) 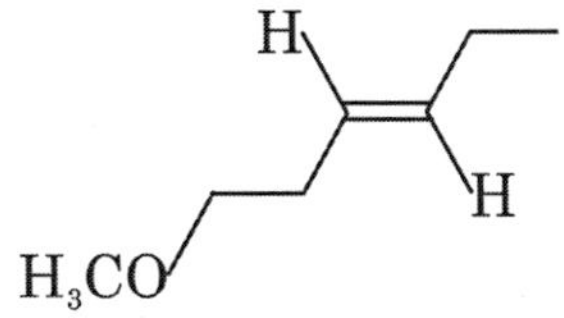

(C) 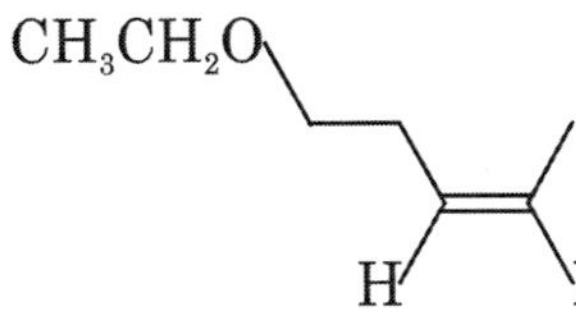

(D) 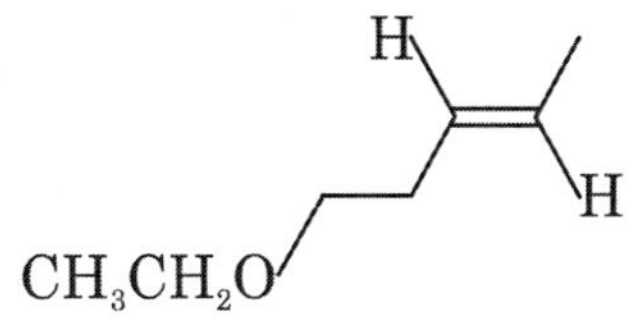

14. The correct statement with respect to product Y is

(A) It gives a positive Tollens test and is a functional isomer of X.
(B) It gives a positive Tollens test and is a geometrical isomer of X.
(C) It gives a positive iodoform test and is a functional isomer of X.
(D) It gives a positive iodoform test and is a geometrical isomer of X.

Paragraph for Questions 15 and 16

An aqueous solution of metal ion M1 reacts separately with reagents Q and R in excess to give tetrahedral and square planar complexes, respectively. An aqueous solution of another metal ion M2 always forms tetrahedral complexes with these reagents. Aqueous solution of M2 on reaction with reagent S gives white precipitate which dissolves in exess of S. The reactions are summarized in the scheme given below.

SCHEME:

$$\text{Tetrahedral} \xleftarrow[\text{excess}]{\text{Q}} \text{M1} \xrightarrow[\text{excess}]{\text{R}} \text{Square planar}$$

$$\text{Tetrahedral} \xleftarrow[\text{excess}]{\text{Q}} \text{M2} \xrightarrow[\text{excess}]{\text{R}} \text{Tetrahedral}$$

M2 $\downarrow$ S, stoichiometric amount

$$\text{White precipitate} \xrightarrow[\text{excess}]{\text{S}} \text{precipitate dissolves}$$

15. M1, Q and R, respectively are

(A) Zn^{2+}, KCN and HCl
(B) Ni^{2+}, HCl and KCN
(C) Cd^{2+}, KCN and HCl
(D) Co^{2+}, HCl and KCN

16. Reagent S is

(A) $K_4[Fe(CN)_6]$
(B) Na_2HPO_4
(C) K_2CrO_4
(D) KOH

SECTION – III (Matching List Type) (Only One Option Correct)

This section contains four questions, each having two matching lists. Choices for the correct combination of elements from List-I and List-II are given as options (A), (B), (C) and (D) out of which only one is correct.

17. Match each coordination compound in List-I with an appropriate pair of characteristics from List-II and select the correct answer using the code given below the lists
(en = $H_2NCH_2CH_2NH_2$; atomic numbers: Ti = 22, Cr = 24; Co = 27; Pt = 78)

List-I	List-II
P. $[Cr(NH_3)_4Cl_2]Cl$	1. Paramagnetic and exhibits ionisation isomerism
Q. $[Ti(H_2O)_5Cl](NO_3)_2$	2. Diamagnetic and exhibits *cis-trans* isomerism
R. $[Pt(en)(NH_3)Cl]NO_3$	3. Paramagnetic and exhibits *cis-trans* isomerism
S. $[Co(NH_3)_4(NO_3)_2]NO_3$	4. Diamagnetic and exhibits ionisation isomerism

Codes:	P	Q	R	S
(A)	4	2	3	1
(B)	3	1	4	2
(C)	2	1	3	4
(D)	1	3	4	2

18. Match the orbital overlap figures shown in List-I with the description given in List-II and select the correct answer using the code given below the lists.

List-I	List-II
P. 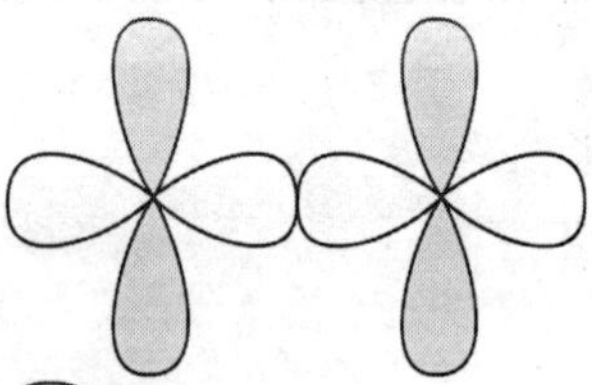	1. *p-d* π antibonding
Q. 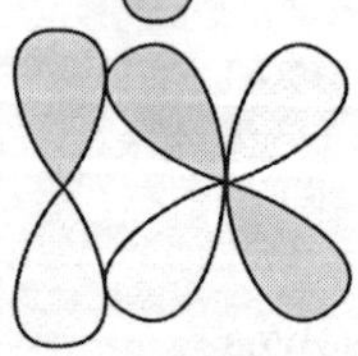	2. *d-d* σ bonding
R. 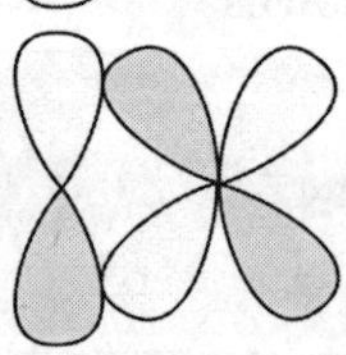	3. *p-d* π bonding
S. 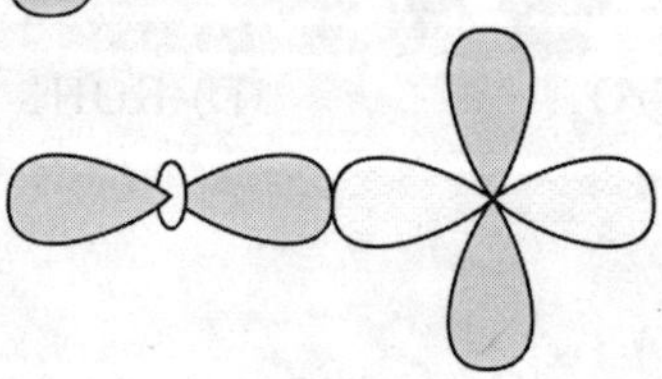	4. *d-d* σ antibonding

Codes:	P	Q	R	S
(A)	2	1	3	4
(B)	4	3	1	2
(C)	2	3	1	4
(D)	4	1	3	2

19. Different possible thermal decomposition pathways for peroxyesters are shown below. Match each pathway from List-I with an appropriate structure from List-II and select the correct answer using the code given below the lists.

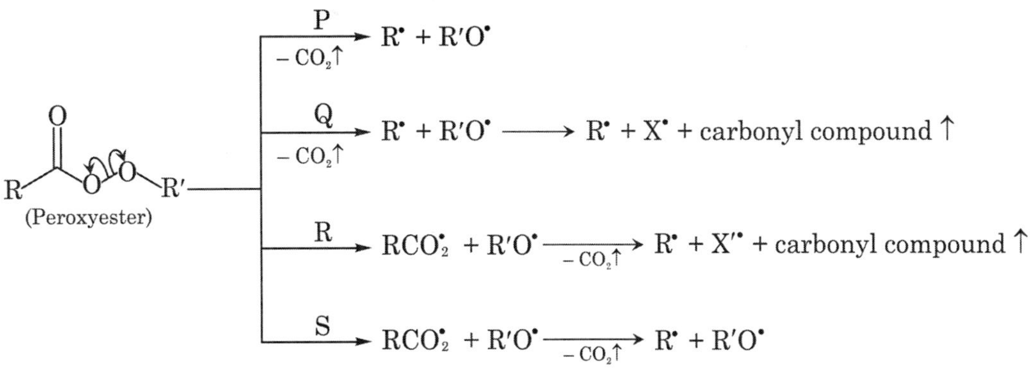

List-I	List-II
P. Pathway P	1. 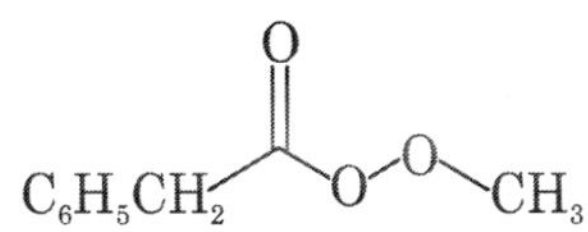
Q. Pathway Q	2. 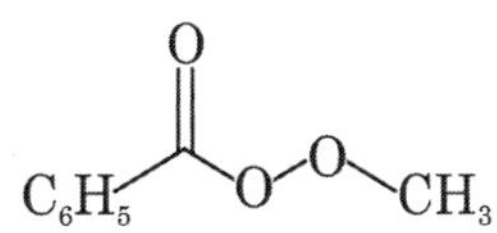
R. Pathway R	3. 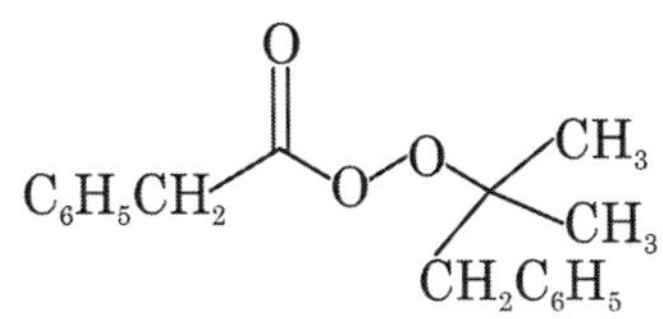
S. Pathway S	4. 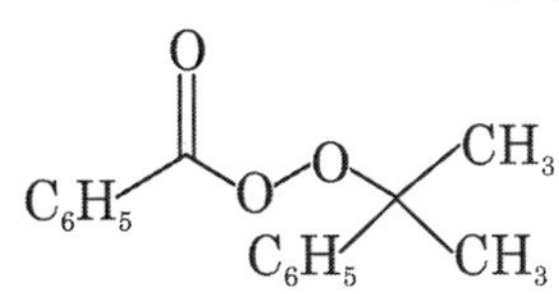

Codes:	P	Q	R	S
(A)	1	3	4	2
(B)	2	4	3	1
(C)	4	1	2	3
(D)	3	2	1	4

20. Match the four starting materials (P, Q, R, S) given in List-I with the corresponding reaction schemes (I, II, III, IV) provided in List-II and select the correct answer using the code given below the lists.

List-I

P. H—≡—H

Q. 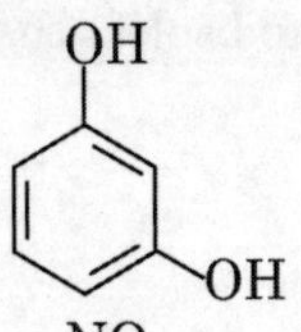

R. 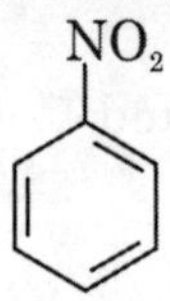

S. 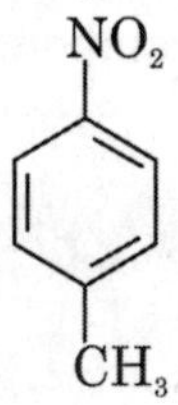

List-II

1. Scheme I

$$? \xrightarrow[\text{(ii) SOCl}_2\text{ (iv) NH}_3]{\text{(i) KMnO}_4\text{, HO}^-\text{, heat (ii) H}^+\text{, H}_2\text{O}} C_7H_6N_2O_3$$

2. Scheme II

$$? \xrightarrow[\text{(ii) HNO}_3\text{ (v) dil. H}_2\text{SO}_4\text{, heat (iv) HO}^-]{\text{(i) Sn/HCl (ii) CH}_3\text{COCl (iii) conc. H}_2\text{SO}_4} C_6H_6N_2O_2$$

3. Scheme III

$$? \xrightarrow[\text{(iii) H}_2\text{S, NH}_3\text{ (iv) NaNO}_2\text{, H}_2\text{SO}_4\text{ (v) hydrolysis}]{\text{(i) red hot iron, 873 K (ii) fuming HNO}_3\text{, H}_2\text{SO}_4\text{, heat}} C_6H_5NO_3$$

4. Scheme IV

$$? \xrightarrow[\text{(ii) conc. HNO}_3\text{, conc. H}_2\text{SO}_4\text{ (iii) dil. H}_2\text{SO}_4\text{, heat}]{\text{(i) conc. H}_2\text{SO}_4\text{, 60°C}} C_6H_5NO_4$$

Codes:	P	Q	R	S
(A)	1	4	2	3
(B)	3	1	4	2
(C)	3	4	2	1
(D)	4	1	3	2

ANSWER KEYS

1. C	2. B	3. B	4. D	5. B	6. D
7. C	8. A	9. A	10. C	11. C	12. D
13. A	14. C	15. B	16. D	17. B	18. C
19. A	20. C				

SOLUTIONS

1. (C)

Assuming that no 2*s*-2*p* mixing takes place

(A) $Be_2 \longrightarrow \sigma 1s^2, \sigma^* 1s^2, \sigma 2s^2, \sigma^* 2s^2$ (diamagnetic)

(B) $B_2 \longrightarrow \sigma 1s^2, \sigma^* 1s^2, \sigma 2s^2, \sigma^* 2s^2, \sigma 2p_z^2, \begin{matrix}\pi 2p_x^0 \\ \pi 2p_y^0\end{matrix}$ (diamagnetic)

(C) $C_2 \longrightarrow \sigma 1s^2, \sigma^* 1s^2, \sigma 2s^2, \sigma^* 2s^2, \sigma 2p_z^2, \begin{matrix}\pi 2p_x^1 \\ \pi 2p_y^1\end{matrix}, \begin{matrix}\pi^* 2p_x^0 \\ \pi^* 2p_y^0\end{matrix}, \sigma^* 2p_z^0$ (paramagnetic)

(D) $N_2 \longrightarrow \sigma 1s^2, \sigma^* 1s^2, \sigma 2s^2, \sigma^* 2s^2, \sigma 2p_z^2, \begin{matrix}\pi 2p_x^2 \\ \pi 2p_y^2\end{matrix}, \begin{matrix}\pi^* 2p_x^0 \\ \pi^* 2p_y^0\end{matrix}, \sigma^* 2p_z^0$ (diamagnetic)

2. (B)

At 100°C and 1 atmosphere pressure $H_2O(\ell) \rightleftharpoons H_2O(g)$ is at equilibrium. For equilibrium $\Delta S_{total} = 0$ and $\Delta S_{system} + \Delta S_{surrounding} = 0$

Hence, $\Delta S_{system} > 0$ and $\Delta S_{surrounding} < 0$

3. (B)

$$\frac{r_1}{r_2} = \frac{1}{8} = \frac{[M]^n}{[2M]^n} \quad \text{or } n = 3$$

4. (D)

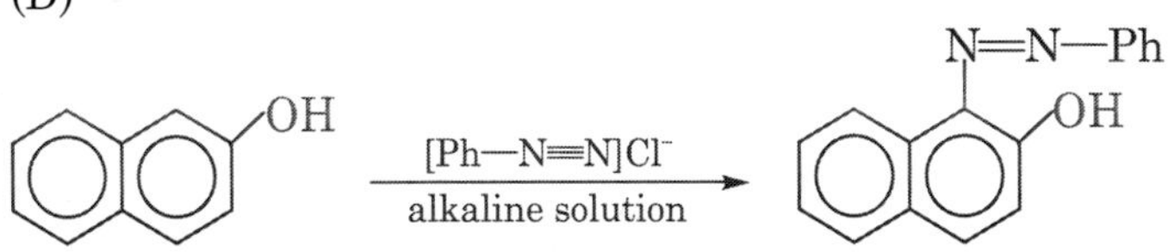

5. (B)

III > II > I

More the branching in an alkane, lesser will be the surface area and lesser will be the boiling point.

6. (D)

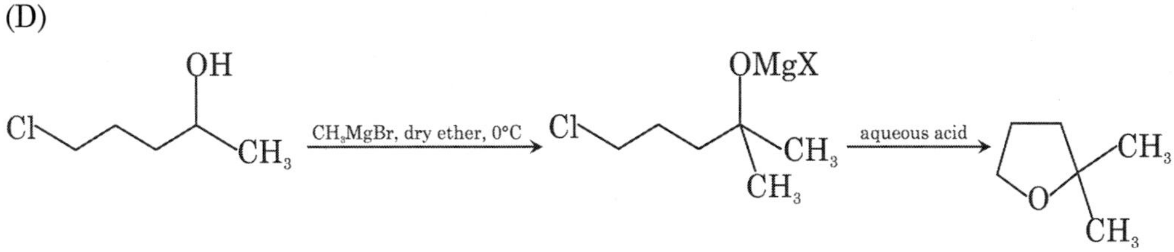

7. (C)

$$XeF_6 + 3H_2O \longrightarrow XeO_3 + 3H_2F_2$$

$$XeO_3 \xrightarrow{OH^-} HXeO_4^-$$

$$HXeO_4^- \xrightarrow{OH^-/H_2O \text{ (disproportionation)}} XeO_6^{4-}(g) + Xe(g) + H_2O(\ell) + O_2(g)$$

8. (A)

$$P_4(s) + 8SOCl(\ell) \longrightarrow 4PCl_3(\ell) + 4SO_2(g) + 2S_2Cl(g)$$

9. (A)

$$KIO_4 + H_2O_2 \longrightarrow KIO_3 + H_2O + O_2$$
$$NH_2OH + 3H_2O_2 \longrightarrow HNO_3 + 4H_2O$$

10. (C)

When two phenyl groups are replaced by two para methoxy group, carbocation formed will be more stable.

11. (C)

$$\frac{r_X}{r_Y} = \frac{d}{24-d} = \sqrt{\frac{40}{10}} = 2$$

or $d = 48 - 2d$ or $3d = 48$ or $d = 16$ cm

12. (D)

As the collision frequency increases then molecular speed decreases than the expected.

13. (A)

14. (C)

Solution for Q. No. 13 to 14

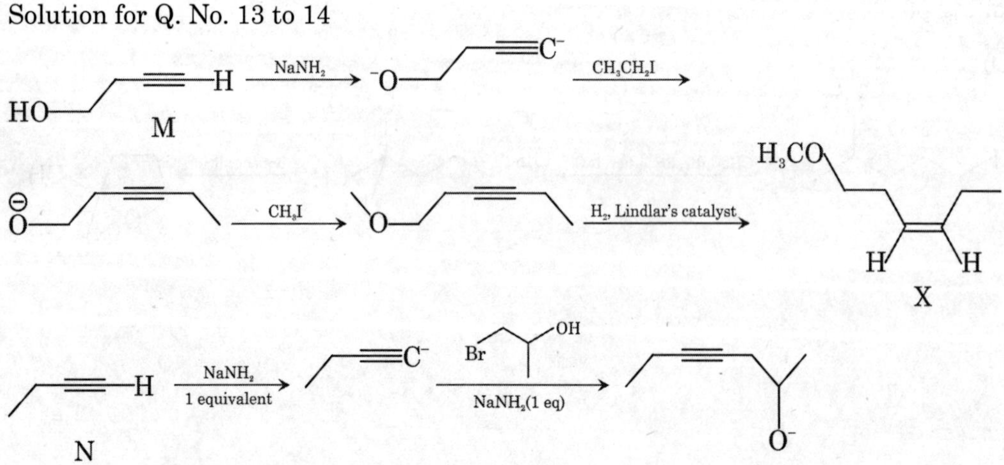

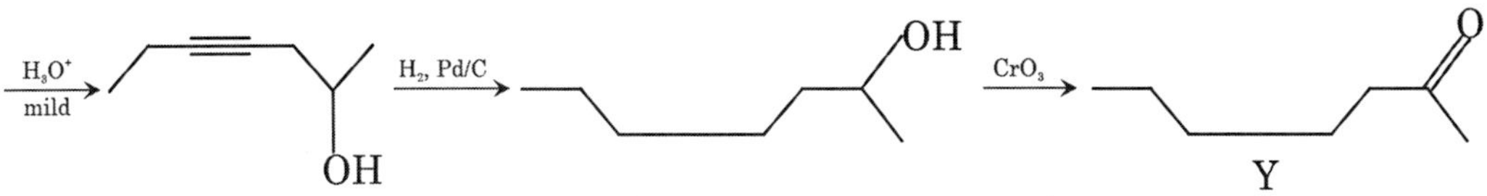

X and Y are functional isomers of each other and Y gives iodoform test.

15. (B)

16. (D)

Solution for the Q. No. 15 to 16.

$$\underset{\text{Tetrahedral}}{[NiCl_4]^{2-}} \xleftarrow[\text{excess}]{HCl(Q)} \underset{(M_1)}{Ni^{2+}} \xrightarrow[\text{excess}]{KCN(R)} \underset{\text{Square planar}}{[Ni(CN)_4]^{2-}}$$

$$\underset{\text{Tetrahedral}}{[ZnCl_4]^{2-}} \xleftarrow[\text{excess}]{HCl(Q)} \underset{(M_2)}{Zn^{2+}} \xrightarrow[\text{excess}]{KCN(R)} \underset{\text{Tetrahedral}}{[Zn(CN)_4]^{2-}}$$

$$Zn^{2+} \xrightarrow{\text{KOS (S)}} \underset{\text{White ppt.}}{Zn(OH)_2} \xrightarrow[\text{excess}]{KOH} \underset{\text{Soluble}}{[Zn(OH)_4]^{2-}}$$

17. (B)

(P) $[Cr(NH_3)_4Cl_2]Cl \longrightarrow$ Paramagnetic and exhibits *cis-trans* isomerism

(Q) $[Ti(H_2O)_5Cl](NO_3)_2 \longrightarrow$ Paramagnetic and exhibits ionization isomerism

(R) $[Pt(en)(NO_3)Cl]NO_3 \longrightarrow$ Diamagnetic and exhibits ionization isomerism

(S) $[Co(NH_3)_4(NO_3)_2]NO_3 \longrightarrow$ Diamagnetic and exhibits *cis-trans* isomerism

18. (C)

P. 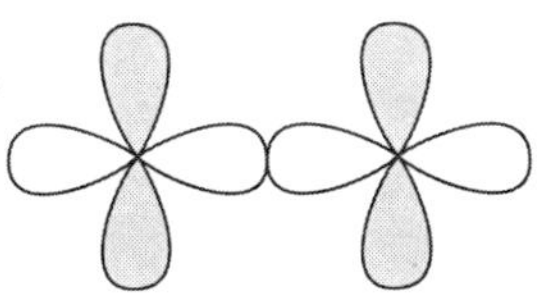 $\longrightarrow$ *d-d* σ bonding

Q. 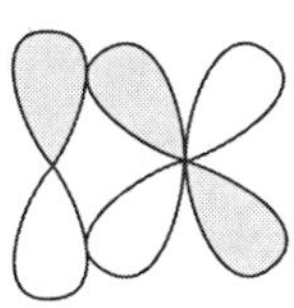 $\longrightarrow$ *p-d* π bonding

R. 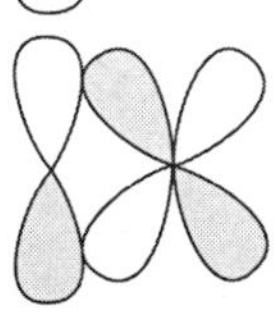 $\longrightarrow$ *p-d* π antibonding

S. 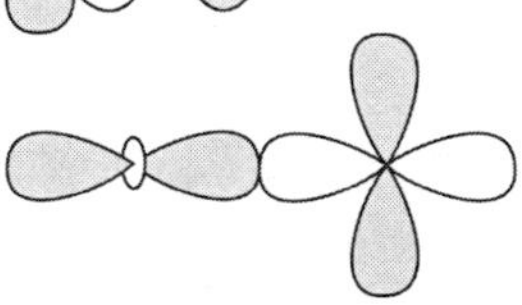 $\longrightarrow$ *d-d* σ antibonding

19. (A)

(P) – 1; (Q) – 3; (R) – 4; (S) – 2

(P)

(Q)

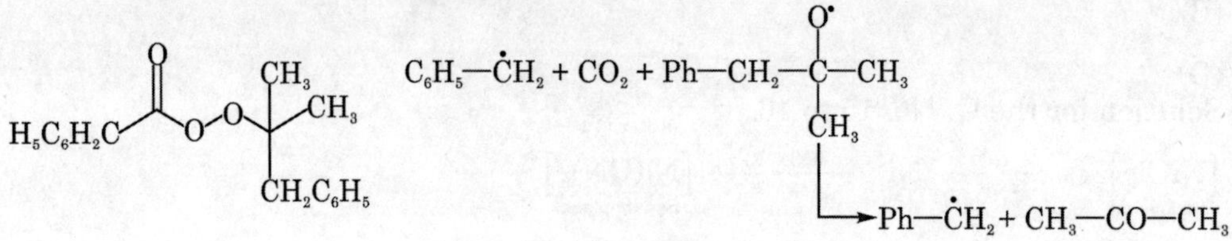

(R)

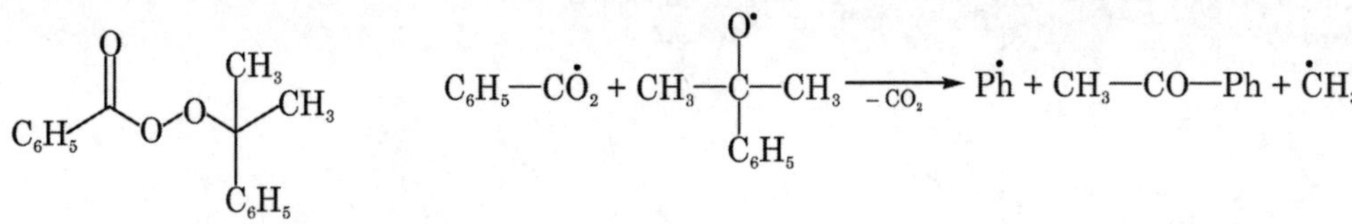

(S)

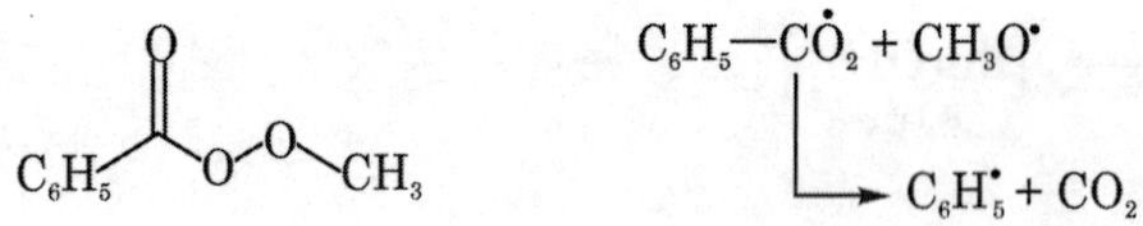

20. (C)

(P)

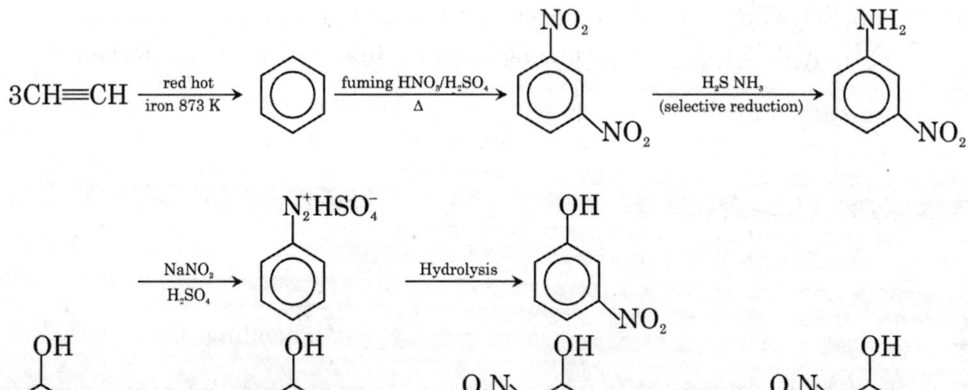

(Q)

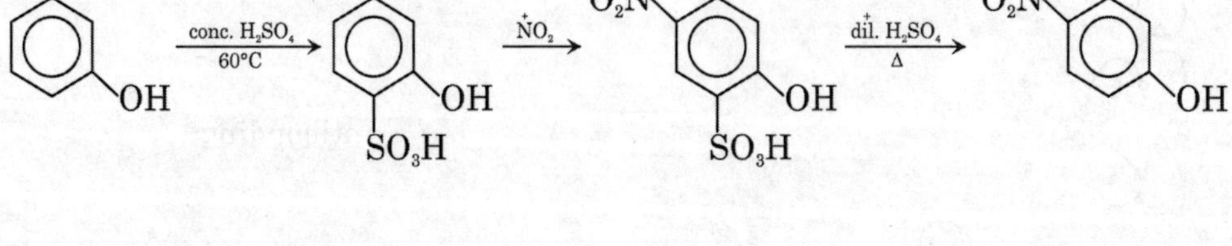

(R)

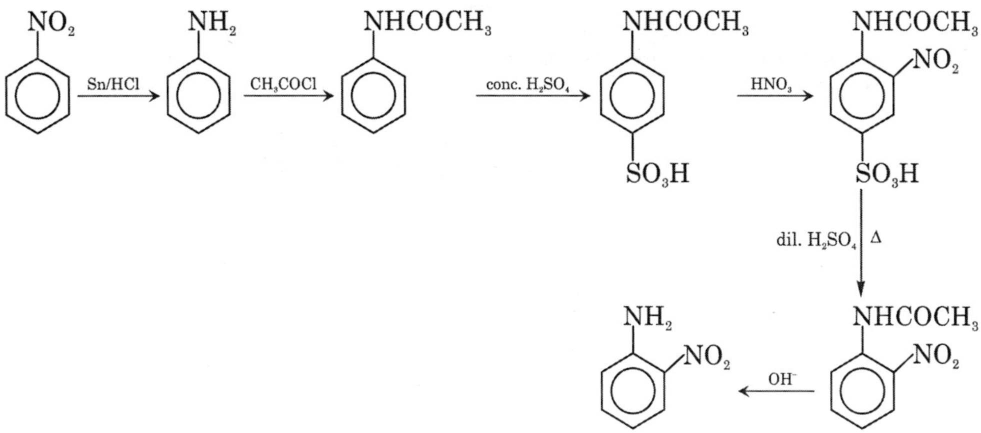

(S)

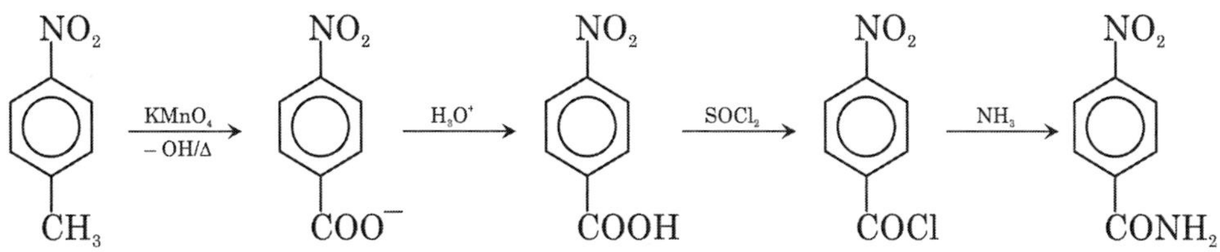

I.I.T. JEE Objective Question Paper—2015 (Main)

1. Which of the following is the energy of a possible excited state of hydrogen?

(A) – 6.8 eV (B) – 3.4 eV (C) + 6.8 eV (D) + 13.6 eV

2. In the following sequence of reactions:

$$\text{Toluene} \xrightarrow{KMnO_4} A \xrightarrow{SOCl_2} B \xrightarrow[BaSO_4]{H_2/Pd} C$$

The product C is

(A) $C_6H_5CH_3$ (B) $C_6H_5CH_2OH$ (C) C_6H_5CHO (D) C_6H_5COOH

3. Which compound would give 5-keto-2-methylhexanal upon ozonolysis?

(A) 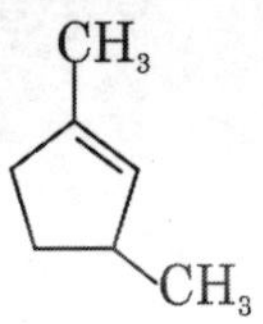

(B) 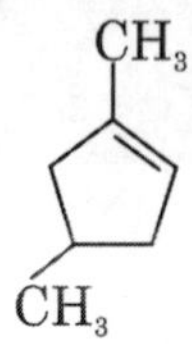

(C) 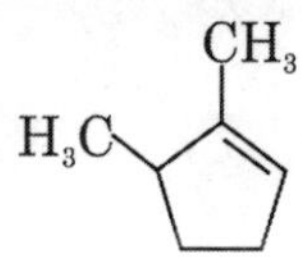

(D) 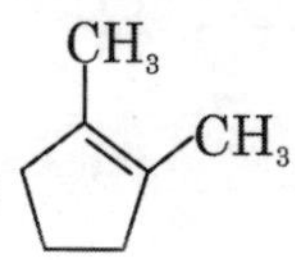

4. The ionic radii (in Å) of N^{3-}, O^{2-} and F^- are respectively

(A) 1.36, 1.71 and 1.40 (B) 1.71, 1.40 and 1.36

(C) 1.71, 1.36 and 1.40 (D) 1.36, 1.40 and 1.71

5. The colour of $KMnO_4$ is due to

(A) *d–d* transition (B) L → M charge transfer transition

(C) σ–σ* transition (D) M → L charge transfer transition.

6. Assertion: Nitrogen and oxygen are the main components in the atmosphere but these do not react to form oxides of nitrogen.

Reason: The reaction between nitrogen and oxygen requires high temperature.

(A) Both assertion and reason are correct, but the reason is not the correct explanation for the assertion

(B) The assertion is incorrect, but the reason is correct

(C) Both the assertion and reason are incorrect

(D) Both assertion and reason are correct, and the reason is the correct explanation for the assertion.

7. Which of the following compounds is not an antacid?

(A) Cimetidine (B) Phenelzine

(C) Ranitidine (D) Aluminium hydroxide

8. In the context of the Hall–Heroult process for the extraction of Al, which of the following statements is false?

(A) Al_2O_3 is mixed with CaF_2 which lowers the melting point of the mixture and brings conductivity

(B) Al^{3+} is reduced at the cathode to form Al

(C) Na_3AlF_6 serves as the electrolyte

(D) CO and CO_2 are produced in this process.

9. Match the catalysts to the correct processes:

Catalyst	Process
(a) $TiCl_3$	(i) Wacker process
(b) $PdCl_2$	(ii) Ziegler–Natta polymerization
(c) $CuCl_2$	(iii) Contact process
(d) V_2O_5	(iv) Deacon's process

(A) (a)–(ii), (b)–(i), (c)–(iv), (d)–(iii)
(B) (a)–(ii), (b)–(iii), (c)–(iv), (d)–(i)
(C) (a)–(iii), (b)–(i), (c)–(ii), (d)–(iv)
(D) (a)–(iii), (b)–(ii), (c)–(iv), (d)–(i)

10. In the reaction,

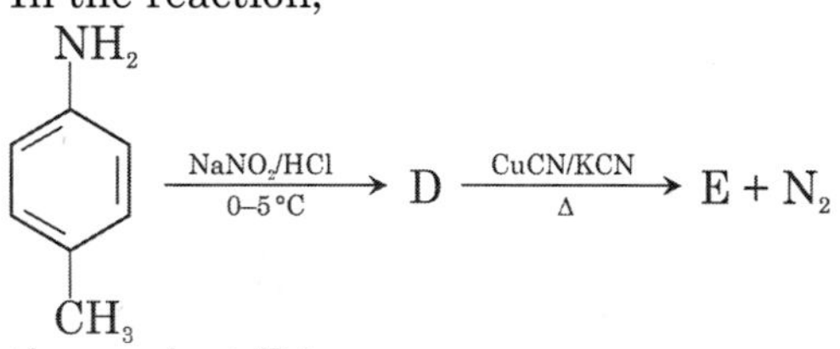

the product E is

(A) 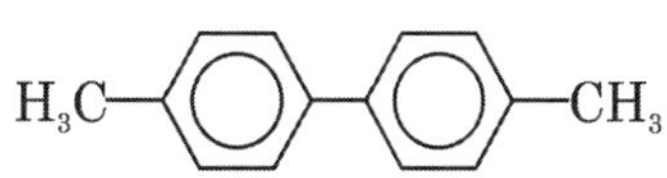

(B) 

(C) 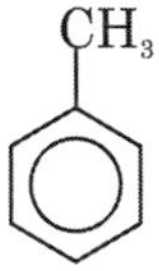

(D) 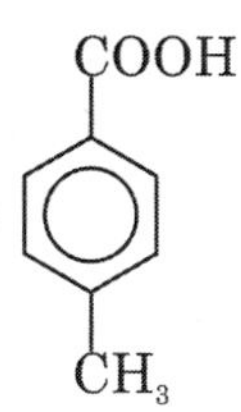

11. Which polymer is used in the manufacture of paints and lacquers?

(A) Glyptal (B) Polypropene (C) Poly vinyl chloride (D) Bakelite

12. The number of geometric isomers that can exist for square planar $[Pt(Cl)(py)(NH_3)(NH_2OH)]^+$ is (py = pyridine):

(A) 3 (B) 4 (C) 6 (D) 2

13. Higher order (> 3) reactions are rare due to

(A) increase in entropy and activation energy as more molecules are involved

(B) shifting of equilibrium towards reactants due to elastic collisions

(C) loss of active species on collision
(D) low probability of simultaneous collision of all the reacting species.

14. Which among the following is the most reactive?
(A) Br_2 (B) I_2 (C) ICl (D) Cl_2

15. Two Faraday of electricity is passed through a solution of $CuSO_4$. The mass of copper deposited at the cathode is (At. Mass of Cu = 63.5 amu)
(A) 63.5 g (B) 2 g (C) 127 g (D) 0 g

16. 3 g of activated charcoal was added to 50 mL of acetic acid solution (0.06*N*) in a flask. After an hour it was filtered and the strength of the filtrate was found to be 0.042*N*. The amount of acetic acid adsorbed (per gram of charcoal) is
(A) 36 mg (B) 42 mg (C) 54 mg (D) 18 mg

17. The synthesis of alkyl fluorides is best accomplished by
(A) Sandmeyer's reaction (B) Finkelstein reaction
(C) Swarts reaction (D) Free radical fluorination.

18. The molecular formula of a commercial resin used for exchanging ions in water softening is $C_8H_7SO_3Na$ (Mol. Wt. 206). What would be the maximum uptake of Ca^{2+} ions by the resin when expressed in mole per gram resin?
(A) $\frac{1}{206}$ (B) $\frac{2}{309}$ (C) $\frac{1}{412}$ (D) $\frac{1}{103}$

19. Which of the vitamins given below is water soluble?
(A) Vitamin D (B) Vitamin E (C) Vitamin K (D) Vitamin C

20. The intermolecular interaction that is dependent on the inverse cube of distance between the molecules is
(A) ion-dipole interaction (B) London force
(C) hydrogen bond (D) ion-ion interaction.

21. The following reaction is performed at 298 K.
$2NO(g) + O_2(g) \rightleftharpoons 2NO_2(g)$
The standard free energy of formation of NO(*g*) is 86.6 kJ/mol at 298 K. What is the standard free energy of formation of $NO_2(g)$ at 298 K? ($K_P = 1.6 \times 10^{12}$)
(A) $86600 + R(298) \ln (1.6 \times 10^{12})$ (B) $86600 \dfrac{\ln (1.6 \times 10^{12})}{R(298)}$
(C) $0.5[2 \times 86,600 - R(298) \ln (1.6 \times 10^{12})]$ (D) $R(298) \ln (1.6 \times 10^{12}) - 86600$

22. Which of the following compounds is not colored yellow?
(A) $K_3[Co(NO_2)_6]$ (B) $(NH_4)_3[As(Mo_3O_{10})_4]$
(C) $BaCrO_4$ (D) $Zn_2[Fe(CN)_6]$

23. In Carius method of estimation of halogens, 250 mg of an organic compound gave 141 mg of AgBr. The percentage of bromine in the compound is (At. Mass Ag = 108; Br = 80)
(A) 36 (B) 48 (C) 60 (D) 24

24. Sodium metal crystallizes in a body centred cubic lattice with a unit cell edge of 4.29 Å. The radius of sodium atom is approximately

(A) 3.22 Å (B) 5.72 Å (C) 0.93 Å (D) 1.86 Å

25. Which of the following compounds will exhibit geometrical isomerism?

(A) 3-Phenyl-1-butene (B) 2-Phenyl-1-butene
(C) 1,1-Diphenyl-1-propane (D) 1-Phenyl-2-butene

26. The vapour pressure of acetone at 20 °C is 185 torr. When 1.2 g of a non-volatile substance was dissolved in 100 g of acetone at 20 °C, its vapour pressure was 183 torr. The molar mass (g mol^{-1}) of the substance is

(A) 64 (B) 128 (C) 488 (D) 32

27. From the following statements regarding H_2O_2, choose the incorrect statement

(A) It decomposes on exposure to light
(B) It has to be stored in plastic or wax lined glass bottles in dark
(C) It has to be kept away from dust
(D) It can act only as an oxidizing agent.

28. Which one of the following alkaline earth metal sulphates has its hydration enthalpy greater than its lattice enthalpy?

(A) $BeSO_4$ (B) $BaSO_4$ (C) $SrSO_4$ (D) $CaSO_4$

29. The standard Gibbs energy change at 300 K for the reaction $2A \rightleftharpoons B + C$ is 2494.2 J. At a given time, the composition of the reaction mixture is $[A] = \frac{1}{2}$, $[B] = 2$ and $[C] = \frac{1}{2}$. The reaction proceeds in the [R = 8.314 J/K/mol, e = 2.718]

(A) reverse direction because $Q > K_c$
(B) forward direction because $Q < K_c$
(C) reverse direction because $Q < K_c$
(D) forward direction because $Q > K_c$

30. Which one has the highest boiling point?

(A) Ne (B) Kr (C) Xe (D) He

ANSWER KEYS

1. B	2. C	3. A	4. B	5. B	6. D
7. B	8. C	9. A	10. B	11. A	12. A
13. D	14. C	15. A	16. D	17. C	18. C
19. D	20. A	21. C	22. D	23. D	24. D
25. D	26. A	27. D	28. A	29. A	30. C

SOLUTIONS

1. (B)

$$E_n = -13.6 \times \frac{Z^2}{n^2} \text{ eV}$$

For hydrogen $Z = 1$

$$E_n = -13.6 \times \frac{1}{n^2} \text{ eV}$$

$$E_2 = \frac{-13.6}{4} = -3.4 \text{ eV}$$

2. (C)

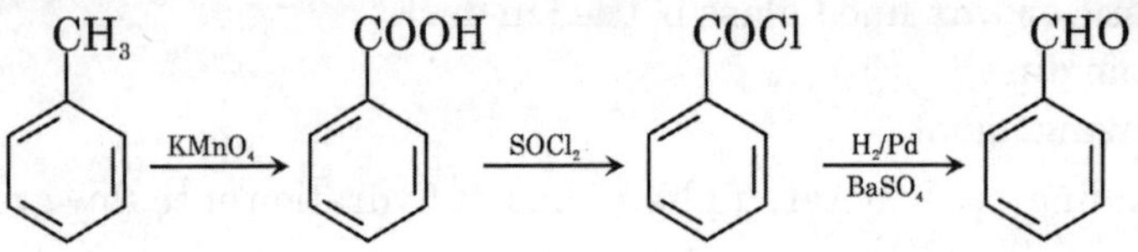

(A) (B) (C)

3. (A)

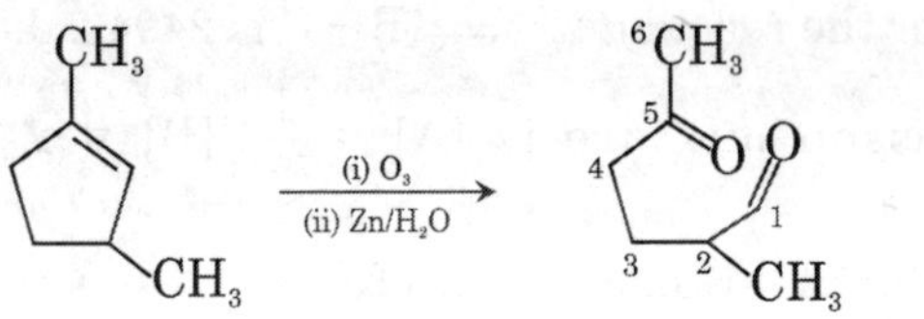

5-keto-2-methylhexanal

4. (B)

Ionic radii

$N^{3-} > O^{2-} > F^-$

5. (B)

L $\longrightarrow$ M charge transfer transition.

6. (D)

7. (B)

Phenelzine is anti-depressant.

8. (C)

$2Al_2O_3 + 3C \longrightarrow 4Al + 3CO_2$

Na_3AlF_6 or CaF_2 is mixed with purified Al_2O_3 to lower the melting point and to bring conductivity. Oxygen liberated at anode reacts with carbon anode to liberate CO and CO_2.

9. (A)

$TiCl_4 + (C_2H_5)_3Al$: Ziegler–Natta catalyst, used for coordination polymerization.

$$CH_2{=}CH_2 \xrightarrow[H_2O\,+\,PdCl_2]{\text{Wacker process}} CH_3{-}CHO + 2HCl$$

$CuCl_2$ is used as catalyst in Deacon's process of production of Cl_2.

V_2O_5 is used as catalyst in contact process of manufacturing of H_2SO_4.

10. (B)

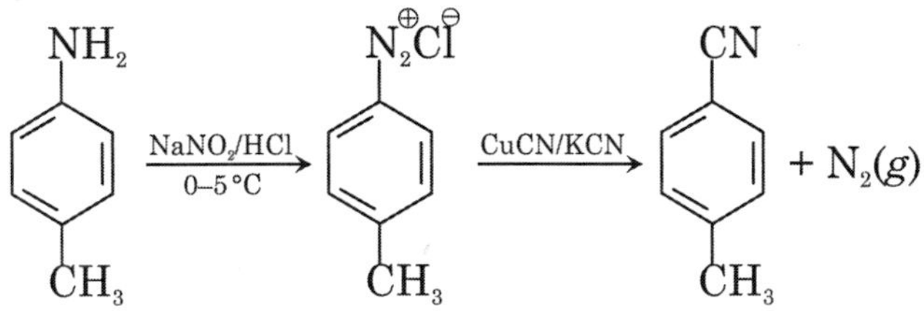

(D) (E)

11. (A)

Glyptal: Manufacture of paints and lacquers

Polypropene: Manufacture of ropes, toys

Poly vinyl chloride: Manufacture of raincoat, handbags

Bakelite: Making combs and electrical switch

12. (A)

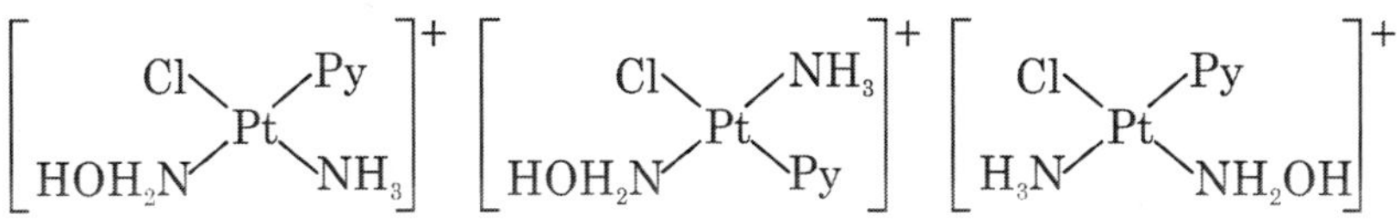

13. (D)

Higher order (> 3) reactions are less probable due to low probability of simultaneous collision of all the reaction species.

14. (C)

Interhalogen compounds are more reactive than the corresponding halogen molecules due to polarity of bond.

15. (A)

$Cu^{2+} + 2e^- \longrightarrow Cu$

2F of electricity will give 1 mole of Cu.

16. (D)

Amount adsorbed

$= (0.060 - 0.042) \times 50 \times 10^{-3} \times 60$

$= 0.018 \times 50 \times 60 \times 10^{-3} = 0.018 \times 3 = 0.054$ gm $= 54$ mg

Amount adsorbed per gram of activated charcoal $= \frac{54}{3} = 18$ mg

17. (C)

$2R\text{—}Cl + Hg_2F_2 \xrightarrow{\Delta} 2F\text{—}F + Hg_2Cl_2$

18. (C)

$2C_8H_7SO_3Na + Ca^{2+} \longrightarrow (C_8H_7SO_3)_2Ca$

2 mole 1 mole

412 gm 1 mole

Maximum uptake of Ca^{2+} ions by the resin = 1/412 (mole per gm resin)

19. (D)

Vitamin C is water soluble.

20. (A)

For ion-dipole interaction,

$F \propto \mu \dfrac{dE}{dr}$ (where μ is dipole moment of dipole and r is distance between ion and dipole)

$\propto \mu \dfrac{d}{dr}\left(\dfrac{1}{r^2}\right) \propto \dfrac{\mu}{r^3}$

21. (C)

$$\Delta G^0_{(\text{Re x}^0)} = 2\Delta G^0_f(NO_2) - 2\Delta G^0_f(NO) - \Delta G^0_f(O_2)$$

$$2\Delta G^0_f(NO_2) = \Delta G^0_{(\text{Re x}^0)} + 2\Delta G^0_f(NO) + \Delta G^0_f(O_2)$$

$$\Delta G = \Delta G^0 + RT \ln k_p$$

At equilibrium,

$$\Delta G = 0,\ Q = k_p,\ \Delta G^0 = -RT\ln k_p,\ \Delta G^0(O_2) = 0$$

$$\Delta G^0_f(NO_2) = 0.5[2 \times 86{,}600 - R(298)\ln(1.6 \times 10^{12})]$$

22. (D)

$Zn_2[Fe(CN)_6]$ is bluish-white and rest are yellow coloured compounds.

23. (D)

Let the organic compound is R—Br.

R—Br + $AgNO_3 \longrightarrow$ AgBr(↓)

Hence, number of moles of AgBr ≡ Number of moles of R—Br ≡ Number of moles of Br

$\dfrac{141 \times 10^{-3}}{188}$ = number of moles of Br = 0.75×10^{-3}

Mass of bromine = $0.75 \times 80 \times 10^{-3}$ g = 60 mg

Percentage of bromine = $\dfrac{60}{250} \times 100 = 24\%$

24. (D)

For body center unit cell, $4r = a\sqrt{3}$

Hence, radius of Na = $\dfrac{\sqrt{3} \times 4.29}{4} = 1.857 \approx 1.86$ Å

25. (D)

Only 1-phenyl-2-butene can show geometrical isomerism.

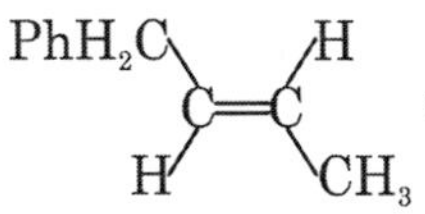

trans

and

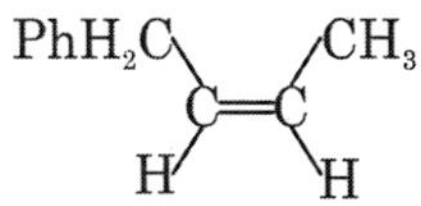

cis

26. (A)

Vapour pressure of pure acetone ($p°$) = 185 torr

Vapour pressure of solution (p) = 183 torr

Then from Raoult's law,

$$\frac{p° - p}{p} = \frac{n_{\text{solute}}}{n_{\text{solvent}}}$$

$$\frac{185 - 183}{183} = \frac{1.2/\text{MW}}{100/58}$$

$$\frac{2}{183} = \frac{1.2 \times 58}{100 \times \text{MW}}$$

$$\text{MW} = \frac{1.2 \times 58 \times 183}{200} = 63.68 \approx 64$$

27. (D)

H_2O_2 can act both as oxidizing agent and reducing agent and H_2O_2 decomposes on exposures to light and dust. Hence, it has to be stored in plastic or wax lined glass bottles in dark.

28. (A)

Because of smallest size of Be^{2+} ion, its hydration energy is maximum and is greater than the lattice energy of $BeSO_4$.

29. (A)

$$\Delta G^0 = -RT \ln K$$

$$2494.2 = -8.314 \times 300 \ln K$$

$$\ln K = -1$$

$$\log_{2.718} K = -1$$

$$K = (2.718)^{-1} = \frac{1}{2.718} = 0.3679 \quad \Rightarrow \quad Q_c = \frac{[\text{B}][\text{C}]}{[\text{A}]^2} = \frac{2 \times \frac{1}{2}}{\left(\frac{1}{2}\right)^2} = 4$$

The reaction proceeds in the reverse direction because $Q_c > K$.

30. (C)

The boiling point of noble gases in increasing order: He < Ne < Ar < Kr < Xe < Rn

Boiling point (K): 4.2 < 27.1 < 87.2 < 119.7 < 165 < 211

I.I.T. JEE Objective Question Paper—2015 (Advanced)

PAPER—1

SECTION–1

(Maximum Marks: 32)

This section contains eight questions.
The answer to each question is a single digit integer ranging from 0 to 9, both inclusive.
For each question, darken the bubble corresponding to the correct integer in the ORS
Marking scheme:
+ 4 if the bubble corresponding to the answer is darkened; 0 in all other cases.

1. If the freezing point of a 0.01 molal aqueous solution of a cobalt(III) chloride-ammonia complex (which behaves as a strong electrolyte) is $-0.0558\ °C$, the number of chloride(s) in the coordination sphere of the complex is (K_f of water $= 1.86\ K\ kg\ mol^{-1}$)

2. The total number of stereoisomers that can exist for M is

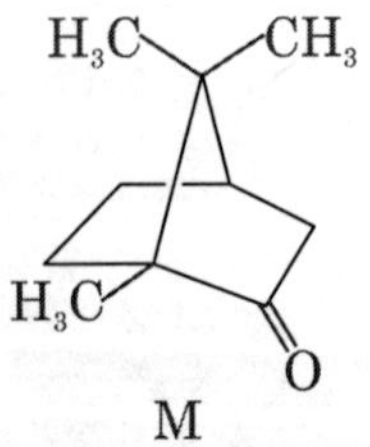

M

3. The number of resonance structures for N is

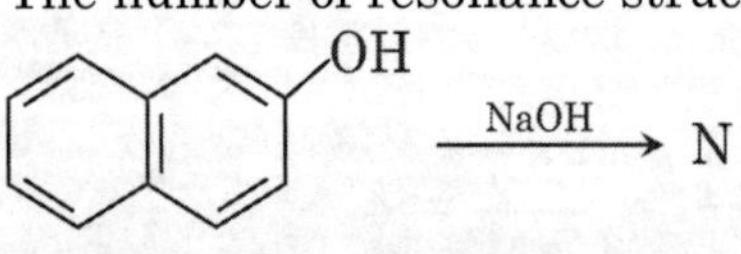

4. The total number of lone pairs of electrons in N_2O_3 is

5. For the octahedral complexes of Fe^{3+} in SCN^- (thiocyanato-S) and in CN^- ligand environments, the difference between the spin-only magnetic moments in Bohr magnetons (when approximated to the nearest integer) is (Atomic number of Fe = 26)

6. Among the triatomic molecules/ions, $BeCl_2$, N_3^-, N_2O, NO_2^+, O_3, SCl_2, ICl_2^-, I_3^- and XeF_2, the total number of linear molecule(s)/ion(s) where the hybridization of the central atom does not have contribution from the d-orbital(s) is (Atomic number: S = 16, Cl = 17, I = 53 and Xe = 54)

7. Not considering the electronic spin, the degeneracy of the second excited state ($n = 3$) of H atom is 9, while the degeneracy of the second excited state of H^- ion is

8. All the energy released from the reaction $X \longrightarrow Y$, $\Delta_r G^0 = -193$ kJ mol^{-1} is used for oxidizing M^+ as $M^+ \longrightarrow M^{3+} + 2e^-$, $E^0 = -0.25$ V.
Under standard conditions, the number of moles of M^+ oxidized when one mole of X is converted to Y is (F = 96500 C mol^{-1})

SECTION–2
(Maximum Marks: 40)

This section contains ten questions.
Each question has four options (A), (B), (C) and (D). One or more than one of these four option(s) is(are) correct.
For each question, darken the bubble(s) corresponding to the correct option(s) in the ORS
Marking scheme:
+ 4 if only the bubble(s) corresponding to all the correct option(s) is(are) darkened.
0 if none of the bubbles is darkened; – 2 in all other cases.

9. If the unit cell of a mineral has cubic close packed (*ccp*) array of oxygen atoms with m fraction of octahedral holes occupied by aluminium ions and n fraction of tetrahedral holes occupied by magnesium ions, m and n, respectively, are

(A) $\frac{1}{2}, \frac{1}{8}$ (B) $1, \frac{1}{4}$ (C) $\frac{1}{2}, \frac{1}{2}$ (D) $\frac{1}{4}, \frac{1}{8}$

10. Compound(s) that on hydrogenation produce(s) optically inactive compound(s) is(are)

(A) 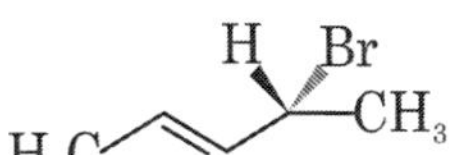

(B) 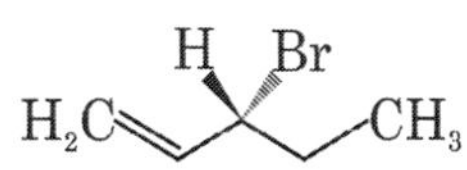

(C) 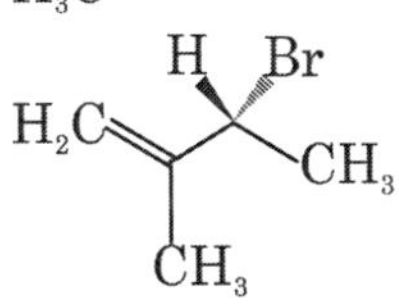

(D) 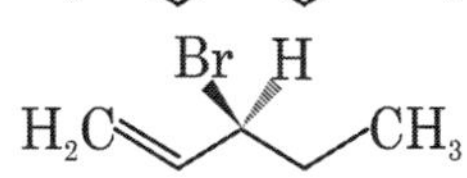

11. Tha major product of the following reaction is

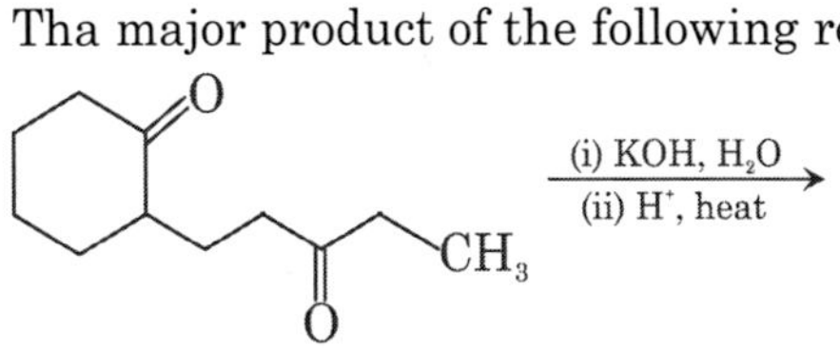

(A) 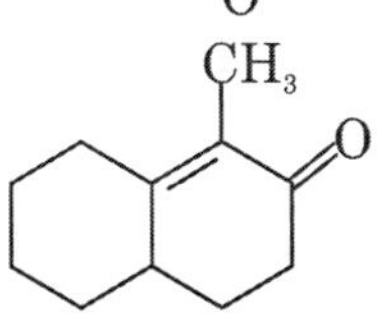

(B) 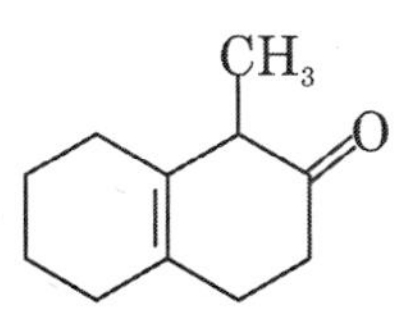

(C) 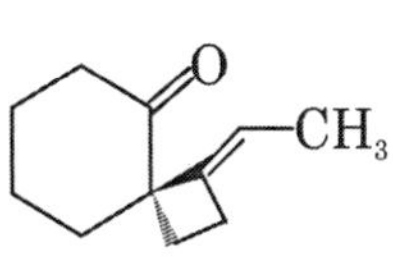

(D) 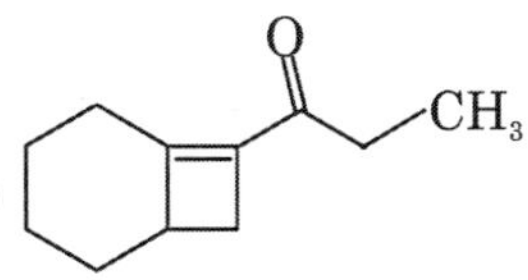

12. In the following reaction, the major product is

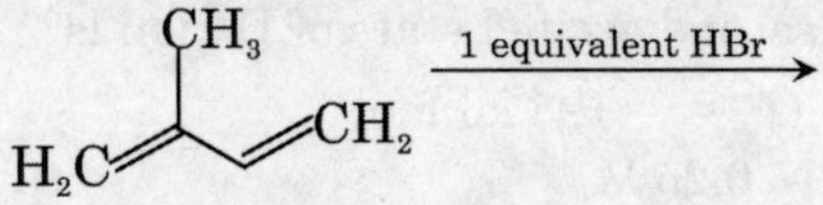

(A) 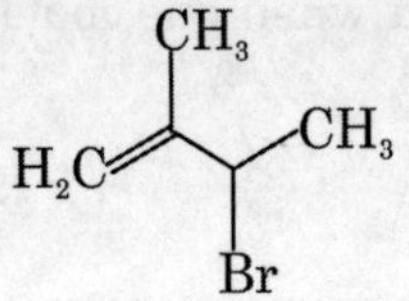

(B) 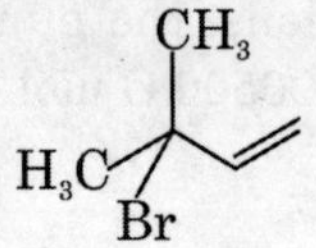

(C) 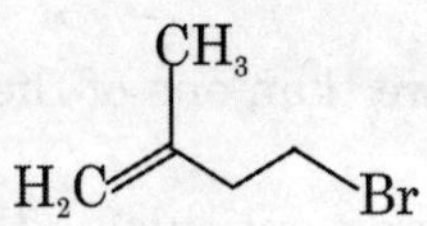

(D) 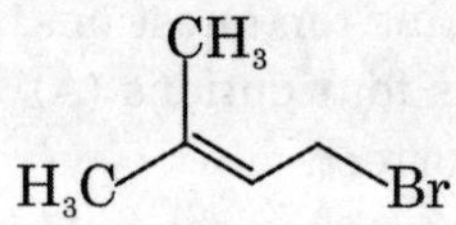

13. The structure of D-(+)-glucose is

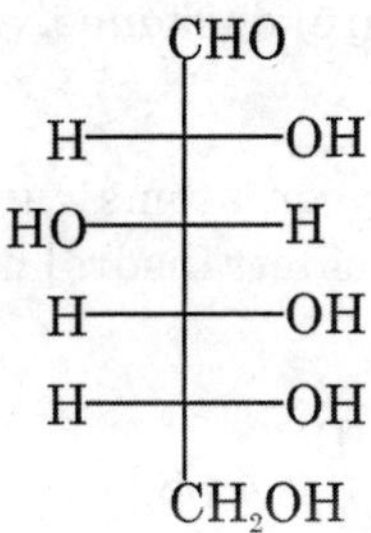

The structure of L-(–)-glucose is

(A) 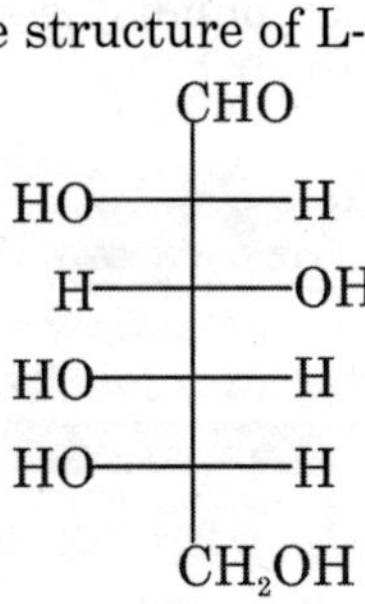

(B) 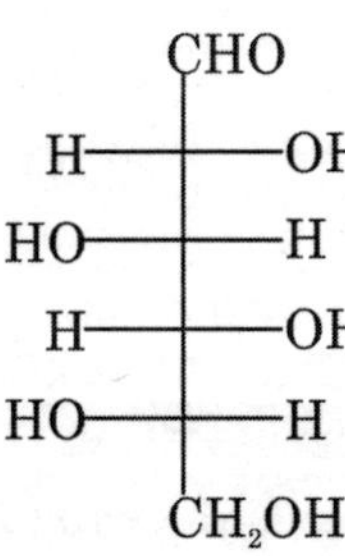

(C) 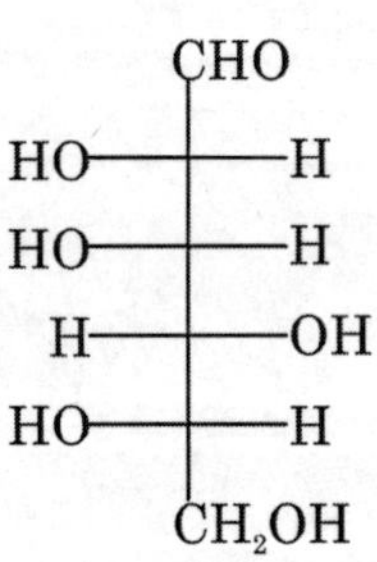

(D) 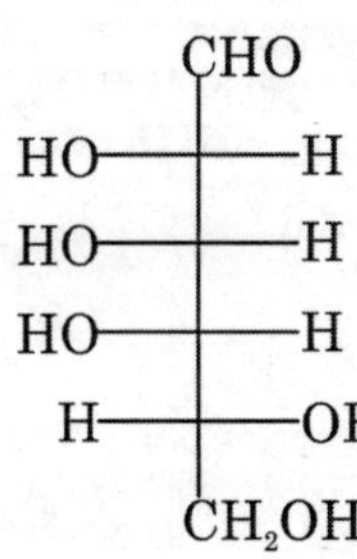

14. The major product of the following reaction is

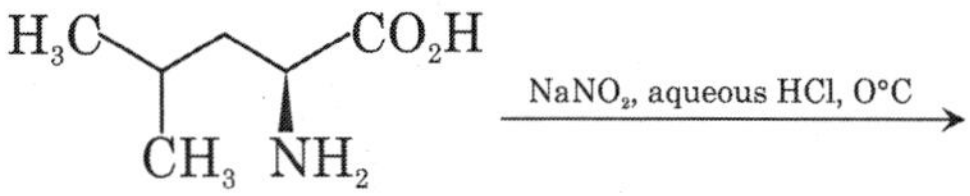

(A) 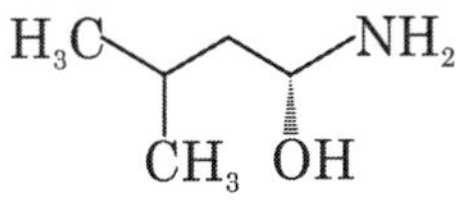

(B) 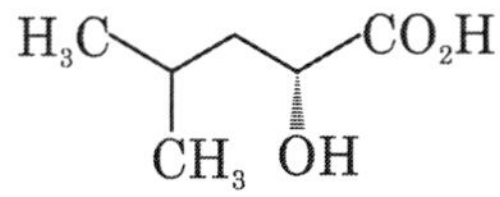

(C) 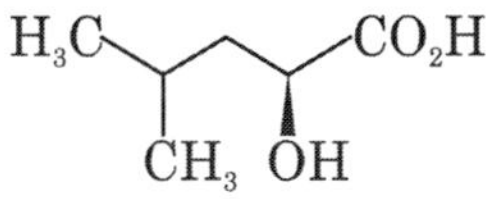

(D) 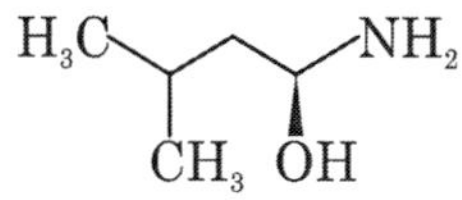

15. The correct statement(s) about Cr^{2+} and Mn^{3+} is (are) (Atomic number of Cr = 24 and Mn = 25)

(A) Cr^{2+} is a reducing agent
(B) Mn^{3+} is an oxidizing agent
(C) Both Cr^{2+} and Mn^{3+} exhibit d^4 electronic configuration
(D) When Cr^{2+} is used as a reducing agent, the chromium ion attains d^5 electronic configuration.

16. Copper is purified by electrolytic refining of blister copper. The correct statement(s) about this process is(are)

(A) Impure Cu strip is used as cathode
(B) Acidified aqueous $CuSO_4$ is used as electrolyte
(C) Pure Cu deposits at cathode
(D) Impurities settle as anode mud.

17. Fe^{3+} is reduced to Fe^{2+} by using

(A) H_2O_2 in presence of NaOH
(B) Na_2O_2 in water
(C) H_2O_2 in presence of H_2SO_4
(D) Na_2O_2 in presence of H_2SO_4

18. The % yield of ammonia as a function of time in the reaction

$N_2(g) + 3H_2(g) \rightleftharpoons 2NH_3(g), \Delta H < 0$

at (P, T_1) is given below:

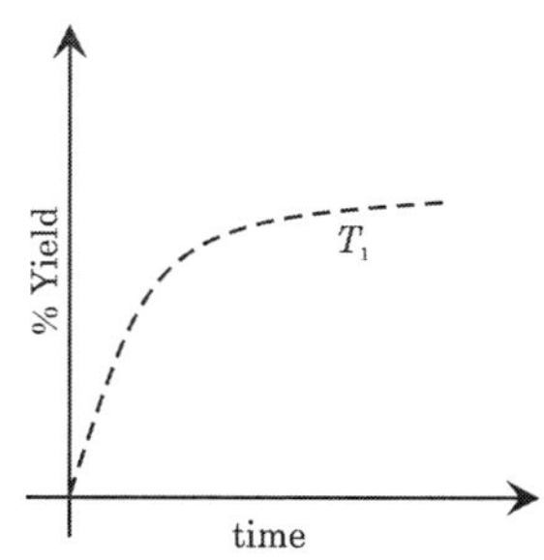

If this reaction is conducted at (P, T_2), with $T_2 > T_1$, the % yield of ammonia as a function of time is represented by

(A) 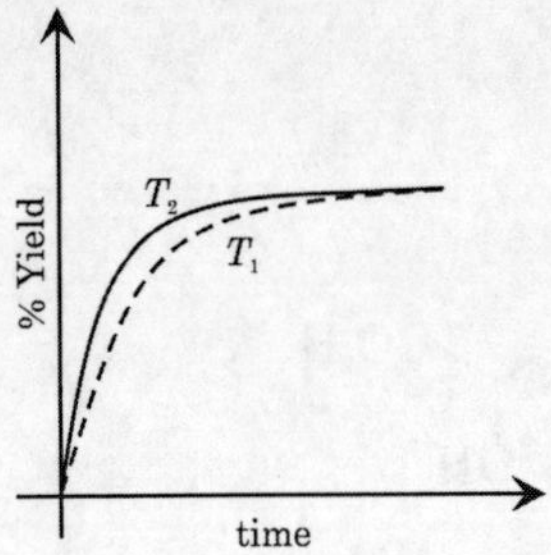

(B) 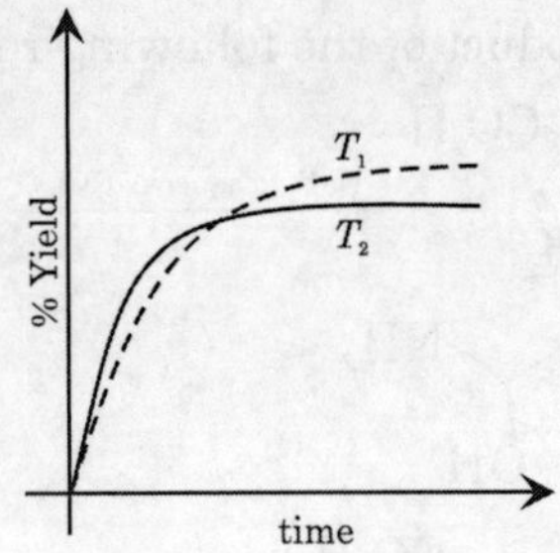

(C) 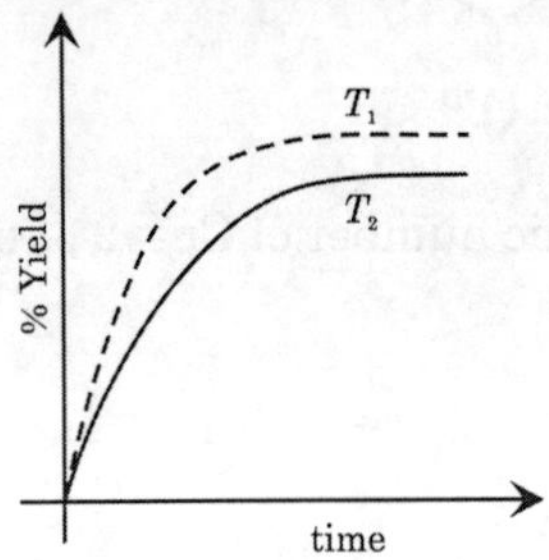

(D) 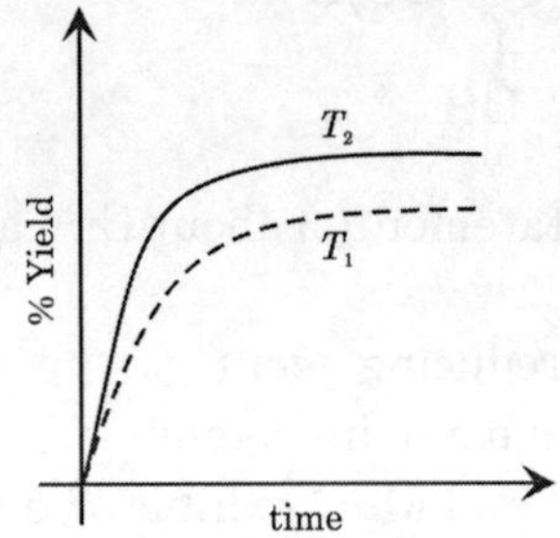

SECTION–3
(Maximum Marks: 16)

This section contain two questions.
Each question contains two columns, **Column I** and **Column II**
Column I has four entries (A), (B), (C) and (D).
Column II has five entries (P), (Q), (R), (S) and (T).
Match the entries in **Column I** with the entries in **Column II.**
One or more entries in **Column I** may match with one or more entries in **Column II.**
The ORS contains a 4 × 5 matrix whose layout will be similar to the one shown below:

(A) (P) (Q) (R) (S) (T)
(B) (P) (Q) (R) (S) (T)
(C) (P) (Q) (R) (S) (T)
(D) (P) (Q) (R) (S) (T)

For each entry in **Column I**, darken the bubbles of all the matching entries. For example, if entry (A) in **Column I** matches with entries (Q), (R) and (T), then darken these three bubbles in the ORS. Similarly, for entries (B), (C) and (D).
Marking scheme:
For each entry in **Column I**,
+ 2 if only the bubble(s) corresponding to all the correct match(es) is(are) darkened.
0 if none of the bubbles is darkened; – 1 in all other cases.

19. Match the anionic species gives in Column I that are present in the ore(s) gives in Column II.

Column I	Column II
(A) Carbonate	(P) Siderite
(B) Sulphide	(Q) Malachite
(C) Hydroxide	(R) Bauxite
(D) Oxide	(S) Calamine
	(T) Argentite

20. Match the thermodynamic processes given under Column I with the expression given under Column II.

Column I	Column II
(A) Freezing of water at 273 K and 1 atm	(P) $q = 0$
(B) Expansion of 1 mol of an ideal gas into a vacuum under isolated conditions	(Q) $w = 0$
(C) Mixing of equal volumes of two ideal gases at constant temperature and pressure in an isolated container	(R) $\Delta S_{sys} < 0$
(D) Reversible heating of $H_2(g)$ at 1 atm from 300 K to 600 K, followed by reversible cooling to 300 K at 1 atm	(S) $\Delta U = 0$
	(T) $\Delta G = 0$

ANSWER KEYS

1. 1	2. 2	3. 9	4. 8	5. 4	6. 4
7. 3	8. 4	9. A	10. B, D	11. A	12. D
13. A	14. C	15. A, B, C	16. B, C, D	17. A, B	18. B

19. (A) $\longrightarrow$ (P, Q, S); (B) $\longrightarrow$ (T); (C) $\longrightarrow$ (Q, R); (D) $\longrightarrow$ (R)

20. (A) $\longrightarrow$ (R, T); (B) $\longrightarrow$ (P, Q, S); (C) $\longrightarrow$ (P, Q, S); (D) $\longrightarrow$ (P, Q, S, T)

SOLUTIONS

1. (1)

$\Delta T_f = iK_f m$

$0.0558 = i \times 1.86 \times 0.01$

$i = 3$

Hence, the complex is $[Co(NH_3)_5Cl]Cl_2$

2. (2)

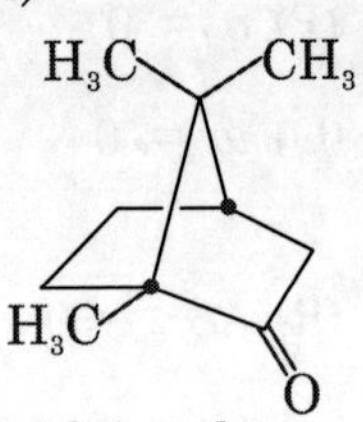

Bridging does not allow the other 2 variants to exist.

Total number of stereoisomers of M = 2

3. (9)

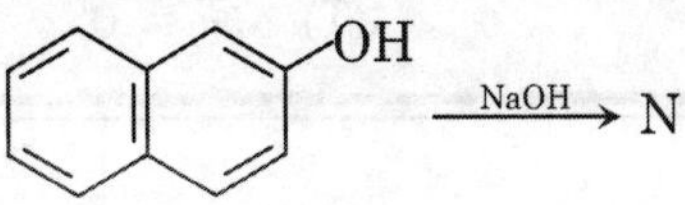

N is 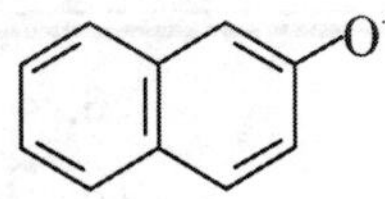

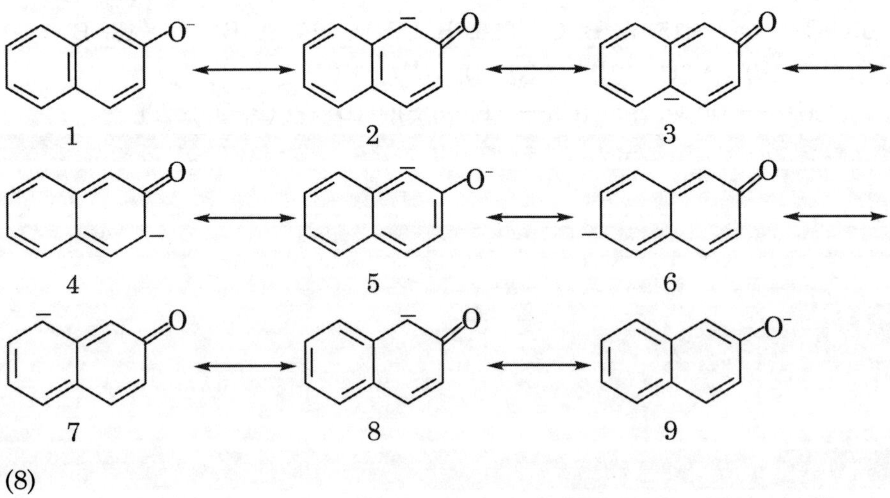

4. (8)

Total number of lone pairs = 8

5. (4)

$[Fe(SCN)_6]^{3-}$ and $[Fe(CN)_6]^{3-}$

In both the cases the electronic configuration of Fe^{3+} will be $1s^2 2s^2 2p^6 3s^2 3p^6 3p^5$.

Since SCN is a weak field ligand and $\bar{C}N$ is a strong field ligand, the pairing will occur only in case of $[Fe(CN)_6]^{3-}$.

Case – 1 $3d^5$

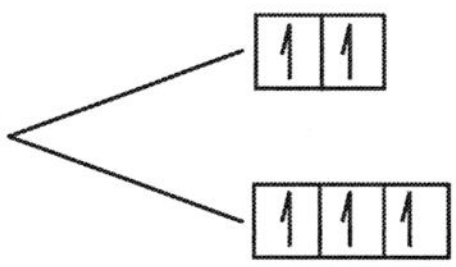

(high spin)
(no pairing)

Case – 2

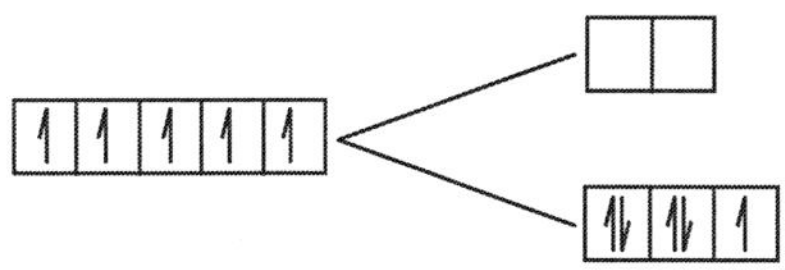

(low spin)
(pairing)

Case – 1 $\mu = \sqrt{n(n+2)} = \sqrt{5(5+2)} = \sqrt{35} = 5.91$ BM

Case – 2 $\mu = \sqrt{n(n+2)} = \sqrt{1(1+2)} = \sqrt{3} = 1.73$ BM

Difference in spin-only magnetic moment $= 5.91 - 1.73 = 4.18 \approx 4$ BM

6. (4)

$BeCl_2$, N_3^-, N_2O, NO_2^+, O_3, SCl_2, ICl_2^-, I_3^-, XeF_2

$BeCl_2 \longrightarrow sp \longrightarrow$ linear

$N_3^- \longrightarrow sp \longrightarrow$ linear

$N_2O \longrightarrow sp \longrightarrow$ linear

$NO_2^+ \longrightarrow sp \longrightarrow$ linear

$O_3 \longrightarrow sp^2 \longrightarrow$ bent

$SCl_2 \longrightarrow sp^3 \longrightarrow$ bent

$I_3^- \longrightarrow sp^3d \longrightarrow$ linear

$ICl_2^- \longrightarrow sp^3d \longrightarrow$ linear

$XeF_2 \longrightarrow sp^3d \longrightarrow$ linear

Hence, only four (4) molecules have linear shape and in these no d-orbital is involved in hybridization.

7. (3)

Single electron species do not follow the $(n + \ell)$ rule but multi electron species do.

Ground state of H^- is $1s^2$.

First excited state of H^- is $1s^1$, $2s^1$.

Second excited state of H^- is $1s^1$, $2s^0$, $2p^1$.

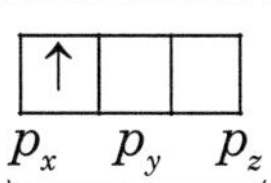

(3 degenerate orbitals)

8. (4)

X $\longrightarrow$ Y $\quad \Delta_r G^0 = -193$ kJ/mol

$M^+ \longrightarrow M^{3+} + 2e^- \quad E^0 = -0.25$ V

ΔG^0 for the this reaction is: $\Delta G^0 = -nFE^0 = -2 \times (-0.25) \times 96500 = 48250$ J/mol = 48.25 kJ/mole

Hence, the number of moles of M^+ ion oxidized using X $\longrightarrow$ Y will be

$$= \frac{193}{48.25} = 4 \text{ moles}$$

9. (A)

In *ccp* lattice:

Number of O atoms: 4

Number of octahedral voids: 4

Number of tatrahedral voids: 8

Number of $Al^{3+} = 4 \times m$

Number of $Mg^{2+} = 8 \times n$

Due to charge neutrality,

$4(-2) + 4m(+3) + 8n(+2) = 0$

Hence, $m = \dfrac{1}{2}$ and $n = \dfrac{1}{8}$

10. (B, D)

(1) 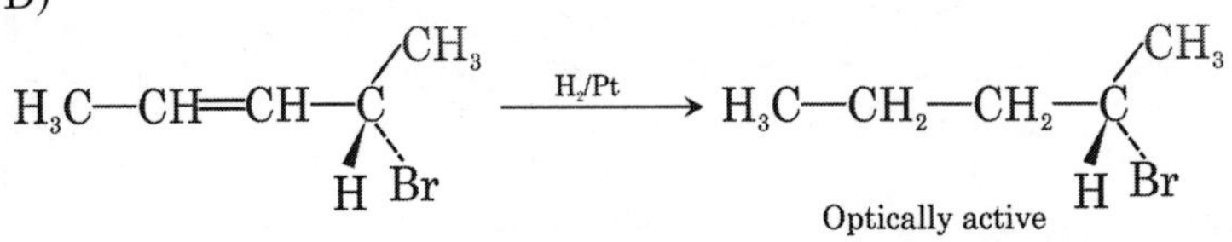

(2) 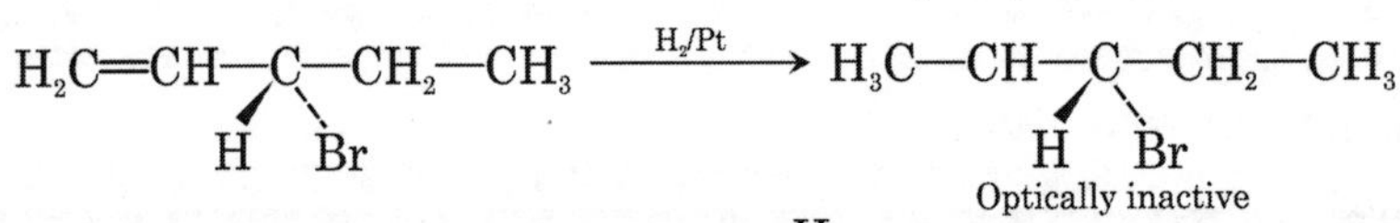

(3) 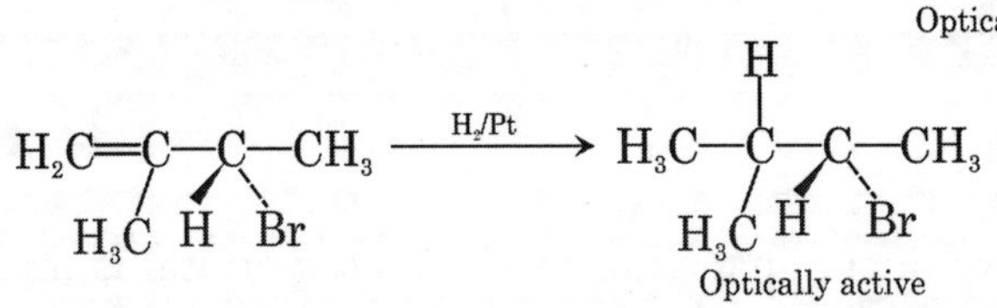

(4) 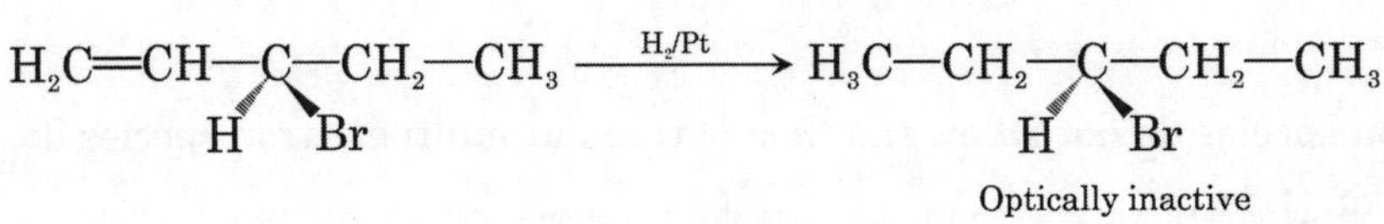

11. (A)

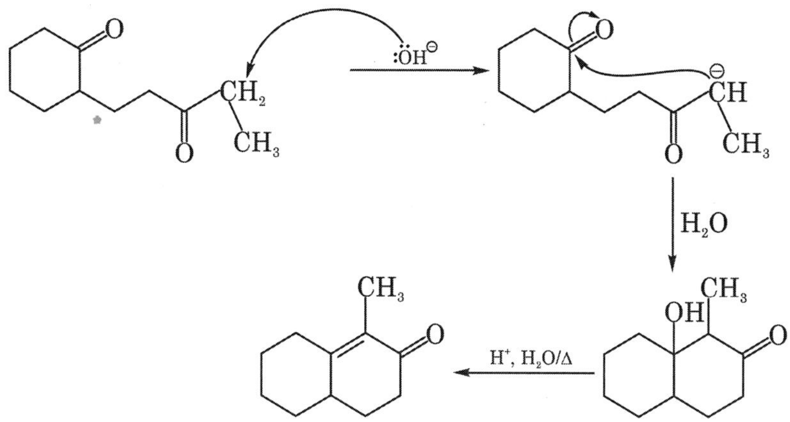

12. (D)

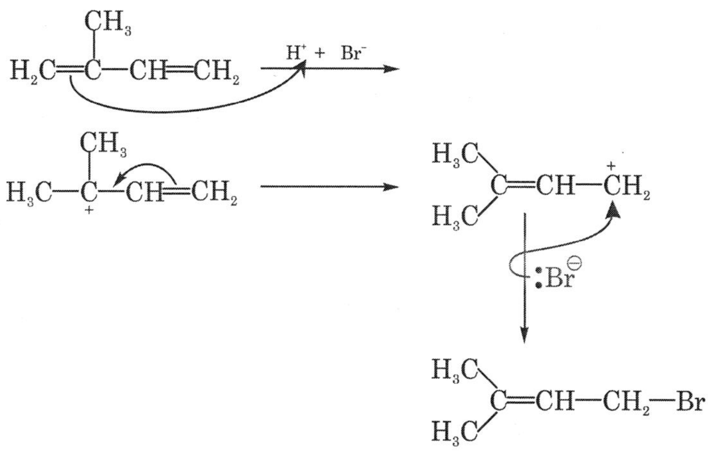

13. (A)

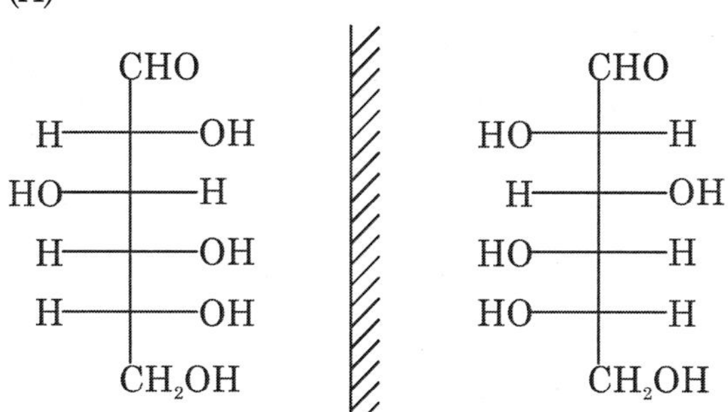

D(+)glucose L(–)glucose

14. (C)

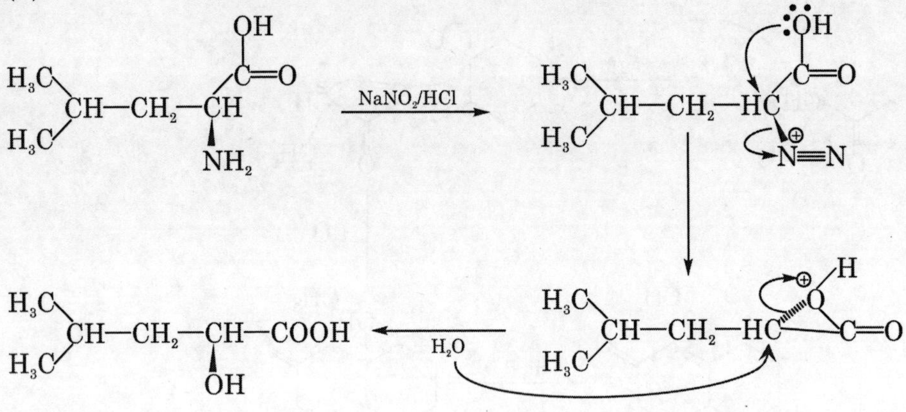

15. (A, B, C)

(1) Cr^{2+} is a reducing agent because Cr^{3+} is more stable.

(2) Mn^{3+} is an oxidizing agent because Mn^{2+} is more stable.

(3) Cr^{2+} and Mn^{3+} have d^4 eletronic configuration.

16. (B, C, D)

(1) Impure Cu strip is used as anode and impurities settle as anode mud.

(2) Pure Cu deposits at the cathode.

(3) Acidified aqueous $CuSO_4$ is used as an electrolyte.

17. (A, B)

$2Fe^{3+} + H_2O_2 + 2OH^- \longrightarrow 2Fe^{2+} + 2H_2O + O_2$

$Na_2O_2 + H_2O \longrightarrow H_2O_2 + NaOH$

18. (B)

$N_2(g) + 3H_2(g) \rightleftharpoons 2NH_3(g); \Delta H < 0$

Increasing the temperature lowers equilibrium yield of ammonia. However, at higher temperature the initial rate of forward reaction would be greater than at lower temperature. That is why the percentage yield of NH_3 would be more initially.

19. (A) $\longrightarrow$ (P, Q, S); (B) $\longrightarrow$ (T); (C) $\longrightarrow$ (Q, R); (D) $\longrightarrow$ (R)

Siderite — $FeCO_3$

Malachite — $CuCO_3 \cdot Cu(OH)_2$

Bauxite — $AlO_x(OH)_{3-2x}$; $0 < x < 1$

Calamine — $ZnCO_3$

Argentite — Ag_2S

20. (A) $\longrightarrow$ (R, T); (B) $\longrightarrow$ (P, Q, S); (C) $\longrightarrow$ (P, Q, S); (D) $\longrightarrow$ (P, Q, S, T)

I.I.T. JEE Objective Question Paper—2015 (Advanced)

PAPER—2

SECTION - I
(Maximum Marks: 32)

This section contains eight questions.
The answer to each question is a single digit integer ranging from 0 to 9, both inclusive.
For each question, darken the bubble corresponding to the correct integer in the ORS.
Marking Scheme: + 4 if the bubble corresponding to the answer is darkened and 0 in all other cases.

1. In dilute aqueous H_2SO_4, the complex diaquodioxalatoferrate(II) is oxidized by MnO_4^-. For this reaction, the ratio of the rate of change of $[H^+]$ to the rate of change of $[MnO_4^-]$ is

2. The number of hydroxyl group(s) in Q is

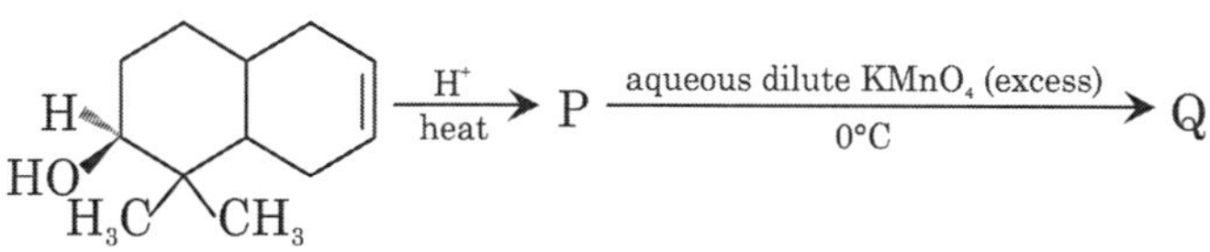

3. Among the following, the number of reaction(s) that produce(s) benzaldehyde is

I. 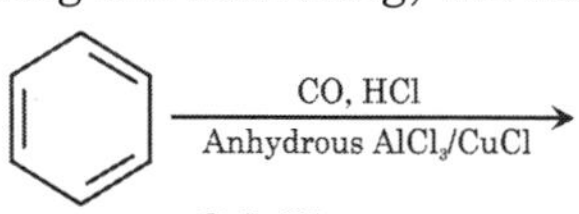

II. 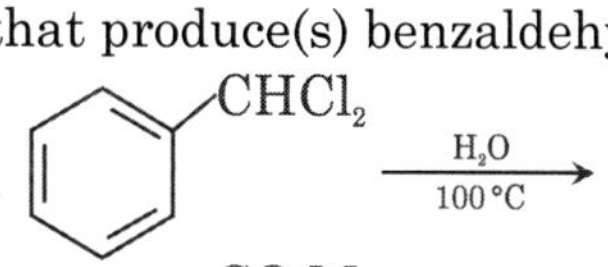

III. 

IV. 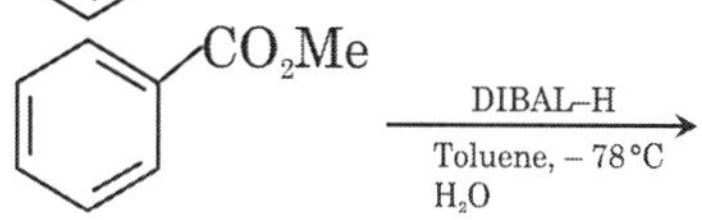

4. In the complex acetylbromidodicarbonylbis(triethylphosphine)iron(II), the number of Fe—C bond(s) is

5. Among the complex ions, $[Co(NH_2—CH_2—CH_2—NH_2)_2Cl_2]^+$, $[CrCl_2(C_2O_4)_2]^{3-}$, $[Fe(H_2O)_4(OH)_2]^+$, $[Fe(NH_3)_2(CN)_4]^-$, $[Co(NH_2—CH_2—CH_2—NH_2)_2(NH_3)Cl]^{2+}$ and $[Co(NH_3)_4(H_2O)Cl]^{2+}$, the number of complex ion(s) that show(s) *cis-trans* isomerism is

6. Three moles of B_2H_6 are completely reacted with methanol. The number of moles of boron containing product formed is

7. The molar conductivity of a solution of a weak acid HX (0.01*M*) is 10 times smaller than the molar conductivity of a solution of a weak acid HY (0.10*M*). If $\lambda^0_{X^-} \approx \lambda^0_{Y^-}$, the difference in their pK_a values, pK_a(HX) – pK_a(HY), is (consider degree of ionisation of both acids to be << 1)

8. A closed vessel with rigid walls contains 1 mol of ${}^{238}_{92}U$ and 1 mol of air at 298 K. Considering complete decay of ${}^{238}_{92}U$ to ${}^{206}_{82}Pb$, the ratio of the final pressure to the initial pressure of the system at 298 K is

SECTION–2
(Maximum Marks: 32)

This section contains eight questions.
Each question has four options (A), (B), (C) and (D). One or more than one of these four option(s) is (are) correct.
For each question, darken the bubble(s) corresponding to the correct option(s) in the ORS.
Marking scheme:
+ 4 if only the bubble(s) corresponding to all the correct option(s) is (are) darkened
0 if none of the bubbles is darkened and – 2 in all other cases.

9. One mole of a monoatomic real gas satisfies the equation $P(V - b) = RT$ where b is a constant. The relationship of interatomic potential $V(r)$ and interatomic distance r for the gas is given by

(A) 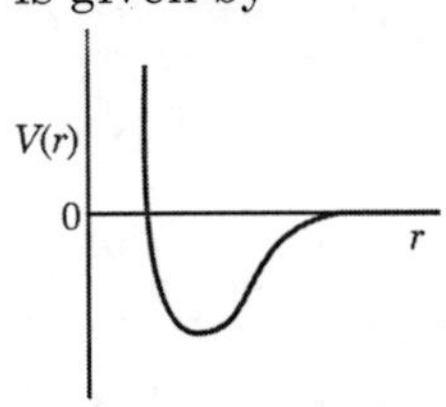

(B) 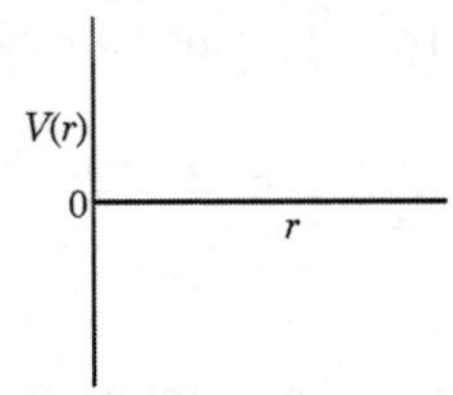

(C) 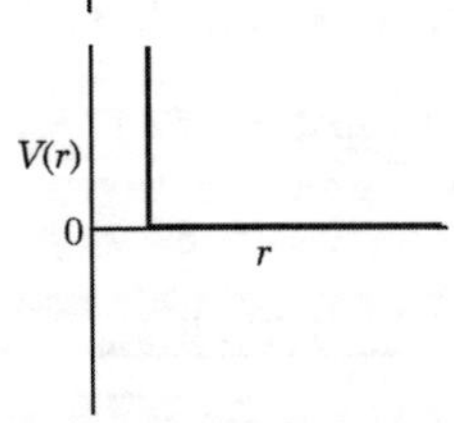

(D) 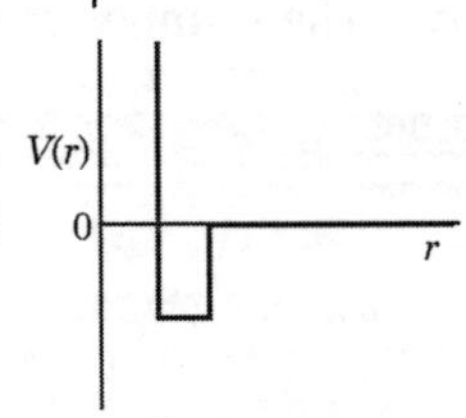

10. In the following reactions, the product S is

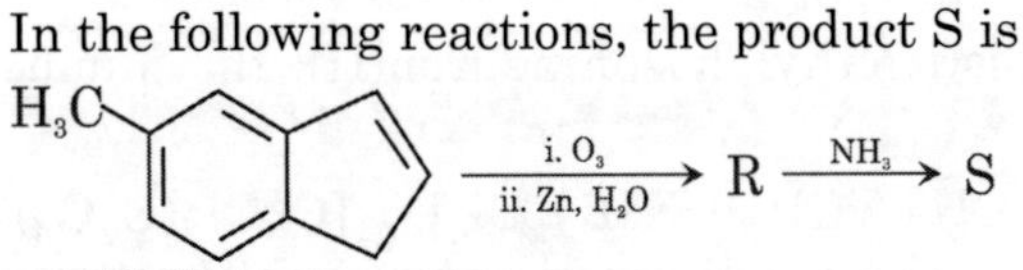

(A) 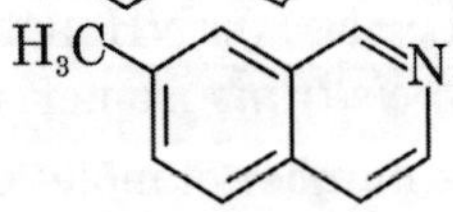

(B) 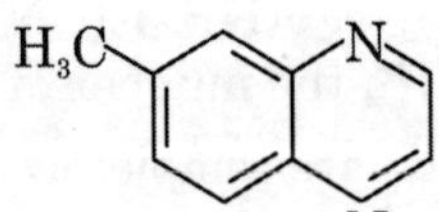

(C) 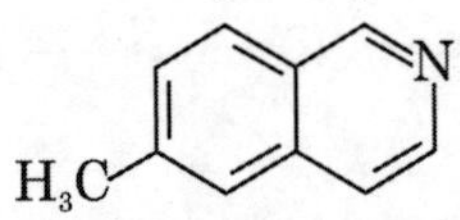

(D) 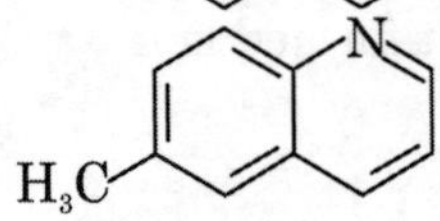

11. The major product U in the following reactions is

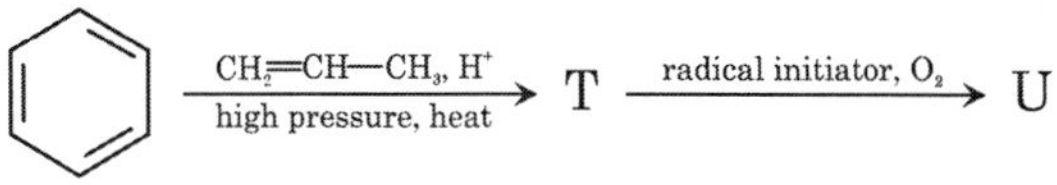

(A) 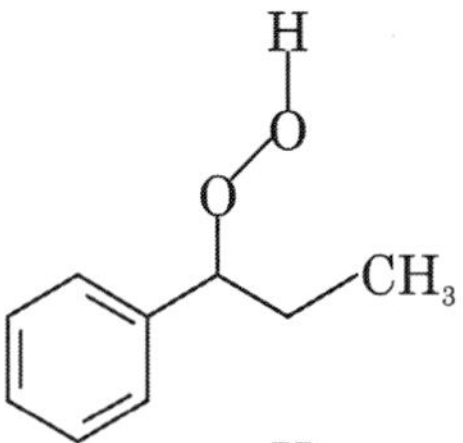

(B) 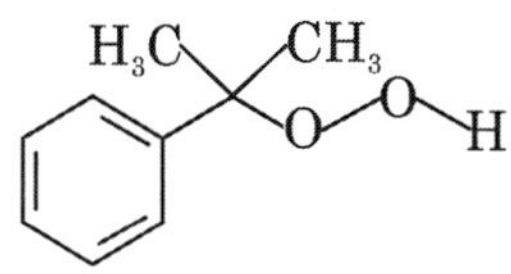

(C) 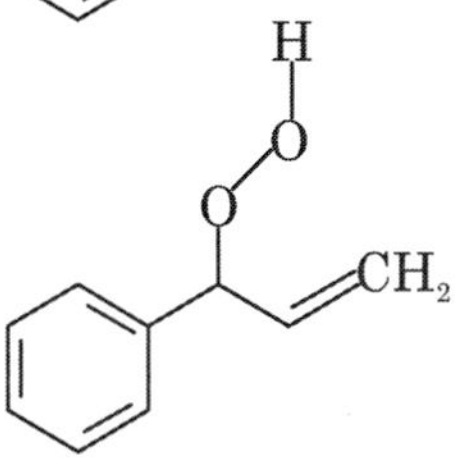

(D) 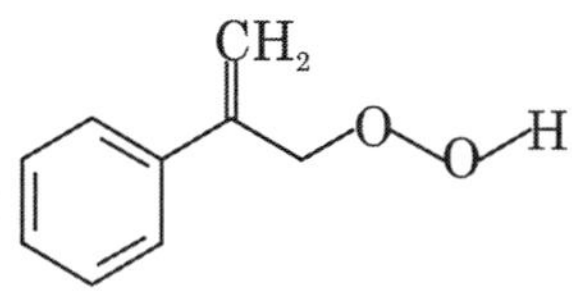

12. In the following reactions, the major product W is

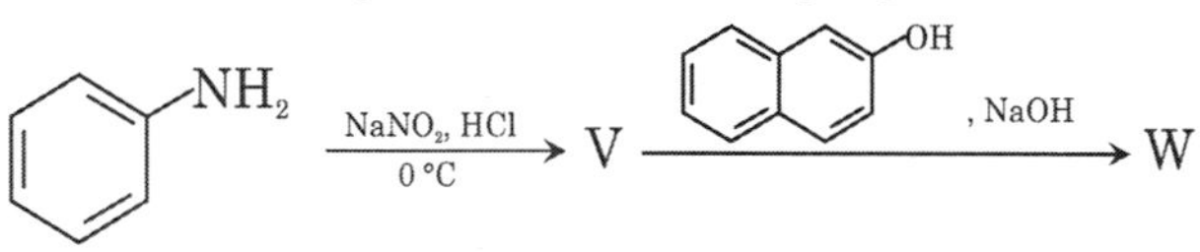

(A) 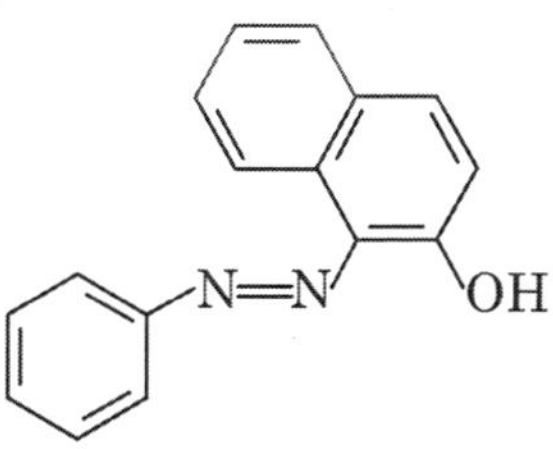

(B) 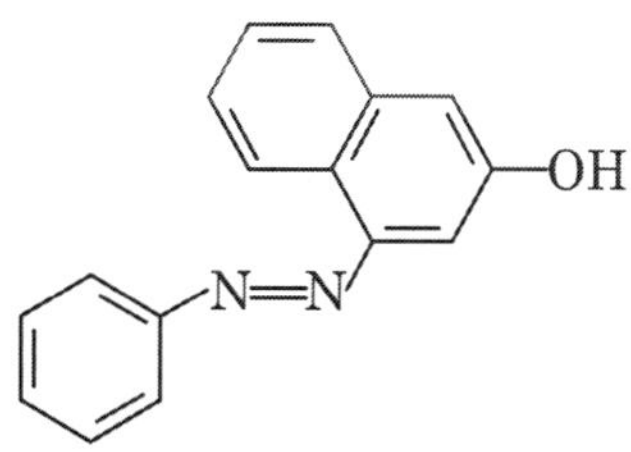

(C) 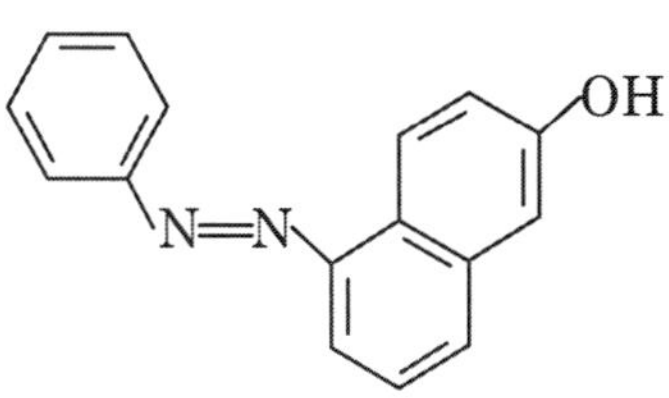

(D) 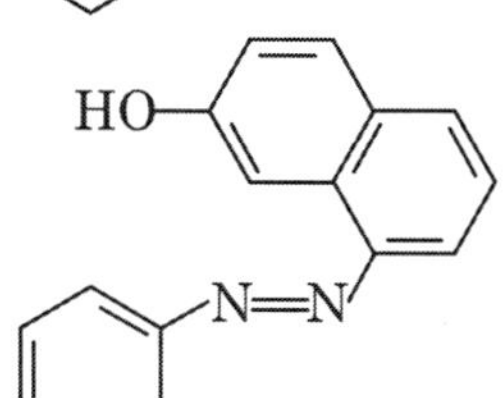

13. The correct statement(s) regarding, (i) HClO, (ii) $HClO_2$, (iii) $HClO_3$, (iv) $HClO_4$, is(are)

(A) The number of Cl═O bonds in (ii) and (iii) together is two.

(B) The number of lone pairs of electrons on Cl in (ii) and (iii) together is three.

(C) The hybridization of Cl in (iv) is sp^3.

(D) Amongst (i) to (iv), the strongest acid is (i).

14. The pair(s) of ions where both the ions are precipitated upon passing H_2S gas in presence of dil. HCl, is(are)

(A) Ba^{2+}, Zn^{2+} (B) Bi^{3+}, Fe^{3+} (C) Cu^{2+}, Pb^{2+} (D) Hg^{2+}, Bi^{3+}

15. Under hydrolytic conditions, the compounds used for preparation of liner polymer and for chain termination, respectively, are

(A) CH_3SiCl_3 and $Si(CH_3)_4$ (B) $(CH_3)_2SiCl_2$ and $(CH_3)_3SiCl$

(C) $(CH_3)_2SiCl_2$ and CH_3SiCl_3 (D) $SiCl_4$ and $(CH_3)_3SiCl$

16. When O_2 is adsorbed on a metallic surface, electron transfer occurs from the metal to O_2. The true statement(s) regarding this adsorption is(are)

(A) O_2 is physisorbed (B) heat is released

(C) occupancy of π^*_{2p} of O_2 is increased (D) bond length of O_2 is increased.

SECTION–3
(Maximum Marks: 16)

This section contains two questions.
Based on each paragraph, there will be two questions.
Each question has four options (A), (B), (C) and (D). One or more than one of these four option(s) is(are) correct. For each question, darken the bubble(s) corresponding to all the correct option(s) in the ORS.
Marking scheme:
+ 4 if only the bubble(s) corresponding to all the correct option(s) is(are) darkened.
0 if none of the bubbles is darkened and – 2 in all other cases.

Paragraph 1

When 100 mL of 1.0*M* HCl was mixed with 100 mL of 1.0*M* NaOH in an insulated beaker at constant pressure, a temperature increase of 5.7 °C was measured for the beaker and its contents (Expt. 1). Because the enthalpy of neutralization of a strong acid with a strong base is a constant (– 57.0 kJ mol^{-1}), this experiment could be used to measure the calorimeter constant. In a second experiment (Expt. 2), 100 mL of 2.0*M* acetic acid ($K_a = 2.0 \times 10^{-5}$) was mixed with 100 mL of 1.0*M* NaOH (under identical conditions to Expt. 1) where a temperature rise of 5.6 °C was measured. (Consider heat capacity of all solutions as 4.2 J g^{-1} K^{-1} and density of all solutions as 1.0 g mL^{-1})

17. Enthalpy of dissociation (in kJ mol^{-1}) of acetic acid obtained from the Expt. 2 is

(A) 1.0 (B) 10.0 (C) 24.5 (D) 51.4

18. The *p*H of the solution after Expt. 2 is

(A) 2.8 (B) 4.7 (C) 5.0 (D) 7.0

Paragraph 2

In the following reactions

$$C_8H_6 \xrightarrow[H_2]{Pd\text{–}BaSO_4} C_8H_8 \xrightarrow[(ii)\ H_2O_2,\ NaOH,\ H_2O]{(i)\ B_2H_6} X$$

$$C_8H_6 \xrightarrow{H_2O,\ HgSO_4,\ H_2SO_4} C_8H_8O$$

$$C_8H_8O \xrightarrow[(ii)\ H^+,\ heat]{(i)\ EtMgBr,\ H_2O} Y$$

19. Compound X is

(A) 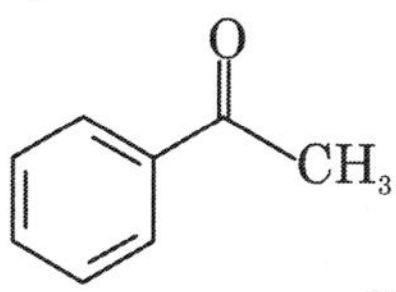

(B) 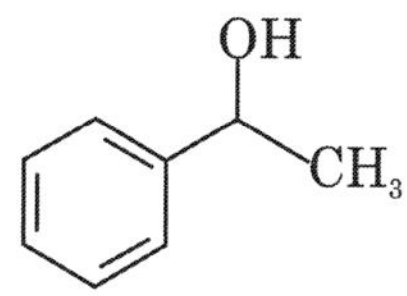

(C) 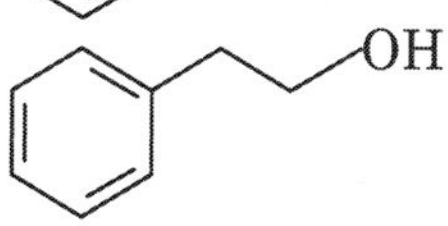

(D) 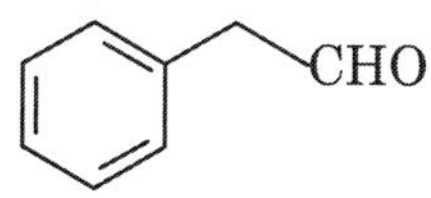

20. The major compound Y is

(A) 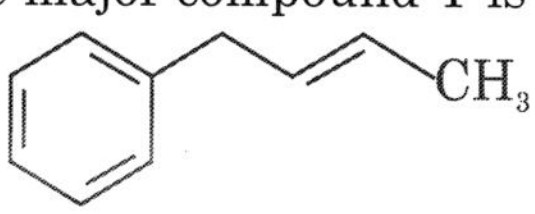

(B) 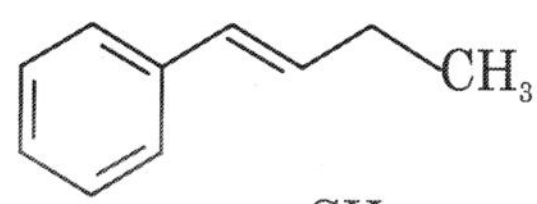

(C) 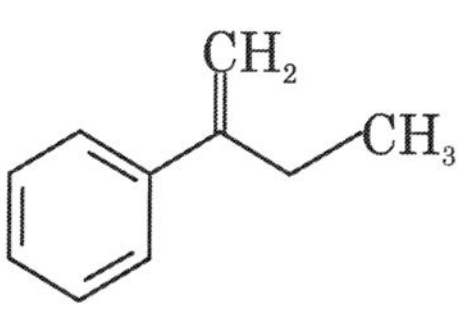

(D) 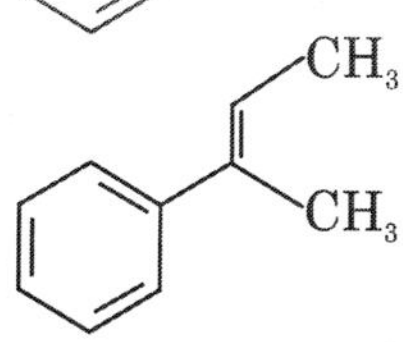

ANSWER KEYS

1. 8	2. 4	3. 4	4. 3	5. 5	6. 6
7. 3	8. 9	9. C	10. A	11. B	12. A
13. B, C	14. C, D	15. B	16. B, C, D	17. A	18. B
19. C	20. D				

SOLUTIONS

1. (8)

$$[Fe(C_2O_4)_2(H_2O)_2]^{2-} + MnO_4^- + 8H^+ \longrightarrow Mn^{2+} + Fe^{3+} + 4CO_2 + 6H_2O$$

Hence, the ratio of rate of change of $[H^+]$ to that of rate of change of $[MnO_4^-]$ is 8.

2. (4)

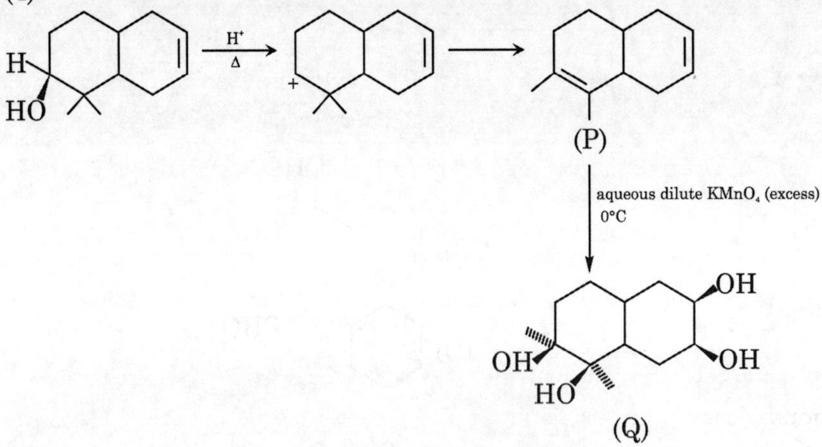

3. (4)

I. 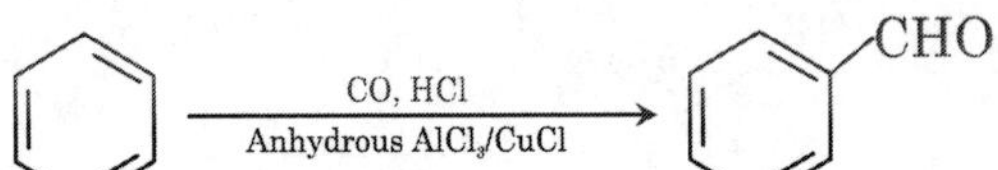

II. 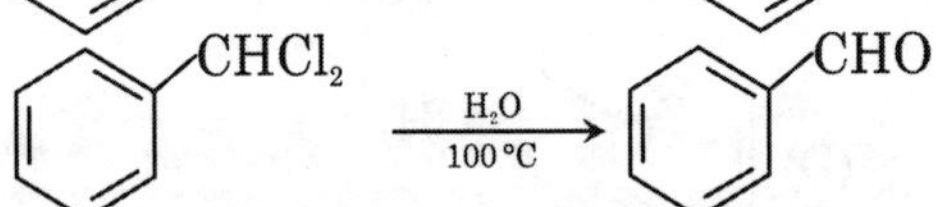

III. 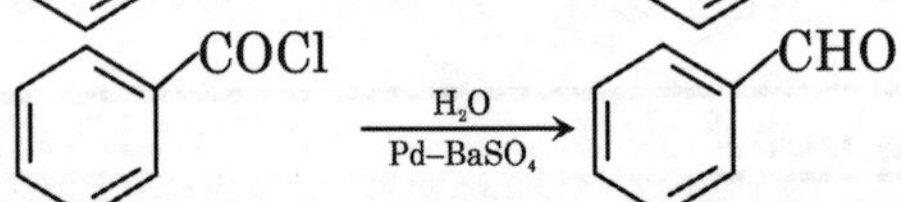

IV. 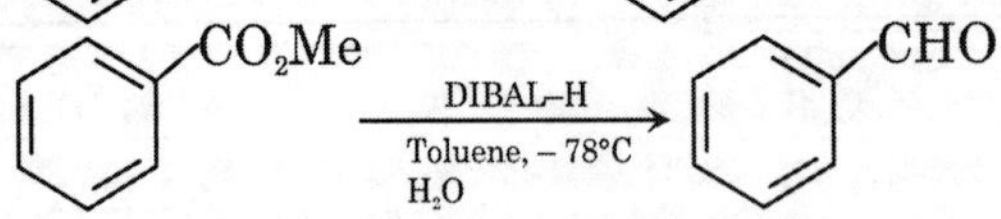

4. (3)

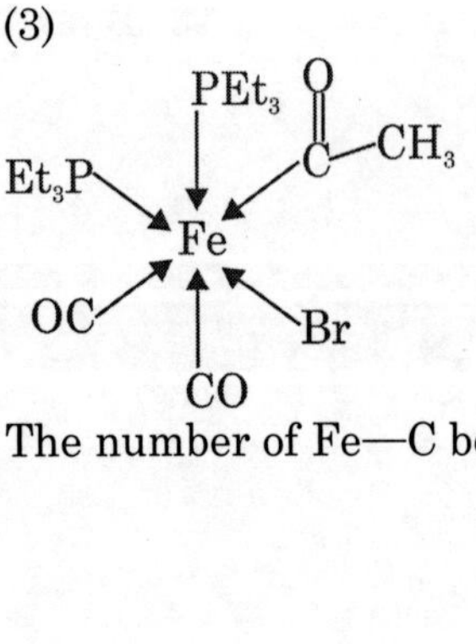

The number of Fe—C bonds is 3.

5. (5)

$[Co(en)_2Cl_2]^+ \longrightarrow$ will show *cis-trans* isomerism

$[CrCl_2(C_2O_4)_2]^{3-} \longrightarrow$ will show *cis-trans* isomerism

$[Fe(H_2O)_4(OH)_2]^+ \longrightarrow$ will show *cis-trans* isomerism

$[Fe(CN)_4(NH_3)_2]^- \longrightarrow$ will show *cis-trans* isomerism

$[Co(en)_2(NH_3)Cl]^{2+} \longrightarrow$ will show *cis-trans* isomerism

$[Co(NH_3)_4(H_2O)Cl]^{2+} \longrightarrow$ will not show *cis-trans* isomerism (Although it will show geometrical isomerism)

6. (6)

$B_2H_6 + 6MeOH \longrightarrow 2B(OMe)_3 + 6H_2$

1 mole of B_2H_6 reacts with 6 mole of MeOH to give 2 moles of $B(OMe)_3$.

3 mole of B_2H_6 will react with 18 mole of MeOH to give 6 moles of $B(OMe)_3$.

7. (3)

$HX \rightleftharpoons H^+ + X^-$ $HY \rightleftharpoons H^+ + Y^-$

$K_a = \dfrac{[H^+][X^-]}{[HX]}$ $K_a = \dfrac{[H^+][Y^-]}{[HY]}$

Λ_m for HX $= \Lambda_{m_1}$ Λ_m for HY $= \Lambda_{m_2}$ $\Lambda_{m_1} = \dfrac{1}{10}\Lambda_{m_2}$

$K_a = C\alpha^2$ $K_{a_1} = C_1 \times \left(\dfrac{\Lambda_{m_1}}{\Lambda^0_{m_1}}\right)^2$ $K_{a_2} = C_2 \times \left(\dfrac{\Lambda_{m_2}}{\Lambda^0_{m_2}}\right)^2$

$$\frac{K_{a_1}}{K_{a_2}} = \frac{C_1}{C_2} \times \left(\frac{\Lambda_{m_1}}{\Lambda_{m_2}}\right)^2 = \frac{0.01}{0.1} \times \left(\frac{1}{10}\right)^2 = 0.001 \quad \Rightarrow \quad pK_{a_1} - pK_{a_2} = 3$$

8. (9)

In the conversion of $^{238}_{92}U$ to $^{206}_{82}Pb$, 8 α-particles and 6 β-particles are ejected.

The number of gaseous moles initially = 1 mol

The number of gaseous moles finally = 1 + 8 mol; (1 mol from air and 8 mol of ^4_2He)

Hence, the ratio = 9/1 = 9

9. (C)

At large interionic distances (because $a \longrightarrow 0$) the P.E. would remain constant. However, when $r \longrightarrow 0$, repulsion would suddenly increase.

10. (A)

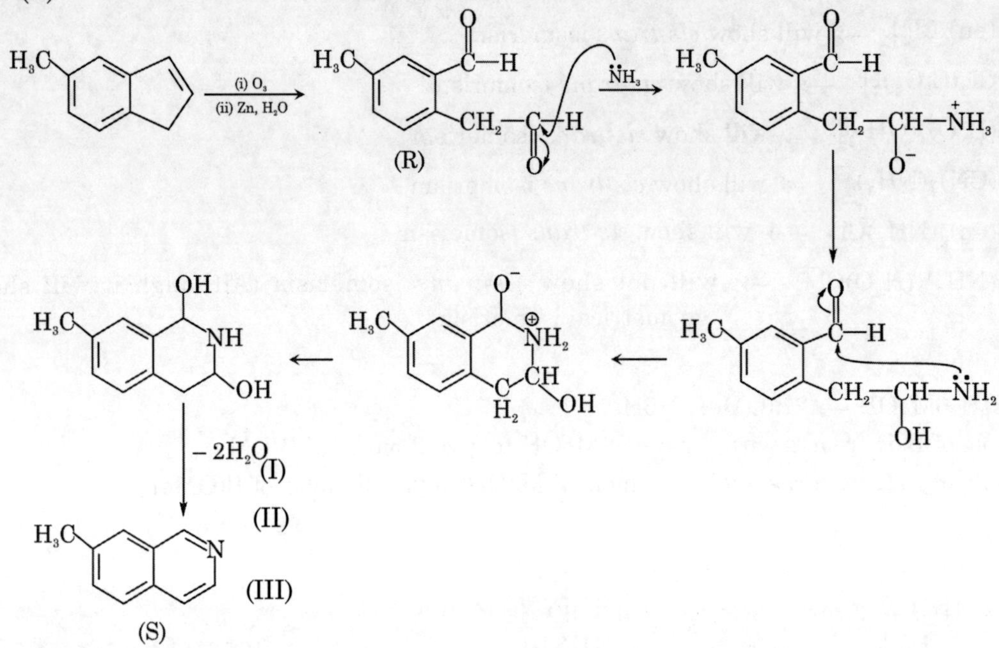

11. (B)

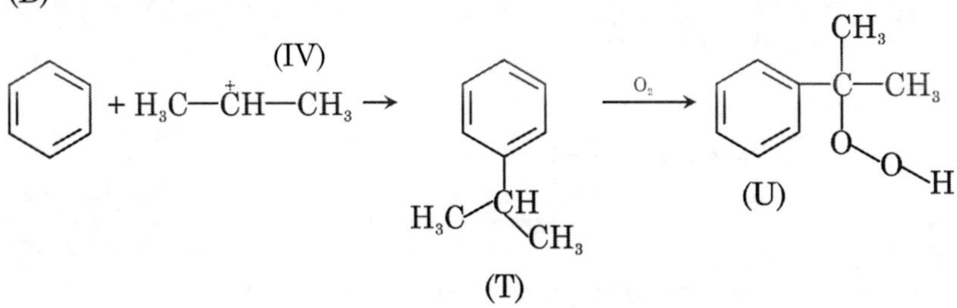

12. (A)

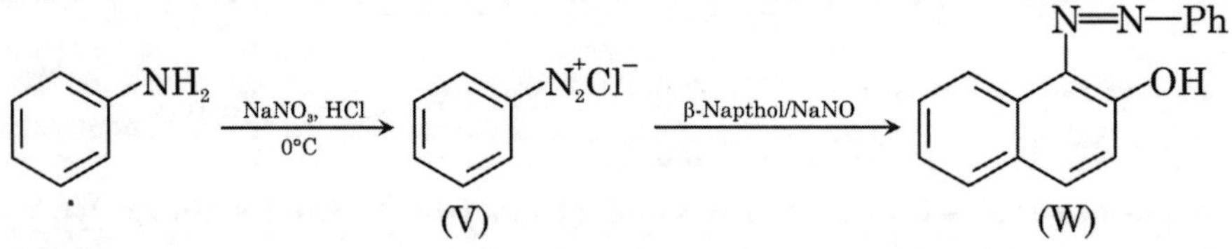

13. (B, C)

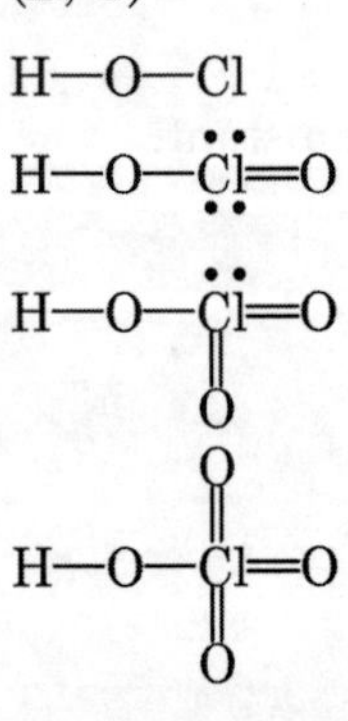

14. (C, D)

Cu^{2+}, Pb^{2+}, Hg^{2+}, Bi^{3+} give precipitate with H_2S in presence of dil. HCl.

15. (B)

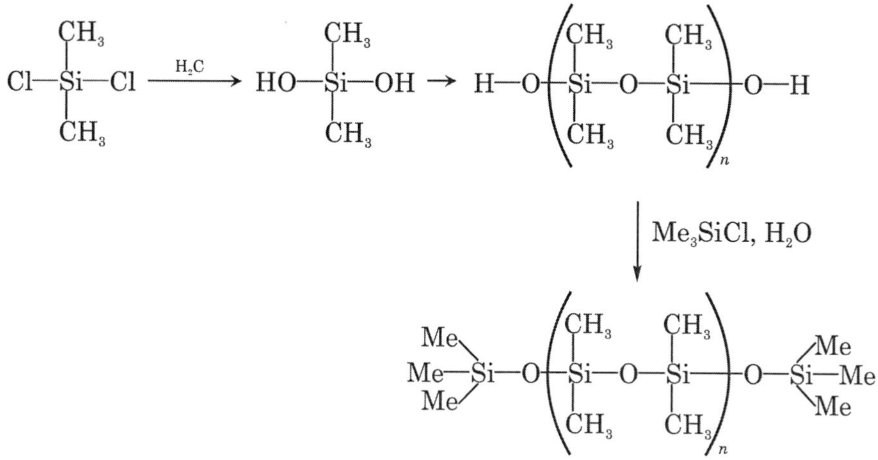

16. (B, C, D)

Adsorption of O_2 on metal surface is exothermic.
During the electron transfer metal to O_2 electron occupies π_{2p} orbital of O_2.
Due to the electron transfer to O_2 the bond order of O_2 decreases and hence bond length increases.

17. (A)

$HCl + NaOH \longrightarrow NaCl + H_2O$

$n = 100 \times 1 = 100$ m mole $= 0.1$ mole

Energy evolved due to neutralization of HCl and NaOH $= 0.1 \times 57 = 5.7$ kJ $= 5700$ J

Energy used to increase temperature of solution $= 200 \times 4.2 \times 5.7 = 4788$ J

Energy used to increase temperature of calorimeter $= 5700 - 4788 = 912$ J

m.s. $\Delta t = 912$

m.s. $\times 5.7 = 912$

ms $= 160$ J/°C [Calorimeter constant]

Energy evolved by neutralization of CH_3COOH and NaOH

$= 200 \times 4.2 \times 5.6 + 160 \times 5.6 = 5600$ J

So energy used in dissociation of 0.1 mole $CH_3COOH = 5700 - 5600 = 100$ J
Enthalpy of dissociation $= 1$ kJ/mole

18. (B)

$$CH_3COOH = \frac{1 \times 100}{200} = \frac{1}{2}$$

$$CH_3CONa = \frac{1 \times 100}{200} = \frac{1}{2}$$

$$pH = pK_a + \log \frac{[\text{salt}]}{[\text{acid}]}$$

$$pH = 5 - \log 2 + \log \frac{1/2}{1/2}$$

$$pH = 4.7$$

19. (C)

C_8H_6: double bond equivalent $= 8 + 1 - \frac{6}{2} = 6$

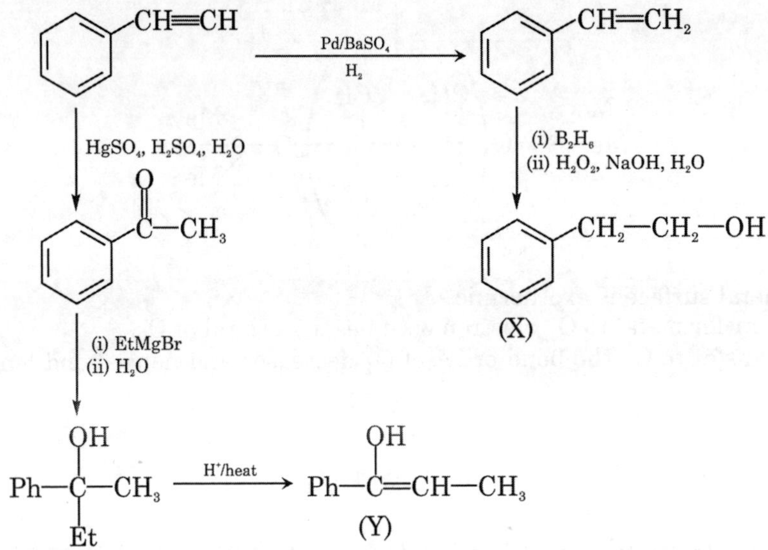

20. (D)

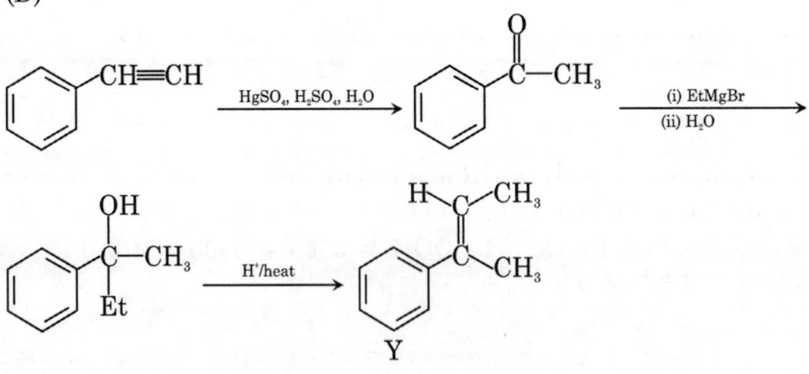

I.I.T. JEE Objective Question Paper—2016 (Main)

1. A stream of electrons from a heated filament was passed between two charged plates kept at a potential difference V esu. If e and m are charge and mass of an electron, respectively, then the value of h/λ (where λ is wavelength associated with electron wave) is given by

   (A) $2meV$ (B) $\sqrt{meV}$ (C) $\sqrt{2meV}$ (D) meV

2. 2-Chloro-2-methylpentane on reaction with sodium methoxide in methanol yields

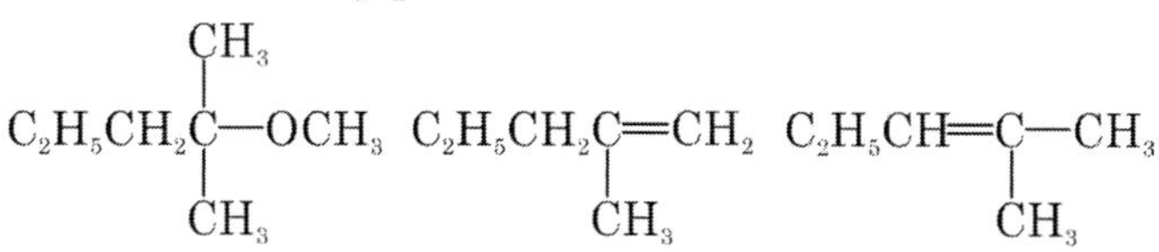

(a) (b) (c)

   (A) (a) and (c) (B) (c) only (C) (a) and (b) (D) All of these.

3. Which of the following compounds is metallic and ferromagnetic?

   (A) CrO_2 (B) VO_2 (C) MnO_2 (D) TiO_2

4. Which of the following statements about low density polythene is false?

   (A) It is a poor conductor of electricity.
   (B) Its synthesis requires dioxygen or a peroxide initiator as a catalyst.
   (C) It is used in the manufacture of buckets, dustbins, etc.
   (D) Its synthesis requires high pressure.

5. For a linear plot of log (x/m) versus log p in a Freundlich adsorption isotherm, which of the following statements is correct? (k and n are constants)

   (A) $1/n$ appears as the intercept. (B) Only $1/n$ appears as the slope.
   (C) log $(1/n)$ appears as the intercept. (D) Both k and $1/n$ appear in the slope term.

6. The heats of combustion of carbon and carbon monoxide are – 393.5 and – 285.3 kJ mol^{-1}, respectively. The heat of formation (in kJ) of carbon monoxide per mole is

   (A) 676.5 (B) – 676.5 (C) – 110.5 (D) 110.5

7. The hottest region of Bunsen flame shown in the figure below is

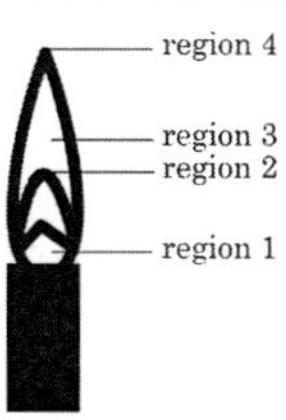

   (A) region 2 (B) region 3 (C) region 4 (D) region 1

8. Which of the following is an anionic detergent?
(A) Sodium lauryl sulphate
(B) Cetyltrimethyl ammonium bromide
(C) Glyceryl oleate
(D) Sodium stearate

9. 18 g glucose ($C_6H_{12}O_6$) is added to 178.2 g water. The vapor pressure of water (in torr) for this aqueous solution is
(A) 76.0
(B) 752.4
(C) 759.0
(D) 7.6

10. The distillation technique most suited for separating glycerol from spent-lye in the soap industry is
(A) Fractional distillation
(B) Steam distillation
(C) Distillation under reduced pressure
(D) Simple distillation.

11. The species in which the N atom is in a state of *sp* hybridization is
(A) NO_2^-
(B) NO_3^-
(C) NO_2
(D) NO_2^+

12. Decomposition of H_2O_2 follows a first order reaction. In fifty minutes the concentration of H_2O_2 decreases from 0.5 to 0.125*M* in one such decomposition. When the concentration of H_2O_2 reaches 0.05*M*, the rate of formation of O_2 will be
(A) 6.93×10^{-4} mol min^{-1}
(B) 2.66 L min^{-1} at STP
(C) 1.34×10^{-2} mol min^{-1}
(D) 6.93×10^{-2} mol min^{-1}

13. The pair having the same magnetic moment is [At. No.: Cr = 24, Mn = 25, Fe = 26, Co = 27]
(A) $[Cr(H_2O)_6]^{2+}$ and $[Fe(H_2O)_6]^{2+}$
(B) $[Mn(H_2O)_6]^{2+}$ and $[Cr(H_2O)_6]^{2+}$
(C) $[CoCl_4]^{2-}$ and $[Fe(H_2O)_6]^{2+}$
(D) $[Cr(H_2O)_6]^{2+}$ and $[CoCl_4]^{2-}$

14. The absolute configuration of 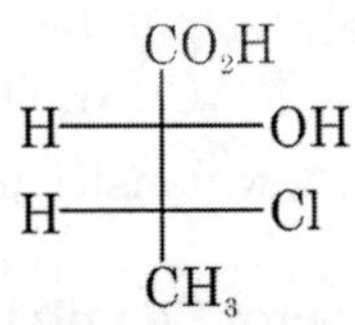

is
(A) (2S, 3R)
(B) (2S, 3S)
(C) (2R, 3R)
(D) (2R, 3S)

15. The equilibrium constant at 298 K for a reaction A + B $\rightleftharpoons$ C + D is 100. If the initial concentration of all the four species were 1*M* each, then equilibrium concentration of D (in mol L^{-1}) will be
(A) 0.818
(B) 1.818
(C) 1.182
(D) 0.182

16. Which one of the following ores is best concentrated by froth floatation method?
(A) Siderite
(B) Galena
(C) Malachite
(D) Magnetite

17. At 300 K and 1 atm, 15 mL of a gaseous hydrocarbon requires 375 mL air containing 20% O_2 by volume for complete combustion. After combustion the gases occupy 330 mL. Assuming that the water formed is in liquid form and the volumes were measured at the same temperature and pressure, the formula of the hydrocarbon is
(A) C_3H_6
(B) C_3H_8
(C) C_4H_8
(D) C_4H_{10}

18. The pair in which phosphorus atoms have a formal oxidation state of + 3 is
(A) Pyrophosphorous and hypophosphoric acids
(B) Orthophosphorous and hypophosphoric acids

(C) Pyrophosphorous and pyrophosphoric acids
(D) Orthophosphorous and pyrophosphorous acids

19. Which one of the following complexes shows optical isomerism?
(A) *cis*-$[Co(en)_2Cl_2]Cl$ (B) *trans*-$[Co(en)_2Cl_2]Cl$
(C) $[Co(NH_3)_4Cl_2]Cl$ (D) $[Co(NH_3)_3Cl_3]$
(en = ethylenediamine)

20. The reaction of zinc with dilute and concentrated nitric acid, respectively, produces
(A) NO_2 and NO (B) NO and N_2O (C) NO_2 and N_2O (D) N_2O and NO_2

21. Which one of the following statements about water is false?
(A) Water can act both as an acid and as a base.
(B) There is extensive intramolecular hydrogen bonding in the condensed phase.
(C) Ice formed by heavy water sinks in normal water.
(D) Water is oxidised to oxygen during photosynthesis.

22. The concentration of fluoride, lead, nitrate and iron in a water sample from an underground lake was found to be 1000 ppb, 40 ppb, 100 ppm and 0.2 ppm, respectively. This water is unsuitable for drinking due to high concentration of
(A) Lead (B) Nitrate (C) Iron (D) Fluoride

23. The main oxides formed on combustion of Li, Na and K in excess of air are, respectively
(A) LiO_2, Na_2O_2 and K_2O (B) Li_2O_2, Na_2O_2 and KO_2
(C) Li_2O, Na_2O_2 and KO_2 (D) Li_2O, Na_2O and KO_2

24. Thiol group is present in
(A) Cystine (B) Cysteine (C) Methionine (D) Cytosine

25. Galvanisation is applying a coating of
(A) Cr (B) Cu (C) Zn (D) Pb

26. Which of the following atoms has the highest first ionization energy?
(A) Na (B) K (C) Sc (D) Rb

27. In the Hofmann bromamide degradation reaction, the number of moles of NaOH and Br_2 used per mole of amine produced are
(A) Four moles of NaOH and two moles of Br_2
(B) Two moles of NaOH and two moles of Br_2
(C) Four moles of NaOH and one mole of Br_2
(D) One mole of NaOH and one mole of Br_2

28. Two closed bulbs of equal volume (V) containing an ideal gas initially at pressure p_i and temperature T_1 are connected through a narrow tube of negligible volume as shown in the figure below. The temperature of one of the bulbs is then raised to T_2. The final pressure p_f is

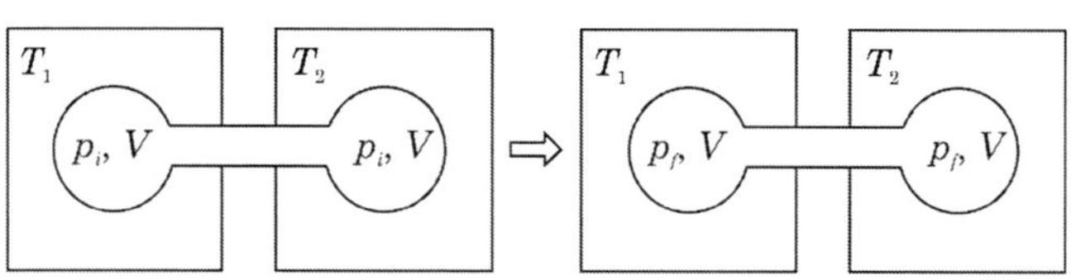

(A) $2p_i\left(\frac{T_1}{T_1+T_2}\right)$ (B) $2p_i\left(\frac{T_2}{T_1+T_2}\right)$ (C) $2p_i\left(\frac{T_1T_2}{T_1+T_2}\right)$ (D) $p_i\left(\frac{T_1T_2}{T_1+T_2}\right)$

29. The reaction of propene with HOCl ($Cl_2 + H_2O$) proceeds through the intermediate

(A) CH_3—CH^+—CH_2—Cl (B) CH_3—CH(OH)—CH_2^+
(C) CH_3—CHCl—CH_2^+ (D) CH_3—CH^+—CH_2—OH

30. The product of the reaction given below is

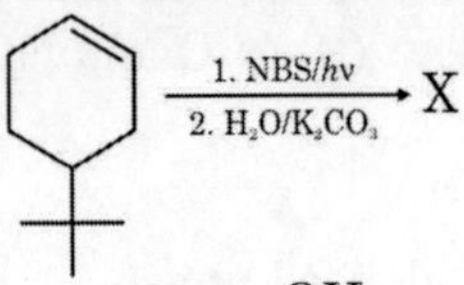

(A) 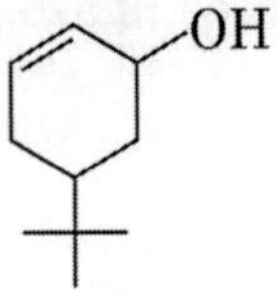

(B) 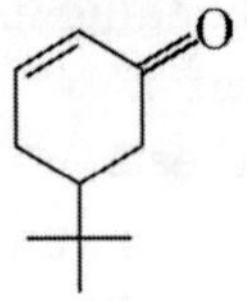

(C) 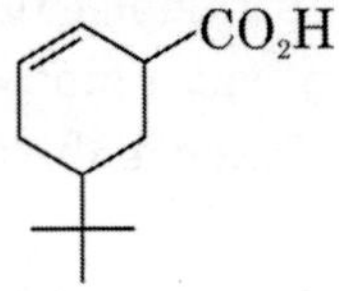

(D) 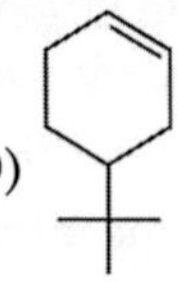

ANSWER KEYS

1. C	2. D	3. A	4. C	5. B	6. C
7. A	8. A	9. B	10. C	11. D	12. A
13. A	14. A	15. B	16. B	17. B	18. D
19. A	20. D	21. B	22. B	23. C	24. B
25. C	26. C	27. C	28. B	29. A	30. A

SOLUTIONS

1. (C)

$$\lambda = \frac{h}{mv} = \frac{h}{\sqrt{2m(\text{KE})}}$$

Hence, $\frac{h}{\lambda} = \sqrt{2meV}$

2. (D)

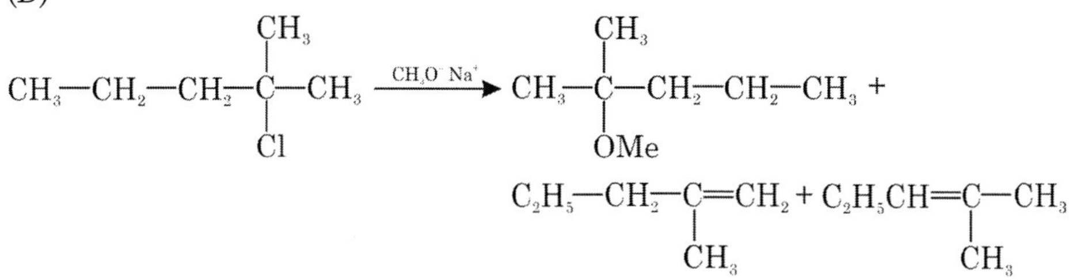

3. (A)

CrO_2 is metallic and ferromagnetic. CrO_2 like metals conducts electricity.

4. (C)

HDPE is used in making buckets and dustbins.

5. (B)

$$\frac{x}{m} = kp^{\frac{1}{n}}$$

$$\log \frac{x}{m} = \log k + \frac{1}{n} \log p$$

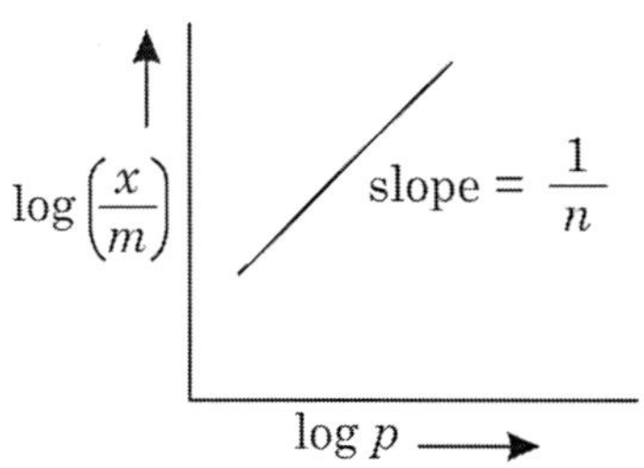

6. (C)

$$C(g) + O_2(g) \longrightarrow CO_2(g) \quad ...(1)$$

$$CO(g) + \frac{1}{2} O_2(g) \longrightarrow CO_2(g) \quad ...(2)$$

$\Delta H_1 = -393.5$ kJ/mol; $\Delta H_2 = -283.5$ kJ/mol

$$C(g) + \frac{1}{2} O_2(g) \;\; CO_2(g)$$

$\Delta_f H_{(CO,\,g)} = \Delta H_1 - \Delta H_2$

$= -393.5 - (-283.5) = -393.5 + 283.5 = -110$ kJ/mol

7. (A)

Region 2 (blue flame)

8. (A)

Sodium lauryl sulphate ($CH_3(CH_2)_{11}OSO_3Na$) is an anionic detergent.

9. (B)

$n_{C_6H_{12}O_6} = 0.1;\ n_{H_2O} = \dfrac{178.2}{18} = 9.9$

$\dfrac{p^0 - p}{p^0} = \dfrac{0.1}{10}$

$p^0 - p = \dfrac{0.1}{10} \times 760 = 7.6$

$p = p^0 - 7.6 = 760 - 7.6 = 752.4$ torr

10. (C)

Glycerol decomposes before its boiling point.

11. (D)

$O{=}\overset{\oplus}{N}{=}O$

sp hybridisation

12. (A)

$H_2O_2 \longrightarrow H_2O + \dfrac{1}{2} O_2$

$2t_{1/2} = 50$ mins; $t_{1/2} = 25$ mins

Hence, $k = \dfrac{0.693}{25} \quad -\dfrac{d[H_2O_2]}{dt} = \dfrac{d[O_2]}{dt} \times 2$

Hence, $\dfrac{d[O_2]}{dt} = \dfrac{1}{2} \times \dfrac{0.693}{25} \times 0.05 = 6.93 \times 10^{-4}$ mol min^{-1}

13. (A)

$Cr^{2+} \rightarrow d^4 \rightarrow$ Number of unpaired electrons 4

$Fe^{2+} \rightarrow d^6 \rightarrow$ Number of unpaired electrons 4

Both configurations have four unpaired electrons due to weak field ligand H_2O.

14. (A)

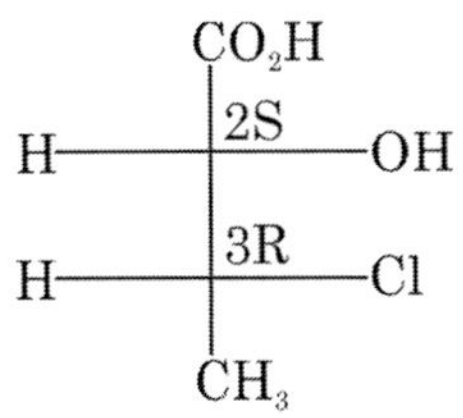

(2S, 3R)

15. (B)

A	+	B	$\rightleftharpoons$	C	+	D
1		1		1		1
$1-x$		$1-x$		$1+x$		$1+x$

Given $K_{eq} = 100 = \dfrac{(1+x)^2}{(1-x)^2} = 100$

or $\dfrac{1+x}{1-x} = 10$ or $1 + x = 10 - 10x$ or $11x = 9$ or $x = \dfrac{9}{11}$

Hence, $[D] = 1 + \dfrac{9}{11} = 1 + 0.818 = 1.818M$

16. (B)

Galena (PbS) is a sulphide ore and it is best concentrated by froth floatation method.

17. (B)

$$C_xH_y + \left(x + \frac{y}{4}\right)O_2 \longrightarrow xCO_2 + \frac{y}{2}H_2O$$

15 mL $\quad 15\left(x + \frac{y}{4}\right)$ mL $\quad 15x$ mL

$$V_{O_2} = \frac{20}{100} \times 375 = 75 \text{ mL} = 15\left(x + \frac{y}{4}\right) \text{ or } x + \frac{y}{4} = 5$$

If we put $x = 3$ and $y = 8$, then the above equation is satisfied and the answer will be C_3H_8.

18. (D)

$H_4P_2O_5$ (pyrophosphorous acid) (+ 3 state)

H_3PO_3 (orthophosphorous acid) (+ 3 state)

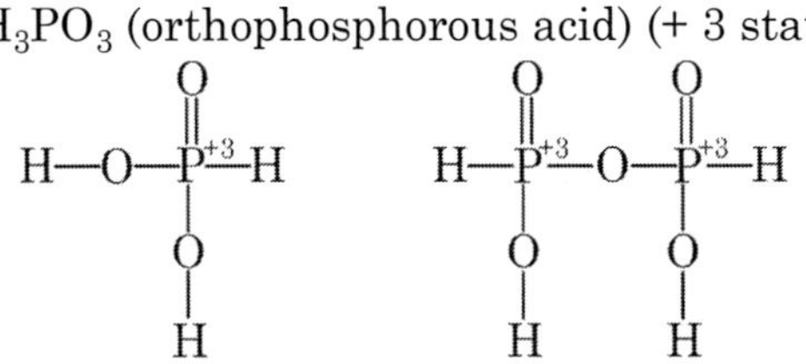

Orthophosphorous acid Pyrophosphorous acid

19. (A)

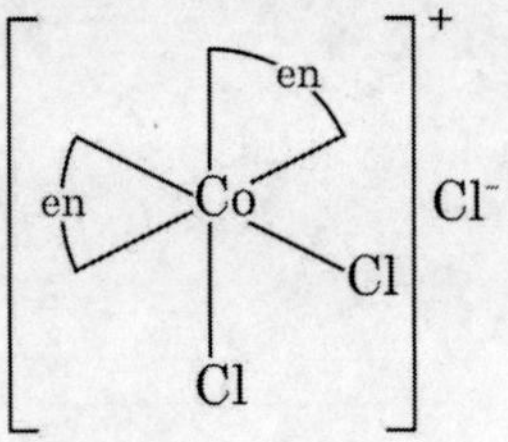

cis-$[CO(en)_2Cl_2]Cl$
(optically active)

20. (D)

$$Zn + 4HNO_3 \text{ (conc.)} \longrightarrow Zn(NO_3)_2 + 2H_2O + 2NO_2(\uparrow)$$

$$4Zn + 10HNO_3 \text{ (dil.)} \longrightarrow 4Zn(NO_3)_2 + N_2O(\uparrow) + 5H_2O$$

21. (B)

Ice shows intermolecular hydrogen bonding.

22. (B)

Maximum limit of nitrate in potable water is 50 ppm.

23. (C)

24. (B)

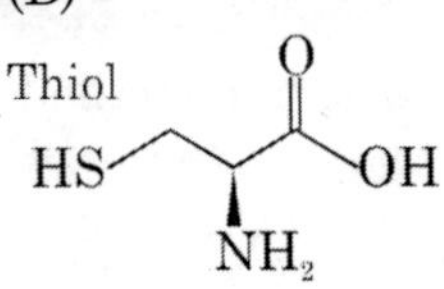

(Cysteine)

25. (C)

26. (C)

Ionisation energy (kJ/mol): Na, 496; K, 419; Rb, 403; Sc, 631.

27. (C)

$$R-\overset{\overset{\large O}{\|}}{C}-NH_2 + 4NaOH + Br_2 \longrightarrow RNH_2 + 2NaBr + Na_2CO_3 + 2H_2O$$

28. (B)

Initial total moles:

$$n = \frac{p_iV}{RT_1} + \frac{p_iV}{RT_1} = \frac{2p_iV}{RT_1} \quad \text{...(1)}$$

Final total moles:

$$n = \frac{p_fV}{RT_1} + \frac{p_fV}{RT_2} \quad \text{...(2)}$$

Equating the two equations, we get

$$\frac{2p_i}{T_1} = \frac{p_f}{T_1} + \frac{p_f}{T_2} \quad \text{or} \quad p_f = \frac{2p_i}{T_1} \times \frac{T_1 T_2}{(T_1 + T_2)} \quad \text{or} \quad p_f = 2p_i \times \frac{T_2}{(T_1 + T_2)}$$

29. (A)

$$CH_3{-}CH{=}CH_2 + Cl^{\oplus} \longrightarrow CH_3{-}\overset{\oplus}{C}H{-}CH_2Cl$$ (Markownikoff's addition

30. (A)

Allylic bromination followed by hydrolysis.

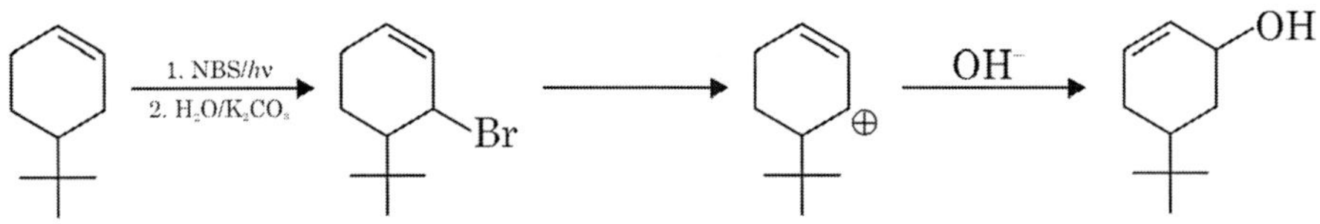

I.I.T. JEE Objective Question Paper—2016 (Advanced)

PAPER—1

SECTION–1

(Maximum Marks: 15)

This section contains five questions.

Each question has four options (A), (B), (C) and (D). Only one of these four options is correct.

For each question, darken the bubbles corresponding to the correct option in the ORS.

For each question, marks will be awarded in one of the following categories:

Full Marks : + 3 If only the bubbles corresponding to the correct option is darkened.

Zero Marks : 0 If none of the bubbles is darkened.

Negative Marks : – 1 In all other cases.

1. P is the probability of finding the 1s electron of hydrogen atom in a spherical shell of infinitesimal thickness, dr, at a distance r from the nucleus. The volume of this shell is $4\pi r^2 dr$. The qualitative sketch of the dependence of P on r is

(A)

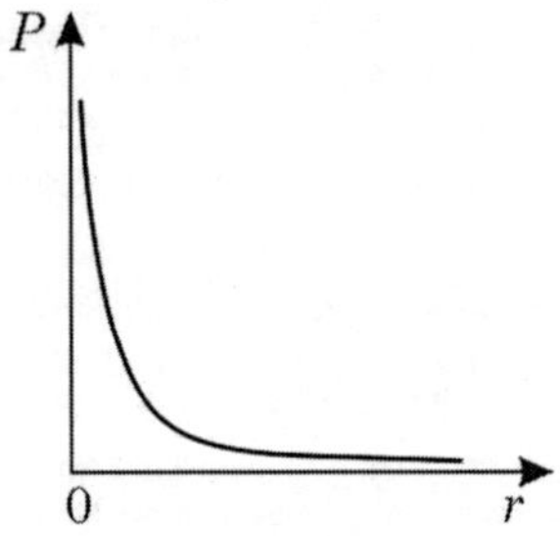

(B)

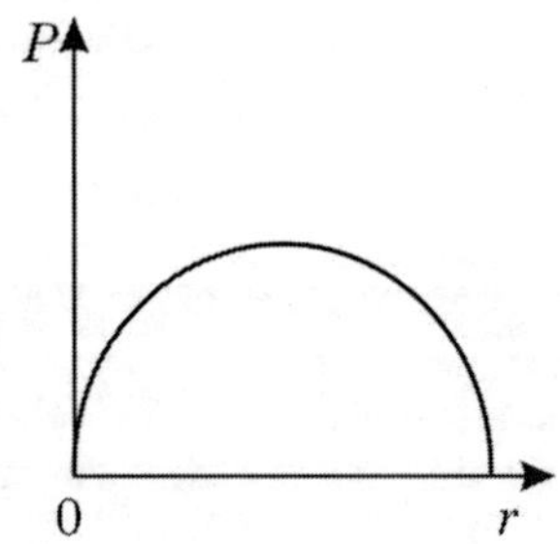

(C)

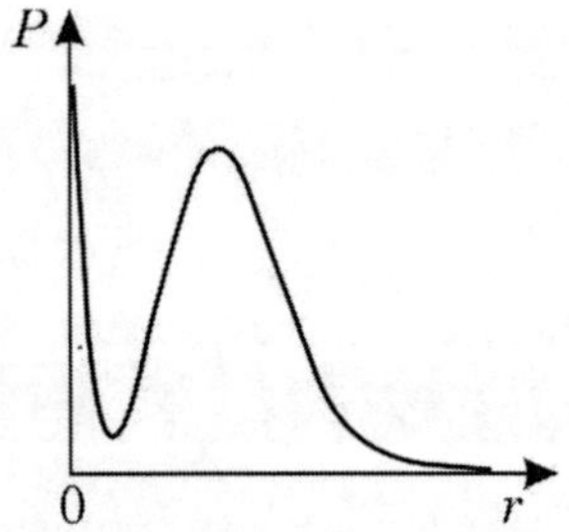

(D)

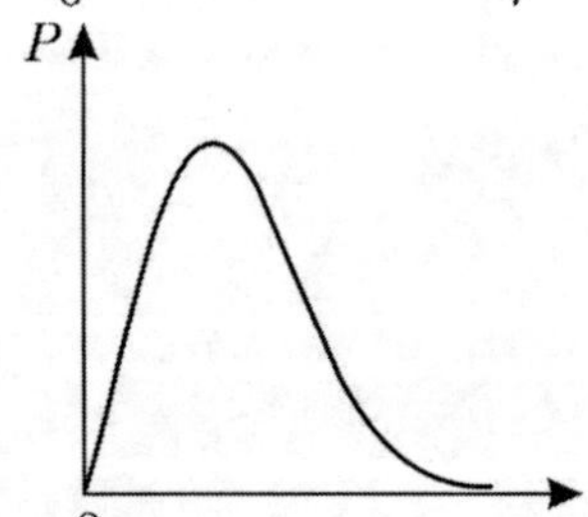

2. One mole of an ideal gas at 300 K in thermal contact with surroundings expands isothermally from 1.0 L to 2.0 L against a constant pressure of 3.0 atm. In this process, the change in entropy of surrounding (ΔS_{surr}) in J K^{-1} is (1 L atm = 101.3 J)
(A) 5.763 (B) 1.013 (C) – 1.013 (D) – 5.763

3. The increasing order of atomic radii of the following Group 13 elements is
(A) Al < Ga < In < T1 (B) Ga < Al < In < T1 (C) Al < In < Ga < T1 (D) Al < Ga < T1 < In

4. Among $[Ni(CO)_4]$, $[NiCl_4]^{2-}$, $[Co(NH_3)_4Cl_2]Cl$, $Na_3[CoF_6]$, Na_2O_2 and CrO_2, the total number of paramagnetic compounds is
(A) 2 (B) 3 (C) 4 (D) 5

5. On complete hydrogenation, natural rubber produces
(A) ethylene-propylene copolymer (B) vulcanised rubber
(C) polypropylene (D) polybutylene

SECTION–2
(Maximum Marks: 32)

This section contains EIGHT questions.
Each question has FOUR options (A), (B), (C) and (D). ONE OR MORE THAN ONE of these four option(s) is(are) correct.
For each question, darken the bubble(s) corresponding to all the correct option(s) in the ORS.
For each question, marks will be awarded in one of the following categories:
Full Marks : + 4 If only the bubble(s) corresponding to all the correct option(s) is(are) darkened.
Partial Marks : + 1 For darkening a bubble corresponding to each correct option, provided NO incorrect option is darkened.
Zero Marks : 0 If none of the bubbles is darkened.
Negative Marks : – 2 In all other cases.
For example, if (A), (C) and (D) are all the correct options for a question, darkening all these three will result in + 4 marks; darkening only (A) and (D) will result in + 2 marks; and darkening (A) and (B) will result in – 2 marks, as a wrong option is also darkened.

6. According to the Arrhenius equation,
(A) a high activation energy usually implies a fast reaction.
(B) rate constant increases with increase in temperature. This is due to a greater number of collisions whose energy exceeds the activation energy.
(C) higher the magnitude of activation energy, stronger is the temperature dependence of the rate constant.
(D) the pre-exponential factor is a measure of the rate at which collisions occur, irrespective of their energy.

7. A plot of the number of neutrons (n) against the number of protons (p) of stable nuclei exhibits upward deviation from linearity for atomic number, $Z > 20$. For an unstable nucleus having n/p ratio less than 1, the possible mode(s) of decay is(are)

(A) β^--decay (β emission) (B) orbital or K-electron capture
(C) neutron emission (D) β^+-decay (positron emission).

8. The crystalline form of borax has
(A) tetranuclear $[B_4O_5(OH)_4]^{2-}$ unit
(B) all boron atoms in the same plane
(C) equal number of sp^2 and sp^3 hybridised boron atoms
(D) one terminal hydroxide per boron atom.

9. The compound(s) with TWO lone pairs of electrons on the central atom is(are)
(A) BrF_5 (B) ClF_3 (C) XeF_4 (D) SF_4

10. The reagent(s) that can selectively precipitate S^{2-} from a mixture of S^{2-} and SO_4^{2-} in aqueous solution is(are)
(A) $CuCl_2$ (B) $BaCl_2$
(C) $Pb(OOCCH_3)_2$ (D) $Na_2[Fe(CN)_5NO]$

11. Positive Tollen's test is observed for

(A) 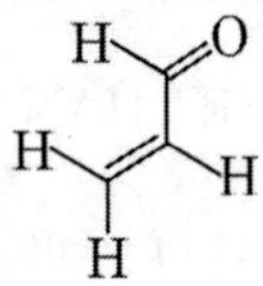 (B) 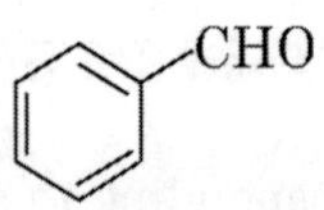 (C) 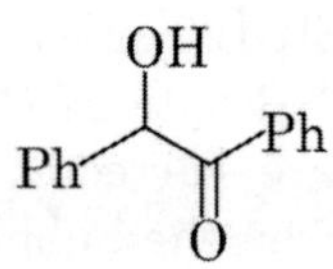 (D) 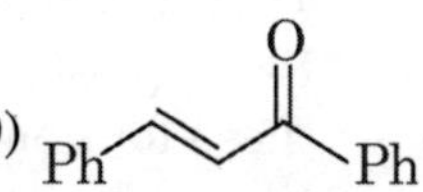

12. The product(s) of the following reaction sequence is(are)

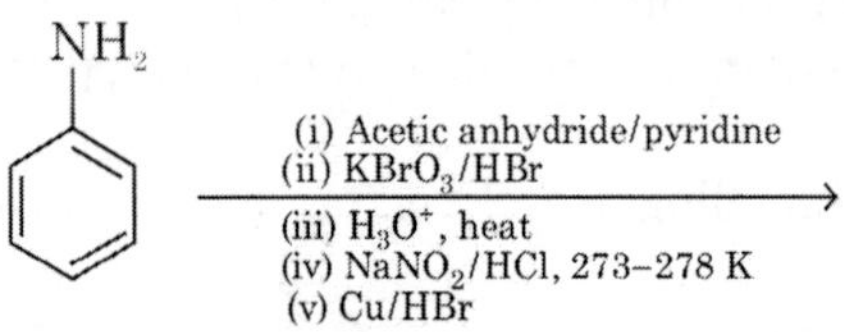

(A) 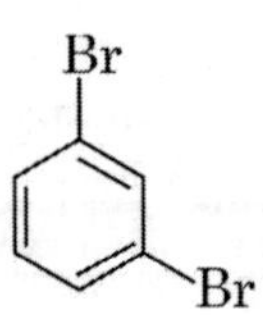 (B) 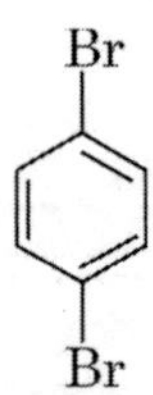 (C) 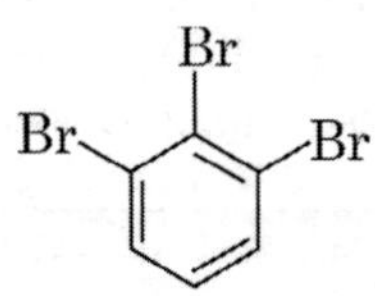 (D) 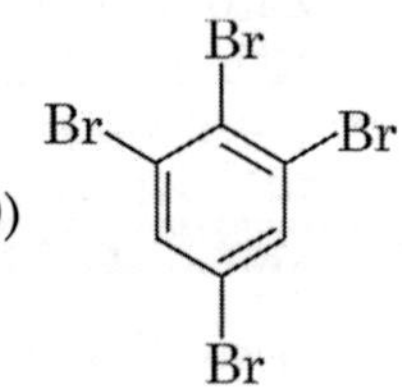

13. The correct statement(s) about the following reaction sequence is(are)

$$\text{Cumene } (C_9H_{12}) \xrightarrow[\text{(ii) } H_3O^+]{\text{(i) } O_2} P \xrightarrow{CHCl_3/NaOH} Q \text{ (major)} + R \text{ (minor)}$$

$$Q \xrightarrow[PhCH_2Br]{NaOH} S$$

(A) R is steam volatile
(B) Q gives dark violet coloration with 1% aqueous $FeCl_3$ solution
(C) S gives yellow precipitate with 2,4-dinitrophenylhydrazine
(D) S gives dark violet coloration with 1% aqueous $FeCl_3$ solution.

SECTION–3
(Maximum Marks: 15)

This section contains five questions.

The answer to each question is a single digit integer ranging from 0 to 9, both inclusive.
For each question, darken the bubble corresponding to the correct integer in the ORS.
For each question, marks will be awarded in one of the following categories:

Full Marks : + 3 If only the bubble corresponding to the correct answer is darkened.

Zero Marks : 0 In all other cases.

14. The mole fraction of a solute in a solution is 0.1. At 298 K, molarity of this solution is the same as its molality. Density of this solution at 298 K is 2.0 g cm^{-3}. The ratio of the molecular weights of the solute and solvent, $\left(\frac{MW_{solute}}{MW_{solvent}}\right)$, is

15. The diffusion coefficient of an ideal gas is proportional to its mean free path and mean speed. The absolute temperature of an ideal gas is increased 4 times and its pressure is increased 2 times. As a result, the diffusion coefficient of this gas increases x times. The value of x is

16. In neutral or faintly alkaline solution, 8 moles of permanganate anion quantitatively oxidise thiosulphate anions to produce X moles of a sulphur containing product. The magnitude of X is

17. The number of geometric isomers possible of the complex $[CoL_2Cl_2]^-$ ($L = H_2NCH_2CH_2O^-$) is

18. In the following monobromination reaction, the number of possible chiral products is

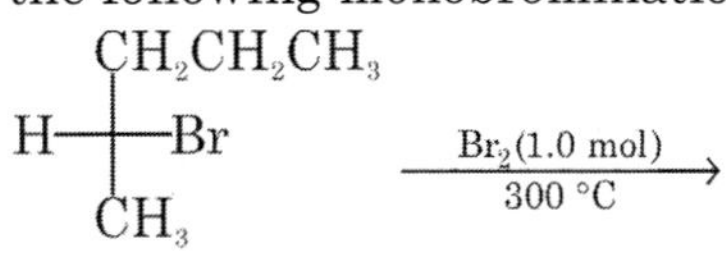

(0.1 mol)
(enantiomerically pure)

ANSWER KEYS

1. D	2. C	3. B	4. B	5. A	6. B, C, D
7. B, D	8. A, C, D	9. B, C	10. A	11. A, B	12. B
13. B, C	14. 9	15. 4	16. 6	17. 5	18. 5

SOLUTIONS

1. (D)

The probability distribution curve for $1s$ electron of hydrogen atom is

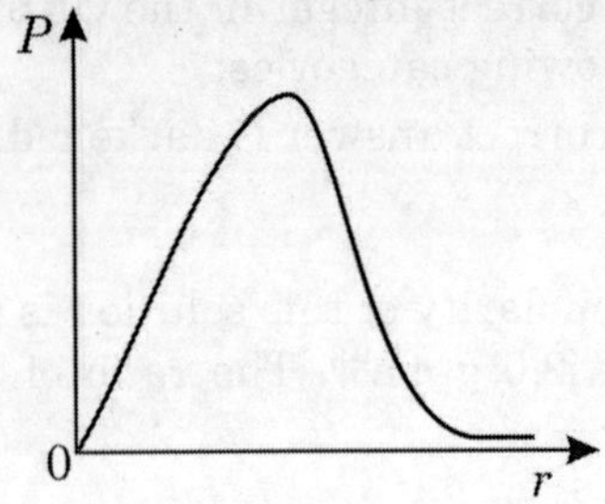

2. (C)

Isothermal process, $\Delta U = 0$

$dq = -dW = P_{ext}(V_2 - V_1) = 3$ L atm $= 3 \times 101.3$ J

$$\Delta S_{surrounding} = -\frac{3 \times 101.3}{300} JK^{-1} = -1.013\ JK^{-1}$$

Hence, $\Delta S_{surr} = -1.013\ JK^{-1}$

3. (B)

Increasing order of atomic radius of group 13 elements: Ga < Al < In < Tl.

Due to poor shielding of d-electrons in Ga, its radius decreases below Al.

4. (B)

Number of paramagnetic compounds are 3. The paramagnetic compounds are:

$[NiCl_4]^{2-}$, $Na_3[CoF_6]$, CrO_2

5. (A)

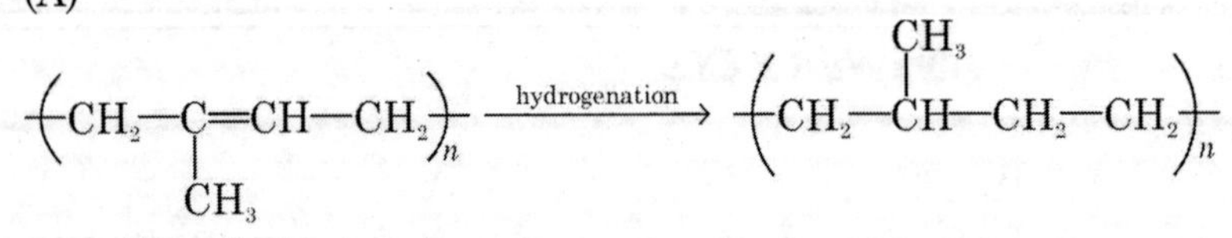

Natural rubber

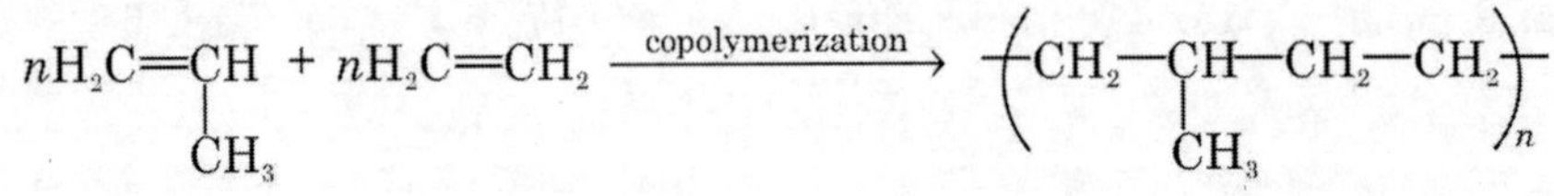

propylene ethylene propylene-ethylene copolymer

6. (B, C, D)

7. (B, D)

$\frac{n}{p} < 1$

Positron, emission

${}_{a}X^{b} \rightarrow {}^{0}_{+1}e + {}_{a-1}Y^{b}$

K-electron capture

${}_{a}X^{b} + {}_{-1}e^{0} \rightarrow {}_{a-1}Y^{b}$

Both the processes cause an increase in n/p ratio towards 1, thus stabilising the nucleus.

8. (A, C, D)

The structure of anion of borax is

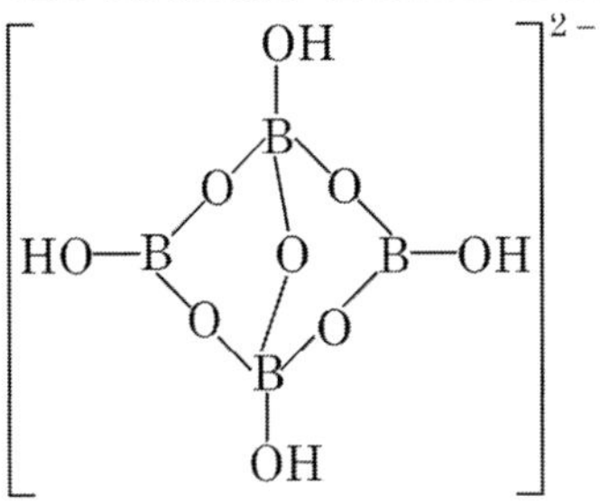

9. (B, C)

ClF_3 and XeF_4 contain two lone pairs of electrons on the central atom.

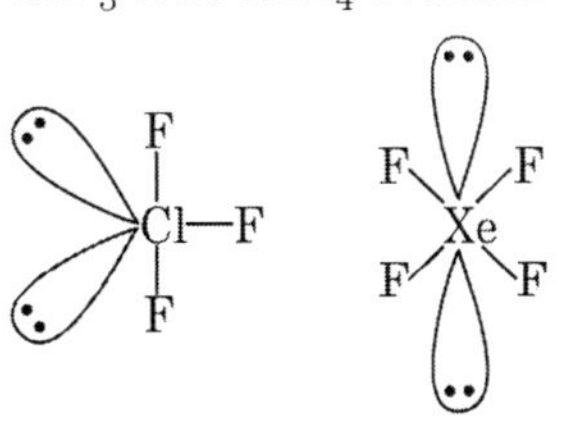

10. (A)

The reagent that can selectively precipitate S^{2-} and SO_4^{2-} in aqueous solution is $CuCl_2$.

$S^{2-} + CuCl_2 \rightarrow CuS(\downarrow) + 2Cl^-$

11. (A, B)

Beside aldehyde α-hydroxy ketones can also exhibit Tollen's test due to rearrangement in aldehyde via ene-diol intermediate. (However, this needs a terminal α-carbon.)

12. (B)

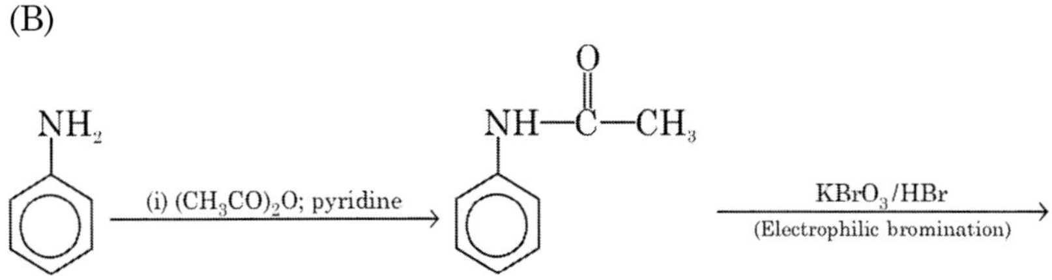

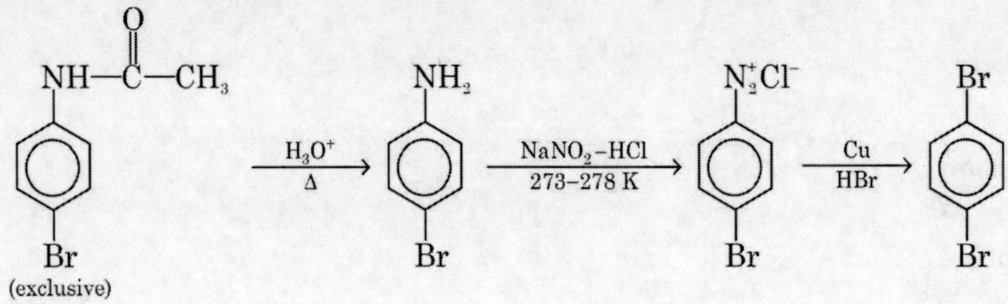

13. (B, C)

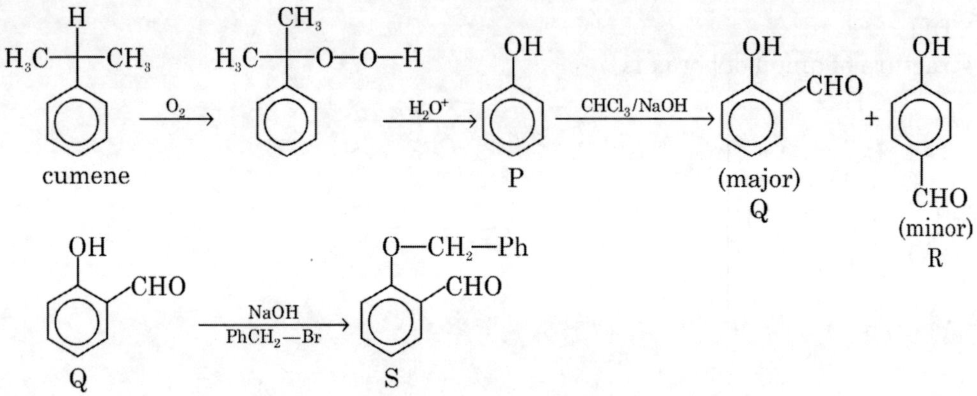

Q (not R) is steam volatile due to intramolecular hydrogen bonding.
Q gives violet colouration due to phenolic functional group.
S gives yellow precipitate with 2,4-DNP due to aldehydic group.
S does not have free phenolic group to respond to $FeCl_3$ test.

14. (9)

$$m = \frac{X_A \times 1000}{X_B \times M_A}$$

$$m = \frac{1000}{9M_A} \qquad \text{...(i)}$$

$$M = \frac{n_B \times 1000 \times d}{n_A \times M_A + n_B \times M_B} = \frac{X_B \times 1000 \times d}{X_A \times M_A + X_B \times M_B}$$

$$= \frac{200}{0.9\,M_A + 0.1\,M_B}$$

$$= \frac{2000}{9\,M_A + M_B} \qquad \text{...(ii)}$$

Since $m = M$, we have

$$\frac{1000}{9M_A} = \frac{2000}{9M_A + M_B} \quad \Rightarrow \quad 9M_A + M_B = 18M_A$$

Hence, $9M_A = M_B$ or $\frac{M_B}{M_A} = 9$

15. (4)

Mean free path $(\ell) \propto \dfrac{T}{P}$; Mean speed $(C_{av}) \propto \sqrt{T}$

Diffusion coefficient (D) is proportional to both mean free path (ℓ) and mean speed (C_{av}).

Hence, $D \propto \dfrac{T^{3/2}}{P}$

At temperature T_1 and P_1,

$$D_1 = \frac{KT_1^{3/2}}{P_1} \qquad ...(1)$$

If T_1 is increased 4 times and P_1 is increased 2 times,

$$D_2 = \frac{K(4T_1)^{3/2}}{2P_1} \qquad ...(2)$$

Hence, $\dfrac{D_2}{D_1} = \dfrac{(4)^{3/2}}{2} = \dfrac{2^3}{2} = 4$

16. (6)

$$8MnO_4^- + 3S_2O_3^{2-} + H_2O \xrightarrow{\text{faintly alkaline}} 8MnO_2 + 6SO_4^{2-} + 2OH^-$$

Hence, 8 moles MnO_4^- produce 6 moles SO_4^{2-}.

17. (5)

Number of geometrical isomers = 5

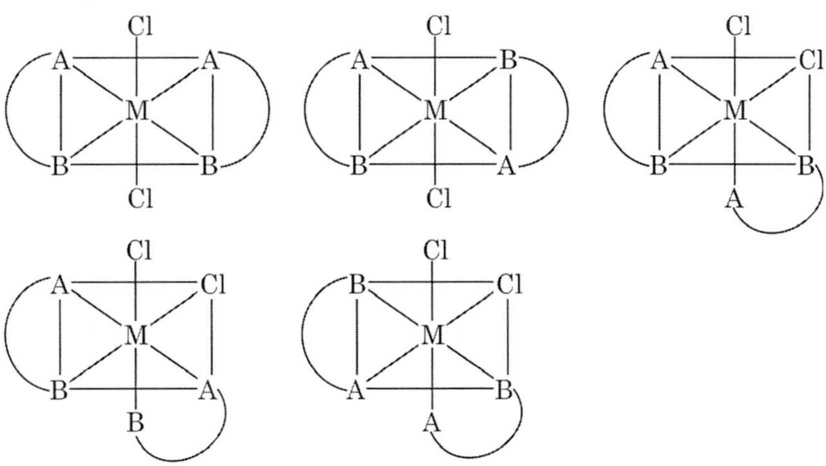

AB = $H_2NCH_2CH_2O^-$; A site is N and B site is O.

18. (5)

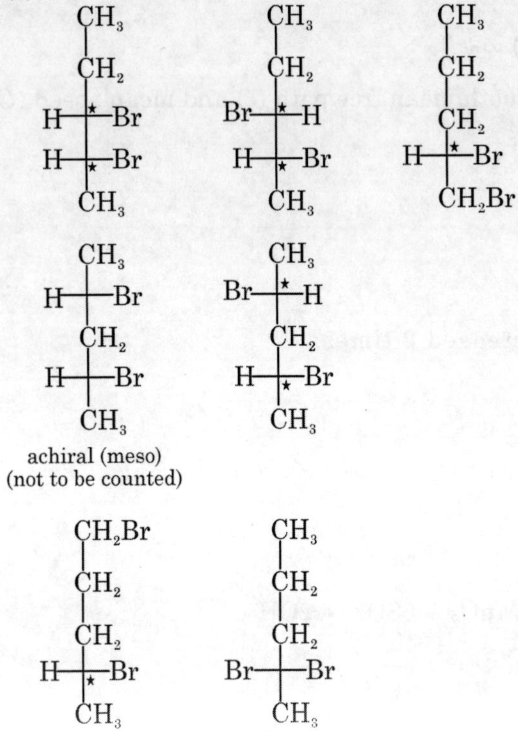

achiral (meso) (not to be counted)

no chiral centre (not to be counted)

I.I.T. JEE Objective Question Paper—2016 (Advanced)

PAPER—2

SECTION–1
(Maximum Marks: 18)

This section contains six questions.
Each question has FOUR options (A), (B), (C) and (D). Only one of these four options is correct.
For each question, darken the bubble corresponding to the correct option in the ORS.
For each question, marks will be awarded in one of the following categories:
Full Marks : + 3 If only the bubble corresponding to the correct option is darkened.
Zero Marks : 0 If none of the bubbles is darkened.
Negative Marks : – 1 In all other cases.

1. The correct order of acidity for the following compounds is

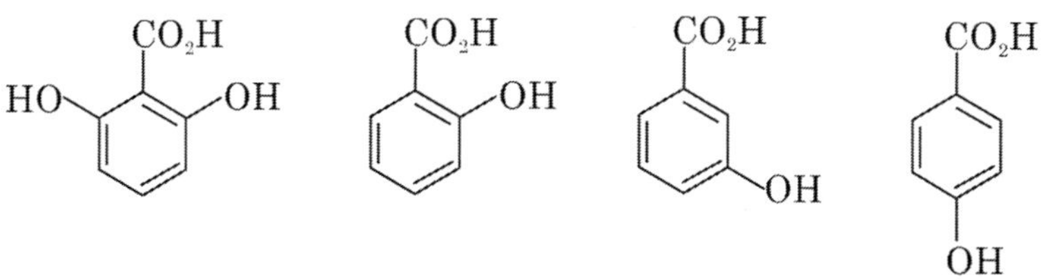

I II III IV

(A) I > II > III > IV (B) III > I > II > IV (C) I > II > IV > II > I (D) I > III > IV > II

2. The geometries of the ammonia complexes of Ni^{2+}, Pt^{2+} and Zn^{2+}, respectively, are
(A) octahedral, square planar and tetrahedral
(B) square planar, octahedral and tetrahedral
(C) tetrahedral, square planar and octahedral
(D) octahedral, tetrahedral and square planar.

3. For the following electrochemical cell at 298 K,

$Pt(s) \mid H_2(g, 1\text{ bar}) \mid H^+(aq, 1M) \mid\mid M^{4+}(aq), M^{2+}(aq) \mid Pt(s)$

$$E_{cell} = 0.092 \text{ V when } \frac{[M^{2+}(aq)]}{[M^{4+}(aq)]} = 10^x$$

Given: $E^0_{M^{4+}/M^{2+}} = 0.151$ V; $2.303\dfrac{RT}{F} = 0.059$ V

The value of x is

(A) – 2 (B) – 1 (C) 1 (D) 2

4. The major product of the following reaction sequence is

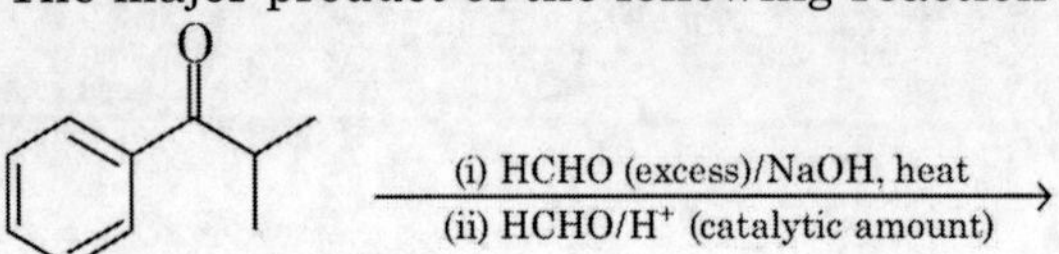

(A) 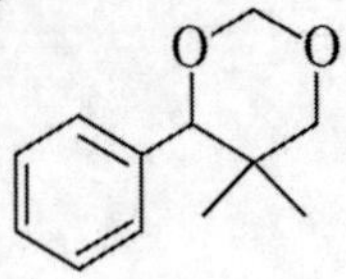

(B) 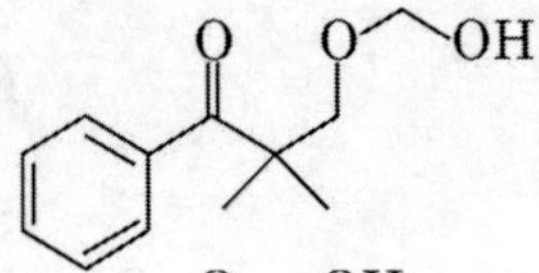

(C) 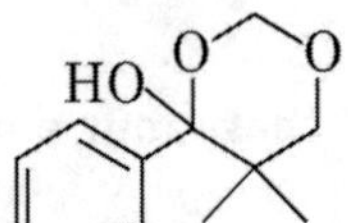

(D) 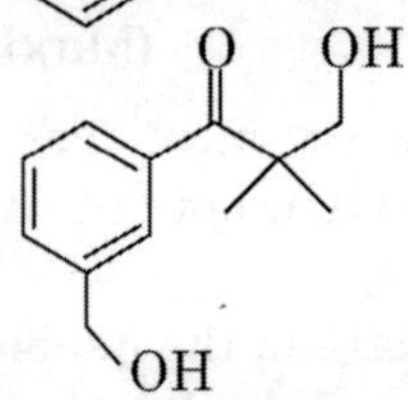

5. In the following reaction sequence in aqueous solution, the species X, Y and Z, respectively, are

$$S_2O_3^{2-} \xrightarrow{Ag^+} \underset{\text{clear solution}}{X} \xrightarrow{Ag^+} \underset{\text{white precipitate}}{Y} \xrightarrow{\text{with time}} \underset{\text{black precipitate}}{Z}$$

(A) $[Ag(S_2O_3)_2]^{3-}$, $Ag_2S_2O_3$, Ag_2S

(B) $[Ag(S_2O_3)_3]^{5-}$, Ag_2SO_3, Ag_2S

(C) $[Ag(SO_3)_2]^{3-}$, $Ag_2S_2O_3$, Ag

(D) $[Ag(SO_3)_3]^{3-}$, Ag_2SO_4, Ag

6. The qualitative sketches I, II and III given below show the variation of surface tension with molar concentration of three different aqueous solutions of KCl, CH_3OH and $CH_3(CH_2)_{11}OSO_3^-Na^+$ at room temperature. The correct assignment of the sketches is

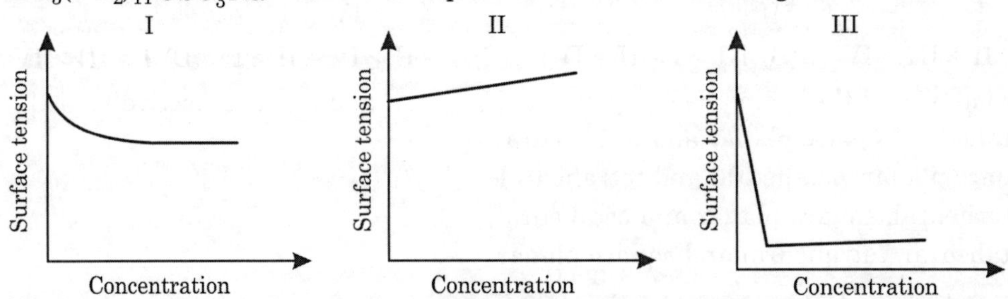

(A) I : KCl II : CH_3OH III : $CH_3(CH_2)_{11}OSO_3^-Na^+$

(B) I : $CH_3(CH_2)_{11}OSO_3^-Na^+$ II : CH_3OH III : KCl

(C) I : KCl   II : $CH_3(CH_2)_{11}OSO_3^-Na^+$ III : CH_3OH

(D) I : CH_3OH II : KCl   III : $CH_3(CH_2)_{11}$ $OSO_3^-Na^+$

SECTION–2
(Maximum Marks: 32)

This section contains eight questions.

Each question has four options (A), (B), (C) and (D). One or more than one of these four option(s) is(are) correct.

For each question, darken the bubble(s) corresponding to all the correct option(s) in the ORS.

For each question, marks will be awarded in one of the following categories:

Full Marks : + 4 If only the bubble(s) corresponding to all the correct option(s) is(are) darkened.

Partial Marks : + 1 For darkening a bubble corresponding to each correct option, provided NO incorrect option is darkened.

Zero Marks : 0 If none of the bubbles is darkened.

Negative Marks : – 2 In all other cases.

For example, if (A), (C) and (D) are all the correct options for a question, darkening all these three will result in + 4 marks; darkening only (A) and (D) will result in + 2 marks; and darkening (A) and (B) will result in – 2 marks, as a wrong option is also darkened.

7. For 'invert sugar', the correct statement(s) is(are)

(Given: specific rotations of (+)-sucrose, (+)-maltose, L-(–)-glucose and L-(+)-fructose in aqueous solution are + 66°, + 140°, – 52° and + 92°, respectively)

(A) 'invert sugar' is prepared by acid catalyzed hydrolysis of maltose

(B) 'invert sugar' is an equimolar mixture of D-(+)-glucose and D-(–)-fructose

(C) specific rotation of 'invert sugar' is – 20°

(D) on reaction with Br_2 water, 'invert sugar' forms saccharic acid as one of the products.

8. Among the following reaction(s) which gives(give) *tert*-butylbenzene as the major product is(are)

(A) 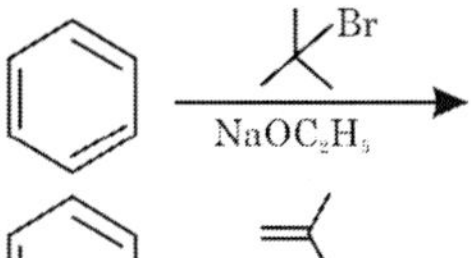

(B) 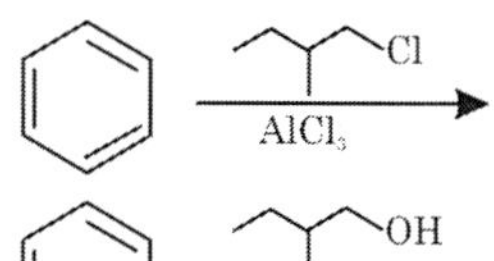

(C) 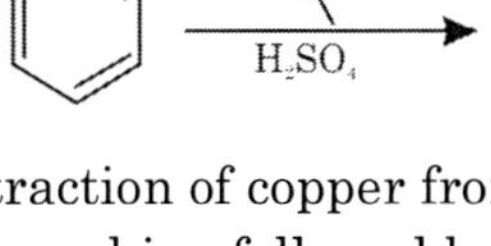

(D) 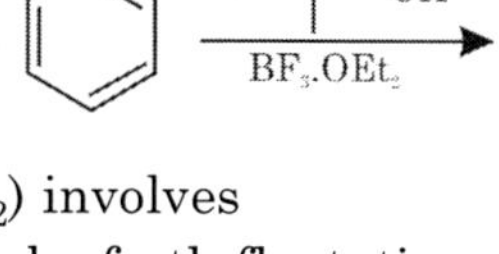

9. Extraction of copper from copper pyrite ($CuFeS_2$) involves

(A) crushing followed by concentration of the ore by froth floatation

(B) removal of iron as slag

(C) self-reduction step to produce 'blister copper' following evolution of SO_2

(D) refining of 'blister copper' by carbon reduction.

10. The correct statement(s) for cubic close packed (*ccp*) three-dimensional structure is(are)

(A) The number of the nearest neighbours of an atom present in the topmost layer is 12

(B) The efficiency of atom packing is 74%

(C) The number of octahedral and tetrahedral voids per atom are 1 and 2, respectively

(D) The unit cell edge length is $2\sqrt{2}$ times the radius of the atom.

11. Reagent(s) which can be used to bring about the following transformation is(are)

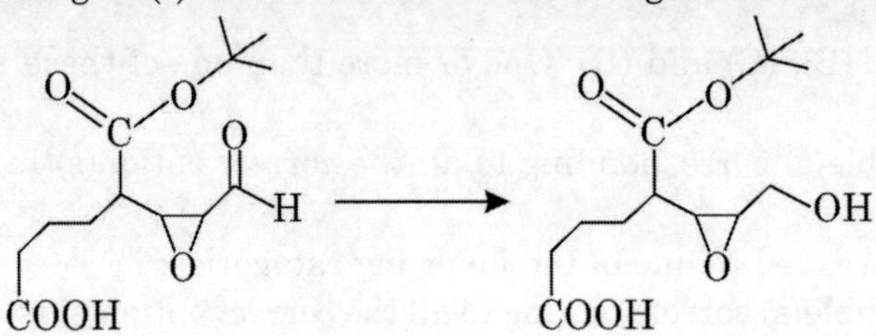

(A) $LiAlH_4$ in $(C_2H_5)_2O$ (B) BH_3 in THF

(C) $NaBH_4$ in C_2H_5OH (D) Raney Ni/H_2 in THF

12. Mixture(s) showing positive deviation from Raoult's law at 35°C is(are)

(A) carbon tetrachloride + methanol (B) carbon disulphide + acetone

(C) benzene + toluene (D) phenol + aniline

13. The nitrogen containing compound produced in the reaction of HNO_3 with P_4O_{10}

(A) can also be prepared by reaction of P_4 and HNO_3

(B) is diamagnetic

(C) contains one N—N bond

(D) reacts with Na metal producing a brown gas.

14. According to Molecular Orbital Theory

(A) C_2^{2-} is expected to be diamagnetic

(B) O_2^{2+} is expected to have a longer bond length than O_2

(C) N_2^+ and N_2^- have the same bond order

(D) He_2^+ has the same energy as two isolated He atoms.

SECTION–3
(Maximum Marks: 12)

This section contains two questions.

Based on each paragraph, there will be two questions.

Each question has four options (A), (B), (C) and (D). Only one of these four options is correct.

For each question, darken the bubble(s) corresponding to all the correct option(s) in the ORS.

For each question, marks will be awarded in one of the following categories:

Full Marks : + 3 If only the bubble corresponding to all the correct option is darkened.
Zero Marks : 0 In all other cases.

Paragraph 1

Thermal decomposition of gaseous X_2 to gaseous X at 298 K takes place according to the following equation:

$$X_2(g) \rightleftharpoons 2X(g)$$

The standard reaction Gibbs energy, $\Delta_r G^0$, of this reaction is positive. At the start of the reaction, there is one mole of X_2 and no X. As the reaction proceeds, the number of moles of X formed is given by β. Thus, $\beta_{\text{equilibrium}}$ is the number of moles of X formed at equilibrium. The reaction is carried out at a constant total pressure of 2 bar. Consider the gases to behave ideally. (Given: $R = 0.083$ L bar K^{-1} mol^{-1})

15. The equilibrium constant K_p for this reaction at 298 K, in terms of $\beta_{\text{equilibrium}}$, is

(A) $\dfrac{8\beta^2_{\text{equilibrium}}}{2-\beta_{\text{equilibrium}}}$ (B) $\dfrac{8\beta^2_{\text{equilibrium}}}{4-\beta^2_{\text{equilibrium}}}$ (C) $\dfrac{4\beta^2_{\text{equilibrium}}}{2-\beta_{\text{equilibrium}}}$ (D) $\dfrac{4\beta^2_{\text{equilibrium}}}{4-\beta^2_{\text{equilibrium}}}$

16. The incorrect statement among the following, for this reaction is

(A) decrease in the total pressure will result in formation of more moles of gaseous X
(B) at the start of the reaction, dissociation of gaseous X_2 takes place spontaneously
(C) $\beta_{\text{equilibrium}} = 0.7$
(D) $K_c < 1$

Paragraph 2

Treatment of compound O with $KMnO_4/H^+$ gave P, which on heating with ammonia gave Q. The compound Q on treatment with Br_2/NaOH produced R. On strong heating, Q gave S, which on further treatment with ethyl-2-bromopropanoate in the presence of KOH followed by acidification, gave a compound T.

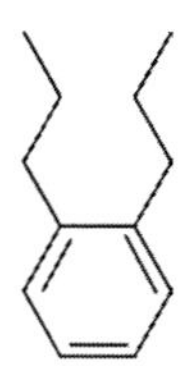

[O]

17. The compound R is

(A) 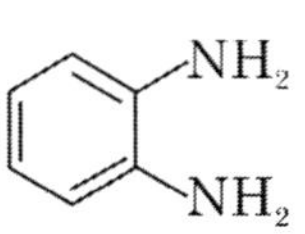

(B) 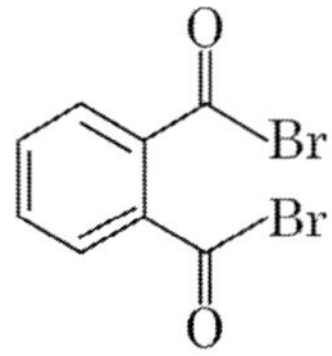

(C) 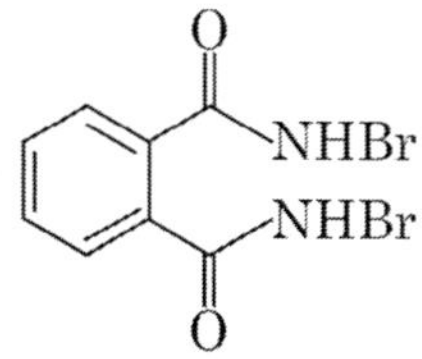

(D) 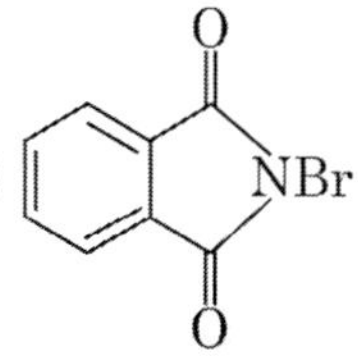

18. The compound T is

(A) glycine (B) alanine (C) valine (D) serine

ANSWER KEYS

1. A	2. A	3. D	4. A	5. A	6. D
7. B, C	8. B, C, D	9. A, B, C	10. B, C, D	11. C, D	12. A, B
13. B, D	14. A, C	15. B	16. C	17. A	18. B

SOLUTIONS

1. (A)

Stabler the conjugate base, stronger the acid.

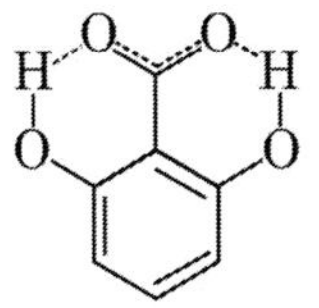

Conjugate base is stabilized by intramolecular H-bond from both the sides.

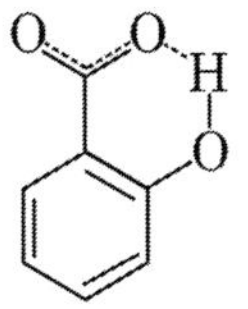

Conjugate base is stabilized by intramolecular H-bond from one side.

2. (A)

$[Ni(NH_3)_6]^{2+}$; $[Pt(NH_3)_4]^{2+}$; $[Zn(NH_3)_4]^{2+}$

octahedral square planer tetrahedral

3. (D)

Anode: $H_2 - 2e \rightleftharpoons 2H^+$

Cathode: $M^{4+} + 2e \rightleftharpoons M^{2+}$

Net cell reaction : $H_2 + M^{4+} \rightleftharpoons 2H^+ + M^{2+}$

$$E_{cell} = E^0_{cell} - \frac{0.059}{2}\log\frac{[H^+]^2[M^{2+}]}{[M^{4+}] \times P_{H_2}}$$

$$0.092 = 0.151 - \frac{0.059}{2}\log 10^x$$

$$\text{or } 0.092 = 0.151 - \frac{0.059}{2}x$$

$$\text{or } \frac{0.059x}{2} = 0.151 - 0.092$$

or $0.059x = 0.059 \times 2$

or $x = 2$

4. (A)

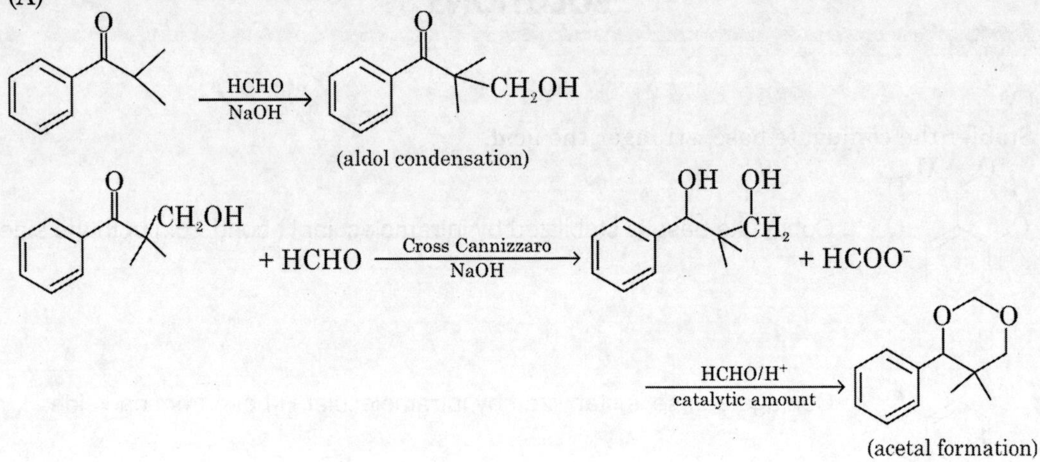

5. (A)

$$Ag^+ + 2S_2O_3^{2-} \rightarrow \underset{\substack{X \\ \text{(clear solution)}}}{[Ag(S_2O_3)_2]^{3-}} \xrightarrow{Ag^+} \underset{\substack{Y \\ \text{(white)}}}{Ag_2S_2O_3} \xrightarrow{\text{with time}} \underset{\substack{Z \\ \text{(black)}}}{Ag_2S}$$

$$Ag_2S_2O_3 + H_2O \rightarrow Ag_2S + H_2SO_4$$

6. (D)

Strong electrolytes such as KCl increase the surface tension slightly. Low molar mass organic compounds usually decrease the surface tension. Surface active organic compounds such as detergents sharply decrease surface tension.

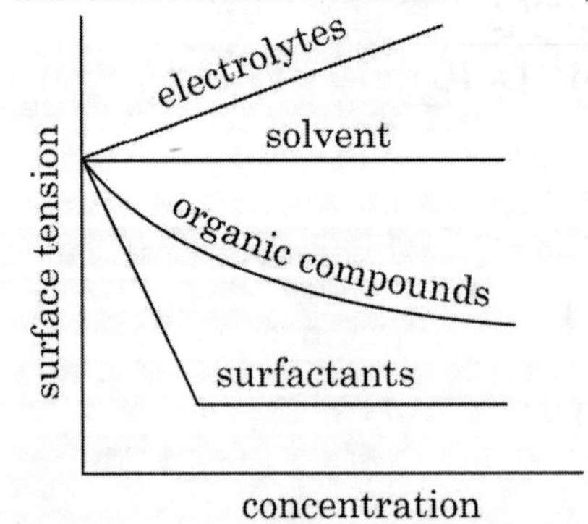

7. (B, C)

$$C_{12}H_{22}O_{11} + H_2O \xrightarrow{H^+} \underset{\substack{\text{D-(+)-glucose} \\ +52^\circ}}{C_6H_{12}O_6} + \underset{\substack{\text{D-(–)-fructose} \\ -92^\circ}}{C_6H_{12}O_6}$$

$$\alpha_{\text{invert sugar}} = \frac{+52^\circ - 92^\circ}{2} = -20^\circ$$ (Average is taken as both monomers are one mole each.)

8. (B, C, D)

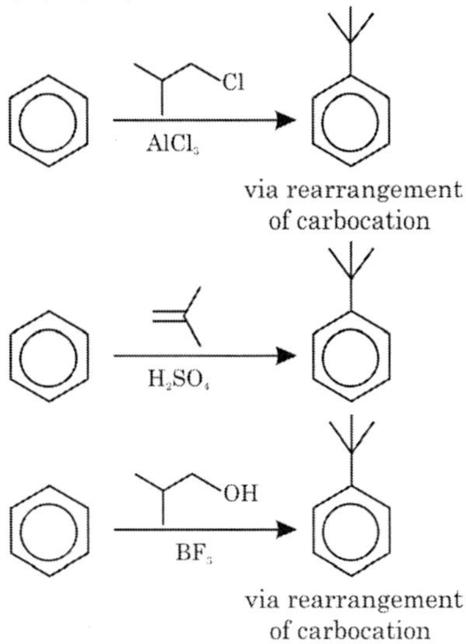

9. (A, B, C)
Refining of blister copper is carried out by poling technique.

10. (B, C, D)
The middle layers will have twelve nearest neighbours. The top-most layer will have nine nearest neighbours. $4r = a\sqrt{2}$, where a = edge length of unit cell and r = radius of atom.

11. (C, D)
$NaBH_4$ and raney Ni/H_2 do not react with acid, ester or epoxide entities of an organic compound.

12. (A, B)
Benzene and toluene will form ideal solution, and phenol and aniline will exhibit negative deviation.

13. (B, D)

$$2HNO_3 \xrightarrow[\text{(dehydration, } - H_2O)]{P_4O_{10}} N_2O_5$$

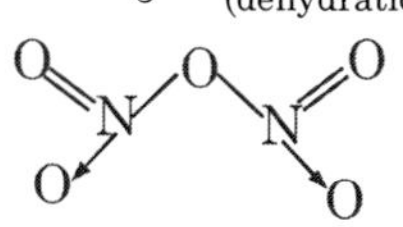

$$Na + N_2O_5 \rightarrow NaNO_3 + \underset{\text{(brown)}}{NO_2(\uparrow)}$$

14. (A, C)

15. (B)

$$X_2(g) \rightleftharpoons 2X(g) \quad \Rightarrow \quad 1-\frac{\beta_e}{2} \quad \beta_e$$

Total number of moles at equilibrium $= 1 - \dfrac{\beta_e}{2} + \beta_e = 1 + \dfrac{\beta_e}{2}$

$$K_p = \frac{(p_x)^2}{p_{x_2}} = \frac{\left(\dfrac{\beta_e \times 2}{1+\dfrac{\beta_e}{2}}\right)^2}{\dfrac{\left(1-\dfrac{\beta_e}{2}\right)\times 2}{1+\dfrac{\beta_e}{2}}} = \frac{2\beta_e^2}{1-\dfrac{\beta_e^2}{4}}$$

$$K_p = \frac{8\beta_e^2}{4-\beta_e^2}$$

16. (C)

No data is given to find the exact value of $\beta_{equilibrium}$

$\Delta G_c^\circ = -2.303RT\log K_c$

$\log K_c = -1$ or $K_c < 1$

17. (A)

18. (B)

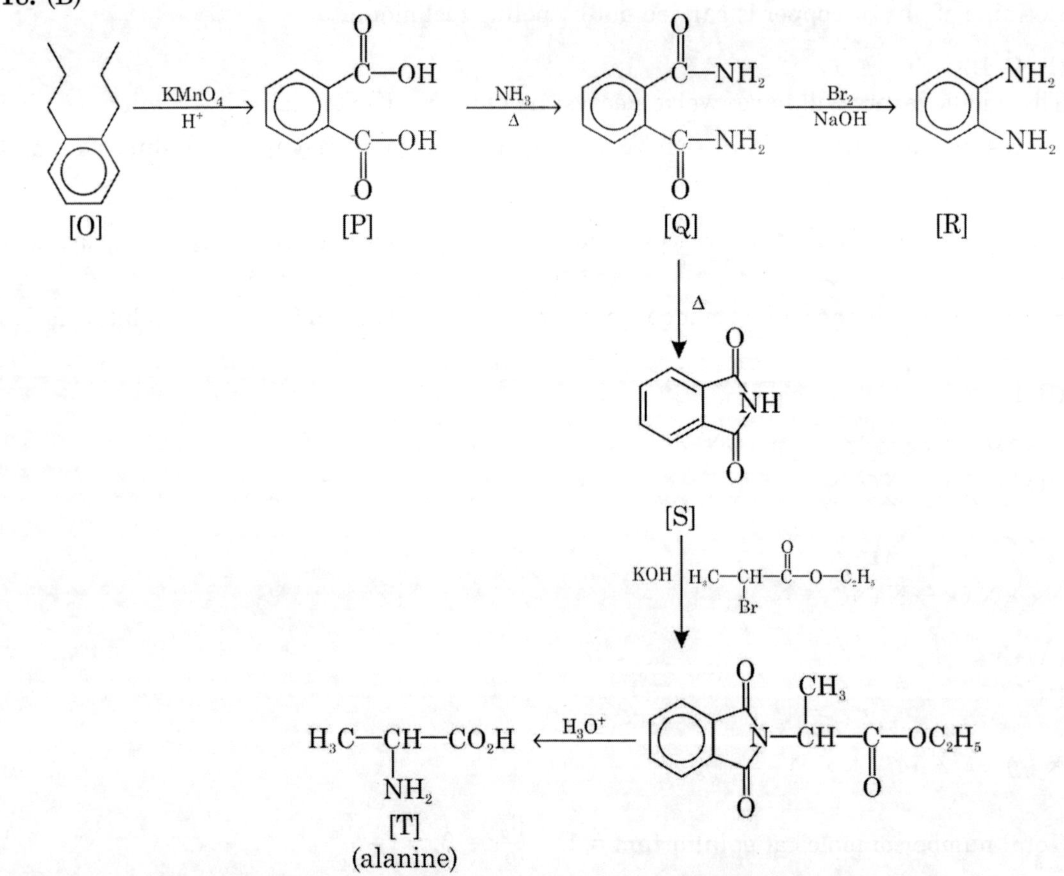

I.I.T. JEE Objective Question Paper—2017 (Main)

1. The freezing point of benzene decreases by 0.45 °C when 0.2 g of acetic acid is added to 20 g of benzene. If acetic acid associates to form a dimer in benzene, percentage association of acetic acid in benzene will be (K_f for benzene = 5.12 K kg mol^{-1})
   (A) 94.6% (B) 64.6% (C) 80.4% (D) 74.6%

2. On treatment of 100 mL of 0.1M solution of $CoCl_3{\cdot}6H_2O$ with excess $AgNO_3$ 1.2×10^{22} ions are precipitated. The complex is:
   (A) $[Co(H_2O)_5Cl]Cl_2{\cdot}H_2O$ (B) $[Co(H_2O)_4Cl_2]Cl{\cdot}2H_2O$
   (C) $[Co(H_2O)_3Cl_3]{\cdot}3H_2O$ (D) $[Co(H_2O)_6]Cl_3$

3. Which of the following compounds will form significant amount of meta product during mono-nitration reaction?

   (A) 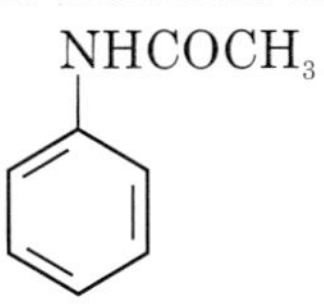 (B) 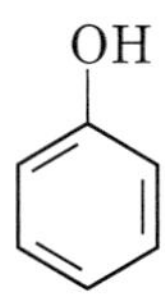 (C) 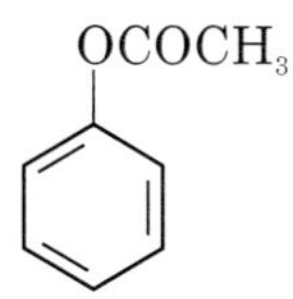 (D) 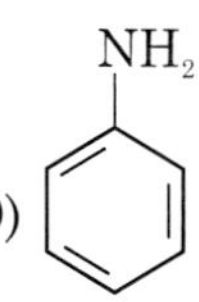

4. The products obtained when chlorine gas reacts with cold and dilute aqueous NaOH are:
   (A) Cl^- and ClO_2^- (B) ClO^- and ClO_3^- (C) ClO_2^- and ClO_3^- (D) Cl^- and ClO^-

5. Both lithium and magnesium display several similar properties due to the diagonal relationship. However, the one which is incorrect, is
   (A) nitrates of both Li and Mg yield NO_2 and O_2 on heating
   (B) both form basic carbonates
   (C) both form soluble hydrogen carbonates
   (D) both form nitrides.

6. A water sample has ppm level concentration of following anions
   $F^- = 10;\ SO_4^{2-} = 100;\ NO_3^- = 50$
   The anion/anions that make/makes the water sample unsuitable for drinking is/are
   (A) Only SO_4^{2-} (B) Only NO_3^- (C) Both SO_4^{2-} and NO_3^- (D) Only F^-

7. The formation of which of the following polymers involves hydrolysis reaction?
   (A) Terylene (B) Nylon 6 (C) Bakelite (D) Nylon 6, 6

8. The Tyndall effect is observed only when the following conditions are satisfied.
   (a) The diameter of the dispersed particles is much smaller than the wavelength of the light used.
   (b) The diameter of the dispersed particle is not much smaller than the wavelength of the light used.
   (c) The refractive indices of the dispersed phase and dispersion medium are almost similar in magnitude.
   (d) The refractive indices of the dispersed phase and dispersion medium differ greatly in magnitude.
   (A) (b) and (c) (B) (a) and (d) (C) (b) and (d) (D) (a) and (c)

9. pK_a of pH_a weak acid (HA) and pK_b of a weak base (BOH) are 3.2 and 3.4, respectively. The pH of their salt (AB) solution is
   (A) 1.0 (B) 7.2 (C) 6.9 (D) 7.0

10. The major product obtained in the following reaction is

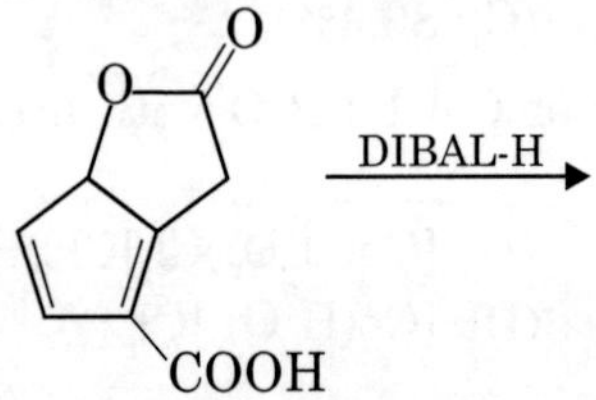

(A) 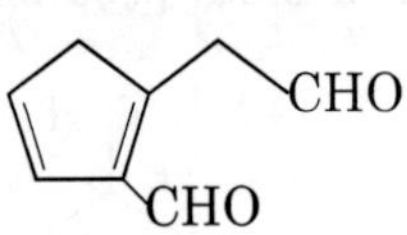 (B) 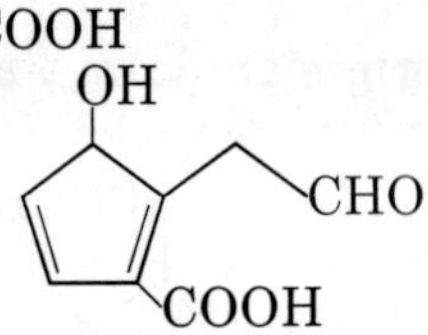 (C) 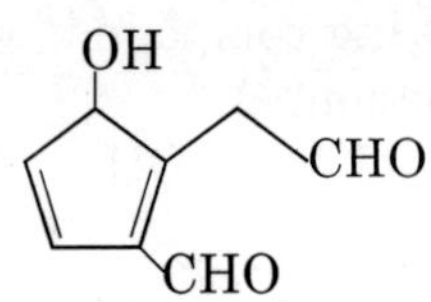 (D) 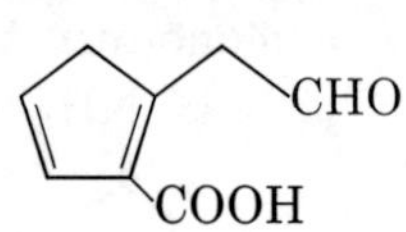

11. Which of the following compounds will behave as a reducing sugar in an aqueous KOH solution?

(A) 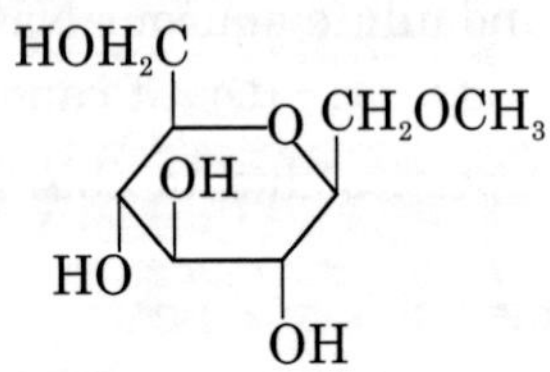

(B) 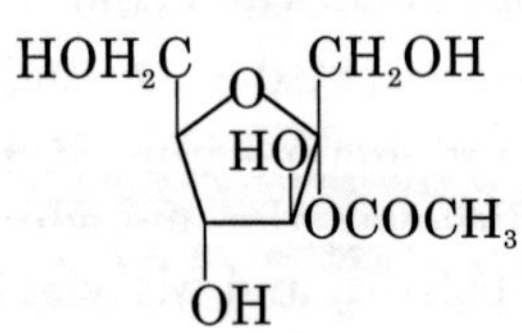

(C) 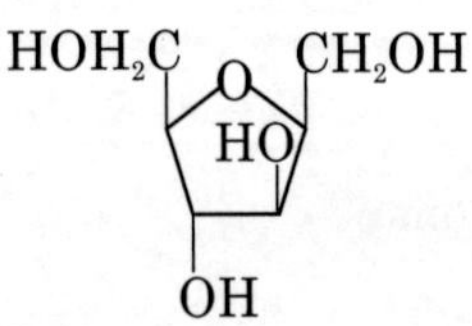

(D) 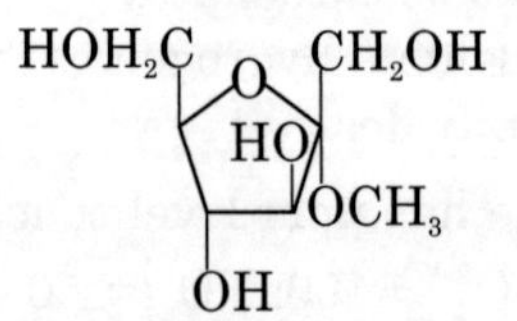

12. The correct sequence of reagents for the following conversion will be

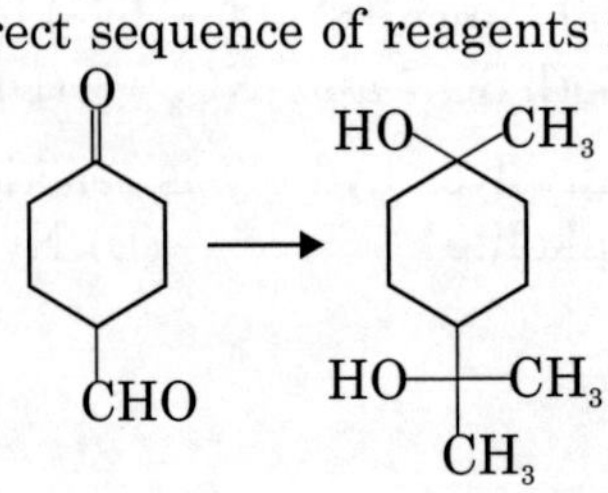

(A) $[Ag(NH_3)_2]^+OH^-$, CH_3MgBr, H^+/CH_3OH (B) $[Ag(NH_3)_2]^+OH^-$, H^+/CH_3OH, CH_3MgBr
(C) CH_3MgBr, H^+/CH_3OH, $[Ag(NH_3)_2]^+OH^-$ (D) CH_3MgBr, $[Ag(NH_3)_2]^+OH^-$, H^+/CH_3OH

13. Which of the following species is not paramagnetic?
(A) B_2 (B) NO (C) CO (D) O_2

14. Which of the following, upon treatment with *tert*-BuONa followed by addition of bromine water, fails to decolourize the colour of bromine?

(A) 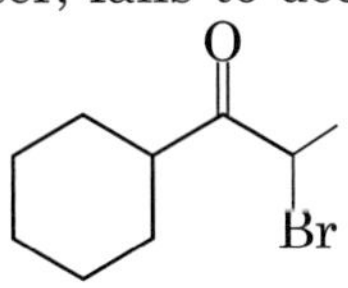

(B) 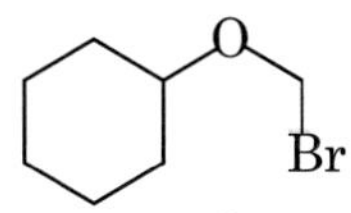

(C) 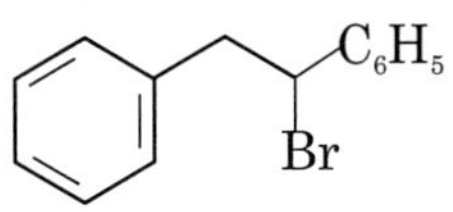

(D) 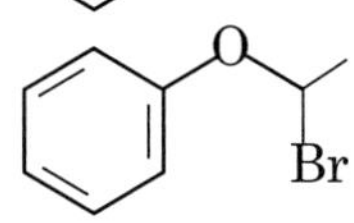

15. Which of the following reactions is an example of a redox reaction?
(A) $XeF_6 + 2H_2O \longrightarrow XeO_2F_2 + 4HF$ (B) $XeF_4 + O_2F_2 \longrightarrow XeF_6 + O_2$
(C) $XeF_2 + PF_5 \longrightarrow [XeF]^+PF_6^-$ (D) $XeF_6 + H_2O \longrightarrow XeOF_4 + 2HF$

16. ΔU is equal to
(A) Isothermal work (B) Isochoric work
(C) Isobaric work (D) Adiabatic work.

17. Which of the following molecules is least resonance stabilized?

(A) 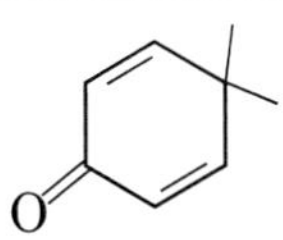 (B)  (C) 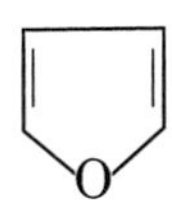 (D) 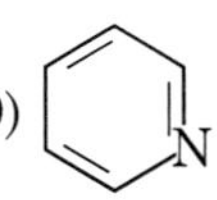

18. The increasing order of the reactivity of the following halides for the S_N1 reaction is
I. $CH_3\underset{\substack{|\\ Cl}}{C}HCH_2CH_3$ II. $CH_3CH_2CH_2Cl$ III. p-$H_3CO—C_6H_4—CH_2Cl$
(A) (II) < (III) < (I) (B) (III) < (II) < (I) (C) (II) < (I) < (III) (D) (I) < (III) < (II)

19. 1 gram of a carbonate (M_2CO_3) on treatment with excess HCl produces 0.01186 mole of CO_2. The molar mass of M_2CO_3 in g mol^{-1} is
(A) 11.86 (B) 1186 (C) 84.3 (D) 118.6

20. Sodium salt of an organic acid X produces effervescence with conc. H_2SO_4. X reacts with the acidified aqueous $CaCl_2$ solution to give a white precipitate which decolourises acidic solution of $KMnO_4$. X is
(A) $Na_2C_2O_4$ (B) C_6H_5COONa (C) HCOONa (D) CH_3COONa

21. The most abundant elements by mass in the body of a healthy human adult are: oxygen (61.4%), carbon (22.9%), hydrogen (10.0%) and nitrogen (2.6%). The weight which a 75 kg person would gain if all ^1H atoms are replaced by ^2H atoms is
(A) 10 kg (B) 15 kg (C) 37.5 kg (D) 7.5 kg

22. The major product obtained in the following reaction is

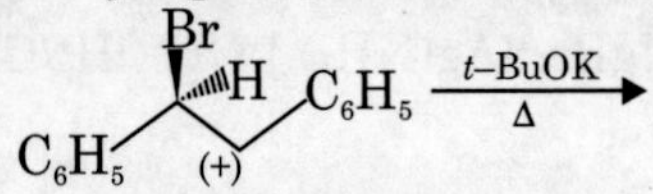

(A) $(-)C_6H_5CH(O^tBu)CH_2C_6H_5$ (B) $(\pm)C_6H_5CH(O^tBu)CH_2C_6H_5$
(C) $C_6H_5CH\ \ CHC_6H_5$ (D) $(+)C_6H_5CH(O^tBu)CH_2C_6H_5$

23. Given: $C_{(graphite)} + O_2(g) \longrightarrow CO_2(g);\ \Delta_rH^0 = -393.5\ kJ\ mol^{-1}$

$$H_2(g) + -O_2(g) \longrightarrow H_2O(l);\ \Delta_rH^0 = -285.8\ kJ\ mol^{-1}$$

$$CO_2(g) + 2H_2O(l) \longrightarrow CH_4(g) + 2O_2(g);\ \Delta_rH^0 = +890.3\ kJ\ mol^{-1}$$

Based on the above thermochemical equations, the value of Δ_rH^0 at 298 K for the reaction $C_{(graphite)} + 2H_2(g) \longrightarrow CH_4(g)$ will be
(A) $-144.0\ kJ\ mol^{-1}$ (B) $+74.8\ kJ\ mol^{-1}$ (C) $+144.0\ kJ\ mol^{-1}$ (D) $-74.8\ kJ\ mol^{-1}$

24. In the following reactions, ZnO is respectively acting as a/an
(a) $ZnO + Na_2O \longrightarrow Na_2ZnO_2$ (b) $ZnO + CO_2 \longrightarrow ZnCO_3$
(A) acid and base (B) base and acid (C) base and base (D) acid and acid

25. The radius of the second Bohr orbit for hydrogen atom is (Planck's const. $h = 6.6262 \times 10^{-34}$ J s, mass of electron $= 9.1091 \times 10^{-31}$ kg, charge of electron, $e = 1.60210 \times 10^{-19}$ C, permittivity of vacuum $\epsilon_0 = 8.854185 \times 10^{-12}\ kg^{-1}\ m^{-3}\ A^2$)
(A) 2.12 Å (B) 1.65 Å (C) 4.76 Å (D) 0.529 Å

26. Two reactions R_1 and R_2 have identical pre-exponential factors. Activation energy of R_1 exceeds that of R_2 by 10 kJ mol^{-1}. If k_1 and k_2 are rate constants for reactions R_1 and R_2 respectively at 300 K, then $\ln(k_2/k_1)$ is equal to ($R = 8.314\ J\ mol^{-1}\ K^{-1}$)
(A) 4 (B) 8 (C) 12 (D) 6

27. A metal crystallizes in a face-centred cubic structure. If the edge length of its unit cell is a, the closest approach between two atoms in metallic crystal will be
(A) $\frac{a}{\sqrt{2}}$ (B) $2a$ (C) $2\sqrt{2}a$ (D) $\sqrt{2}a$

28. The group having isoelectronic species is
(A) O^-, F^-, Na^+, Mg^{2+} (B) $O^{2-}, F^-, Na^+, Mg^{2+}$
(C) O^-, F^-, Na, Mg^+ (D) O^{2-}, F^-, Na, Mg^{2+}

29. Given $E^0_{Cl_2/Cl^-} = 1.36\ V$, $E^0_{Cr^{3+}/Cr} = -0.74\ V$

$$E^0_{Cr_2O_7^{2-}/Cr^{3+}} = 1.33\ V,\ E^0_{MnO_4^-/Mn^{2+}} = 1.51\ V$$

Among the following, the strongest reducing agent is
(A) Cl^- (B) Cr (C) Mn^{2+} (D) Cr^{3+}

30. 3-Methylpent-2-ene on reaction with HBr in presence of peroxide forms an addition product. The number of possible stereoisomers for the product is
(A) Four (B) Six (C) Zero (D) Two

ANSWER KEYS

1. A	2. A	3. D	4. D	5. B	6. D
7. B	8. C	9. C	10. B	11. B	12. B
13. C	14. B	15. B	16. D	17. A	18. C
19. C	20. A	21. D	22. C	23. D	24. A
25. A	26. A	27. A	28. B	29. B	30. A

SOLUTIONS

1. (A)

$\Delta T_f = i \times K_f \times m$

or $0.45 = i \times 5.12 \times \frac{0.2}{60} \times \frac{1000}{20}$

$i = 0.527$

$\underset{1-\alpha}{2\,\text{AcOH}} \rightleftharpoons \underset{\alpha/2}{(\text{AcOH})_2}$

$i = 1 - \frac{\alpha}{2}$ (where α is degree of association)

$0.527 = 1 - \frac{\alpha}{2}$

$\alpha = 0.946$ or 94.6%

2. (A)

$\underset{0.01M}{CoCl_3 \cdot 6H_2O} + AgNO_3 \longrightarrow nAgCl \downarrow$

$\frac{1.2 \times 10^{22}}{6.0 \times 10^{23}} = 2 \times 10^{-2}$ moles $= 0.02$ moles

$0.01 \times n = 0.02$ or $n = 2$

Hence, complex is $[Co(H_2O)_5Cl]Cl_2 \cdot H_2O$.

3. (D)

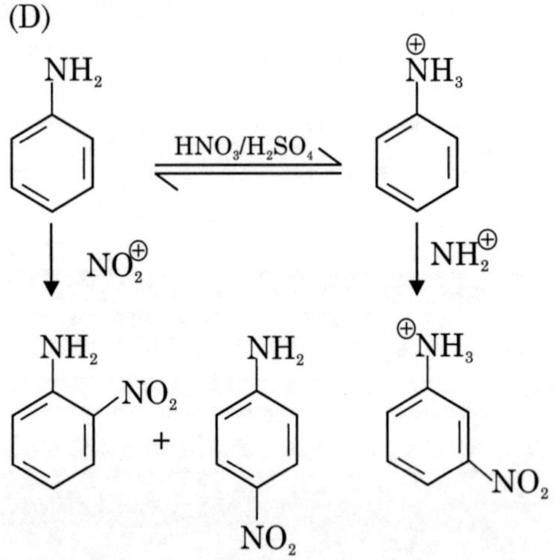

(2%) (51%) (47%)

Anilinium ion is strongly deactivating group and is a meta directing in nature. Hence, it gives meta nitration product in significant amount (47%).

4. (D)

$Cl_2 + \underset{\text{(cold \& dil.)}}{NaOH} \longrightarrow NaCl + NaClO + H_2O$

5. (B)

A, C and D are correct statements.

Li forms only carbonate (Li_2CO_3) and not basic carbonate. But Mg forms basic carbonate.

$$5Mg^{2+} + 6CO_3^{2-} + 7H_2O \longrightarrow 4MgCO_3 \cdot Mg(OH)_2 \cdot 5H_2O \downarrow + 2HCO_3^-$$

6. (D)

Permissible limit of F^- ion is 1 ppm.

7. (B)

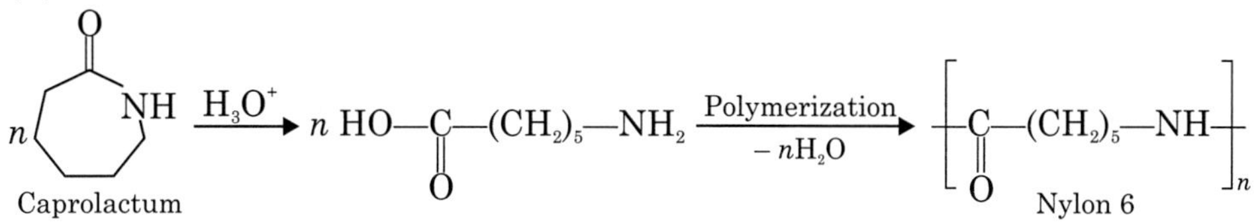

8. (C)

Tyndall effect is observed only when the following two conditions are satisfied.

(i) The diameter of the dispersed particle is not much smaller than the wavelength of the light used.

(ii) The refractive indices of the dispersed phase and dispersion medium differ greatly in magnitude.

9. (C)

pH of a salt made up of weak acid and weak base is calculated by using the expression

$$pH = 7 + \frac{1}{2}pK_a - \frac{1}{2}pK_b = 7 + \frac{3.2}{2} - \frac{3.4}{2} = 6.9$$

10. (B)

DIBAL-H reduces only ester group to an aldehyde and alcohol.

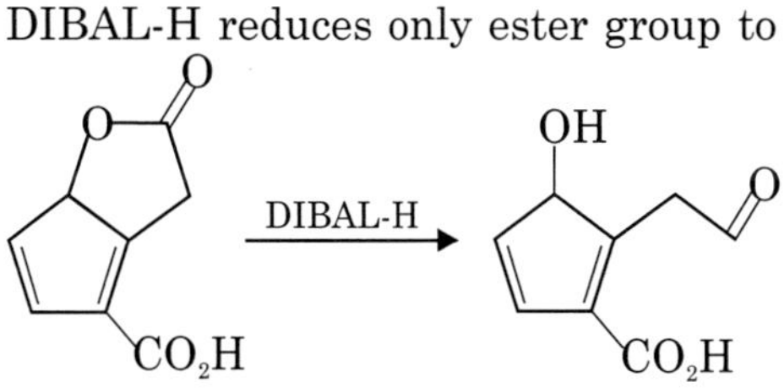

11. (B)

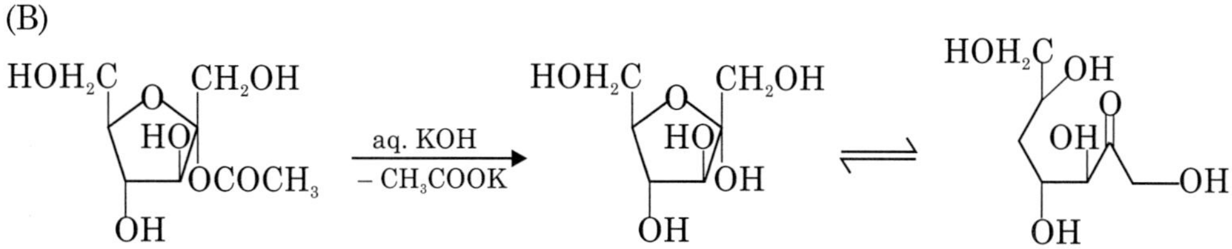

(α-Hydroxyketone)
(a reducing sugar)

12. (B)

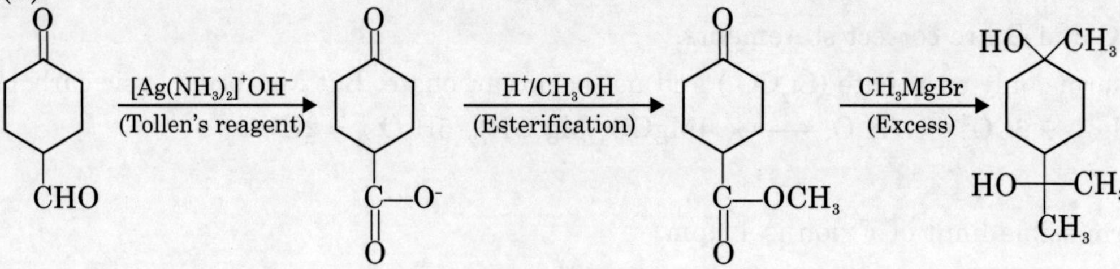

13. (C)

(1) $B_2 : \sigma_{1s}^2, \sigma_{1s}^{*2}, \sigma_{2s}^2, \sigma_{2s}^{*2}, \pi_{2p_x}^1, \pi_{2p_y}^1$

B_2 contains two unpaired electrons. Hence, it is paramagnetic.

(2) $\dot{N}O$ — odd electron species. Hence, it is paramagnetic.

(3) $CO: \sigma_{1s}^2, \sigma_{1s}^{*2}, \sigma_{2s}^2, \sigma_{2s}^{*2}, \pi_{2p_x}^2, \pi_{2p_y}^2, \sigma_{2p_z}^2$

CO contains no unpaired electron. Hence, it is diamagnetic.

(4) $O_2 : \sigma_{1s}^2, \sigma_{1s}^{*2}, \sigma_{2s}^2, \sigma_{2s}^{*2}, \sigma_{2p_z}^2, \pi_{2p_x}^2 \pi_{2p_y}^2, \pi_{2p_x}^{*1} \pi_{2p_y}^{*1}$

O_2 contains two unpaired electrons. Hence, it is paramagnetic.

14. (B)

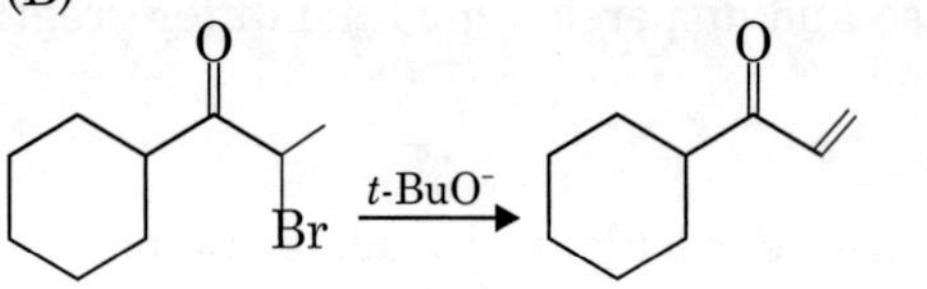

(decolourizes Br_2 water)

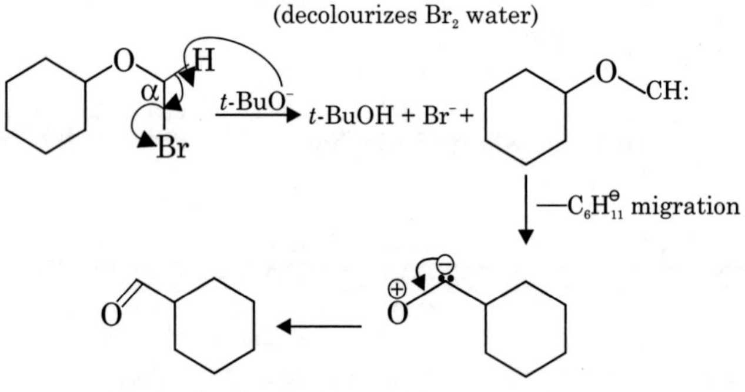

(cannot decolourize Br_2 water)

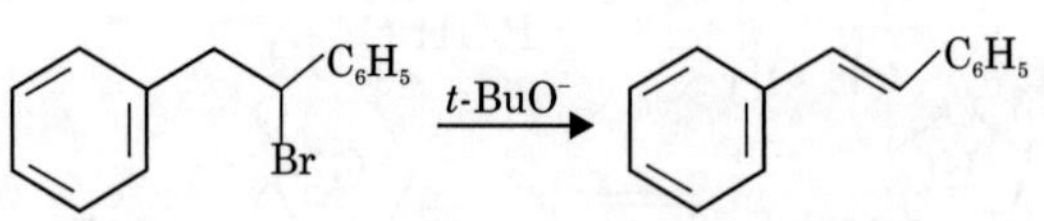

(decolourizes Br_2 water)

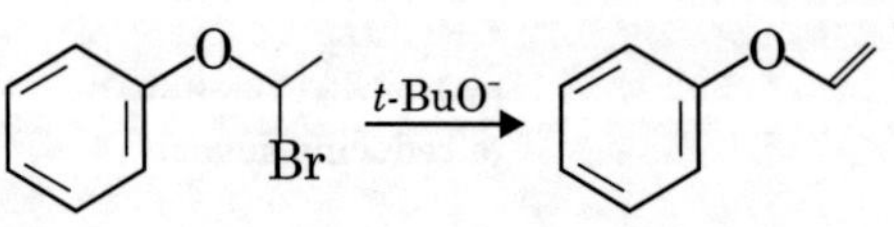

(decolourizes Br_2 water)

15. (B)

Any reaction having a substance in its elemental form is a redox reaction.

$$\overset{+4}{X}eF_4 + \overset{+1}{O_2}F_2 \longrightarrow \overset{+6}{X}eF_6 + \overset{0}{O_2}$$

Xenon undergoes oxidation while oxygen undergoes reduction.

16. (D)

For adiabatic process: $q = 0$

From 1st law $\Delta U = q + w$, we can write $\Delta U = w$

Work involved in adiabatic process is at the expense of change in interval energy of the system.

17. (A)

, 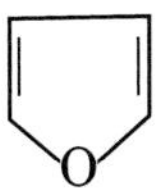, 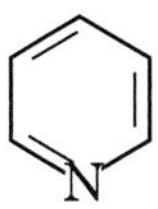 are aromatic and stabilized according to Huckel's rule.

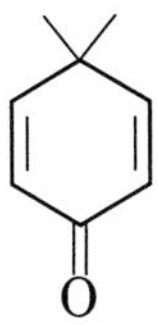 is non-aromatic. Hence, it is least stabilized by resonance.

18. (C)

Order of reactivity for S_N1 mechanism is in accordance with order of stability of carbocation involved.

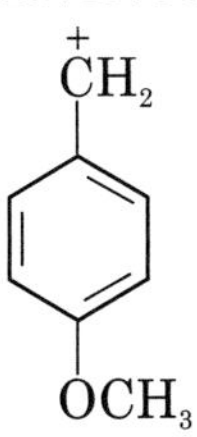 $> CH_3—\overset{+}{C}H—CH_2—CH_3 > CH_3CH_2\overset{+}{C}H_2$

Hence, the order of reactivity is as follows:

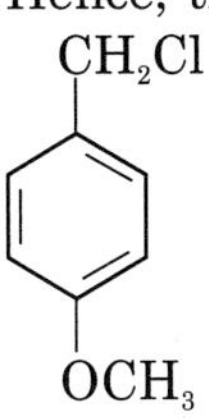 $> CH_3—\underset{}{\overset{Cl}{\overset{|}{C}}}H—CH_2—CH_3 > CH_3CH_2CH_2Cl$

19. (C)

$$M_2CO_3 + 2HCl \longrightarrow 2MCl + H_2O + CO_2$$

No. of moles of M_2CO_3 = No. of moles of CO_2 evolved

$\frac{1}{M} = 0.01186$ (M = molar mass of M_2CO_3)

Or $M = \frac{1}{0.01186} = \frac{10^5}{1186} = 84.3$ g/mol

20. (A)

$$\underset{(X)}{Na_2C_2O_4} + H_2SO_4\text{ (conc.)} \longrightarrow Na_2SO_4 + H_2O + CO_2\uparrow + CO\uparrow$$

$$CaCl_2 + Na_2C_2O_4 \longrightarrow \underset{\text{white precipitate}}{CaC_2O_4} + 2NaCl$$

$$C_2O_4^{2-} + MnO_4^- \xrightarrow{H^+} CO_2 + Mn^{2+}$$

21. (D)

Mass of ^1H in body $= 75 \times 10^3 \times \dfrac{10}{100}$ g $= 7.5 \times 10^3$ g

No. of moles ^1H replaced by $^2H = 7.5 \times 10^3$

Hence, mass increased $= 7.5 \times 10^3$ g $= 7.5$ kg

22. (C)

Elimination reaction is highly favoured if (i) bulkier base is used, (ii) higher temperature is used. Hence, 1,2-elimination provides the major product.

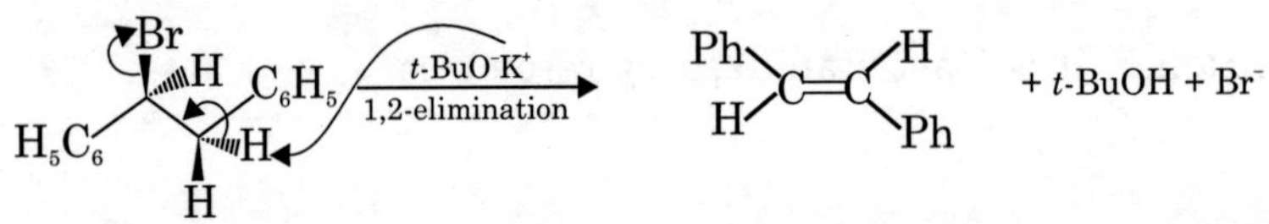

23. (D)

$$C_{(graphite)} + O_2(g) \longrightarrow CO_2(g);\ \Delta_r H^0 = -393.5\text{ kJ mol}^{-1} \quad \text{...(i)}$$

$$H_2(g) + \frac{1}{2}O_2(g) \longrightarrow H_2O(l);\ \Delta_r H^0 = -285.8\text{ kJ mol}^{-1} \quad \text{...(ii)}$$

$$CO_2(g) + 2H_2O(l) \longrightarrow CH_4(g) + 2O_2(g);\ \Delta_r H^0 = +890.3\text{ kJ mol}^{-1} \quad \text{...(iii)}$$

$$CO_2(g) + 2H_2O(l) \longrightarrow CH_4(g) + 2O_2(g) \quad \text{...(iv)}$$

$$\Delta_r H_4 = \Delta_r H_1 + 2\Delta_r H_2 + \Delta_r H_3$$

$$= -393.5 + (-285.8 \times 2) + 890.3 = -74.8\text{ kJ/mol}$$

24. (A)

$$\underset{\text{Acid}}{ZnO} + \underset{\text{Base}}{Na_2O} \rightarrow \underset{\text{Salt}}{Na_2ZnO_2}$$

$$\underset{\text{Base}}{ZnO} + \underset{\text{Acid}}{CO_2} \rightarrow \underset{\text{Salt}}{ZnCO_3}$$

Non-metal oxides are generally acidic while alkali metal oxides are basic. ZnO is an amphoteric oxide.

25. (A)

$$r_n = 0.529 \times \frac{n^2}{Z}$$

$$r_2 = 0.529 \times 4 = 2.116\text{ Å}$$

26. (A)

Arrhenius equation, $k = Ae^{-E_a/RT}$

$$\ln\left(\frac{k_2}{k_1}\right) = \frac{1}{RT}[(E_a)_1 - (E_a)_2] = \frac{1 \times 10 \times 10^3}{8.314 \times 300} = 4$$

27. (A)

In FCC, distance of closest approach between two atoms = $2r$

In FCC, atoms are in close contact along face diagonal of FCC unit cell.

$$4r = \sqrt{2}a$$

Or $2r = \frac{\sqrt{2}a}{2} = \left(\frac{a}{\sqrt{2}}\right)$

28. (B)

Isoelectronic species have same number of electrons. All these species contain ten electrons.

29. (B)

$$E^0_{\text{Red}} \propto \frac{1}{\text{Reducing strength}}$$

$E^0_{Cl_2/Cl^-} = 1.36$ V; $E^0_{Cr^{3+}/Cr} = -0.74$ V;

$E^0_{MnO_4^-/Mn^{2+}} = 1.51$ V; $E^0_{Cr_2O_7^{2-}/Cr^{3+}} = 1.33$ V

30. (A)

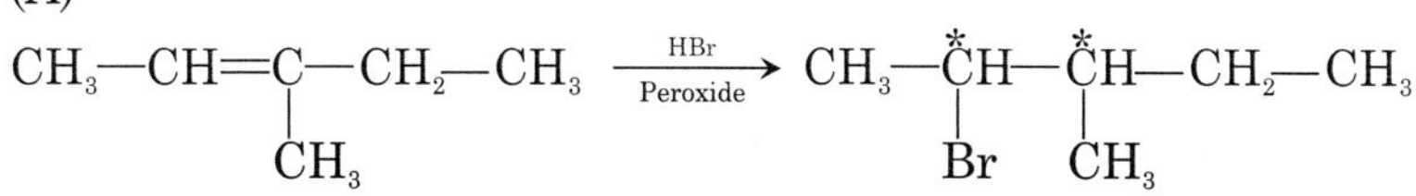

No. of stereoisomers = $2^n = 2^2 = 4$ [n = No. of chiral carbon atoms]

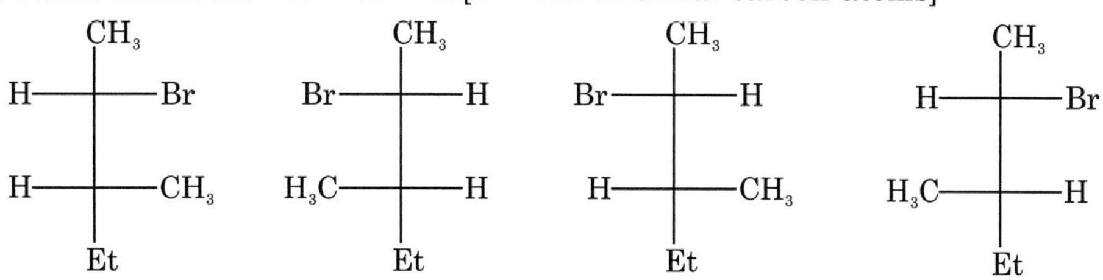

(I) (II) (III) (IV)

I.I.T. JEE Objective Question Paper—2017 (Advanced)

PAPER–1

SECTION–1
(Maximum Marks: 28)

- This section contains SEVEN questions.
- Each question has FOUR options (A), (B), (C) and (D). ONE OR MORE THAN ONE of these four options is(are) correct.
- For each question, darken the bubble(s) corresponding to the correct option(s) in the ORS.
- For each question, marks will be awarded in one of the following categories:

  Full Marks : + 4 If only the bubbles corresponding to the correct option(s) is(are) darkened.

  Partial Marks : + 1 For darkening a bubble corresponding to each correct option, provided NO incorrect option is darkened.

  Zero Marks : 0 If none of the bubbles is darkened.

  Negative Marks : – 2 In all other cases.
- For example, if (A), (C) and (D) are all the correct options for a question, darkening all these three will get + 4 marks; darkening only (A) and (D) will get + 2 marks; and darkening (A) and (B) will get – 2 marks, as a wrong option is also darkened.

1. The IUPAC name(s) of the following compound is(are)

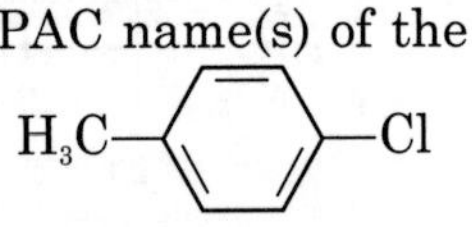

(A) 4-methylchlorobenzene
(B) 4-chlorotoluene
(C) 1-chloro-4-methylbenzene
(D) 1-methyl-4-chlorobenzene

2. The correct statement(s) for the following addition reactions is(are)

(i) 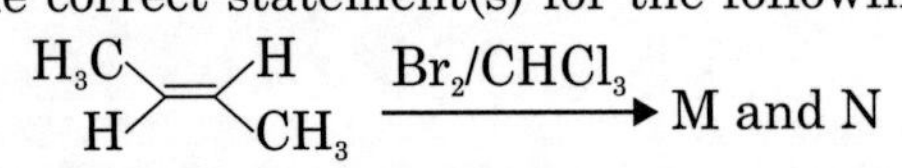

(ii) 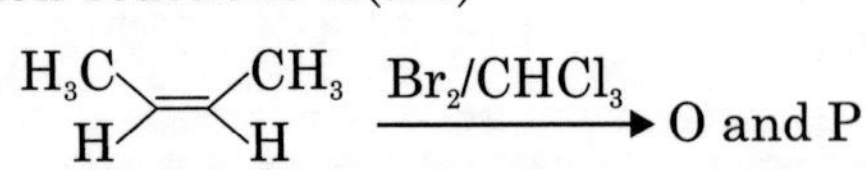

(A) (M and O) and (N and P) are two pairs of diastereomers
(B) Bromination proceeds through *trans*-addition in both the reactions
(C) O and P are identical molecules
(D) (M and O) and (N and P) are two pairs of enantiomers.

3. Addition of excess aqueous ammonia to a pink coloured aqueous solution of $MCl_2 \cdot 6H_2O$ (X) and NH_4Cl gives an octahedral complex Y in the presence of air. In aqueous solution, complex Y behaves as 1 : 3 electrolyte. The reaction of X with excess HCl at room temperature results in the formation of a blue coloured complex Z. The calculated spin-only magnetic moment of X and Z is 3.87 B.M., whereas it is zero for the complex Y. Among the following options, which statements is(are) correct?
   (A) The hybridization of the central metal ion in Y is d^2sp^3
   (B) Z is a tetrahedral complex
   (C) Addition of silver nitrate to Y gives only two equivalents of silver chloride
   (D) When X and Z are in equilibrium at 0°C, the colour of the solution is pink.

4. For a solution formed by mixing liquids L and M, the vapour pressure of L plotted against the mole fraction of M in solution is shown in the following figure. Here P_L and x_M represent mole fractions of L and M, respectively, in the solution. The correct statement(s) applicable to this system is(are)

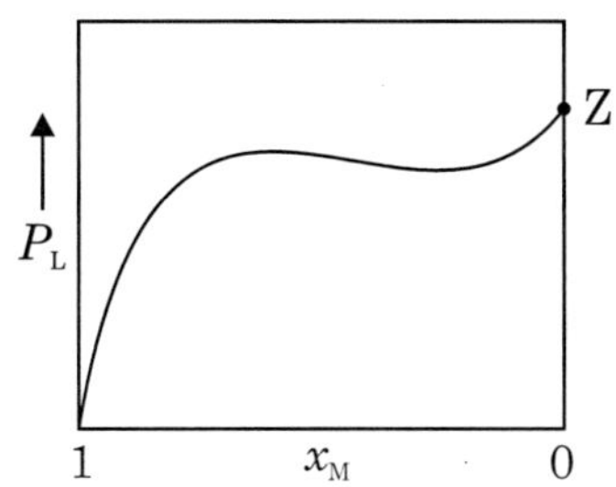

   (A) Attractive intramolecular interactions between L–L in pure liquid L and M–M in pure liquid M are stronger than those between L–M when mixed in solution
   (B) The point Z represents vapour pressure of pure liquid M and Raoult's law is obeyed when $P_L \longrightarrow 0$
   (C) The point Z represents vapour pressure of pure liquid L and Raoult's law is obeyed when $P_L \longrightarrow 1$
   (D) The point Z represents vapour pressure of pure liquid M and Raoult's law is obeyed from $P_L = 0$ to $P_L = 1$.

5. An ideal gas is expanded from (P_1, V_1, T_1) to (P_2, V_2, T_2) under different conditions. The correct statement(s) among the following is(are)
   (A) The work done on the gas is maximum when it is compressed irreversibly from (P_2, V_2) to (P_1, V_1) against constant pressure P_1
   (B) The work done on the gas is less when it is expanded reversibly from V_1 to V_2 under adiabatic conditions as compared to that when expanded reversibly from V_1 to V_2 under isothermal conditions
   (C) The change in internal energy of the gas (i) is zero, if it is expanded reversibly with $T_1 = T_2$, and (ii) is positive, if it is expanded reversibly under adiabatic conditions with $T_1 \neq T_2$
   (D) If the expansion is carried out freely, it is simultaneously both isothermal as well as adiabatic.

6. The correct statement(s) about the oxoacids, $HClO_4$ and HClO, is(are)
(A) $HClO_4$ is more acidic than HClO because of the resonance stabilization of its anion
(B) $HClO_4$ is formed in the reaction between Cl_2 and H_2O
(C) The central atom in both $HClO_4$ and HClO is sp^3 hybridized
(D) The conjugate base of $HClO_4$ is a weaker base than H_2O.

7. The colour of the X_2 molecules of group 17 elements changes gradually from yellow to violet down the group. This is due to
(A) the physical state of X_2 at room temperature changes from gas to solid down the group
(B) decrease in HOMO-LUMO gap down the group
(C) decrease in π^*-σ^* down the group
(D) decrease in ionization energy down the group.

SECTION–2
(Maximum Marks: 15)

- This section contains FIVE questions.
- The answer to each question is a SINGLE DIGIT INTEGER ranging from 0 to 9, both inclusive.
- For each question, darken the bubble corresponding to the correct integer in the ORS.
- For each question, marks will be awarded in one of the following categories:

Full Marks : + 3 If only the bubble corresponding to the correct answer is darkened.

Zero Marks : 0 In all other cases.

8. Among H_2, He_2^+, Li_2, Be_2, B_2, C_2, N_2, O_2^- and F_2, the number of diamagnetic species is
(Atomic number: H = 1, He = 2, Li = 3, Be = 4, B = 5, C = 6, N = 7, O = 8, F = 9)

9. Among the following, the number of aromatic compound(s) is

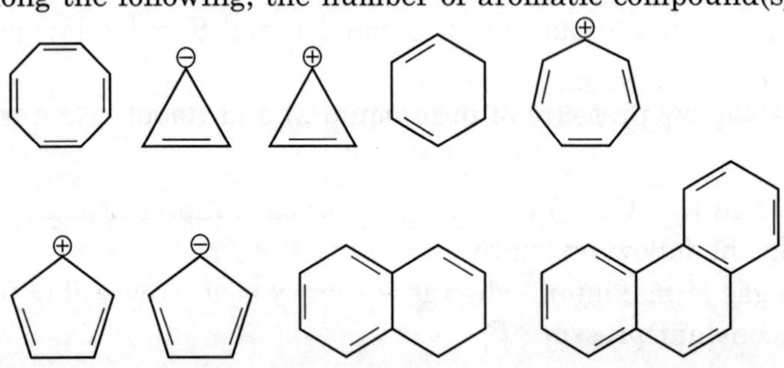

10. The conductance of a 0.0015*M* aqueous solution of a weak monobasic acid was determined by using a conductivity cell consisting of platinized Pt electrodes. The distance between the electrodes is 120 cm with an area of cross section of 1 cm². The conductance of this solution was found to be 5×10^{-7} S. The *p*H of the solution is 4. The value of limiting molar conductivity (Λ_m^0) of this weak monobasic acid in aqueous solution is Z × 10^2 S cm^{-1} mol^{-1}. The value of Z is

11. The sum of the number of lone pairs of electrons on each central atom in the following species is $[TeBr_6]^{2-}$, $[BrF_2]^+$, SNF_3 and $[XeF_3]^-$
[Atomic number: N = 7, F = 9, S = 16, Br = 35, Te = 52, Xe = 54]

12. A crystalline solid of a pure substance has a face-centred cubic structure with a cell edge of 400 pm. If the density of the substance in the crystal is 8 g cm^{-3}, then the number of atoms present in 256 g of the crystal is $N \times 10^{24}$. The value of N is

SECTION–3
(Maximum Marks: 18)

- This section contains SIX questions of matching type.
- This section contains TWO tables (each having 3 columns and 4 rows).
- Based on each table, there are THREE questions.
- Each question has FOUR options (A), (B), (C) and (D). ONLY ONE of these four options is correct.
- For each question, darken the bubble corresponding to the correct option in the ORS.
- For each question, marks will be awarded in one of the following categories:

Full Marks : + 3 If only the bubble corresponding to the correct option is darkened.
Zero Marks : 0 If none of the bubbles is darkened.
Negative Marks : – 1 In all other cases.

Answer Q. 13, Q. 14 and Q. 15 by appropriately matching the information given in the three columns of the following table.

The wave function ψ_{n,l,m_l} is a mathematical function whose value depends upon spherical polar coordinates (r, θ, ϕ) of the electron and characterized by the quantum numbers n, l and m_1. Here r is distance from nucleus, θ is colatitude and ϕ is azimuth. In the mathematical functions given in the table, Z is atomic number and a_0 is Bohr radius.

Column–1	Column–2	Column–3
(I) 1s orbital	(i) $\psi_{n,l,m_l} \propto \left(\frac{Z}{a_0}\right)^{\frac{3}{2}} e^{-\left(\frac{Zr}{a_e}\right)}$	(P) 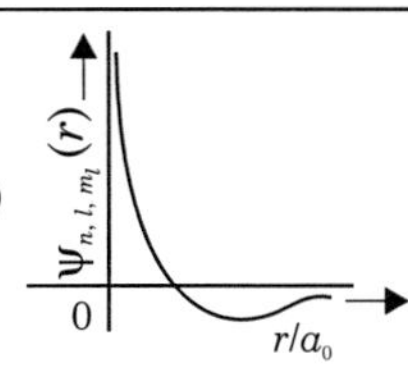
(II) 2s orbital	(ii) One radial node	(Q) Probability density at nucleus $\propto \frac{1}{a_0^3}$
(III) 2p_z orbital	(iii) $\psi_{n,l,m_l} \propto \left(\frac{Z}{a_0}\right)^{\frac{5}{2}} re^{-\left(\frac{Zr}{2a_0}\right)} \cos\theta$	(R) Probability density is maximum at nucleus.

(IV) $3d_z^2$ orbital	(iv) xy-plane is a nodal plane.	(S) Energy needed to excite an electron from $n = 2$ state to $n = 4$ state is 27/32 times the energy needed to excite an electron from $n = 2$ state to $n = 6$ state.

13. For the given orbital in column 1, the only correct combination for any hydrogen-like species is
(A) (IV) (iv) (R) (B) (II) (ii) (P) (C) (III) (iii) (P) (D) (I) (ii) (S)

14. For He^+ ion, the only incorrect combination is
(A) (II) (ii) (Q) (B) (I) (i) (S) (C) (I) (i) (R) (D) (I) (iii) (R)

15. For hydrogen atom, the only correct combination is
(A) (I) (iv) (R) (B) (I) (i) (P) (C) (II) (i) (Q) (D) (I) (i) (S)

Answer Q. 16, Q. 17 and Q. 18 by appropriately matching the information given in the three columns of the following table.

Columns 1, 2 and 3 contain starting materials, reaction conditions and type of reactions, respectively		
Column–1	**Column–2**	**Column–3**
(I) Toluene	(i) $NaOH/Br_2$	(P) Condensation
(II) Acetophenone	(ii) $Br_2/h\varpi v$	(Q) Carboxylation
(III) Banzaldehyde	(iii) $(CH_3CO)_2O/CH_3COOK$	(R) Substitution
(IV) Phenol	(iv) $NaOH/CO_2$	(S) Haloform

16. For the synthesis of benzoic acid, the only correct combination is
(A) (III) (iv) (R) (B) (IV) (ii) (P) (C) (I) (iv) (Q) (D) (II) (i) (S)

17. The only correct combination in which the reaction proceeds through radical mechanism is
(A) (I) (ii) (R) (B) (II) (iii) (R) (C) (III) (ii) (P) (D) (IV) (i) (Q)

18. The only correct combination that gives two different carboxylic acids is
(A) (IV) (iii) (Q) (B) (III) (iii) (P) (C) (II) (iv) (R) (D) (I) (i) (S)

ANSWER KEYS

1. B, C	2. A, B	3. A, B, D	4. A, C	5. A, B, D	6. A, C, D
7. B, C	8. 6	9. 5	10. 6	11. 6	12. 2
13. B	14. D	15. D	16. D	17. A	18. B

SOLUTIONS

1. (B, C)

Toluene is accepted by IUPAC as a name of parent carbon atom. Hence, the IUPAC name of the compound is 4-chlorotoluene and 1-chloro-4-methyl benzene.

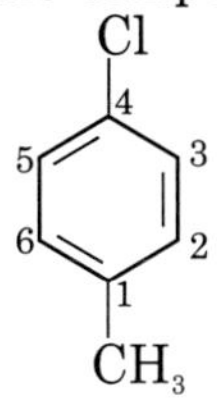

2. (A, B)

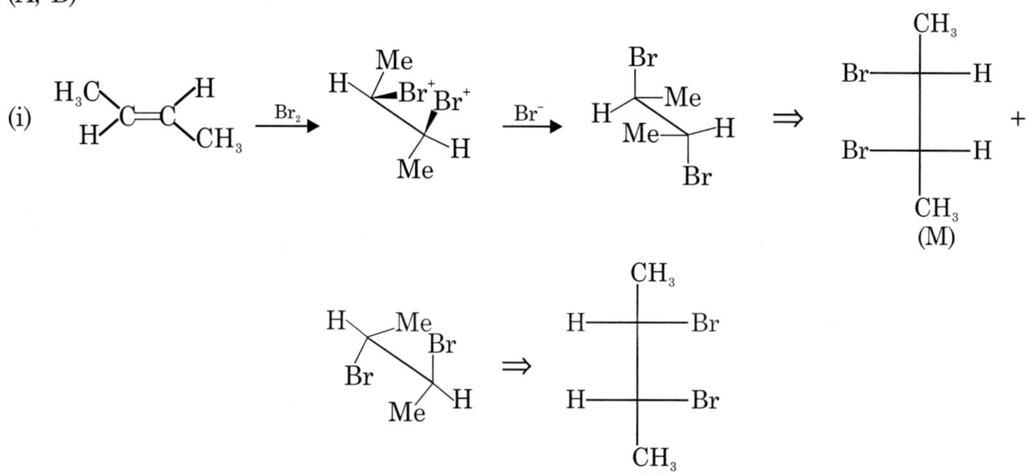

(M) (N)

(M) and (N) are identical meso compounds.

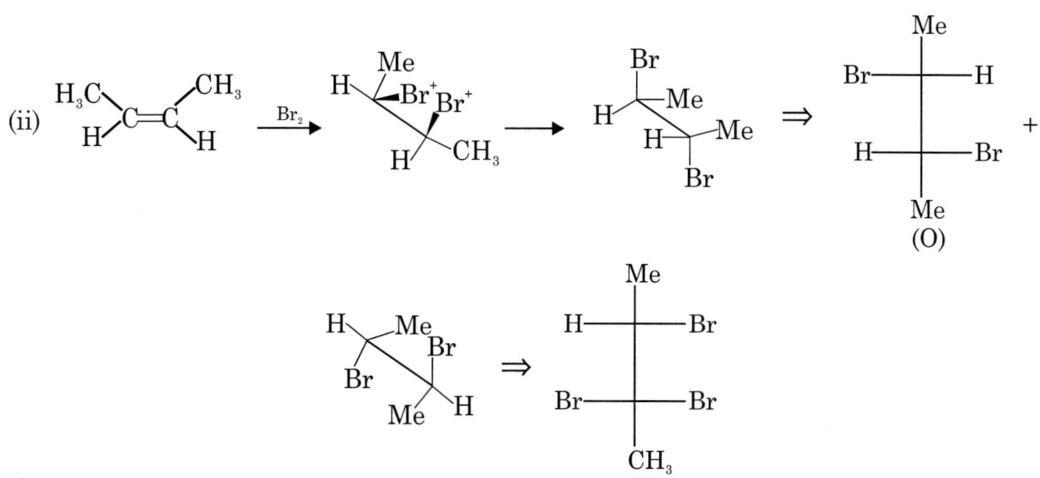

(O) (P)

3. (A, B, D)

$$\underset{\text{Pink (X)}}{[Co^{II}(H_2O)_6]Cl_2} \xrightarrow[\text{Air}/O_2]{\text{Excess } NH_4OH/NH_4Cl} \underset{(Y)}{[Co^{III}(NH_3)_6]Cl_3}$$

$$\underset{(X)}{[Co(H_2O)_6]^{2+}} + \underset{(\text{excess})}{4Cl^-} \longrightarrow \underset{(Z)\text{ blue}}{[CoCl_4]^{2-}}$$

(A) Hybridization of Y is d^2sp^3 as NH_3 is a strong field ligand.

(B) $[CoCl_4]^{2-}$ has sp^3 hybridization as Cl^- is a weak field ligand.

(C) $\underset{(Y)}{[Co(NH_3)_6]Cl_3} + 3AgNO_3(aq) \rightarrow 3AgCl$

(D) $\underset{(\text{Blue})}{[CoCl_4]^{2-}} + 6H_2O \rightleftharpoons \underset{(\text{Pink})}{[Co(H_2O)_6]^{2+}} + 4Cl^- \quad \Delta H = (-)\text{ve (exothermic)}$

When ice is added to the solution, the equilibrium shifts to the right. Hence pink colour will remain predominant. Hence, the correct answer is (A, B, D).

4. (A, C)

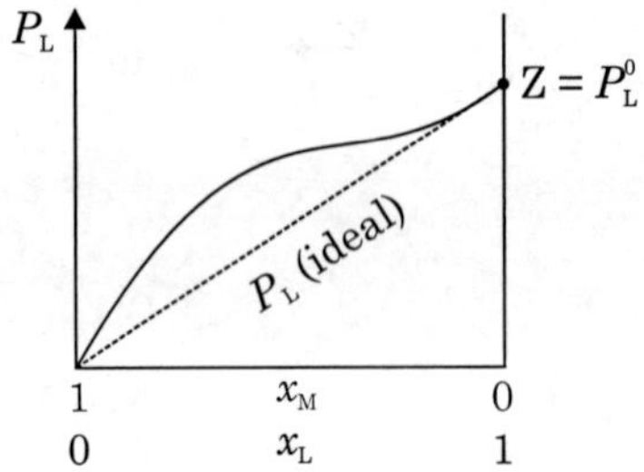

(A) This is a case of positive deviation. Hence, F_{L-L}, $F_{M-M} > F_{L-M}$

(C) $P_L \geq P_L^0 x_L$

But when $x_L \rightarrow 1$, the mixture contains almost pure liquid L. Hence, $P_L \rightarrow P_L^0$

5. (A, B, D)

(A)

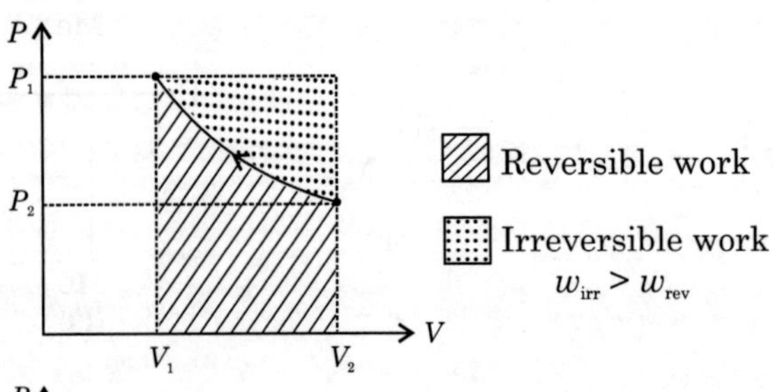

(B)

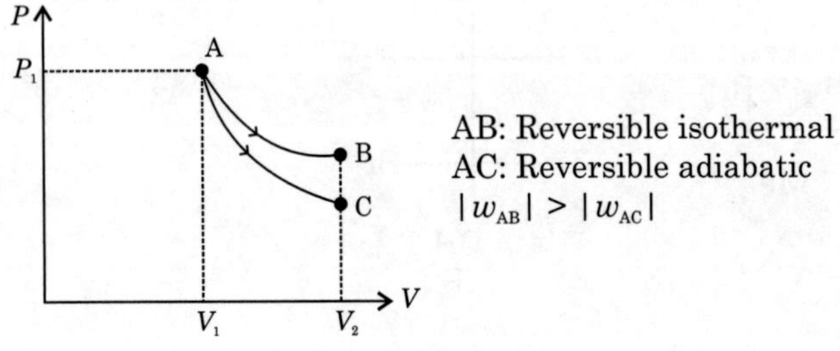

(C) (i) $\Delta U = nC_V\Delta T = 0$ (isothermal, hence $\Delta T = 0$)

(ii) $\Delta U = q + w = -\text{ve}$ ($q = 0$, $w < 0$)

$\Delta U = nC_V\Delta T \Rightarrow \Delta T < 0$

(D) $q = 0$ (adiabatic), $w = 0$ (free expansion)

$\Delta U = 0 \Rightarrow \Delta T = 0$ (isothermal)

6. (A, C, D)

(A) $HClO_4 \rightleftharpoons H^+ + ClO_4^-$ ClO_4^- is a resonance stablized anion

$HClO \rightleftharpoons H^+ + ClO^-$

Hence, $HClO_4$ is more acidic than HClO.

(B) $Cl_2 + H_2O \longrightarrow HClO + HCl$

(C) $HClO_4$

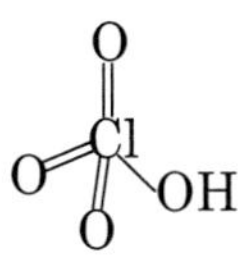

Central atom is sp^3 hybridised.

HClO 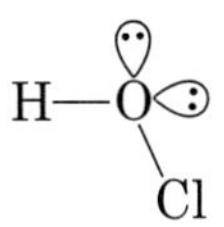

Central atom is sp^3 hybridised.

(D) $HClO_4 + H_2O \rightarrow H_3O^{\oplus} + ClO_4^-$. Since H_2O accepts H^+ from $HClO_4$, H_2O is stronger base compared to ClO_4^-.

7. (B, C)

Halogens are coloured due to HOMO-LUMO transition of electrons.

M.O. (configuration) : $F_2 = \sigma 1s^2, \sigma^* 1s^2, \sigma 2s^2, \sigma^* 2s^2, \sigma 2p_z^2, \pi 2p_x^2 = \pi 2p_y^2, \pi^* 2p_x^2 = \pi^* 2p_y^2, \sigma^* 2p_z$

$\pi^* 2p_x^2$ ↗ filled (HOMO); $\sigma^* 2p_z$ ↗ vacant (LUMO)

On moving down the group HOMO-LUMO energy gap decreases. Hence, the transition of electrons becomes easier from $\pi^* 2p$ to $\sigma^* 2p$. Therefore, colour intensifies.

8. (6)

H_2	$\Rightarrow$	$\sigma 1s^2$	(Diamagnetic)
He_2^+	$\Rightarrow$	$\sigma 1s^2 \sigma * 1s^1$	(Paramagnetic)
Li_2	$\Rightarrow$	$\sigma 1s^2 \sigma * 1s^2 \sigma 2s^2$	(Diamagnetic)
Be_2	$\Rightarrow$	$\sigma 1s^2 \sigma * 1s^2 \sigma 2s^2 \sigma * 2s^2$	(Diamagnetic)
B_2	$\Rightarrow$	$\sigma 1s^2 \sigma * 1s^2 \sigma 2s^2 \sigma * 2s^2 \pi 2p_x^1 = \pi 2p_y^1$	(Paramagnetic)
C_2	$\Rightarrow$	$\sigma 1s^2 \sigma * 1s^2 \sigma 2s^2 \sigma * 2s^2 \pi 2p_x^2 = \pi 2p_y^2$	(Diamagnetic)
N_2	$\Rightarrow$	$\sigma 1s^2 \sigma * 1s^2 \sigma 2s^2 \sigma * 2s^2 \pi 2p_x^2 = \pi 2p_y^2 \sigma 2p_z^2$	(Diamagnetic)

$O_2^- \Rightarrow \sigma 1s^2 \sigma * 1s^2 \sigma 2s^2 \sigma * 2s^2 \pi 2p_z^2 \pi 2p_x^2 = \pi 2p_y^2 \pi * 2p_x^2 = \pi * 2p_y^1$ (Paramagnetic)

$F_2 \Rightarrow \sigma 1s^2 \sigma * 1s^2 \sigma 2s^2 \sigma * 2s^2 \sigma 2p_z^2 \pi 2p_x^2 = \pi 2p_y^2 \pi * 2p_x^2 = \pi * 2p_y^2$ (Diamagnetic)

9. (5)

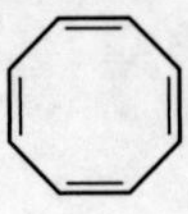

Cyclooctatetraene, non-aromatic

Due to the non-planarity of ring, the π-electrons are not delocalized.

Cyclopropenyl anion, non-aromatic

4π-electrons are delocalized.

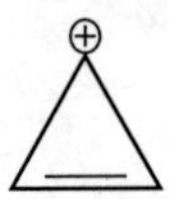

Cyclopropenyl cation; aromatic

2π-electrons are delocalized.

Cyclohexadiene, non-aromatic

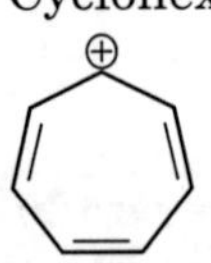

Tropylium ion: aromatic

6π-electrons are delocalized.

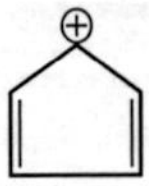

Cyclopentadienyl cation; non-aromatic

4π-electrons are delocalized.

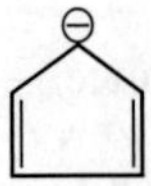

Cyclopentadienyl anion; aromatic

6π-electrons are delocalized.

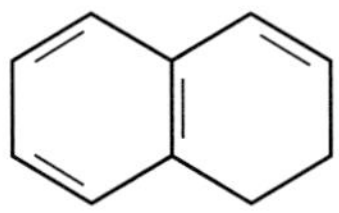

aromatic

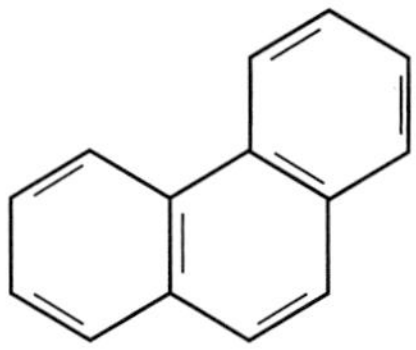

aromatic
14 πe^-s
(delocalized)

10. (6)

For weak acid $[H^+] = c\alpha = c\frac{\Lambda_m^C}{\Lambda_m^\infty} = c \times \frac{\kappa \times \frac{1000}{c}}{\Lambda_m^\infty} = \frac{\kappa \times 1000}{\Lambda_m^\infty} = \frac{G \times \left(\frac{l}{a}\right) \times 1000}{\Lambda_m^\infty}$

$$10^{-4} = \frac{5 \times 10^{-7} \times \left(\frac{120}{1}\right) \times 1000}{Z \times 10^2} \Rightarrow Z = 6$$

11. (6)

	Number of σ-bonds formed by central atom	Number of lone pairs on central atom
(i) In $[TeBr_6]^{2-}$	6	1
(ii) In $[BrF_2]^+$	2	2
(iii) In SNF_3	4	0
(iv) In $[XeF_3]^-$	3	3

Hence, total number of lone pairs of electrons = 1 + 2 + 0 + 3 = 6

12. (2)

$$d = \frac{Z \times M}{N_A \times a^3}$$

For FCC unit cell $Z = 4$. Edge length $a = 4 \times 10^{-8}$ cm

Hence, $M = \frac{d \times N_A \times a^3}{Z} = \frac{8 \times 6 \times 10^{23} \times 64 \times 10^{-24}}{4}$ g/mol

No. of atoms = $\frac{\text{wt (g)}}{\text{molar mass}} \times N_A = \frac{256 \times 10 \times 6 \times 10^{23}}{8 \times 6 \times 16} = 2 \times 10^{24}$ (value of $N = 2$)

13. (B)

(A) (IV) (iv) (R) ⇒ incorrect, because, d_{z^2} has no nodal plane.

(B) (II) (ii) (P) ⇒ correct, because 2*s* orbital has 1 radial node.

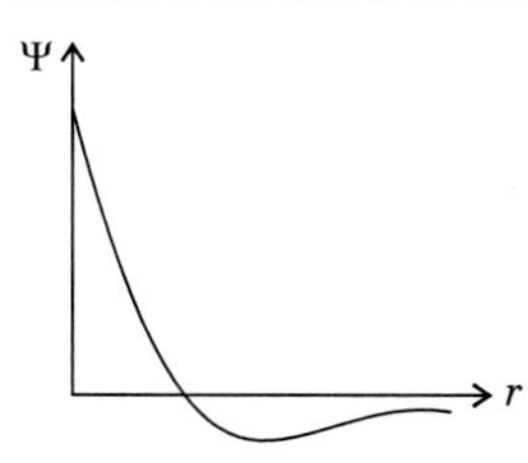

(C) (III) (iii) (P) ⇒ incorrect, because probability density for $2p$ at nucleus is zero.

(D) (I) (ii) (S) ⇒ incorrect, because $1s$ orbital has no radial node.

14. (D)

The option (D) is incorrect because in the wave function of $1s$ orbital, no angular function is present.

15. (D)

For $1s$ orbital: $\Psi_{n, l, m_l} \; \alpha \left(\dfrac{Z}{a_0}\right)^{3/2 - zr} e^{a_0}$

Energy needed to excite an electron from $n = 2$ to $n = 4$,

$$\Delta E_{2\text{-}4} = 13.6Z^2 \times \frac{3}{16} \text{ eV}$$

Energy needed to excite an electron from $n = 2$ to $n = 6$,

$$\Delta E_{2\text{-}6} = 13.6Z^2 \times \frac{8}{36} \text{ eV}$$

$$\Delta E_{2\text{-}4} = \frac{27}{32} E_{2\text{-}6} \text{ (hence, true)}$$

16. (D)

(II) (i) (S)

(A) 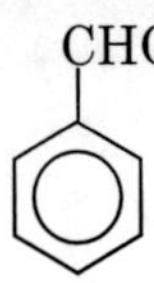 $\xrightarrow{NaOH/CO_2}$ Benzoic acid is not formed.

(III) ↑ (iv) (R)

(B) 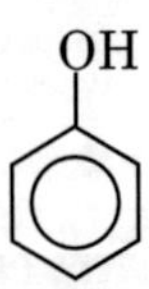 $\xrightarrow{Br_2/hv}$ Benzoic acid is not formed.

(I) ↑ (ii) (P)

(C) 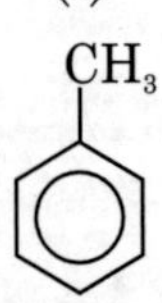 $\xrightarrow{NaOH/CO_2}$ Benzoic acid is not formed.

(IV) ↑ (iv) (Q)

(D) 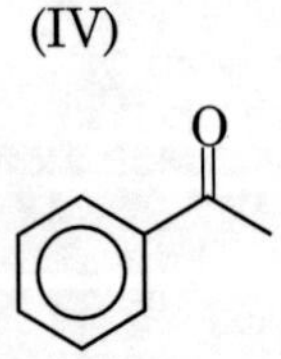 $\xrightarrow{NaOH/Br_2}$ 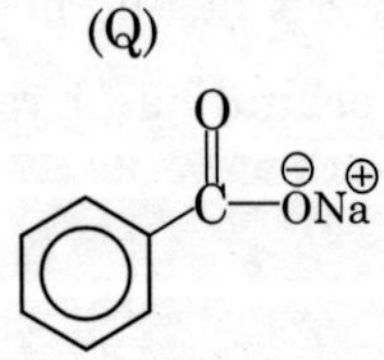 + $CHBr_3$ (haloform reaction)

(II) ↑ (i) (S)

17. (A)

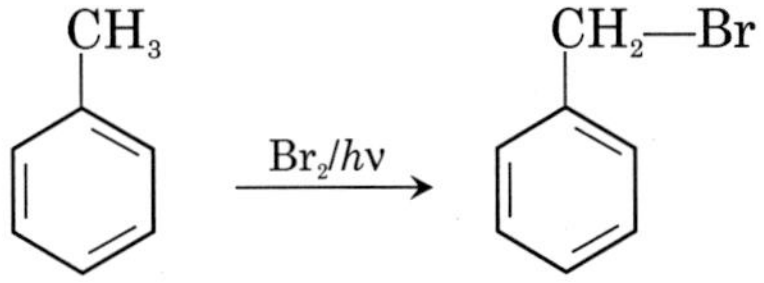

The mechanism involved is free radical substitution.

(B) 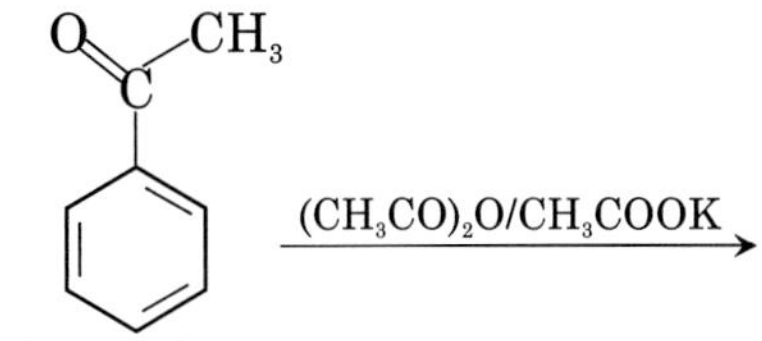 Condensation reaction.

Acetophenone

(No radical mechanism is involved.)

(C) 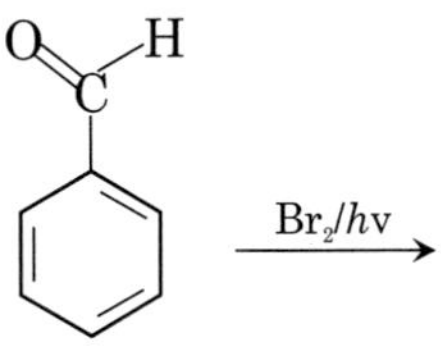 Condensation does not occur.

(D) 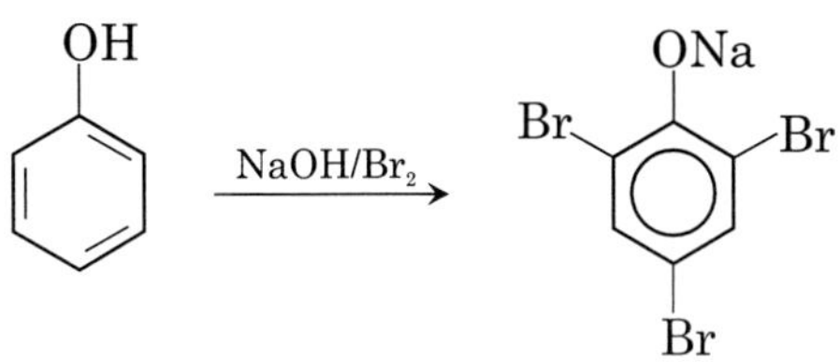

(No free radical mechanism is involved.)

18. (B)

$$C_6H_5CHO + CH_3-\underset{\underset{O}{\|}}{C}-O-\underset{\underset{O}{\|}}{C}-CH_3 \xrightarrow{CH_3COOK} \underset{\text{Cinnamic acid (cis and trans)}}{C_6H_5CH{=}CH-COOH}$$

It is a Perkin condensation reaction.

Mechanism

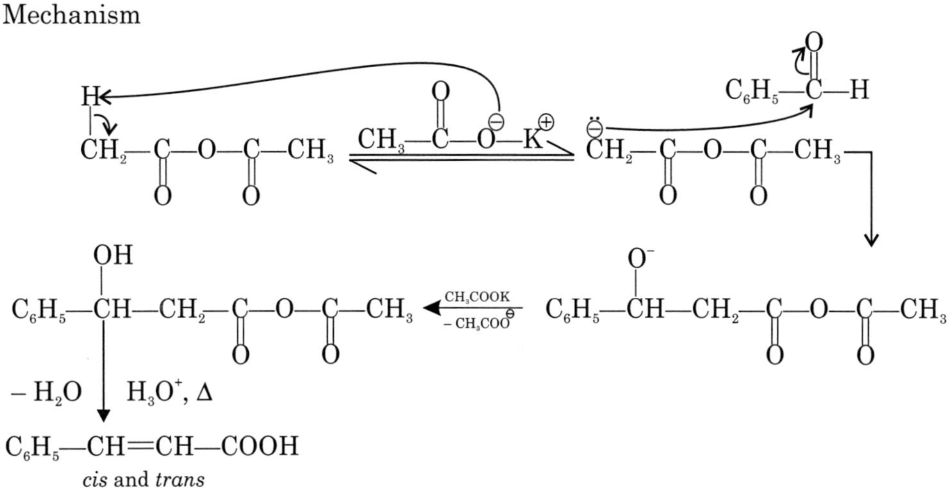

I.I.T. JEE Objective Question Paper—2017 (Advanced)

PAPER–2

SECTION–1

(Maximum Marks: 21)

- This section contains SEVEN questions.
- Each question has FOUR options (A), (B), (C) and (D). ONLY ONE of these four options is correct.
- For each question, darken the bubble corresponding to the correct option in the ORS.
- For each question, marks will be awarded in one of the following categories:

  Full Marks : + 3 If only the bubble corresponding to the correct option is darkened.

  Zero Marks : 0 If none of the bubbles is darkened.

  Negative Marks : – 1 In all other cases.

1. Which of the following combinations will produce H_2 gas?

(A) Zn metal and NaOH(*aq*)

(B) Au metal and NaCN(*aq*) in the presence of air

(C) Cu metal and conc. HNO_3

(D) Fe metal and conc. HNO_3

2. Pure water freezes at 273 K and 1 bar. The addition of 34.5 g of ethanol to 500 g of water changes the freezing point of the solution. Use the freezing point depression constant of water as 2 K kg mol^{-1}. The figures shown below represent plots of vapour pressure (V. P.) versus temperature (T). [Molecular weight of ethanol is 46 g mol^{-1}.]

Among the following, the option representing change in the freezing point is

(A) 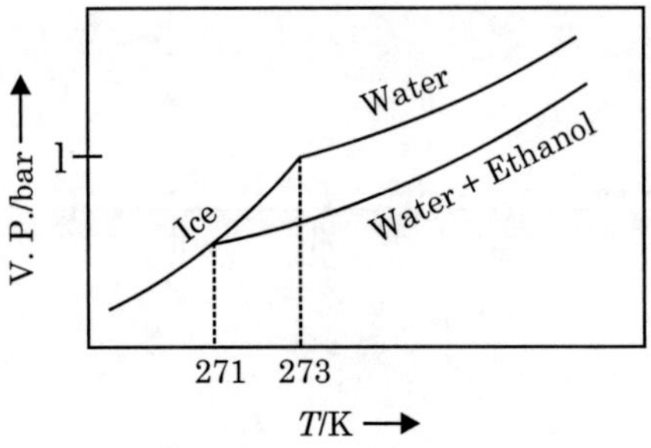

(B) 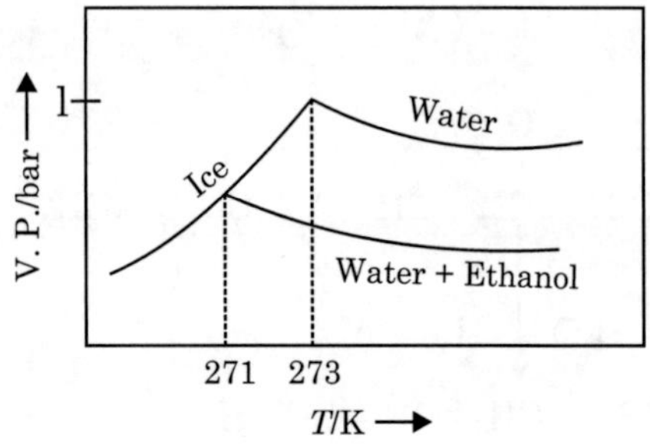

(C) 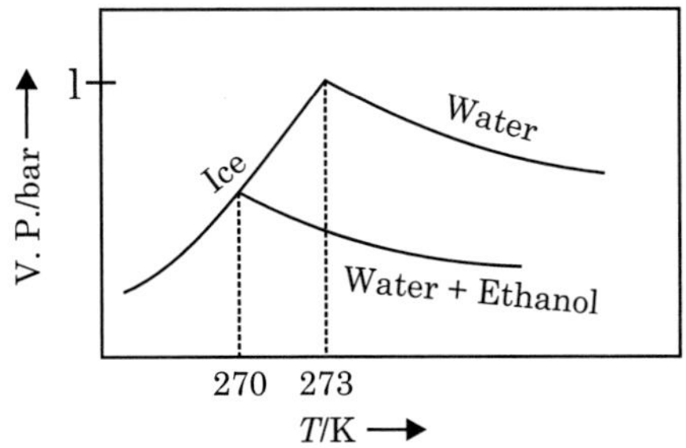

(D) 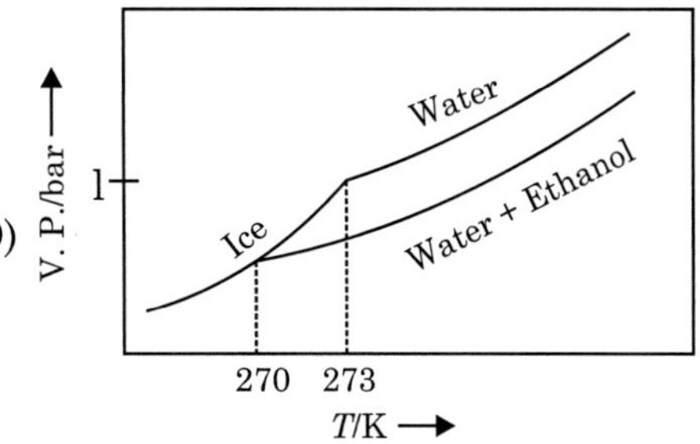

3. The order of the oxidation state of the phosphorus atom in H_3PO_2, H_3PO_4, H_3PO_3 and $H_4P_2O_6$ is

(A) $H_3PO_4 > H_4P_2O_6 > H_3PO_3 > H_3PO_2$
(B) $H_3PO_3 > H_3PO_2 > H_3PO_4 > H_4P_2O_6$
(C) $H_3PO_2 > H_3PO_3 > H_4P_2O_6 > H_3PO_4$
(D) $H_3PO_4 > H_3PO_2 > H_3PO_3 > H_4P_2O_6$

4. The standard state Gibbs free energies of formation of C (graphite) and C (diamond) at $T = 298$ K are

$\Delta_f G^0$ [C (graphite)] = 0 kJ mol^{-1}; $\Delta_f G^0$ [C (diamond)] = 2.9 kJ mol^{-1}

The standard state means that the pressure should be 1 bar and the substance should be pure at a given temperature. The conversion of graphite [C (graphite)] to diamond [C (diamond)] reduces its volume by 2×10^{-6} m^3 mol^{-1}. If C (graphite) is converted to C (diamond) isothermally at $T = 298$ K, the pressure at which C (graphite) is in equilibrium with C (diamond), is

[Useful information: 1 J = 1 kg m^2 s^{-2}; 1 Pa = 1 kg m^{-1} s^{-2}; 1 bar = 10^5 Pa]

(A) 14501 bar
(B) 29001 bar
(C) 58001 bar
(D) 1405 bar

5. The major product of the following reaction is

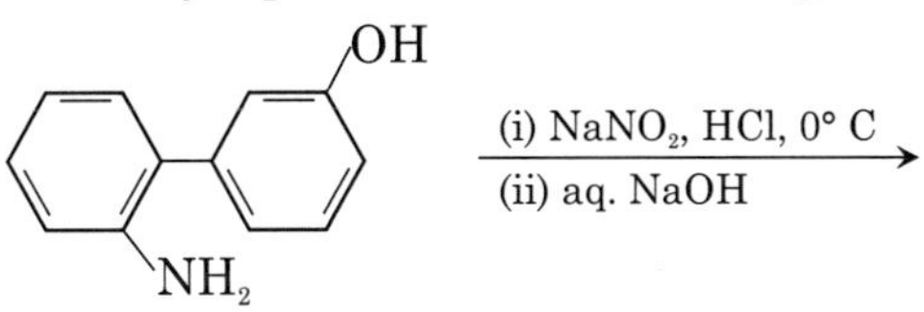

(A) 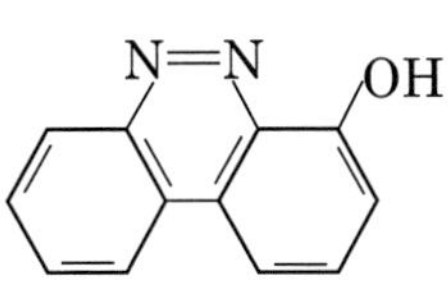

(B) 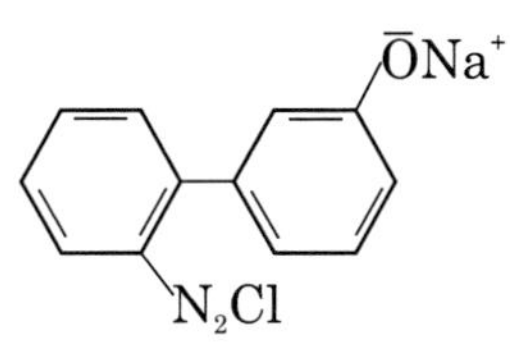

(C) 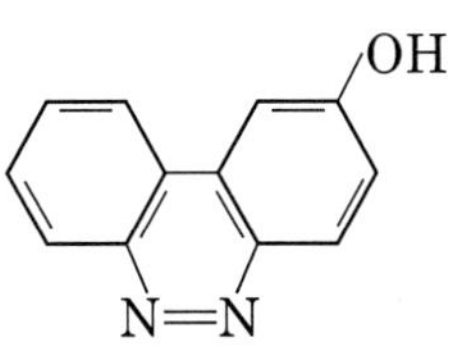

(D) 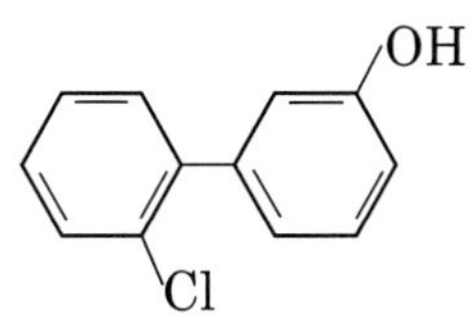

6. The order of basicity among the following compounds is

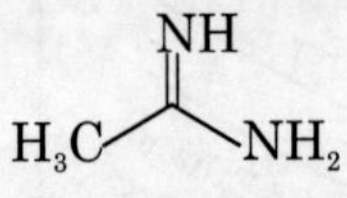

(I)

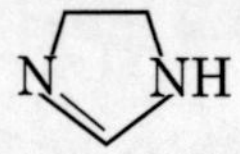

(II)

(III)

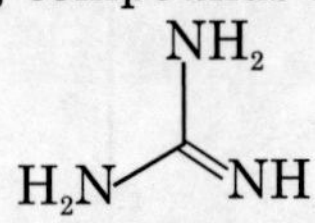

(IV)

(A) II > I > IV > III (B) IV > II > III > I (C) I > IV > III > II (D) IV > I > II > III

7. For the following cell:

$Zn(s) \mid ZnSO_4(aq) \parallel CuSO_4(aq) \mid Cu(s)$

When the concentration of Zn^{2+} is 10 times the concentration of Cu^{2+}, the expression of ΔG (in J mol^{-1}) is

[F is Faraday constant, R is gas constant, T is temperature, E^0(cell) = 1.1 V]

(A) $2.303RT + 1.1$ F (B) $2.303RT - 2.2$ F (C) 1.1 F (D) $-$ 2.2 F

SECTION–2
(Maximum Marks: 28)

- This section contains SEVEN questions.
- Each question has FOUR options (A), (B), (C) and (D). ONE OR MORE THAN ONE of these four options is(are) correct.
- For each question, darken the bubble(s) corresponding to all the correct option(s) in the ORS.
- For each question, marks will be awarded in one of the following categories:

  Full Marks : + 4 If only the bubble(s) corresponding to all the correct option(s) is(are) darkened.

  Partial Marks : + 1 For darkening a bubble corresponding to each correct option, provided NO incorrect option is darkened.

  Zero Marks : 0 If none of the bubbles is darkened.

  Negative Marks : – 2 In all other cases.
- For example, if (A), (C) and (D) are all the correct options for a question, darkening all these three will result in + 4 marks; darkening only (A) and (D) will get in + 2 marks; and darkening (A) and (B) will get in – 2 marks, as a wrong option is also darkened.

8. In a bimolecular reaction, the steric factor P was experimentally determined to be 4.5. The correct option(s) among the following is(are)

(A) The value of frequency factor predicted by Arrhenius equation is higher than that determined experimentally

(B) The activation energy of the reaction is unaffected by the value of the steric factor

(C) Since P = 4.5, the reaction will not proceed unless an effective catalyst is used

(D) Experimentally determined value of frequency factor is higher than that predicted by Arrhenius equation.

9. Compounds P and R upon ozonolysis produce Q and S, respectively. The molecular formula of Q and S is C_8H_8O. Q undergoes Cannizzaro reaction but not haloform reaction, whereas S undergoes haloform reaction but not Cannizzaro reaction.

(i) $P \xrightarrow[\text{(ii) Zn/H}_2\text{O}]{\text{(i) O}_3/\text{CH}_2\text{Cl}_2} Q$ (C_8H_8O) (ii) $R \xrightarrow[\text{(ii) Zn/H}_2\text{O}]{\text{(i) O}_3/\text{CH}_2\text{Cl}_2} S$ (C_8H_8O)

The option(s) with suitable combination of P and R, respectively, is(are)

(A) 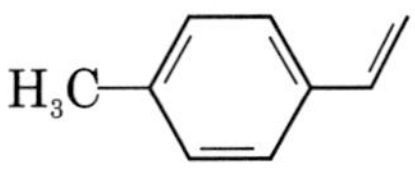 and 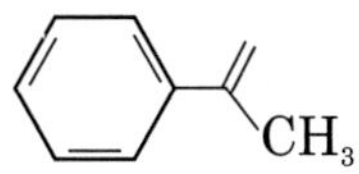

(B) 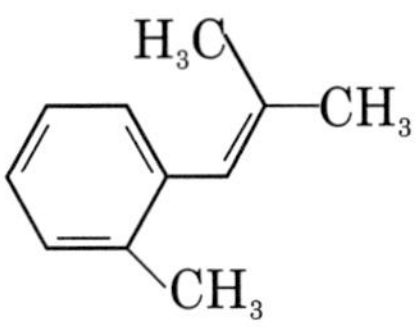 and 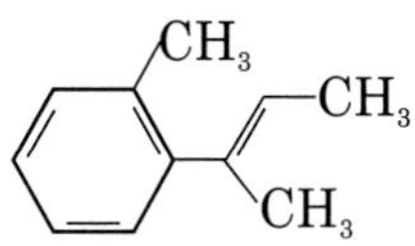

(C) 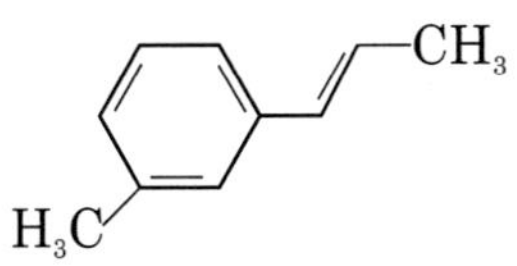 and 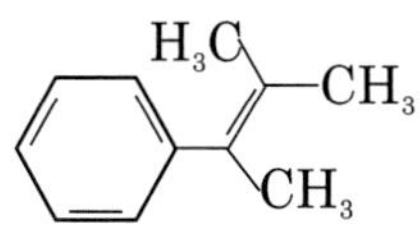

(D) 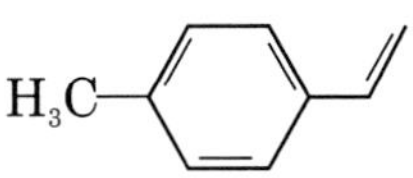 and 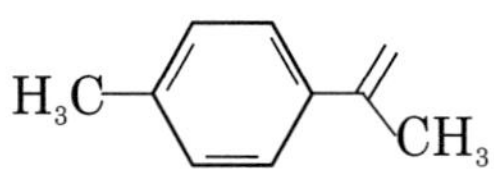

10. For a reaction taking place in a container in equilibrium with its surroundings, the effect of temperature on its equilibrium constant K in terms of change in entropy is described by

(A) With increase in temperature, the value of K for exothermic reaction decreases because the entropy change of the system is positive

(B) With increase in temperature, the value of K for endothermic reaction increases because unfavourable change in entropy of the surroundings decreases

(C) With increase in temperature, the value of K for exothermic reaction decreases because favourable change in entropy of the surroundings decreases

(D) With increase in temperature, the value of K for endothermic reaction increases because the entropy change of the system is negative.

11. The option(s) with only amphoteric oxides is(are)

(A) Cr_2O_3, CrO, SnO, PbO
(B) NO, B_2O_3, PbO, SnO_2
(C) Cr_2O_3, BeO, SnO, SnO_2
(D) ZnO, Al_2O_3, PbO, PbO_2

12. Among the following, the correct statement(s) is(are)

(A) $Al(CH_3)_3$ has the three-centre two-electron bonds in its dimeric structure

(B) $AlCl_3$ has the three-centre two-electron bonds in its dimeric structure

(C) BH_3 has the three-centre two-electron bonds in its dimeric structure

(D) The Lewis acidity of BCl_3 is greater than that of $AlCl_3$.

13. The correct statement(s) about surface properties is(are)

(A) Cloud is an emulsion type of colloid in which liquid is dispersed phase and gas is dispersion medium

(B) Adsorption is accompanied by decrease in enthalpy and decrease in entropy of the system

(C) Brownian motion of colloidal particles does not depend on the size of the particles but depends on viscosity of the solution

(D) The critical temperatures of ethane and nitrogen are 563 K and 126 K, respectively. The adsorption of ethane will be more than that of nitrogen on same amount of activated charcoal at a given temperature.

14. For the following compounds, the correct statement(s) with respect of nucleophilic substitution reactions is(are)

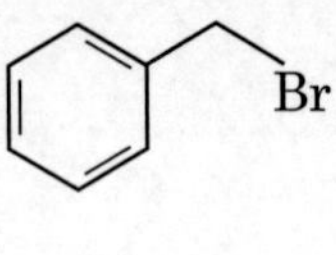

(I)

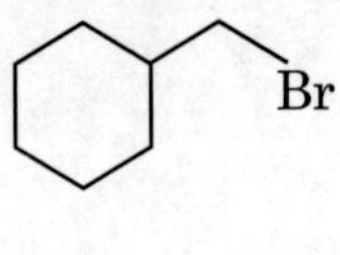

(II)

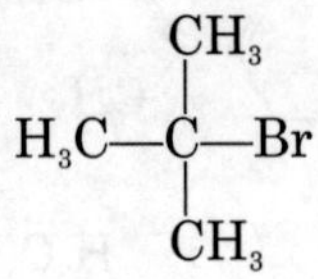

(III)

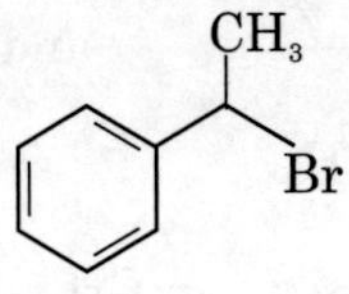

(IV)

(A) I and II follow S_N2 mechanism

(B) The order of reactivity for I, III and IV is: IV > I > III

(C) I and III follow S_N1 mechanism

(D) Compound IV undergoes inversion of configuration.

SECTION–3
(Maximum Marks: 12)

- This section contains TWO paragraphs.
- Based on each paragraph, there are TWO questions.
- Each question has FOUR options (A), (B), (C) and (D). ONLY ONE of these four options is correct.
- For each question, darken the bubble corresponding to the correct option in the ORS.
- For each question, marks will be awarded in one of the following categories:
  Full Marks : + 3 If only the bubble corresponding to all the correct option is darkened.
  Zero Marks : 0 In all other cases.

Paragraph for Questions 15 & 16

Upon heating $KClO_3$ in the presence of catalytic amount of MnO_2, a gas W is formed. Excess amount of W reacts with white phosphorus to give X. The reaction of X with pure HNO_3 gives Y and Z.

15. W and X are, respectively

(A) O_3 and P_4O_6 (B) O_2 and P_4O_{10} (C) O_3 and P_4O_{10} (D) O_2 and P_4O_6

16. Y and Z are, respectively

(A) N_2O_4 and H_3PO_3 (B) N_2O_4 and HPO_3

(C) N_2O_5 and HPO_3 (D) N_2O_3 and H_3PO_4

Paragraph for Questions 17 & 18

The reaction of compound P with CH_3MgBr (excess) in $(C_2H_5)_2O$ followed by addition of H_2O gives Q. The compound Q on treatment with H_2SO_4 at 0 °C gives R. The reaction of R with CH_3COCl in the presence of anhydrous $AlCl_3$ in CH_2Cl_2 followed by treatment with H_2O produces compound S. [Et in compound P is ethyl group]

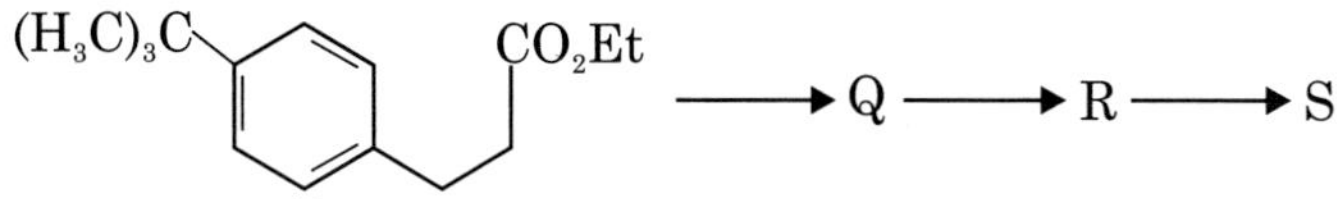

P

17. The reactions, Q to R and S to S, are
(A) Dehydration and Friedel–Crafts acylation
(B) Friedel–Crafts alkylation, dehydration and Friedel–Crafts acylation
(C) Aromatic sulfonation and Friedel–Crafts acylation
(D) Friedel–Crafts alkylation and Friedel–Crafts acylation.

18. The product S is

(A) 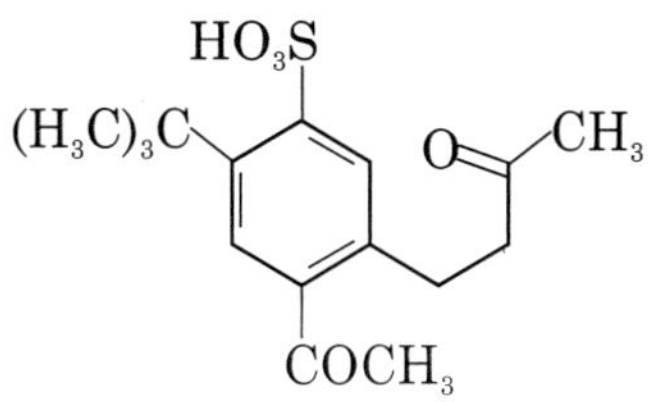

(B) 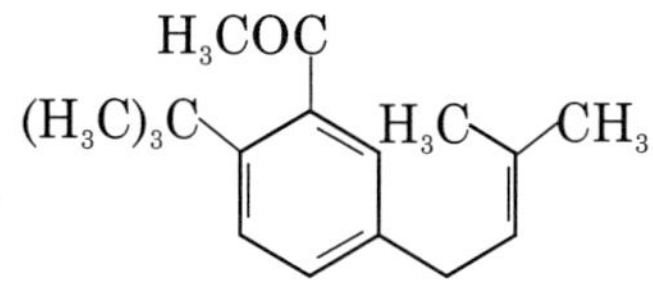

(C) 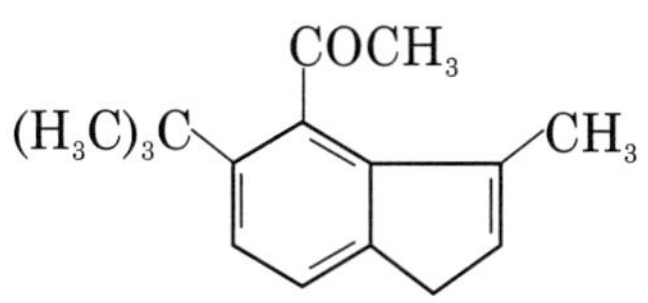

(D) 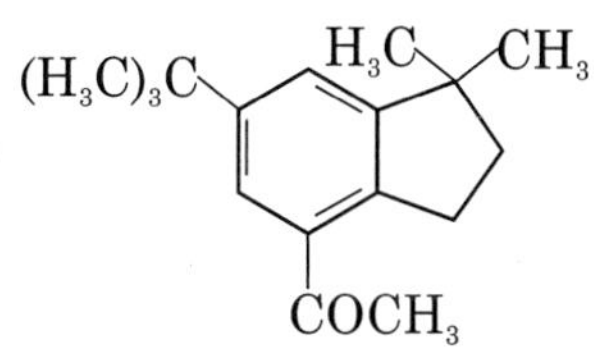

ANSWER KEYS

1. A	2. D	3. A	4. A	5. C	6. D
7. B	8. B, D	9. A, C	10. B, C	11. C, D	12. A, C, D
13. B, D	14. A, B, C, D	15. B	16. C	17. B	18. D

SOLUTIONS

1. (A)

(A) $Zn + 2NaOH \longrightarrow Na_2ZnO_2 + H_2$

(B) $4Au + 8NaCN + O_2 + 2H_2O \longrightarrow 4Na[Au(CN)_2] + 4NaOH$

(C) $Cu + \underset{\text{(conc.)}}{4HNO_3} \longrightarrow Cu(NO_3)_2 + 2NO_2 + 2H_2O$

(D) Formation of passive layer of Fe_2O_3 on the surface of Fe and NO_2 gas is evolved.

2. (D)

Ethanol should be considered non-volatile as per given option.

$\Delta T_f = K_f \times m$

$$\Delta T_f = 2 \times \frac{34.5}{46 \times 0.5} = 3 \text{ K}$$

$T_f^0 = 273$ K

$T_f = 270$ K

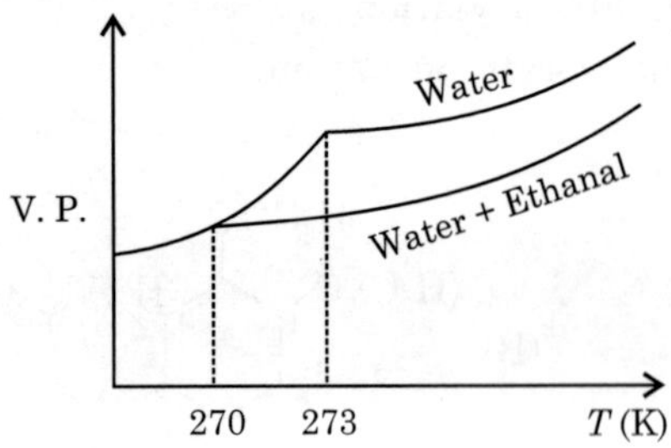

3. (A)

$H_3PO_4 \rightarrow$ 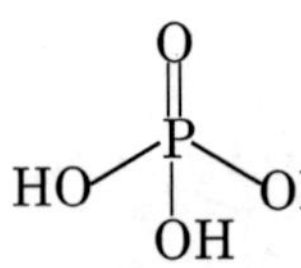 OH; oxidation state of P = + 5

$H_4P_2O_6 \rightarrow$ 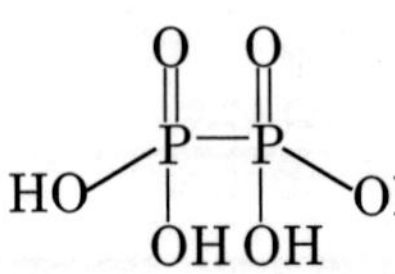 OH; oxidation state of P = + 4

$H_3PO_4 \rightarrow$ 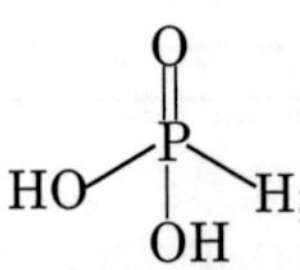 H; oxidation state of P = + 3

$H_3PO_2 \rightarrow$ 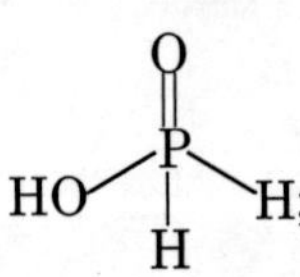 H; oxidation state of P = + 1

Hence, Ans. (A).

4. (A)

C(graphite) $\longrightarrow$ C(diamond); $\Delta G^0 = \Delta_f G^0_{\text{diamond}} - \Delta_f G^0_{\text{graphite}} = 2.9$ kJ/mol at 1 bar.

As $dG_T = V.dP$,

$$\int_{\Delta G_1}^{\Delta G_2} d(\Delta G_T) = \int_{P_1}^{P_2} \Delta V.dP$$

$$\Delta G_2 - \Delta G_1 = \Delta V.(P_2 - P_1)$$

$$(2.9 \times 10^3 - 0) = (-2 \times 10^{-6})(1 - P_2)$$

$$P_2 - 1 = \frac{2.9 \times 10^3}{2 \times 10^{-6}} \text{Pa} = 1.45 \times 10^4 \text{ bar}$$

$$P_2 = 14501 \text{ bar}$$

5. (C)

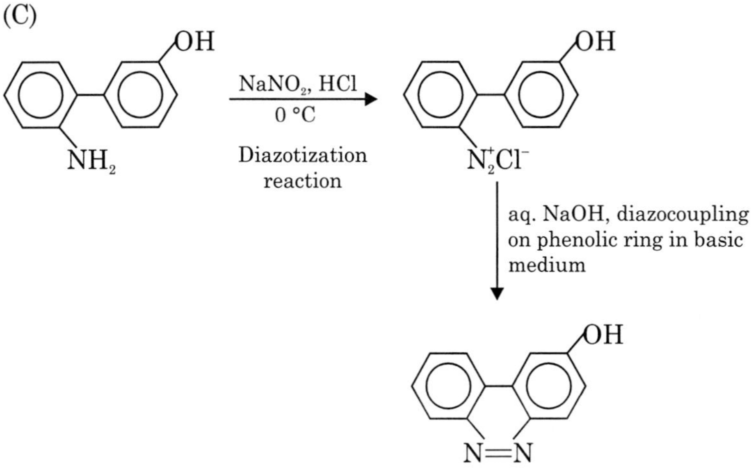

Para is less hindered.

6. (D) (IV > I > II > III)

Basic strength ∝ stability of conjugated acid

∝ + M / + H / + I

(I) 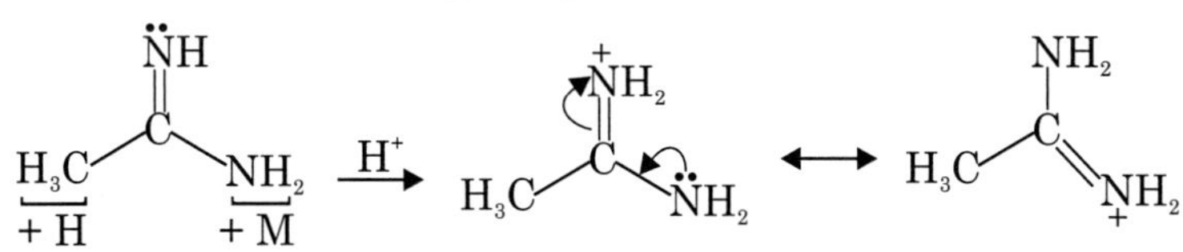

Conjugated acid stabilized by 2 equivalent resonating structures.

(II) 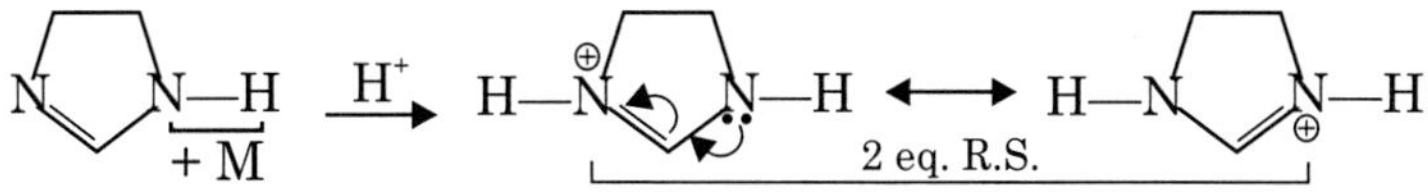

(III) 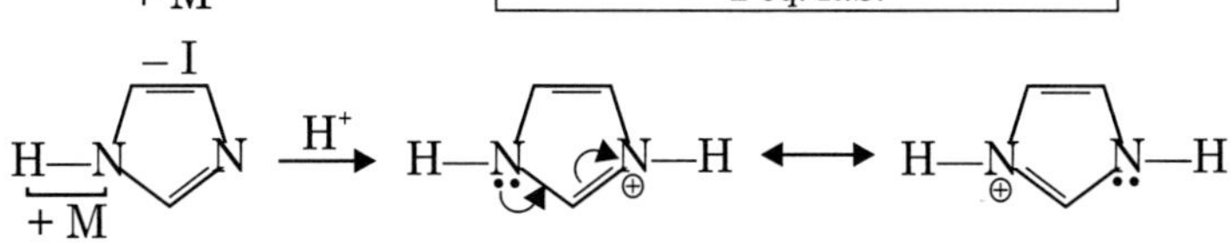

(IV)

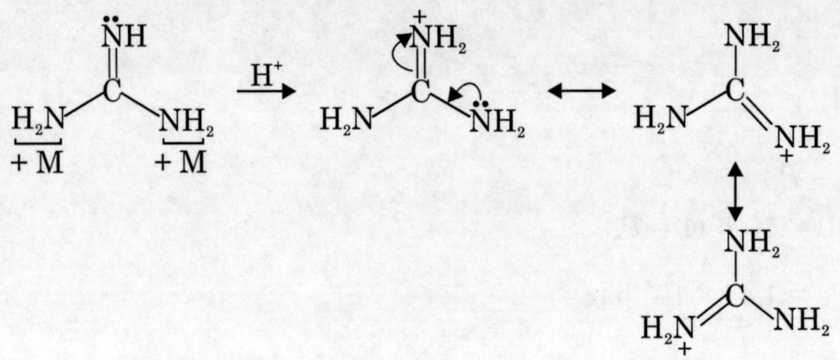

Conjugated acid stabilized by 3 equivalent resonating structures.

7. (B)

$\Delta G = \Delta G^0 + 2.303RT \log Q$

$\Delta G = -nFE^0 + 2.303RT \log Q$

Given: $E^0 = 1.1$ V and $n = 2$

$$\Delta G = (-2 \times 1.1 \times F) + 2.303RT \log \frac{[Zn^{2+}]}{[Cu^{2+}]}$$

$\Delta G = -2.2\, F + 2.303RT$

8. (B, D)

$K = P.\,A.\,e^{-E_a/RT}$

(A) If $P < 1$, $A_{arr} > A_{expt}$

(D) If $P > 1$, $A_{arr} < A_{expt}$

(C) If P is very small, then catalyst is required to carry out the reaction at measurable rate.

9. (A, C)

(A)

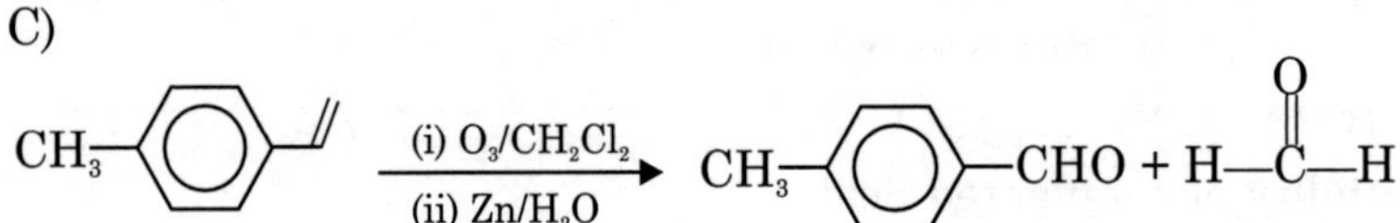

P — Q

give Cannizzaro reaction
but no haloform reaction

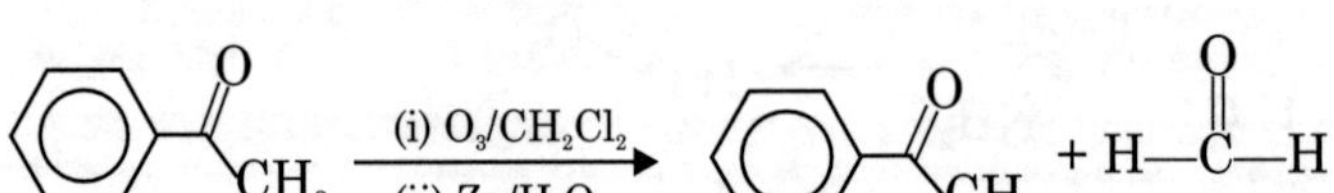

R — S

no Cannizzaro reaction
but give haloform reaction

(B) Products of ozonolysis of R are having 9 carbon atoms.

(C)

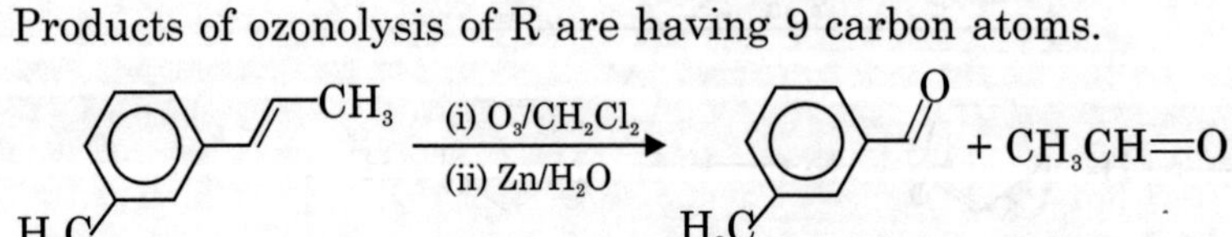

P — Q

gives Cannizzaro reaction
but no haloform reaction

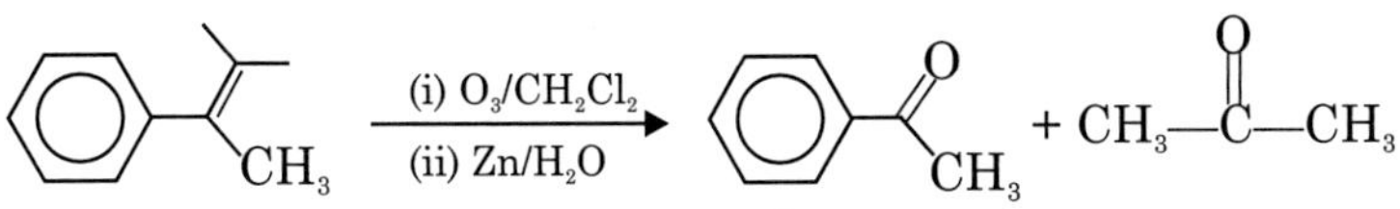

R

S

no Cannizzaro reaction
but give haloform reaction

(D) Products of ozonolysis of R are having 9 carbon atoms.

10. (B, C)

$$\Delta S_{surr.} = \frac{-q_{process}}{T_{surr.}}$$

If $\Delta H > 0$ on $T \uparrow K_{eq} \uparrow$, $\Delta S_{surr.} < 0$ (Surrounding is unfavourable.)
If $\Delta H < 0$ on $T \uparrow K_{eq} \downarrow$, $\Delta S_{surr.} > 0$ (Surrounding is favourable.)

11. (C, D)

(C) Cr_2O_3, BeO, SnO, SnO_2
All are amphoteric oxides.

(D) ZnO, Al_2O_3, PbO, PbO_2
All are amphoteric oxides.

12. (A, C, D)

(A) 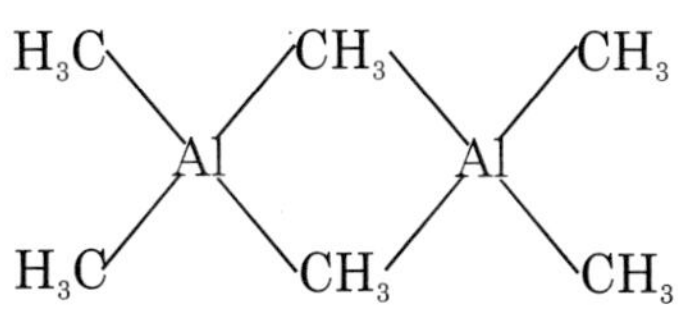 Two 3c – $2e^-$ bond (three-centre-two-electron bond)

(B) 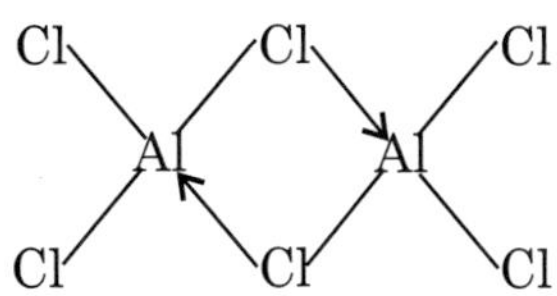 Two 3c – $4e^-$ bond (three-centre-four-electron bond)

(C) 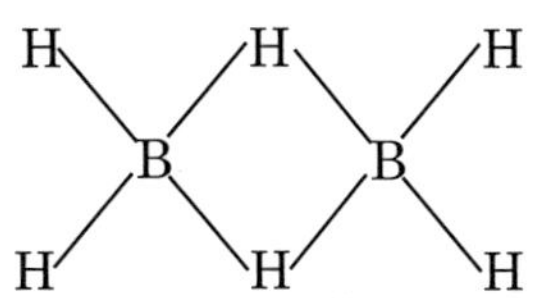 Two 3c – $2e^-$ bond (three-centre-two-electron bond)

(D) Lewis acidic strength decreases down the group. The decrease in acid strength occurs because as size increases, the attraction between the incoming electron pair and the nucleus weakens. Hence, Lewis acidic strength of BCl_3 is more than that of $AlCl_3$.

13. (B, D)

(A) Emulsion is liquid in liquid type colloid.

(B) For adsorption, $\Delta H < 0$ and $\Delta S < 0$.

(C) Smaller the size and less viscous the dispersion medium, more will be the Brownian motion.

(D) Higher the T_C, greater will be the extent of adsorption.

14. (A, B, C, D)

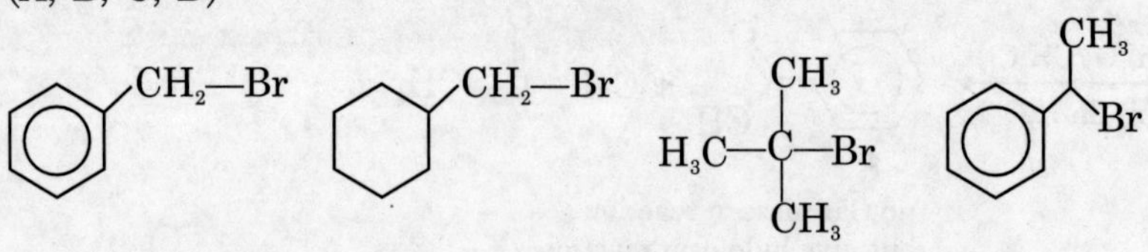

(A) I and II follow S_N2 mechanism as they are primary.
(B) Reactivity order: IV > I > III.
(C) I and III follow S_N1 mechanism as they form stable carbocation.
(D) Compound IV undergoes inversion of configuration.

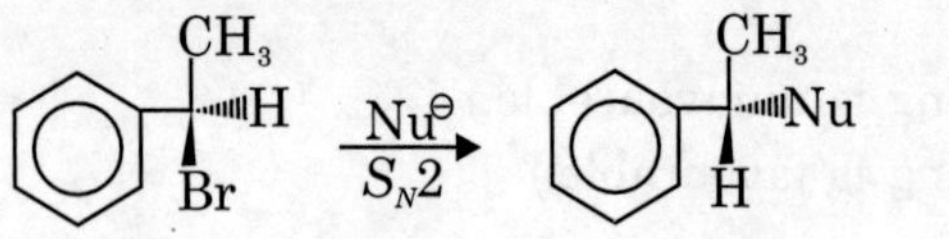

(inverted product)

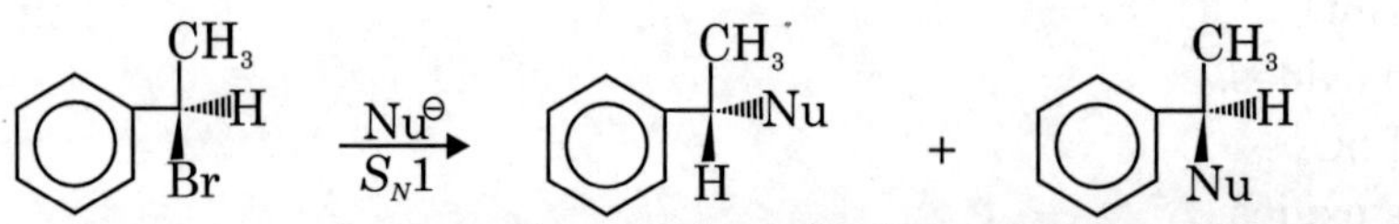

(inverted product) (retained product)

Solutions for paragraph Q. 15 & 16

15. (B)

W and X are, respectively, $W = O_2$ and $X = P_4O_{10}$

16. (C)

Y and Z are, respectively, $Y = N_2O_5$, $Z = HPO_3$

$$2KClO_3 \xrightarrow[\Delta]{MnO_2} 2KCl + \underset{(W)}{3O_2}$$

$$\underset{\substack{\text{white}\\\text{phosphorus}}}{P_4} + \underset{\substack{\text{(excess)}\\(W)}}{5O_2} \rightarrow \underset{(X)}{P_4O_{10}}$$

$$\underset{(X)}{P_4O_{10}} + 4HNO_3 \rightarrow \underset{(Y)}{2N_2O_5} + \underset{(Z)}{(HPO_3)_4}$$

Solutions for paragraph Q. 17 & 18

17. (B)

The reaction Q to R is Friedel–Crafts alkylation and hydration.

The reaction R to S is Friedel–Crafts alkylation.

18. (D)

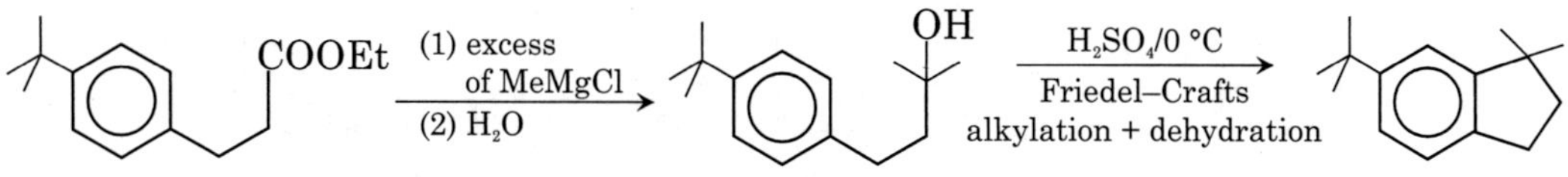

P Q R

(1) $Me-\overset{O}{\overset{\|}{C}}-Cl/AlCl_3$ in CH_2Cl_2 (Friedel–Crafts acylation)
(2) H_2O

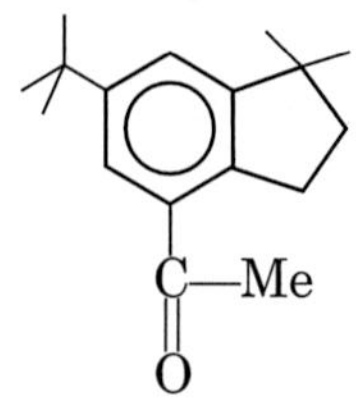

S

I.I.T. JEE Objective Question Paper—2018 (Main)

1. Total number of lone pair of electron in I_3^- ion is

(A) 9 (B) 12 (C) 3 (D) 6

2. Which of the following salts is the most basic in aqueous solution?

(A) $FeCl_3$ (B) $Pb(CH_3COO)_2$ (C) $Al(CN)_3$ (D) CH_3COOK

3. Phenol reacts with methyl chloroformate in the presence of NaOH to form product A. A reacts with Br_2 to form product B. A and B are respectively

(A) 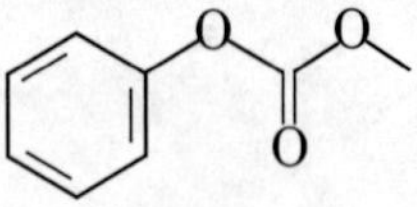 and 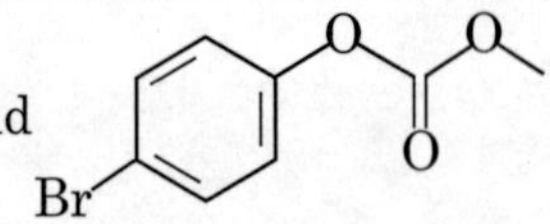

(B) 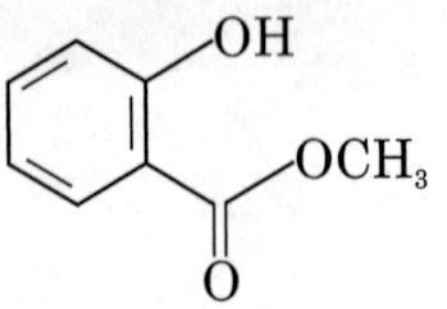 and 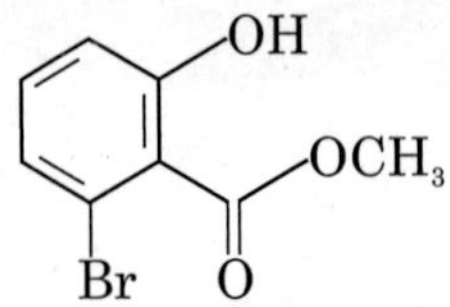

(C) 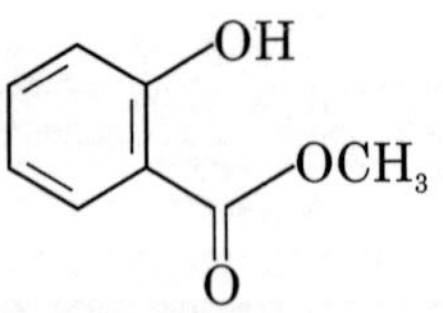 and 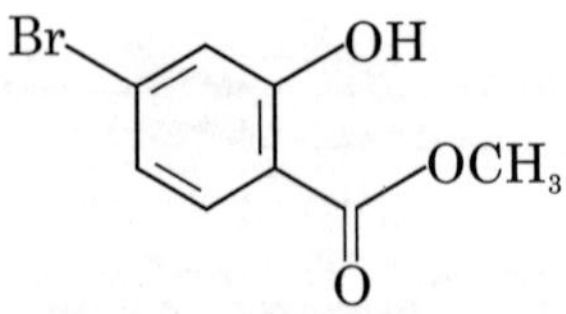

(D) 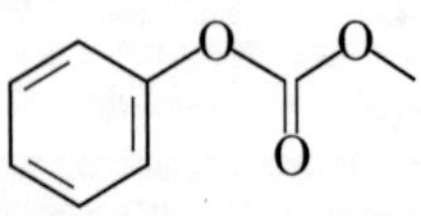 and 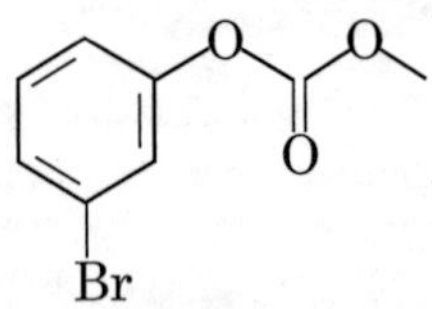

4. The increasing order of basicity of the following compounds is

(a)  (b)  (c) 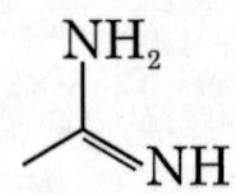 (d) 

(A) (b) < (a) < (d) < (c) (B) (d) < (b) < (a) < (c)

(C) (a) < (b) < (c) < (d) (D) (b) < (a) < (c) < (d)

5. An alkali is titrated against an acid with methyl orange as indicator. Which of the following is a correct combination?

	Base	Acid	End Point
(A)	Weak	Strong	Yellow to pinkish red
(B)	Strong	Strong	Pink to colourless
(C)	Weak	Strong	Colourless to pink
(D)	Strong	Strong	Pinkish red to yellow

6. The trans-alkenes are formed by the reduction of alkynes with

(A) Na/liq.NH_3 (B) Sn–HCl (C) H_2–Pd/C, $BaSO_4$ (D) $NaBH_4$

7. The ratio of mass percent of C and H of an organic compound ($C_XH_YO_Z$) is 6 : 1. If one molecule of the above compound ($C_XH_YO_Z$) contains half as much oxygen as required to burn one molecule of compound C_XH_Y completely to CO_2 and H_2O. The empirical formula of compound $C_XH_YO_Z$ is

(A) $C_3H_4O_2$ (B) $C_2H_4O_3$ (C) $C_3H_6O_3$ (D) C_2H_4O

8. Hydrogen peroxide oxidises $[Fe(CN)_6]^{4-}$ to $[Fe(CN)_6]^{3-}$ in acidic medium but reduces $[Fe(CN)_6]^{3-}$ to $[Fe(CN)_6]^{4-}$ in alkaline medium. The other products formed are, respectively

(A) H_2O and ($H_2O + O_2$) (B) H_2O and ($H_2O + OH^-$)

(C) ($H_2O + O_2$) and H_2O (D) ($H_2O + O_2$) and ($H_2O + OH^-$)

9. The major product formed in the following reaction is

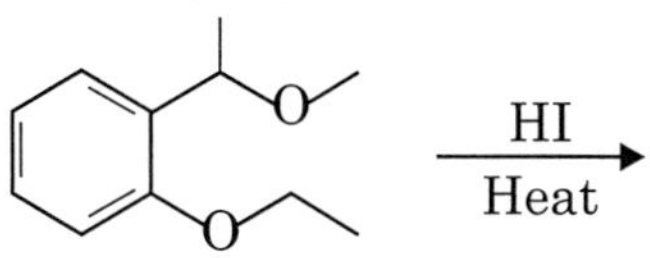

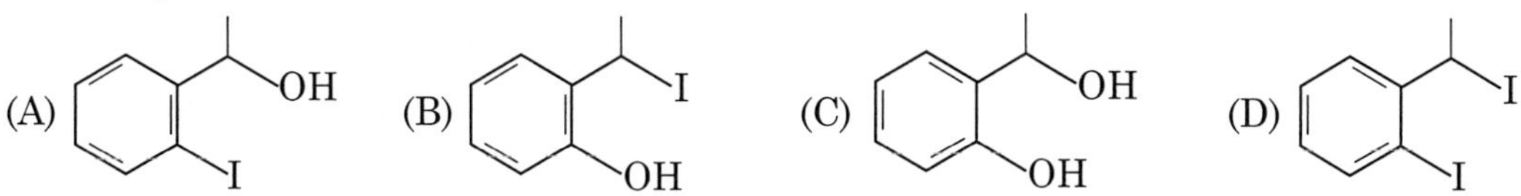

10. How long (approximate) should water be electrolysed by passing through 100 amperes current so that the oxygen released can completely burn 27.66 g of diborane? (Atomic weight of B = 10.8 u)

(A) 3.2 hours (B) 1.6 hours (C) 6.4 hours (D) 0.8 hours

11. Which of the following lines correctly show the temperature dependence of equilibrium constant K, for an exothermic reaction?

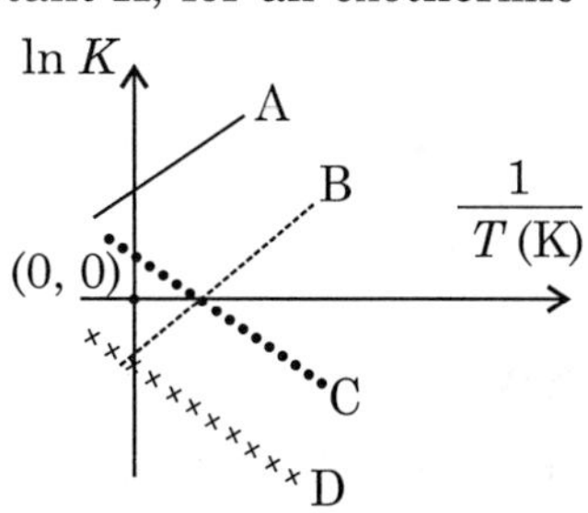

(A) C and D (B) A and D (C) A and B (D) B and C

12. At 518°C, the rate of decomposition of a sample of gaseous acetaldehyde, initially at a pressure of 363 Torr, was 1.00 Torr s^{-1} when 5% had reacted and 0.5 Torr s^{-1} when 33% had reacted. The order of the reaction is

(A) 1 (B) 0 (C) 2 (D) 3

13. Glucose on prolonged heating with HI gives

(A) Hexanoic acid (B) 6-iodohexanal (C) *n*-Hexane (D) 1-Hexene

14. Consider the following reaction and statements:

$[Co(NH_3)_4Br_2]^+ + Br^- \rightarrow [Co(NH_3)_3Br_3] + NH_3$

(I) Two isomers are produced if the reactant complex ion is a *cis*-isomer

(II) Two isomers are produced if the reactant complex ion is a *trans*-isomer

(III) Only one isomer is produced if the reactant complex ion is a *trans*-isomer

(IV) Only one isomer is produced if the reactant complex ion is a *cis*-isomer.

The correct statements are:

(A) (III) and (IV) (B) (II) and (IV) (C) (I) and (II) (D) (I) and (III)

15. The major product of the following reaction is

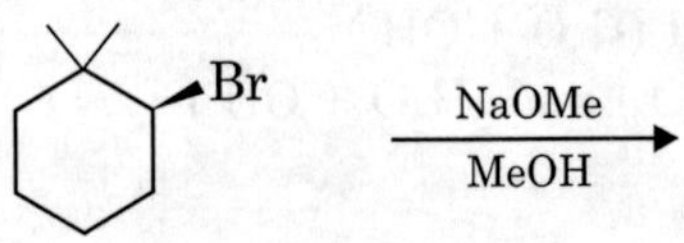

(A) 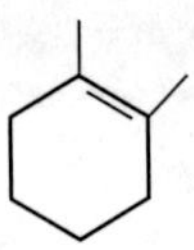 (B) 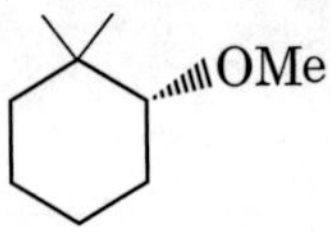 (C) 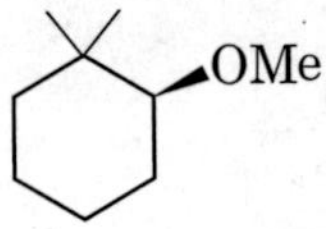 (D) 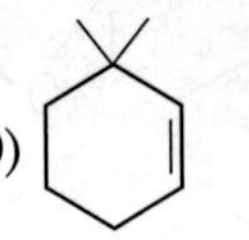

16. Phenol on treatment with CO_2 in the presence of NaOH followed by acidification produces compound X as the major product. X on treatment with $(CH_3CO)_2O$ in the presence of catalytic amount of H_2SO_4 produces

(A) 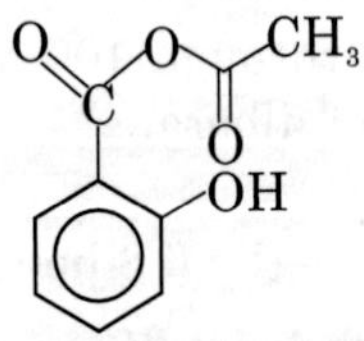 (B) 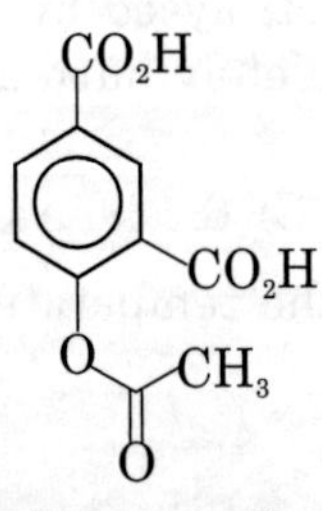 (C) 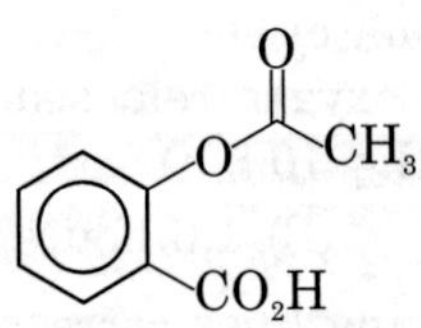 (D) 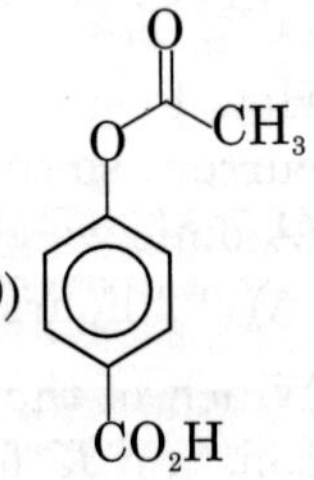

17. An aqueous solution contains an unknown concentration of Ba^{2+}. When 50 mL of a 1*M* solution of Na_2SO_4 is added, $BaSO_4$ just begins to precipitate. The final volume is 500 mL. The solubility product of $BaSO_4$ is 1×10^{-10}. What is the original concentration of Ba^{2+}?

(A) $1.1 \times 10^{-9}M$ (B) $1.0 \times 10^{-10}M$ (C) $5 \times 10^{-9}M$ (D) $2 \times 10^{-9}M$

18. Which of the following compounds will be suitable for Kjeldahl's method for nitrogen estimation?

(A) 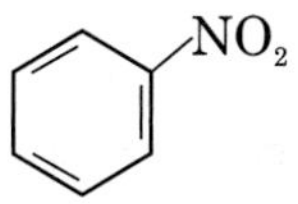

(B) 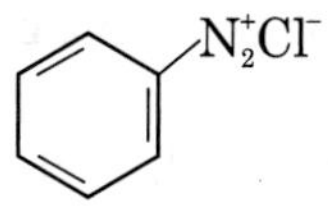

(C) 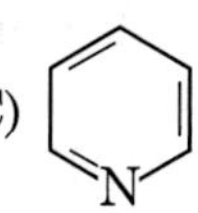

(D) 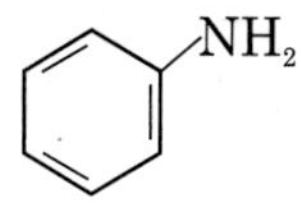

19. When metal M is treated with NaOH, a white gelatinous precipitate X is obtained, which is soluble in excess of NaOH. Compound X when heated strongly gives an oxide which is used in chromatography as an adsorbent. The metal M is

(A) Al (B) Fe (C) Zn (D) Ca

20. An aqueous solution contains $0.10M$ H_2S and $0.20M$ HCl. If the equilibrium constants for the formation of HS^- from H_2S is 1.0×10^{-7} and that of S^{2-} from HS^- ions is 1.2×10^{-13}, then the concentration of S^{2-} ions in aqueous solution is

(A) $6 \times 10^{-21}M$ (B) $5 \times 10^{-19}M$ (C) $5 \times 10^{-8}M$ (D) $3 \times 10^{-20}M$

21. The recommended concentration of fluoride ion in drinking water is up to 1 ppm as fluoride ion is required to make teeth enamel harder by converting $[3Ca_3(PO_4)_2 \cdot Ca(OH)_2]$ to

(A) $[3Ca_3(PO_4)_2 \cdot CaF_2]$ (B) $[3\{Ca(OH)_2\} \cdot CaF_2]$

(C) $[CaF_2]$ (D) $[3(CaF_2) \cdot Ca(OH)_2]$

22. The compound that does not produce nitrogen gas by the thermal decomposition is

(A) NH_4NO_2 (B) $(NH_4)_2SO_4$ (C) $Ba(N_3)_2$ (D) $(NH_4)_2Cr_2O_7$

23. The predominant form of histamine present in human blood is (pK_a, histidine = 6.0)

(A) 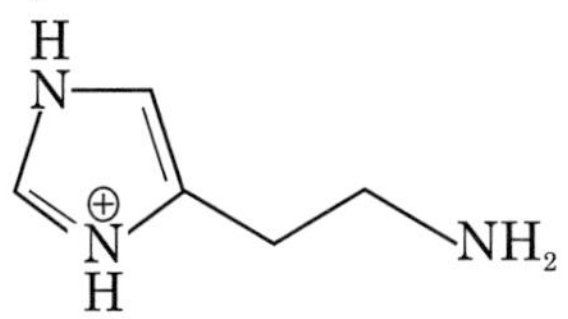

(B) 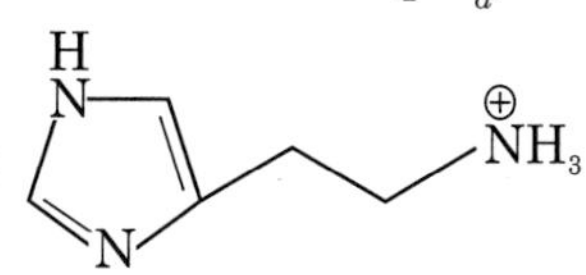

(C) 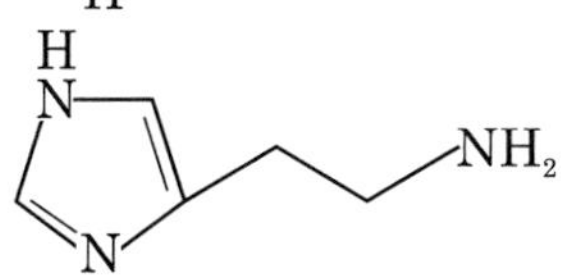

(D) 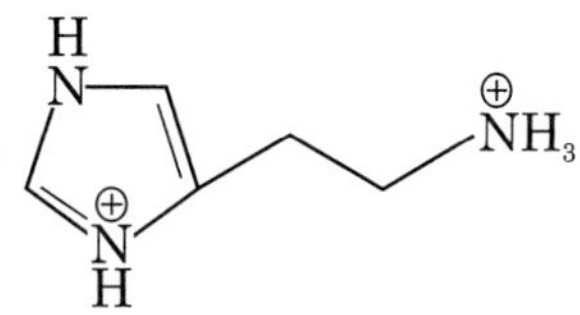

24. The oxidation states of Cr in $[Cr(H_2O)_6]Cl_3$, $[Cr(C_6H_6)_2]$, and $K_2[Cr(CN)_2(O)_2(O_2)(NH_3)]$ respectively are

(A) + 3, 0, and + 6 (B) + 3, 0, and + 4

(C) + 3, + 4, and + 6 (D) + 3, + 2 and + 4

25. Which type of 'defect' has the presence of cations in the interstitial sites?

(A) Frenkel defect (B) Metal deficiency defect

(C) Schottky defect (D) Vacancy defect.

26. The combustion of benzene (ℓ) gives $CO_2(g)$ and $H_2O(\ell)$. Given that heat of combustion of benzene at constant volume is – 3263.9 kJ mol^{-1} at 25°C. Heat of combustion (in kJ mol^{-1}) of benzene at constant pressure will be (R = 8.314 J K^{-1} mol^{-1})

(A) 3260 (B) – 3267.6 (C) 4152.6 (D) – 452.46

27. Which of the following are Lewis acids?

(A) PH_3 and $SiCl_4$ (B) BCl_3 and $AlCl_3$ (C) PH_3 and BCl_3 (D) $AlCl_3$ and $SiCl_4$

28. Which of the following compounds contain(s) no covalent bond(s)?
KCl, PH_3, O_2, B_2H_6, H_2SO_4

(A) KCl (B) KCl, B_2H_6 (C) KCl, B_2H_6, PH_3 (D) KCl, H_2SO_4

29. For 1 molal aqueous solution of the following compounds, which one will show the highest freezing point?

(A) $[Co(H_2O)_4Cl_2]Cl\cdot 2H_2O$ (B) $[Co(H_2O)_3Cl_3]\cdot 3H_2O$
(C) $[Co(H_2O)_6]Cl_3$ (D) $[Co(H_2O)_5Cl]Cl_2\cdot H_2O$

30. According to molecular orbital theory, which of the following will not be a viable molecule?

(A) H_2^- (B) H_2^{2-} (C) He_2^{2+} (D) He_2^+

ANSWER KEYS

1. A	2. D	3. A	4. A	5. A	6. A
7. B	8. A	9. B	10. A	11. C	12. C
13. C	14. D	15. D	16. C	17. A	18. D
19. A	20. D	21. A	22. B	23. B	24. A
25. A	26. B	27. B	28. A	29. B	30. B

SOLUTIONS

1. (A)

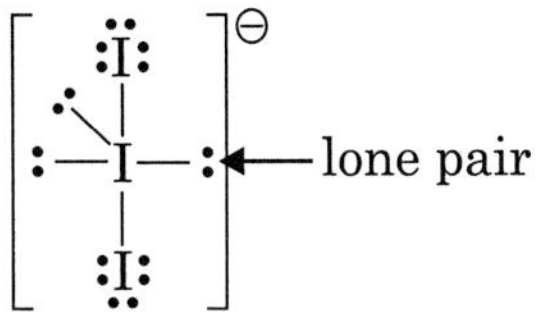

2. (D)

CH_3COOK is most basic among the given options.

3. (A)

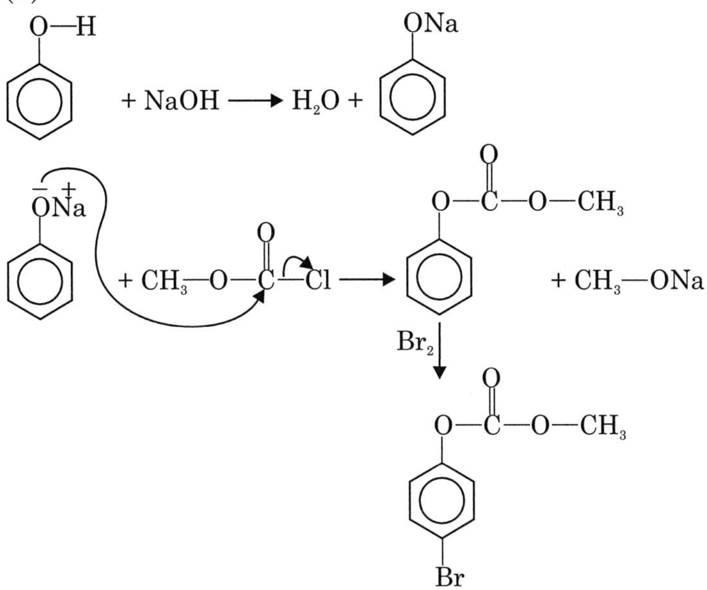

4. (A)

Basic Nature: (b) < (a) < (d) < (c)

5. (A)

When a strong acid is titrated against a weak base using methyl orange as an indicator, the end point is yellow to pinkish red.

6. (A)

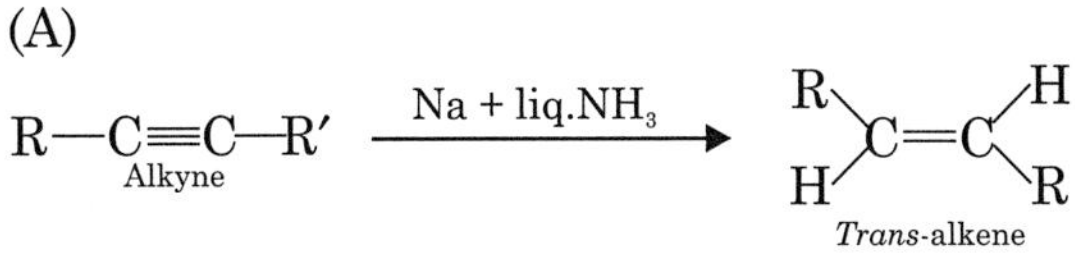

7. (B)

$C_XH_YO_Z$

$$\frac{w_C}{w_H} = \frac{12X}{Y} = \frac{6}{1} \quad \text{or} \quad \frac{X}{Y} = \frac{1}{2}$$

$$C_XH_Y + \left(X + \frac{Y}{4}\right) O_2 \quad XCO_2 + \frac{Y}{2} H_2O$$

$X : Y : Z = 2 : 4 : 3$

8. (A)

$$\underset{(OA)}{H_2\overset{-1}{O}_2} + [Fe(CN)_6]^{4-} \xrightarrow{H^+} [Fe(CN)_6]^{3-} + H_2\overset{-2}{O}$$

$$\underset{(RA)}{H_2\overset{-1}{O}_2} + [Fe(CN)_6]^{3-} \longrightarrow [Fe(CN)_6]^{4-} + \overset{0}{O}_2$$

9. (B)

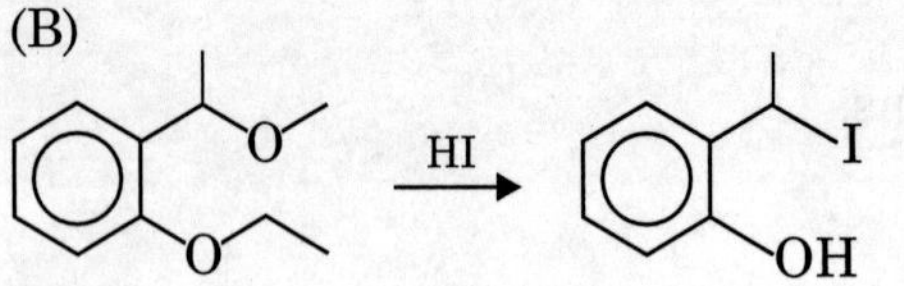

10. (A)

$$B_2H_6 + 3O_2 \longrightarrow B_2O_3 + 3H_2O$$

$\frac{27.6}{27.6} = 3 \text{ mol}$

1 mol

$\frac{It}{F} = \frac{\theta}{F} = \frac{w}{E} = n_{O_2} \times 4$

$t = 965 \times 12 \text{ sec} = 3.2 \text{ hr}$

11. (C)

$$\log_{10} K_{eq} = \text{constant} - \left(\frac{\Delta H}{2.303R}\right)\left(\frac{1}{T}\right)$$

Given: = ΔH = –ve i.e. slope = + ve

12. (C)

Rate of a reaction = $k[P]^n$

$$R_1 = 1 \text{ torr/sec} = k\left[365 \times \frac{95}{100}\right]^n$$

$$R_2 = 0.5 \text{ torr/sec} = k\left[365 \times \frac{67}{100}\right]^n$$

$$\frac{1}{0.5} = \left(\frac{95}{67}\right)^n$$

$$2 = \left(\frac{96}{67}\right)^n = (1.43)^n \quad \text{or} \quad n = 2$$

13. (C)

$$\text{Glucose} + \text{HI} \xrightarrow{\Delta} n\text{-Hexane}$$

14. (D)

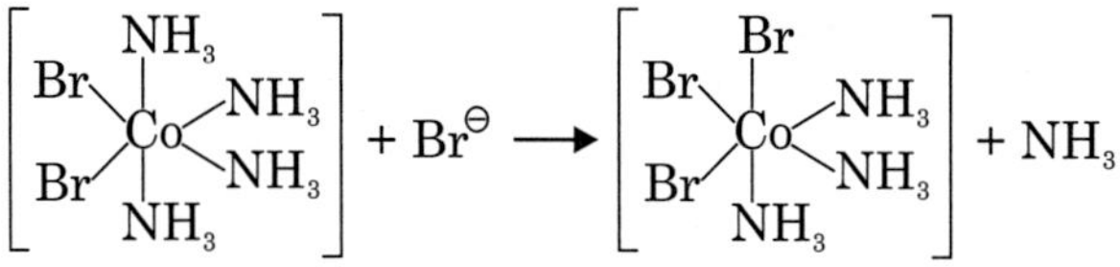

cis

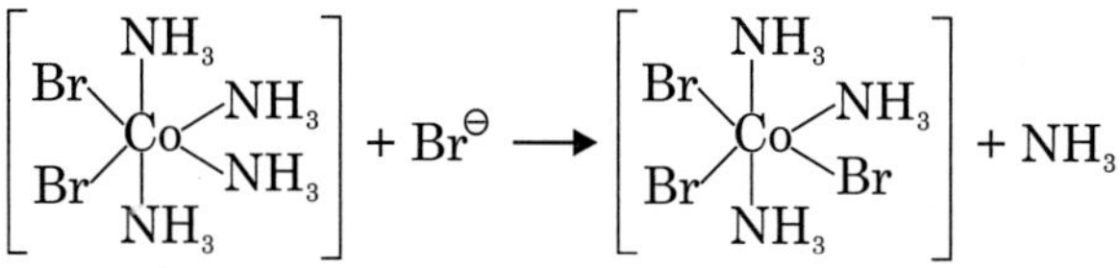

cis

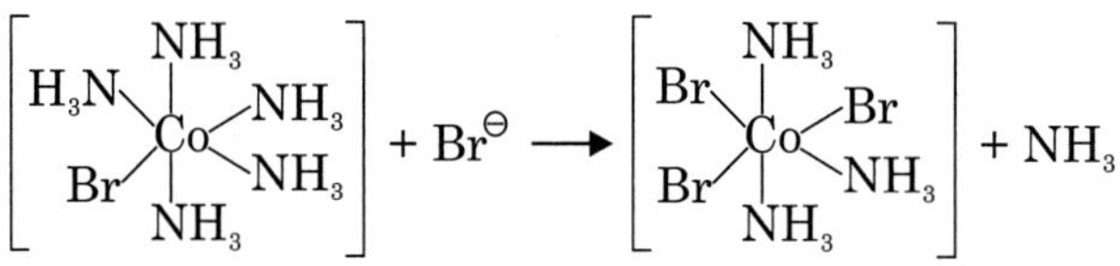

trans one product

15. (D)

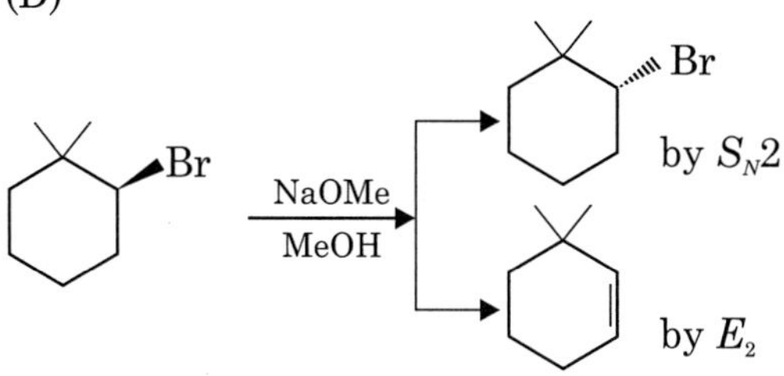

16. (C)

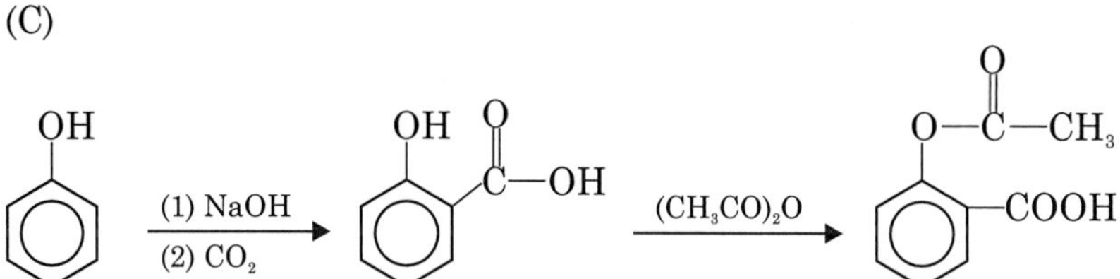

Aspirin

17. (A)

$$\left(\frac{MV}{V+50}\right)\left(\frac{1\times 50}{V+50}\right) = 10^{-10} \qquad ...(1)$$

$V = 450$ mL as total volume = 500 mL

$$\left(\frac{M\times 450}{500}\right)\left(\frac{50}{100}\right) = 10^{-10}$$

$$M = \frac{50}{45}\times 10\times 10^{-10} = \frac{50}{45}\times 10^{-9} = 1.1\times 10^{-9}\,M$$

18. (D)

A compound containing primary amine N atom is suitable for estimation of nitrogen by Kjeldahl's method.

19. (A)
Alumina is used in chromatography as an adsorbent.

20. (D)
$K_1 = 10^{-7}$, $K_2 = 1.2 \times 10^{-13}$

$$K_1 K_2 = \frac{[H^+]^2\,[S^{2-}]}{[H_2S]}$$

$$10^{-7} \times 1.2 \times 10^{-13} = \frac{(0.2)^2\,(S^{2-})}{(0.1)}$$

$$[S^{2-}] = \frac{1.2 \times 10^{-13} \times 10^{-7} \times 0.1}{4 \times 10^{-2}} = \frac{12}{4} \times 10^{-20} = 3 \times 10^{-20} M$$

21. (A)
Solid $[3Ca_3(PO_4)_2 \cdot Ca(F_2)]$ is formed.

22. (B)
$(NH_4)_2SO_4 \longrightarrow NH_3\uparrow$

23. (B)
Blood is slightly basic in nature (*p*H 7.35) and at this *p*H terminal NH_2 will get + charge due to more basic nature.

24. (A)
The oxidation states of Cr in $[Cr(H_2O)_6]Cl_3$, $[Cr(C_6H_6)_2]$ and $K_2[Cr(CN)(O_2)(O_2)]$ respectively are + 3, 0 and + 6.

25. (A)
In Frenkel defect, cations are present in interstitial sites.

26. (B)

$C_6H_6(\ell) + 7.5O_2(g) \longrightarrow 6CO_2(g) + 3H_2O(\ell)$

$\Delta n_g = 6 - 7.5 = -1.5$

$\Delta H = \Delta U + (\Delta_{ng})RT$

$$= -3263.9 + \frac{(-1.5) \times 8.314 \times 298}{1000} = -3267.6 \text{ kJ / mol}^{-1}$$

27. (B)
In both BCl_3 and $AlCl_3$, the octet is not fulfilled and hence they behave as Lewis acid.

28. (A)
KCl contains ionic bond.

29. (B)
1 molal aqueous solution of the non-electrolytic molecule will exhibit the highest freezing point.

30. (B)
Bond order of $H_2^{2-} = 0$
Hence, according to the molecular orbital theory, H_2^{2-} is not a viable molecule.

I.I.T. JEE Objective Question Paper—2018 (Advanced)

Paper–1

INSTRUCTIONS

- This question paper has three (3) parts: PART-I: Physics, PART-II: Chemistry and PART-III: Mathematics.
- Each part has total of eighteen (18) questions divided into three (3) sections (Section-1, Section-2 and Section-3).
- Total number of questions in Paper-1: Fifty-four (54).
- Paper-1 Maximum Marks: One Hundred Eighty (180).

Instructions for Section-1: Questions and Marking Scheme

SECTION–1 (Maximum Marks: 24)

- This section contains **SIX (6)** questions.
- Each question has **FOUR options** for correct answer(s). **ONE OR MORE THAN ONE** of these four option(s) is (are) correct option(s).
- For each question, choose the correct option(s) to answer the question.
- Answer to each question will be evaluated according to the following marking scheme:

Full Marks	:	+ 4	If only (all) the correct option(s) is (are) chosen.
Partial Marks	:	+ 3	If all the four options are correct but ONLY three options are chosen.
Partial Marks	:	+ 2	If three or more options are correct but ONLY two options are chosen, both of which are correct options.
Partial Marks	:	+ 1	If two or more options are correct but ONLY one option is chosen and it is a correct option.
Zero Marks	:	0	If none of the options is chosen (i.e. the question is unanswered).
Negative Marks	:	− 2	In all other cases.

For Example: If first, third and fourth are the ONLY three correct options for a question with second option being an incorrect option; selecting only two of the three correct options (e.g., the first and fourth options), without selecting any incorrect option (second option in this case), will result in + 2 marks. Selecting only one of the three correct options (either first or third or fourth option), without selecting any incorrect option (second option in this case), will result in + 1 marks. Selecting any incorrect option(s) (second option in this case), with or without selection of any correct option(s) will result in – 2 marks.

Answering Section-1 Questions:

- To select the option(s), **using the mouse click** on the corresponding button(s) of the option(s).
- To deselect chosen option(s), click on the button(s) of the chosen options(s) again or click on the **Clear Response** button to clear all the chosen options.
- To change the option(s) of a previously answered question, if required, first click on the **Clear Response** button to clear all the chosen options and then select the new option(s).
- To mark a question ONLY for review (i.e. without answering it), click on the **Mark for Review & Next button**.
- To mark a question for review (after answering it), click on **Mark for Review & Next button** – answered question which is also marked for review will be evaluated.
- To save the answer, click on the **Save & Next button** – the answered question will be evaluated.

SECTION–2: (Maximum Marks: 44)

- This section contains **ELEVEN (11)** questions.
- Each question has **FOUR options** for correct answer(s). **ONE OR MORE THAN ONE** of these four option(s) is (are) correct option(s).
- For each question, choose the correct option(s) to answer the question.
- Answer to each question will be evaluated according to the following marking scheme.

Full Marks	: + 4	If only (all) the correct option(s) is (are) chosen.
Partial Marks	: + 3	If all the four options are correct but ONLY three options are chosen.
Partial Marks	: + 2	If three or more options are correct but ONLY two options are chosen, both of which are correct options.
Partial Marks	: + 1	If two or more options are correct but ONLY one option is chosen and it is a correct option.
Zero Marks	: 0	If none of the options is chosen (i.e. the question is unanswered).
Negative Marks	: – 2	In all other cases.

For Example: If first, third and fourth are the ONLY three correct options for a question with second option being an incorrect option; selecting only two of the three correct options (e.g., the first and fourth options), without selecting any incorrect option (second option in this case), will result in + 2 marks. Selecting only one of the three correct options (either first or third or fourth option), without selecting any incorrect option (second option in this case), will result in + 1 marks. Selecting any incorrect option(s) (second option in this case), with or without selection of any correct option(s) will result in – 2 marks.

1. The compound(s) which generate(s) N_2 gas upon thermal decomposition below 300°C is (are)
(A) NH_4NO_3 (B) $(NH_4)_2Cr_2O_7$ (C) $Ba(N_3)_2$ (D) Mg_3N_2

2. The correct statement(s) regarding the binary transition metal carbonyl compounds is (are)
(Atomic numbers: Fe = 26, Ni = 28)
(A) Total number of valence shell electrons at metal centre in $Fe(CO)_5$ or $Ni(CO)_4$ is 16
(B) These are predominantly low spin in nature
(C) Metal—carbon bond strengthens when the oxidation state of the metal is lowered
(D) The carbonyl C—O bond weakens when the oxidation state of the metal is increased

3. Based on the compounds of group 15 elements, the correct statement(s) is (are)
(A) Bi_2O_5 is more basic than N_2O_5
(B) NF_3 is more covalent than BiF_3
(C) PH_3 boils at lower temperature than NH_3
(D) The N—N single bond is stronger than the P—P single bond

4. In the following reaction sequence, the correct structure(s) of X is (are)

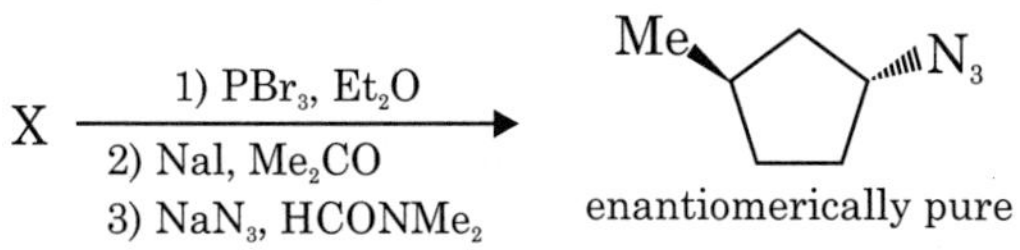

(A) 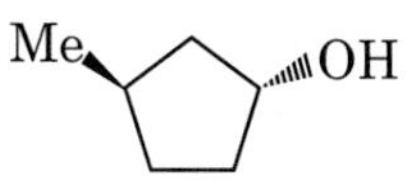

(B) 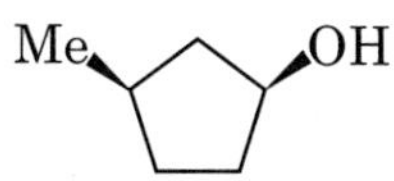

(C) 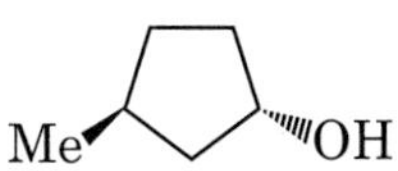

(D) 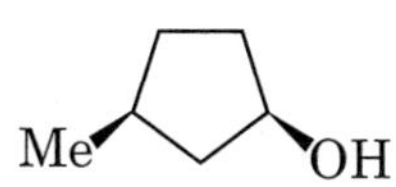

5. The reaction(s) leading to the formation of 1,3,5-trimethylbenzene is (are)

(A) 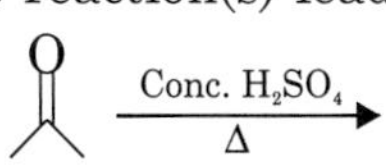

(B) $Me \equiv H \xrightarrow[\text{873 K}]{\text{heated iron tube}}$

(C) 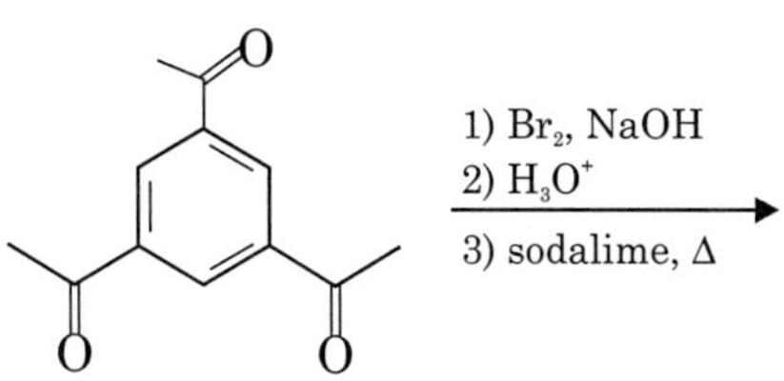

(D) 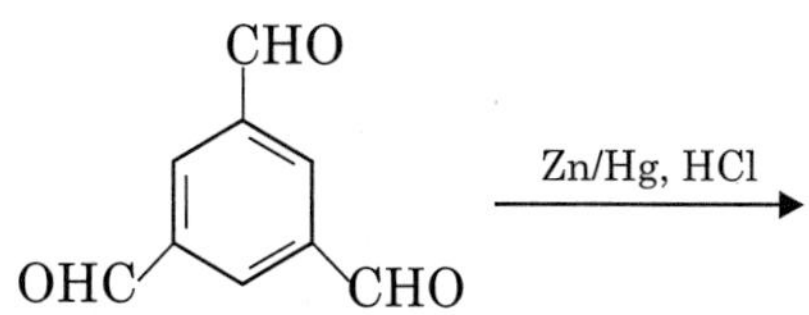

6. A reversible cyclic process for an ideal gas is shown below. Here, *P*, *V*, and *T* are pressure, volume and temperature, respectively. The thermodynamic parameters *q*, *w*, *H* and *U* are heat, work, enthalpy and internal energy, respectively.

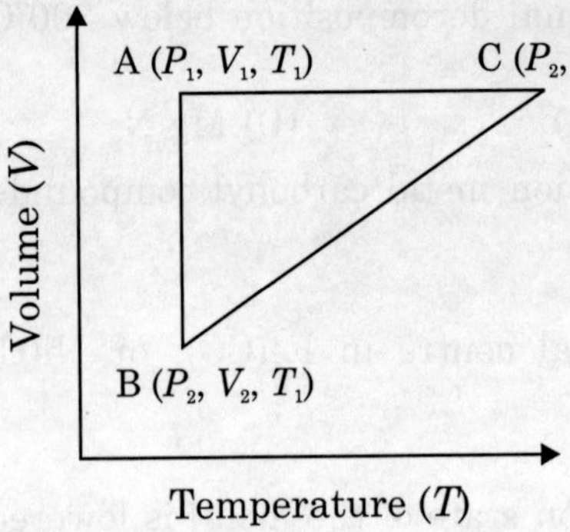

The correct option(s) is (are)

(A) $q_{AC} = \Delta U_{BC}$ and $w_{AB} = P_2(V_2 - V_1)$

(B) $w_{BC} = P_2(V_2 - V_1)$ and $q_{BC} = \Delta H_{AC}$

(C) $\Delta H_{CA} < \Delta U_{CA}$ and $q_{AC} = \Delta U_{BC}$

(D) $q_{BC} = \Delta H_{AC}$ and $\Delta H_{CA} > \Delta U_{CA}$

7. Among the species given below, the total number of diamagnetic species is _____.
H atom, NO_2 monomer, O_2^- (superoxide), dimeric sulphur in vapour phase, Mn_3O_4, $(NH_4)_2[FeCl_4]$, $(NH_4)_2[NiCl_4]$, K_2MnO_4, K_2CrO_4

8. The ammonia prepared by treating ammonium sulphate with calcium hydroxide is completely used by $NiCl_2 \cdot 6H_2O$ to form a stable coordination compound. Assume that both the reactions are 100% complete. If 1584 g of ammonium sulphate and 952 g of $NiCl_2 \cdot 6H_2O$ are used in the preparation, the combined weight (in gram) of gypsum and the nickel-ammonia coordination compound thus produced is ______.
(Atomic weights in g mol^{-1}: H = 1, N = 14, O = 16, S = 32, Cl = 35.5, Ca = 40, Ni = 59)

9. Consider an ionic solid MX with NaCl structure. Construct a new structure (Z) whose unit cell is constructed from the unit cell of MX following the sequential instructions given below. Neglect the charge balance.

(i) Remove all the anions (X) except the central one

(ii) Replace all the face centered cations (M) by anions (X)

(iii) Remove all the corner cations (M)

(iv) Replace the central anion (X) with cation (M).

The value of $\left(\dfrac{\text{number of anions}}{\text{number of cations}}\right)$ in Z is______.

10. For the electrochemical cell,

$$Mg(s) \mid Mg^{2+}\ (aq,\ 1M) \parallel Cu^{2+}\ (aq,\ 1M) \mid Cu(s)$$

the standard emf of the cell is 2.70 V at 300 K. When the concentration of Mg^{2+} is changed to xM, the cell potential changes to 2.67 V at 300 K. The value of x is ____.
[Given, $\dfrac{F}{R} = 11500\ KV^{-1}$, where F is the Faraday constant and R is the gas constant, ln(10) = 2.30.]

11. A closed tank has two compartments A and B, both filled with oxygen (assumed to be ideal gas). The partition separating the two compartments is fixed and is a perfect heat insulator (Figure 1). If the old partition is replaced by a new partition which can slide and conduct heat but does NOT allow the gas to leak across (Figure 2), the volume (in m^3) of the compartment A after the system attains equilibrium is ____.

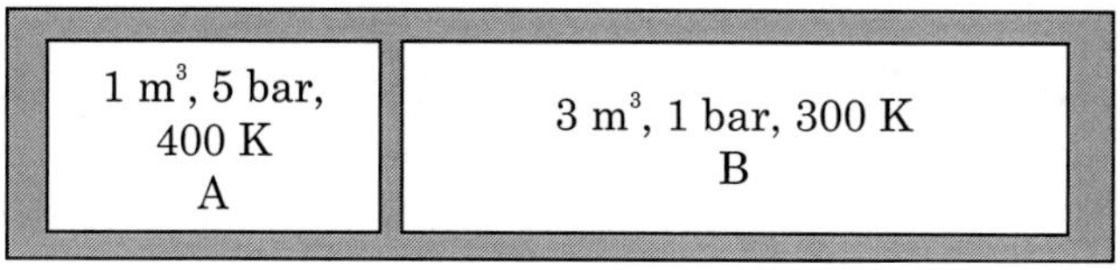

Figure 1

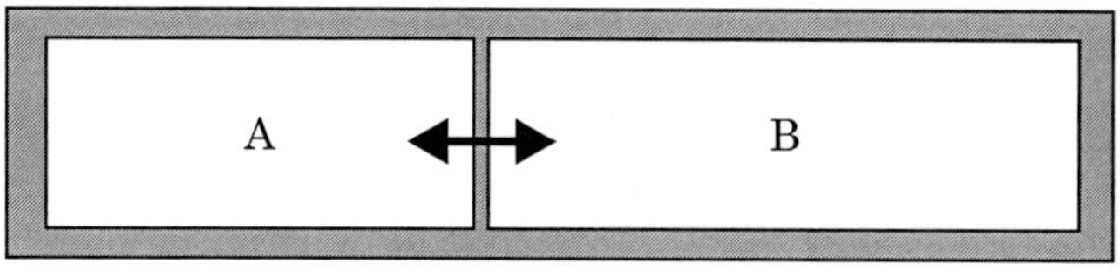

Figure 2

12. Liquids A and B form ideal solution over the entire range of composition. At temperature T, equimolar binary solution of liquids A and B has vapour pressure 45 Torr. At the same temperature, a new solution of A and B having mole fractions x_A and x_B, respectively, has vapour pressure of 22.5 Torr. The value of x_A/x_B in the new solution is _____.

(Given that the vapour pressure of pure liquid A is 20 Torr at temperature T)

13. The solubility of a salt of weak acid (AB) at pH 3 is Y $\times 10^{-3}$ mol L^{-1}. The value of Y is _____.

(Given that the value of solubility product of AB (K_{sp}) = 2×10^{-10} and the value of ionization constant of HB (K_a) = 1×10^{-8})

14. The plot given below shows $P - T$ curves (where P is the pressure and T is the temperature) for two solvents X and Y and isomolal solutions of NaCl in these solvents. NaCl completely dissociates in both the solvents.

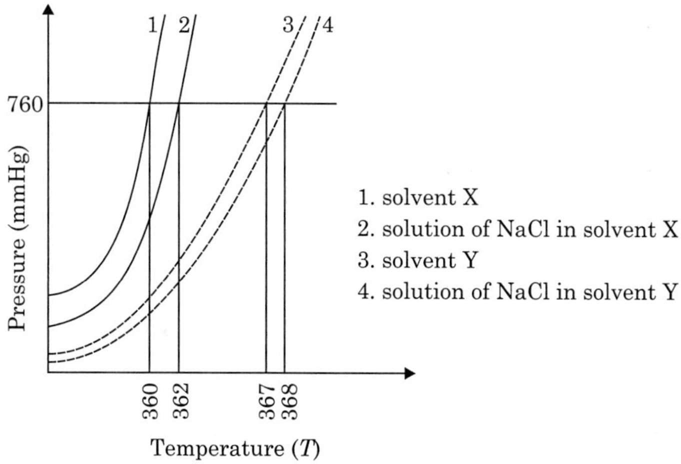

On addition of equal number of moles of a non-volatile solute S in equal amount (in kg) of these solvents, the elevation of boiling point of solvent X is three times that of solvent Y. Solute S is known to undergo dimerization in these solvents. If the degree of dimerization is 0.7 in solvent Y, the degree of dimerization in solvent X is _____.

Paragraph "X"

Treatment of benzene with CO/HCl in the presence of anhydrous $AlCl_3$/CuCl followed by reaction with Ac_2O/NaOAc gives compound **X** as the major product. Compound **X** upon reaction with Br_2/Na_2CO_3, followed by heating at 473 K with moist KOH furnishes **Y** as the major product. Reaction of **X** with H_2/Pd-C, followed by H_3PO_4 treatment gives **Z** as the major product.

(There are two questions based on Paragraph "X", the question given below is one of them.)

15. The compound **Y** is

(A) 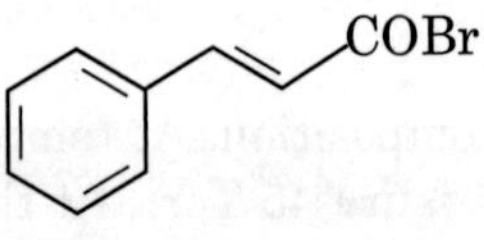

(B) 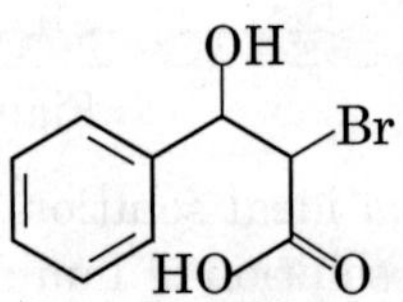

(C) 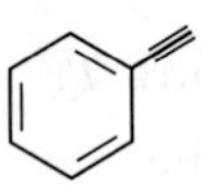

(D) 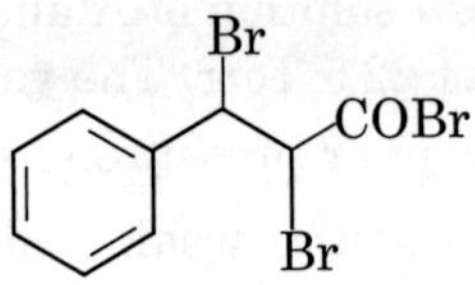

16. The compound **Z** is

(A) 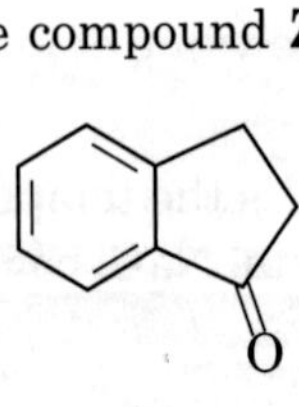

(B) 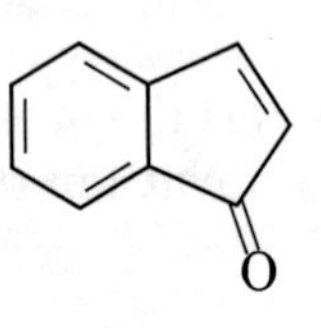

(C) 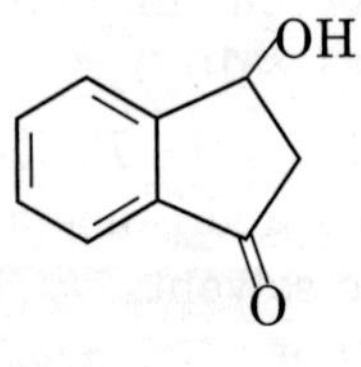

(D) 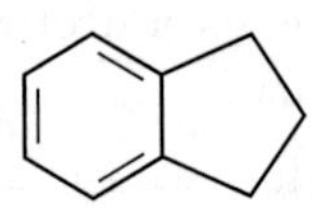

Paragraph "A"

An organic acid $P(C_{11}H_{12}O_2)$ can easily be oxidized to a dibasic acid which reacts with ethyleneglycol to produce a polymer dacron. Upon ozonolysis, **P** gives an aliphatic ketone as one of the products. **P** undergoes the following reaction sequences to furnish **R** via **Q**. The compound **P** also undergoes another set of reactions to produce **S**.

S ← [(1) H_2/Pd-C; (2) NH_3/Δ; (3) Br_2/NaOH; (4) $CHCl_3$, KOH, Δ; (5) H_2/Pd-C] — P → [(1) H_2/Pd-C; (2) $SOCl_2$; (3) MeMgBr, $CdCl_2$; (4) $NaBH_4$] Q → [(1) HCl; (2) Mg/Et_2O; (3) CO_2 (dry ice); (4) H_3O^+] R

(There are two questions based on Paragraph "A". The question given on next page is one of them.)

17. The compound **R** is

(A) 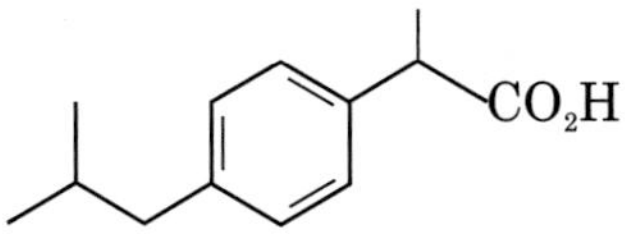

(B) 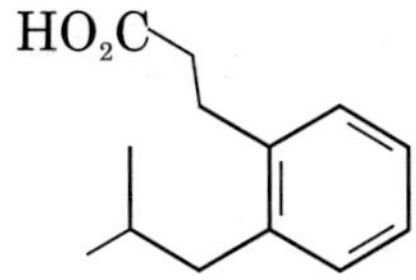

(C) 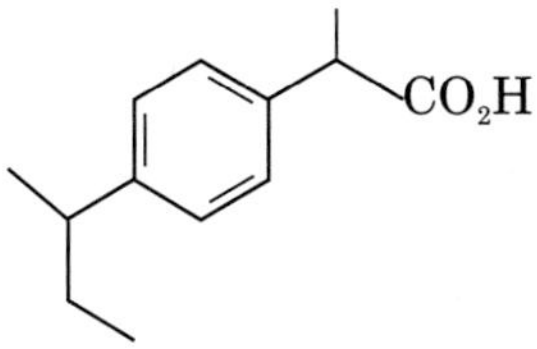

(D) 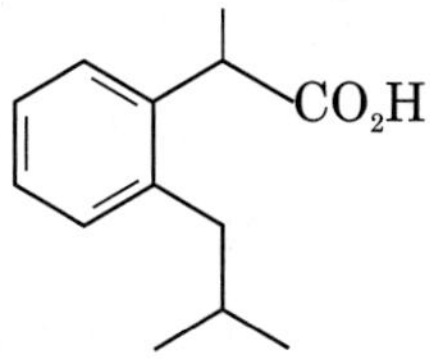

18. The compound **S** is

(A) 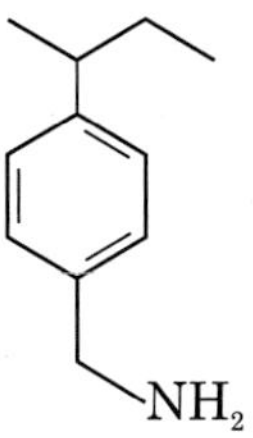

(B) 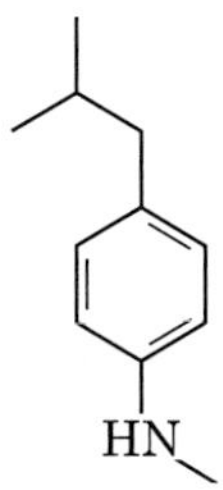

(C) 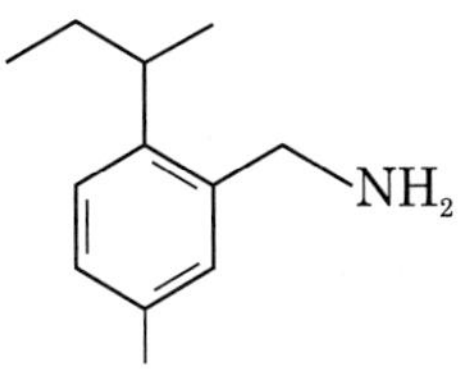

(D) 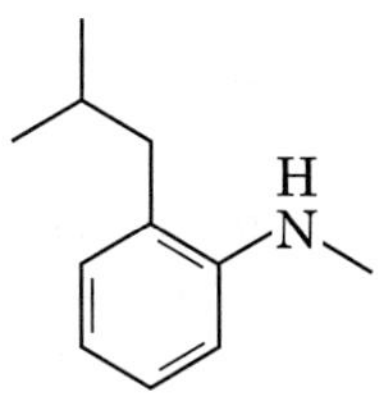

ANSWER KEYS

1. B, C	2. B, C	3. A, B, C	4. B	5. A, B, D	6. B, C
7. 1	8. 2992	9. 3	10. 10	11. 2.22	12. 19
13. 4.47	14. 0.05	15. C	16. A	17. A	18. B

SOLUTIONS

1. (B, C)

(A) $NH_4NO_3 \xrightarrow[\text{below } 300°C]{\Delta} N_2O + 2H_2O$

(B) $(NH_4)_2Cr_2O_7 \xrightarrow{\Delta} N_2 + Cr_2O_3 + 4H_2O$

(C) $Ba(N_3)_2 \xrightarrow{\Delta} Ba + 3N_2$

(D) Mg_3N_2 (it does not decompose into N_2)

2. (B, C)

(A) $[Fe(CO)_5]$ and $[Ni(CO)_4]$ complexes have 18-electrons in their valence shell.

(B) Carbonyl complexes are predominantly low spin complexes due to strong ligand field.

(C) As electron density increases on metals (with lowering oxidation state on metals), the extent of synergic bonding increases. Hence M—C bond strength increases.

(D) While positive charge on metals increases and the extent of synergic bond decreases, the C—O bond becomes stronger.

3. (A, B, C)

(A) Bi_2O_5 is metallic oxide but N_2O_5 is non-metallic oxide. Therefore, Bi_2O_5 is basic but N_2O_5 is acidic.

(B) In NF_3, N and F are non-metals but in BiF_3, Bi is metal but F is non-metal. Therefore, NF_3 is more covalent than BiF_3.

(C) In PH_3, hydrogen bonding is absent but in NH_3 hydrogen bonding is present. Therefore, PH_3 boils at lower temperature than NH_3.

(D) Due to small size in N—N single bond *lp-lp* repulsion is more than P—P single bond. Therefore, N—N single bond is weaker than the P—P single bond.

4. (B)

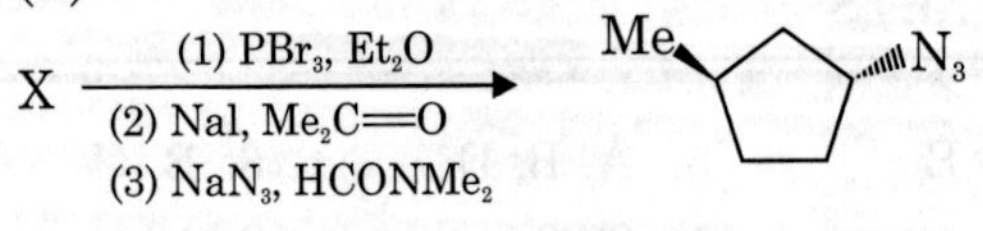

All the three reactions are S_N2. Hence, X is 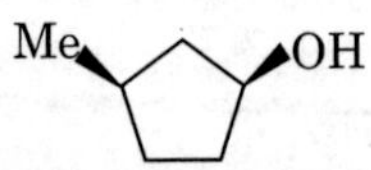

5. (A, B, D)

(A) 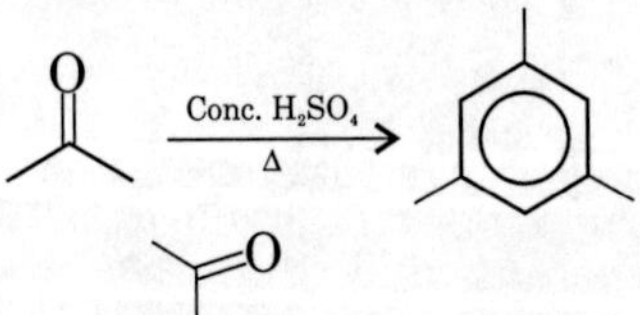

(B) 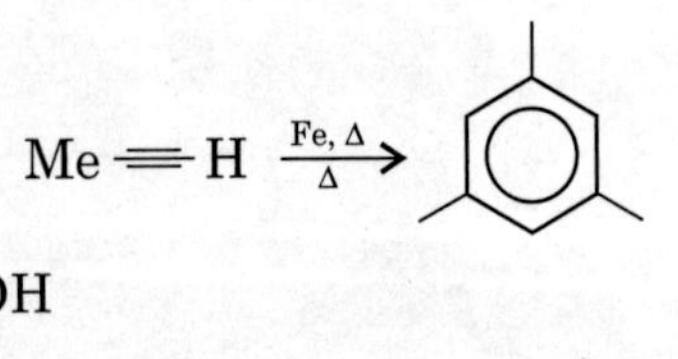

(C) 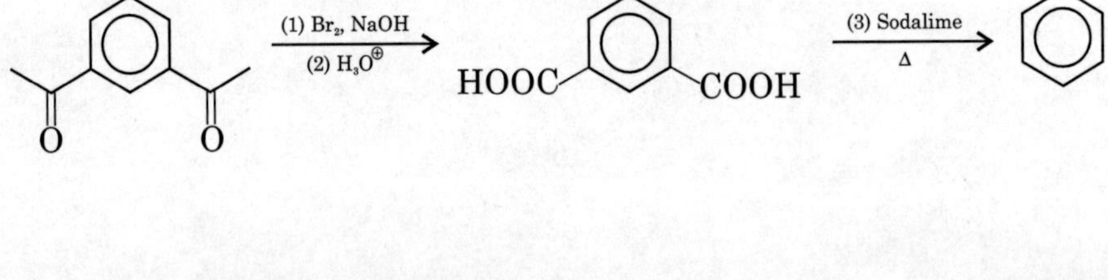

(D) 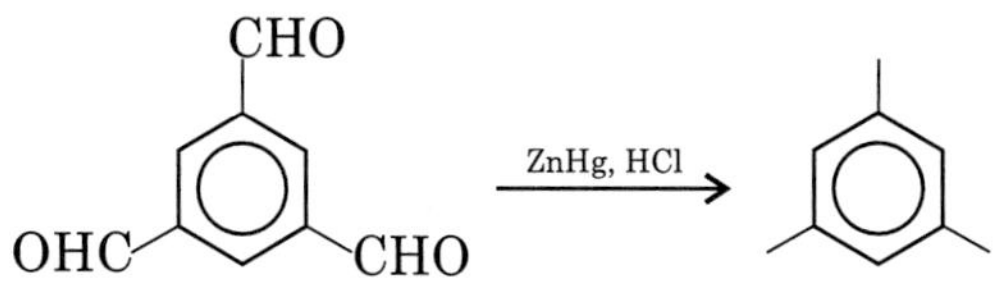

6. (B, C)

AC → Isochoric

AB → Isothermal

BC → Isobaric

$q_{AC} = \Delta U_{BC} = nC_V (T_2 - T_1)$

$W_{AB} = nRT_1 \quad \ln\left(\frac{V_2}{V_1}\right)$ A (wrong)

$q_{BC} = \Delta H_{AC} = nC_P (T_2 - T_1)$

$W_{BC} = - P_2 (V_1 - V_2)$ B (correct)

$nC_P (T_1 - T_2) < nC_V (T_1 - T_2)$ C (correct)

$\Delta H_{CA} < \Delta U_{CA}$

D (wrong)

7. (1)

H-atom = $\boxed{1}$ $1s^1$ Paramagnetic

NO_2 = 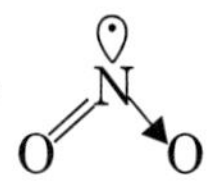 odd electron species Paramagnetic

O_2^- (superoxide) = One unpaired electrons in π^* M.O. Paramagnetic

S_2 (in vapour phase) = same as O_2, two unpaired electrons are present in π^* M.O. Paramagnetic

$Mn_3O_4 = 2\overset{2+}{Mn}O.\overset{4+}{Mn}O_2$ Paramagnetic

$(NH_4)_2[FeCl_4]$ = $Fe^{2+} = 3d^6 4s^0$

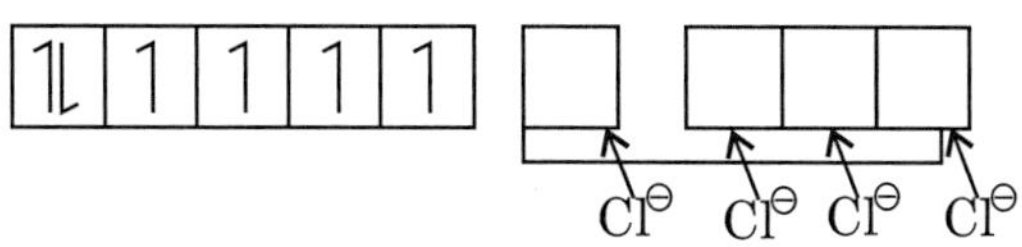

Paramagnetic

$(NH_4)_2[NiCl_4]$: $Ni = 3d^8 4s^2$; $Ni^{2+} = 3d^8 4s^0$ sp^3

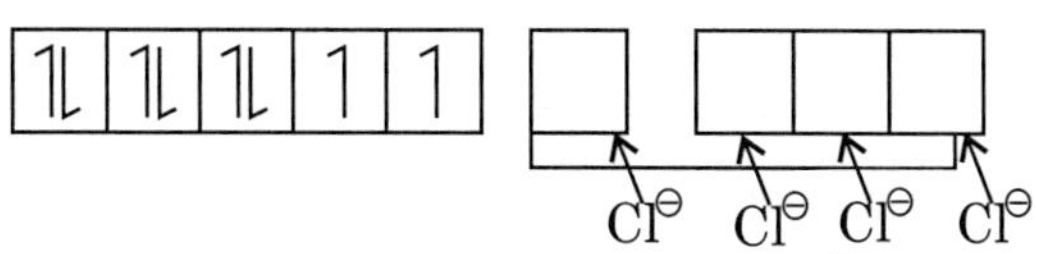

sp^3

Paramagnetic

$K_2MnO_4 = 2K^+$ 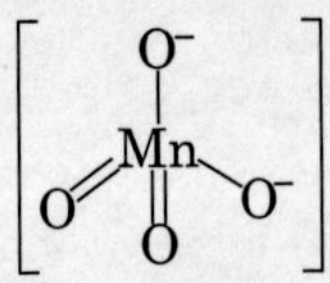, $Mn^{6+} = [Ar]\ 3d^1$ Paramagnetic

$K_2CrO_4 = 2K^+$ 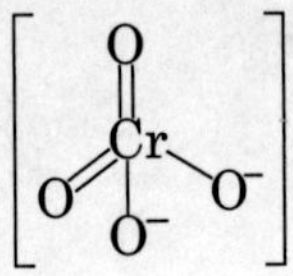, $Cr^{6+} = [Ar]\ 3d^0$ Diamagnetic

8. (2992)

$\underset{\substack{1584\,g \\ = 12\,mol}}{(NH_4)_2SO_4} + Ca(OH)_2 \qquad \underset{\substack{\text{gypsum } (M = 172) \\ 12\,mol}}{CaSO_4 \cdot 2H_2O} + \underset{24\,mol}{2NH_3}$

$\underset{952\,g = 4\,mol}{NiCl_2 \cdot 6H_2O} + \underset{24\,mol}{6NH_3} \qquad \underset{(M = 232);\,4\,mol}{Ni(NH_3)_6]Cl_2} + 6H_2O$

Total mass = 12 × 172 + 4 × 232 = 2992 g

9. (3)

$X^{\ominus}$: O.V.

M^+: FCC

M^+	X^-
(i) 4	1
(ii) 4 – 3	3 + 1
(iii) 4 – 3 – 1	3 + 1
(iv) 1	3

$Z = \frac{3}{1} = 3$

10. (10)

$Mg(s) + Cu^{2+}(aq) \rightarrow Mg^{2+}(aq) + Cu(s)$

$E^0_{Cell} = 2.70 \qquad Mg^{2+} = xM$

$E_{Cell} = 2.67 \qquad Cu^{2+} = 1M$

$$E_{Cell} = E^0_{Cell} - \frac{RT}{nF}\ln x$$

$$2.67 = 2.70 - \frac{RT}{2F}\ln x$$

$$-0.03 = -\frac{R \times 300}{2F} \times \ln x$$

$$\ln x = \frac{0.03 \times 2}{300} \times \frac{F}{R} = \frac{0.03 \times 2 \times 11500}{300 \times 1}$$

$\ln x = 2.30 = \ln(10)$

$x = 10$

11. (2.22)

$P_1 = 5$ $\quad\quad$ $P_2 = 1$

$v_1 = 1$ $\quad\quad$ $v_2 = 3$

$T_1 = 400$ $\quad\quad$ $T_2 = 300$

$n_1 = \dfrac{5}{400R}$ $\quad\quad$ $n_2 = \dfrac{3}{300R}$

Let volume be $(v + x)$ $\quad$ $v = (3 - x)$ $\quad$ $15 - 5x = 4 + 4x$

$$\frac{P_A}{T_A} = \frac{P_B}{T_B}$$

$$\frac{n_{b_1} \times R}{v_{b_1}} = \frac{n_{b_2} \times R}{v_{b_2}}$$

$$\frac{5}{400(4 + x)} = \frac{3}{300R(3 - x)}$$

or $5(3 - x) = 4 + 4x$

or $x = \dfrac{11}{9}$

$$v = 1 + x = 1 + \frac{11}{9} = \left(\frac{20}{9}\right) = 2.22$$

12. (19)

$$45 = P_A^0 \times \frac{1}{2} + P_B^0 \times \frac{1}{2}$$

$P_A^0 + P_B^0 = 90$ ...(1)

Given $P_A^0 = 20$ torr; $P_B^0 = 70$ torr

$22.5 \text{ torr} = 20\, x_A + 70\,(1 - x_A) = 70 - 50\, x_A$

or $x_A = \left(\dfrac{70 - 22.5}{50}\right) = 0.95$

$x_B = (1 - 0.95) = 0.05$

Hence, $\dfrac{x_A}{x_B} = \dfrac{0.95}{0.05} = 19$

13. (4.47)

$$S = \sqrt{K_{sp}\left(\frac{[H^+]}{K_a} + 1\right)} = \sqrt{2 \times 10^{-10}\left(\frac{10^{-3}}{10^{-8}} + 1\right)} \simeq \sqrt{2 \times 10^{-5}} = 4.47 \times 10^{-3} M$$

14. (0.05)

From graph

For solvent X $\quad\quad$ $\Delta T_{bx} = 2$

$\Delta T_{bx} = m_{NaCl} \times K_{b(x)}$ ...(1)

For solvent Y $\quad\quad$ $\Delta T_{by} = 1$

$\Delta T_{b(y)} = m_{NaCl} \times K_{b(y)}$ ...(2)

Dividing equation (1) by equation (2), we have

$$\frac{K_{b(x)}}{K_{b(y)}} = 2$$

For solute S

$$2(S) \rightarrow S_2$$
$$1-\alpha \quad \alpha/2$$

$i = (1 - \alpha/2)$

$$\Delta T_{b(x)(s)} = \left(1 - \frac{\alpha_1}{2}\right) K_{b(x)}$$

$$\Delta T_{b(y)(s)} = \left(1 - \frac{\alpha_2}{2}\right) K_{b(y)}$$

Given $\Delta T_{b(x)(s)} = 3\Delta T_{b(y)(s)}$

$$\left(1 - \frac{\alpha_1}{2}\right) K_{b(x)} = 3 \times \left(1 - \frac{\alpha_2}{2}\right) \times K_{b(y)}$$

or $2\left(1 - \frac{\alpha_1}{2}\right) = 3\left(1 - \frac{\alpha_2}{2}\right)$

$\alpha_2 = 0.7$

Hence, $\alpha_1 = 0.05$

15. (C)

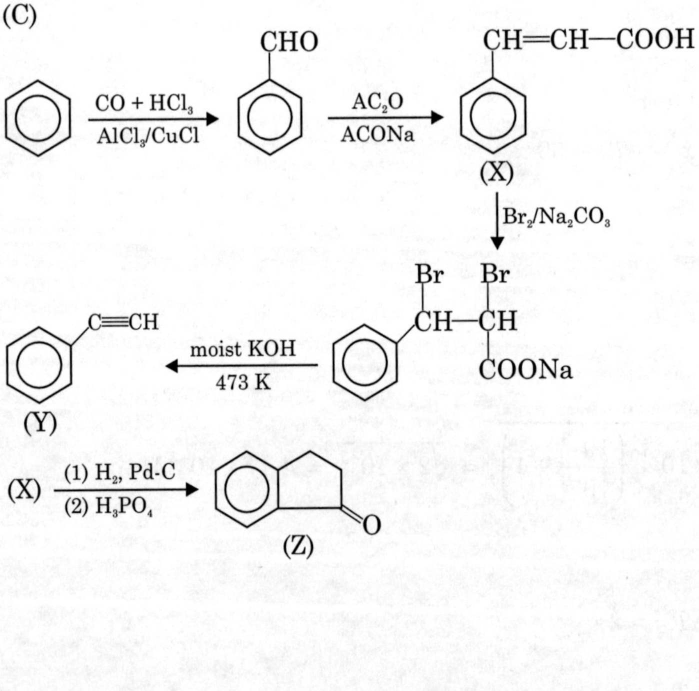

16. (A)

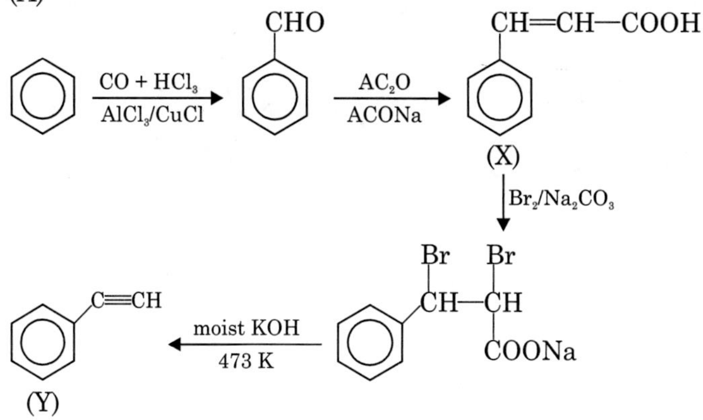

17. (A)

18. (B) **Solution 17 and 18:**

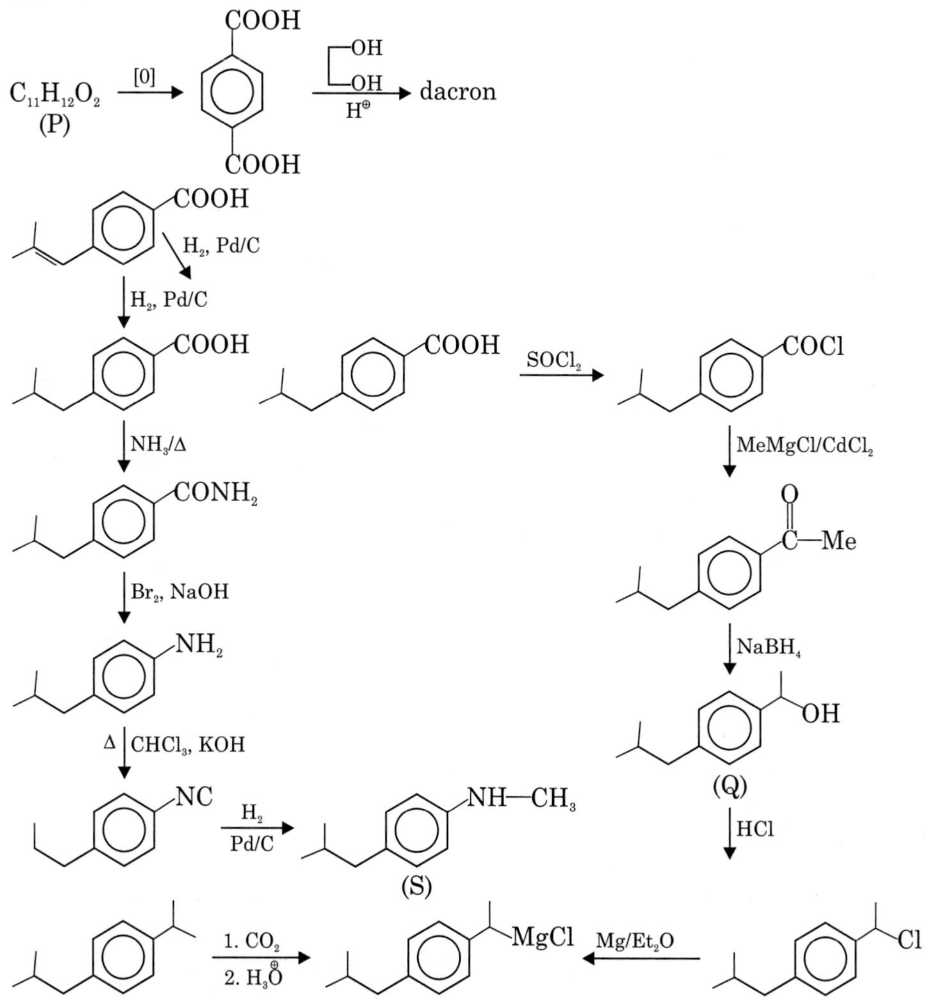

I.I.T. JEE Objective Question Paper—2018 (Advanced)

Paper–2

INSTRUCTIONS TO CANDIDATES

- This question paper has three (3) parts: PART-I: Physics, PART-II: Chemistry and PART-III: Mathematics.
- Each part has total of eighteen (18) questions divided into three (3) sections (Section-1, Section-2 and Section-3).
- Total number of questions in Paper-2: Fifty-four (54).
- Paper-2 Maximum Marks: One Hundred Eighty (180).

Instructions for Section-1: Questions and Marking Scheme

SECTION–1 (Maximum Marks: 24)

- This section contains **SIX (6)** questions.
- Each question has **FOUR options** for correct answer(s). **ONE OR MORE THAN ONE** of these four option(s) is (are) correct option(s).
- For each question, choose the correct option(s) to answer the question.
- Answer to each question will be evaluated according to the following marking scheme:

Full Marks	: + 4	If only (all) the correct option(s) is (are) chosen.
Partial Marks	: + 3	If all the four options are correct but ONLY three options are chosen.
Partial Marks	: + 2	If three or more options are correct but ONLY two options are chosen, both of which are correct options.
Partial Marks	: + 1	If two or more options are correct but ONLY one option is chosen, it is a correct option.
Zero Marks	: 0	If none of the options is chosen (i.e. the question is unanswered).
Negative Marks	: − 2	In all other cases.

For Example: If first, third and fourth are the ONLY three correct options for a question with second option being an incorrect option; selecting only two of the three correct options (e.g., the first and fourth options), without selecting any incorrect option (second option in this case), will result in + 2 marks. Selecting only one of the three correct options (either first or third or fourth option), without selecting any incorrect option (second option in this case), will result in + 1 marks. Selecting any incorrect option(s) (second option in this case), with or without selection of any correct option(s) will result in – 2 marks.

Answering Section-1 Questions

- To select the option(s), **using the mouse click** on the corresponding button(s) of the option(s).
- To deselect chosen option(s), click on the button(s) of the chosen option(s) again or click on the **Clear Response** button to clear all the chosen options.
- To change the option(s) of a previously answered question, if required, first click on the **Clear Response** button to clear all the chosen options and then select the new option(s).
- To mark a question ONLY for review (i.e. without answering it), click on the **Mark for Review & Next button**.
- To mark a question for review (after answering it), click on **Mark for Review & Next button** – answered question which is also marked for review will be evaluated.
- To save the answer, click on the **Save & Next button** – the answered question will be evaluated.

Instructions for Section-2: Questions and Marking Scheme

SECTION–2 (Maximum Marks: 24)

- This section contains **EIGHT (8)** questions. The answer to each question is **NUMERICAL VALUE**.
- For each question, enter the correct numerical value (in decimal notation, truncated/ rounded-off to the second decimal place; e.g., 6.25, 7.00, – 0.33, – 0.30, 30.27, – 127.30) using the mouse and the on-screen virtual numeric keypad in the place designated to enter the answer.
- Answer to each question will be evaluated according to the following marking scheme.

  Full Marks : + 3 If ONLY the correct numerical value is entered as answer.

  Zero Marks : 0 In all other cases.

Answering Section-2 Questions

- Using the attached computer mouse, click on numbers (and/or symbols) on the on-screen virtual numeric keypad to enter the numerical value as answer in the space provided for answer.
- To change the answer, if required, first click on the **Clear Response** button to clear the entered answer and then enter the new numerical value.
- To mark a question ONLY for review (i.e. answering it), click on the **Mark for Review & Next button** – the answered question which is also marked for review will be evaluated.
- To mark a question for review (after answering it), click **Mark for Review & Next button** – the answered question which is also marked for review will be evaluated.
- To save the answer, click on the **Save & Next button** – the answered question will be evaluated.

Instructions for Section-3: Questions and Marking Scheme

SECTION–3 (Maximum Marks: 12)

- This section contains **TWO (2)** paragraphs. Based on each paragraph, there are **TWO (2)** questions.
- Each question has **FOUR** options. **ONLY ONE** of these four options corresponds to the correct answer.
- For each question, choose the option corresponding to the correct answer.
- Answer to each question will be evaluated according to the following marking scheme:

Full Marks : + 3 If ONLY the correct option is chosen.

Zero Marks : 0 If none of the options is chosen (i.e. the question is unanswered).

Negative Marks : – 1 In all other cases.

Answering Section-3 Questions

- To select an option, using the mouse click on the corresponding button of the option.
- To deselect the chosen answer, click on the button of the chosen option again or click on the **Clear Response button**.
- To change the chosen answer, click on the button of another option.
- To mark a question ONLY for review (i.e. without answering it), click on **Mark for Review & Next button.**
- To mark a question for review (after answering it), click on **Mark for Review & Next button** – the answered question which is also marked for review will be evaluated.
- To save the answer, click on the **Save & Next button** – the answered question will be evaluated.

1. The correct option(s) regarding the complex $[Co(en)(NH_3)_3(H_2O)]^{3+}$ (en = $H_2NCH_2CH_2NH_2$) is (are)
   (A) It has two geometrical isomers
   (B) It will have three geometrical isomers if bidentate 'en' is replaced by two cyanide ligands
   (C) It is paramagnetic
   (D) It absorbs light at longer wavelength as compared to $[Co(en)(NH_3)_4]^{3+}$
2. The correct option(s) to distinguish nitrate salts of Mn^{2+} and Cu^{2+} taken separately is (are):
   (A) Mn^{2+} shows the characteristic green colour in the flame test
   (B) Only Cu^{2+} shows the formation of precipitate by passing H_2S in acidic medium
   (C) Only Mn^{2+} shows the formation of precipitate by passing H_2S in faintly basic medium
   (D) Cu^{2+}/Cu has higher reduction potential than Mn^{2+}/Mn (measured under similar conditions).
3. Aniline reacts with mixed acid (conc. HNO_3 and conc. H_2SO_4) at 288 K to give P (51%), Q (47%) and R (2%). The major product(s) of the following reaction sequence is (are)

$$\mathrm{R} \xrightarrow[\text{(3) } H_3O^+ \text{ (4) } NaNO_2,\ HCl/273\text{-}278\ K \text{ (5) } EtOH,\ \Delta]{\text{(1) } Ac_2O,\ \text{pyridine (2) } Br_2,\ CH_3CO_2H} \mathrm{S} \xrightarrow[\text{(3) } NaNO_2,\ HCl/273\text{-}278\ K \text{ (4) } H_3PO_2]{\text{(1) } Sn/HCl \text{ (2) } Br_2/H_2O \text{ (excess)}} \text{major product(s)}$$

(A) 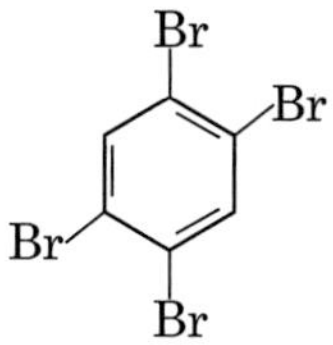

(B) 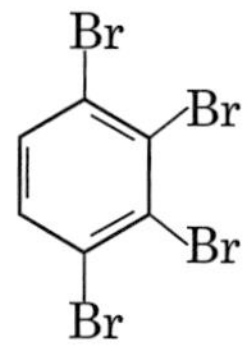

(C) 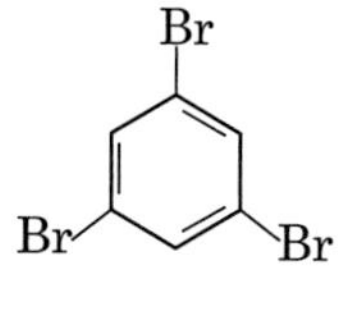

(D) 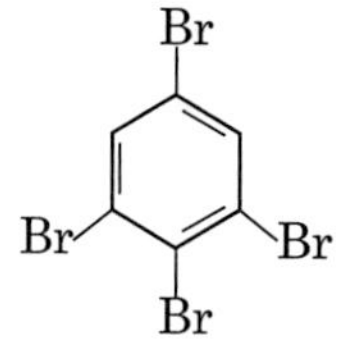

4. The Fischer presentation of D-glucose is given below.

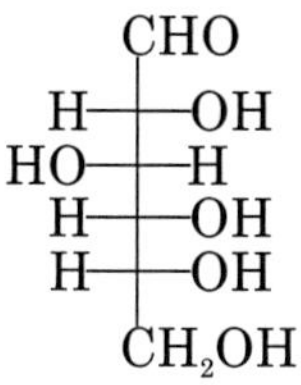

D-glucose

The correct structure(s) of β-glucopyranose is (are)

(A) 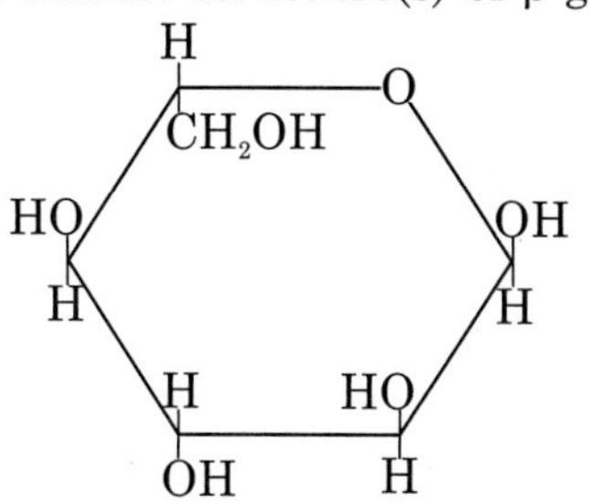

(B) 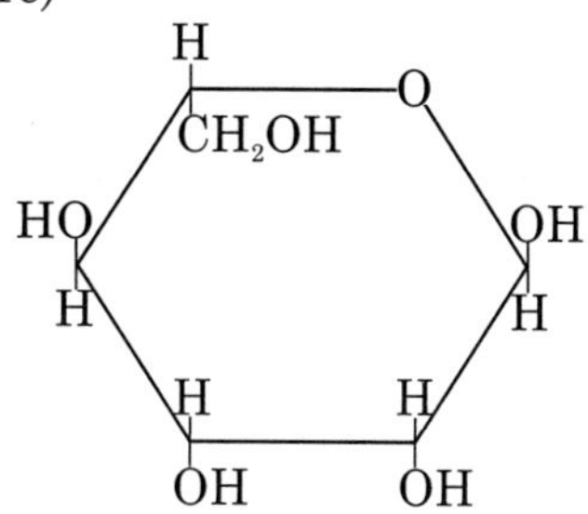

(C) 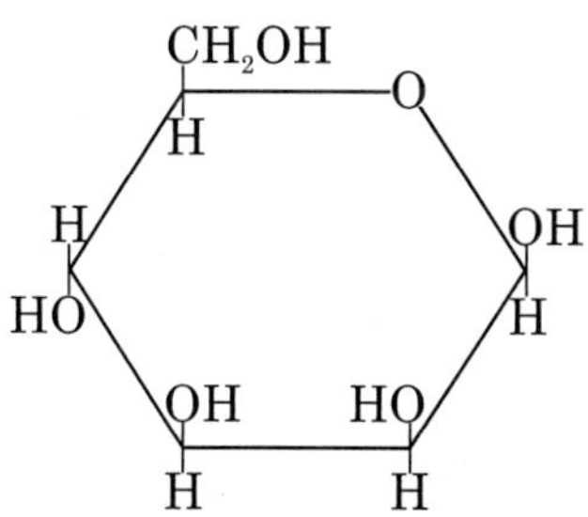

(D) 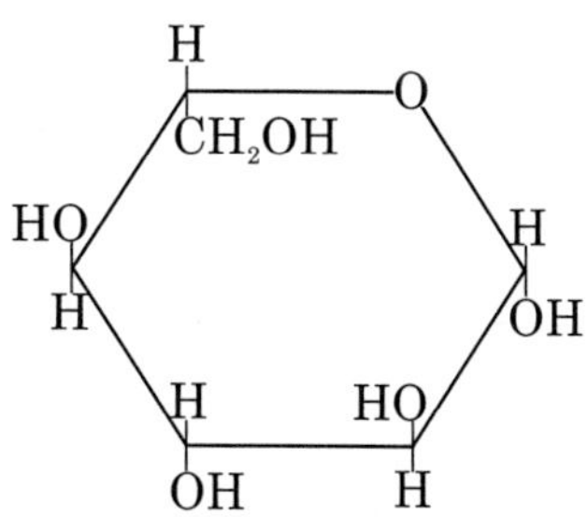

5. For a first order reaction A(g) → 2B(g) + C(g) at constant volume and 300 K, the total pressure at the beginning (t = 0) and at time t are P_0 and P_t, respectively. Initially, only A is present with concentration $[A]_0$, and $t_{1/3}$ is the time required for the partial pressure of A to reach 1/3rd of its initial value. The correct option(s) is (are)
(Assume that all these gases behave as ideal gases.)

(A) 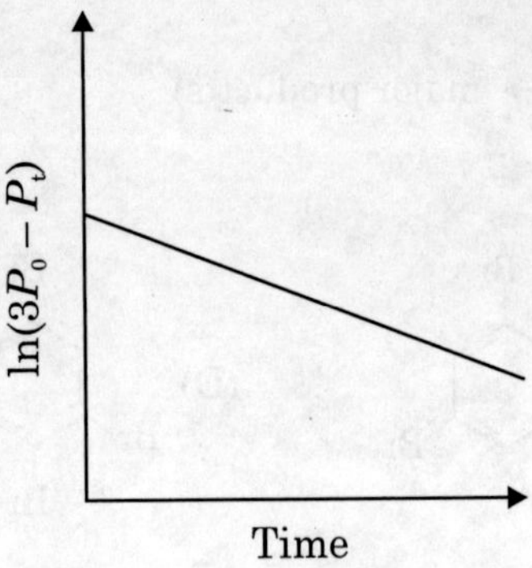

(B) 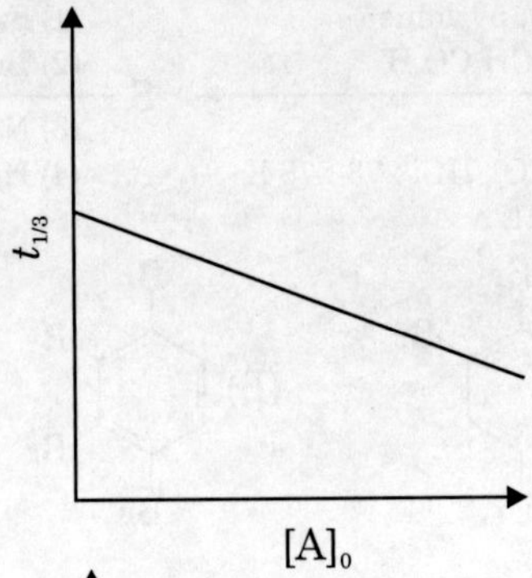

(C) 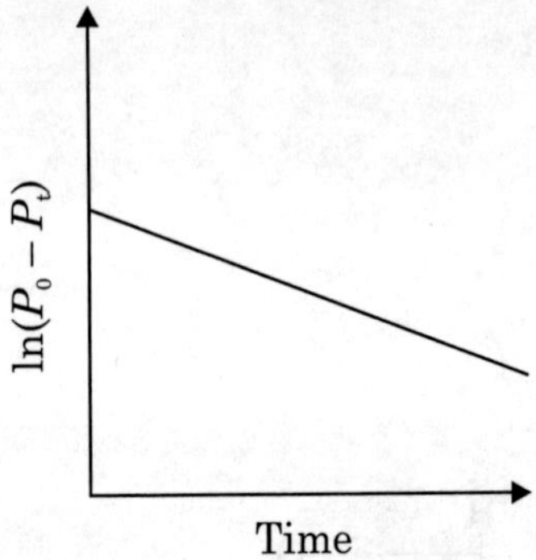

(D) 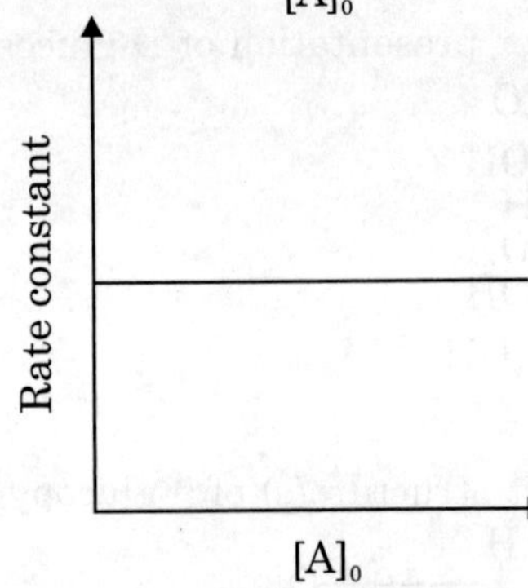

6. For a reaction, $A \rightleftharpoons P$, the plots of [A] and [P] with time at temperatures T_1 and T_2 are given below.

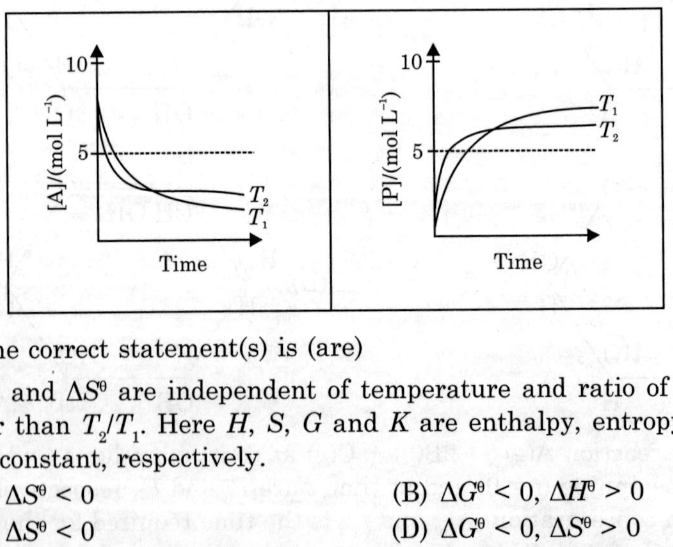

If $T_2 > T_1$, the correct statement(s) is (are)

Assume ΔH^θ and ΔS^θ are independent of temperature and ratio of $\ln K$ at T_1 to $\ln K$ at T_2 is greater than T_2/T_1. Here H, S, G and K are enthalpy, entropy, Gibbs energy and equilibrium constant, respectively.

(A) $\Delta H^\theta < 0, \Delta S^\theta < 0$
(B) $\Delta G^\theta < 0, \Delta H^\theta > 0$
(C) $\Delta G^\theta < 0, \Delta S^\theta < 0$
(D) $\Delta G^\theta < 0, \Delta S^\theta > 0$

7. The total number of compounds having at least one bridging oxo group among the molecules given below is ______.

N_2O_3, N_2O_5, P_4O_6, P_4O_7, $H_4P_2O_5$, $H_5P_3O_{10}$, $H_2S_2O_3$, $H_2S_2O_5$

8. Galena (an ore) is partially oxidized by passing air through it at high temperature. After some time, the passage of air is stopped, but the heating is continued in a closed furnace such that the contents undergo self-reduction. The weight (in kg) of Pb produced per kg of O_2 consumed is ______.

(Atomic weights in g mol^{-1}: O = 16, S = 32, Pb = 207)

9. To measure the quantity of $MnCl_2$ dissolved in an aqueous solution, it was completely converted to $KMnO_4$ using the reaction,

$MnCl_2 + K_2S_2O_8 + H_2O \rightarrow KMnO_4 + H_2SO_4 + HCl$ (equation not balanced).

Few drops of concentrated HCl were added to this solution and gently warmed. Further, oxalic acid (225 g) was added in portions till the colour of the permanganate ion disappeared. The quantity of $MnCl_2$ (in mg) present in the initial solution is _____.

(Atomic weights in g mol^{-1}: Mn = 55, Cl = 35.5)

10. For the given compound X, the total number of optically active stereoisomers is ____.

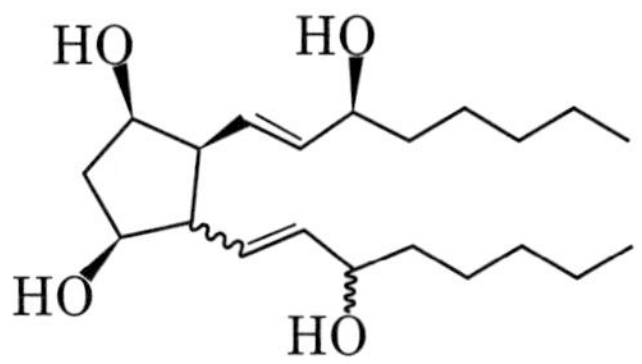

X

This type of bond indicates that the configuration at the specific carbon and the geometry of the double bond is fixed.

This type of bond indicates that the configuration at the specific carbon and the geometry of the double bond is NOT fixed.

11. In the following reaction sequence, the amount of D (in g) formed from 10 moles of acetophenone is _____.

(Atomic weight in g mol^{-1}: H = 1, C = 12, N = 14, O = 16, Br = 80. The yield (%) corresponding to the product in each step is given in the parenthesis.)

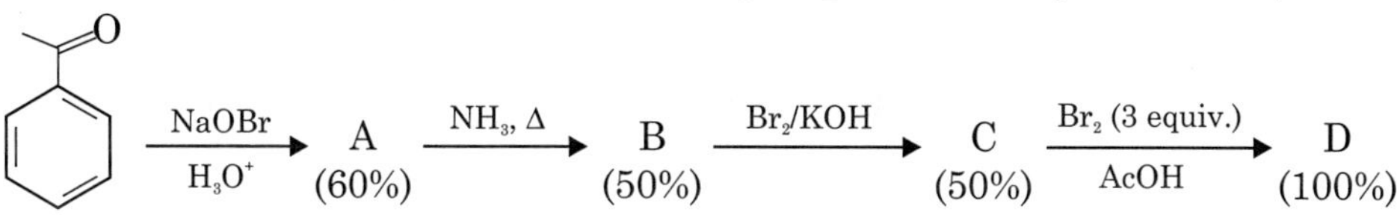

12. The surface of copper gets tarnished by the formation of copper oxide. N_2 gas was passed to prevent the oxide formation during heating of copper at 1250 K. However, the N_2 gas contains 1 mole % of water vapour as impurity. The water vapour oxidises copper as per the reaction given below:

$2Cu(s) + H_2O(g) \rightarrow Cu_2O(s) + H_2(g)$

p_{H_2} is the minimum partial pressure of H_2 (in bar) needed to prevent the oxidation at 1250 K. The value of ln (p_{H_2}) is ____.

(Given: total pressure = 1 bar, R (universal gas constant) = 8 J K^{-1} mol^{-1}, ln(10) = 2.3. Cu(*s*) and $Cu_2O(s)$ are mutually immiscible.)

At 1250 K: $2Cu(s) + 1/2O_2(g) \rightarrow Cu_2O(s)$; $\Delta G^\theta = -$ 78,000 J mol^{-1}

$H_2(g) + 1/2O_2(g) \rightarrow H_2O(g)$; $\Delta G^\theta = -$ 1,78,000 J mol^{-1}; G is the Gibbs energy.

13. Consider the following reversible reaction,

$A(g) + B(g) \rightleftharpoons AB(g)$.

The activation energy of the backward reaction exceeds that of the forward reaction by $2RT$ (in J mol^{-1}). If the pre-exponential factor of the forward reaction is 4 times that of the reverse reaction, the absolute value of $\Delta G^{\ominus}$ (in J mol^{-1}) for the reaction at 300 K is ____.

(Given: ln(2) = 0.7, RT = 2500 J mol^{-1} at 300 K and G is the Gibbs energy.)

14. Consider an electrochemical cell: $A(s) \mid A^{n+}(aq, 2M) \parallel B^{2n+}(aq, 1M) \mid B(s)$. The value of $\Delta H^{\ominus}$ for the cell reaction is twice that of $\Delta G^{\ominus}$ at 300 K. If the emf of the cell is zero, the $\Delta S^{\ominus}$ (in J K^{-1} mol^{-1}) of the cell reaction per mole of B formed at 300 K is ____.

(Given: ln(2) = 0.7, R (universal gas constant) = 8.3 J K^{-1} mol^{-1}. H, S and G are enthalpy, entropy and Gibbs energy, respectively.)

15. Match each set of hybrid orbitals from LIST–I with complex(es) given in LIST–II.

LIST–I	LIST–II
P. dsp^2	1. $[FeF_6]^{4-}$
Q. sp^3	2. $[Ti(H_2O)_3Cl_3]$
R. sp^3d^2	3. $[Cr(NH_3)_6]^{3+}$
S. d^2sp^3	4. $[FeCl_4]^{2-}$
	5. $Ni(CO)_4$
	6. $[Ni(CN)_4]^{2-}$

The correct option is

(A) P → 5; Q → 4, 6; R → 2, 3; S → 1

(B) P → 5, 6; Q → 4; R → 3; S → 1, 2

(C) P → 6; Q → 4, 5; R → 1; S → 2, 3

(D) P → 4, 6; Q → 5, 6; R → 1, 2; S → 3

16. The desired product X can be prepared by reacting the major product of the reactions in LIST–I with one or more appropriate reagents in LIST–II. (Given: order of migratory aptitude: aryl > alkyl > hydrogen)

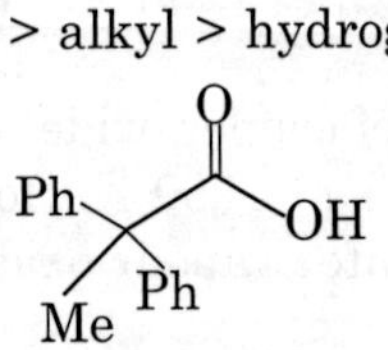

LIST–I	LIST–II
P. 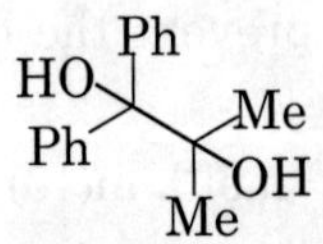 $+ H_2SO_4$	1. I_2, NaOH
Q. 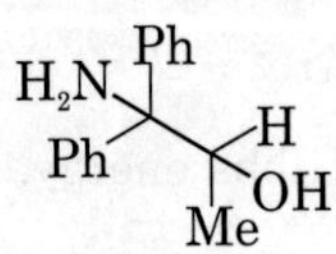 $+ HNO_2$	2. $[Ag(NH_3)_2]OH$

R. 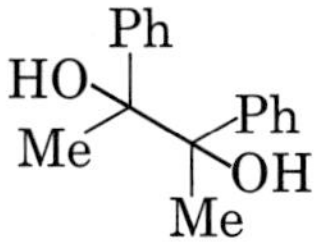 + H_2SO_4 3. Fehling solution

S. 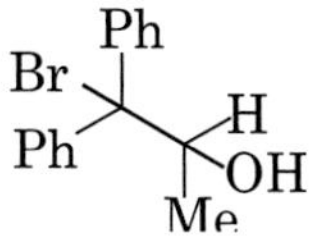 + $AgNO_3$ 4. HCHO, NaOH

5. NaOBr

The correct option is

(A) P → 1; Q → 2, 3; R → 1, 4; S → 2, 4   (B) P → 1, 5; Q → 3, 4; R → 4, 5; S → 3

(C) P → 1, 5; Q → 3, 4; R → 5; S → 2, 4   (D) P → 1, 5; Q → 2, 3; R → 1, 5; S → 2, 3

17. LIST–I contains reactions and LIST–II contains major products.

LIST–I	LIST–II
P. 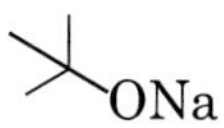 + 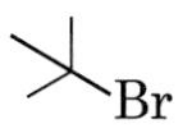 ⟶	1. 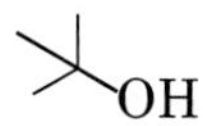
Q. 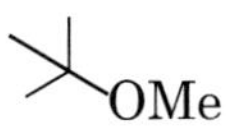 + HBr ⟶	2. 
R. 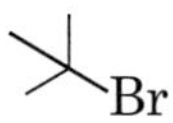 + NaOMe ⟶	3. 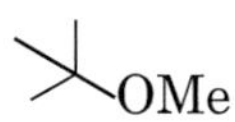
S. 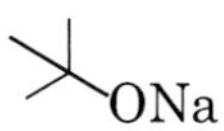 + MeBr ⟶	4. 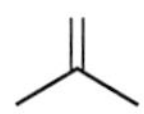
	5. 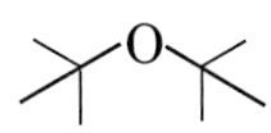

Match each reaction in LIST–I with one or more products in LIST–II and choose the correct option.

(A) P → 1, 5; Q → 2; R → 3; S → 4   (B) P → 1, 4; Q → 2; R → 4; S → 3

(C) P → 1, 4; Q → 1, 2; R → 3, 4; S → 4   (D) P → 4, 5; Q → 4; R → 4; S → 3, 4

18. Dilution process of different aqueous solutions, with water, is given in LIST–I. The effects of dilution of the solutions on [H^+] are given in LIST–II.

(Note: Degree of dissociation (α) of weak acid and weak base is << 1; degree of hydrolysis of salt << 1; [H^+] represents the concentration of H^+ ions)

LIST–I	LIST–II
P. (10 mL of 0.1M NaOH + 20 mL of 0.1M acetic acid) diluted to 60 mL	1. the value of [H^+] does not change on dilution
Q. (20 mL of 0.1M NaOH + 20 mL of 0.1M acetic acid) diluted to 80 mL	2. the value of [H^+] changes to half of its initial value on dilution
R. (20 mL of 0.1M HCl + 20 mL of 0.1M ammonia solution) diluted to 80 mL	3. the value of [H^+] changes to two times of its initial value on dilution

S. 10 mL saturated solution of $Ni(OH)_2$ in equilibrium with excess solid $Ni(OH)_2$ is diluted to 20 mL (solid $Ni(OH)_2$ is still present after dilution).

4. the value of $[H^+]$ changes to $\frac{1}{\sqrt{2}}$ times of its initial value on dilution

5. the value of $[H^+]$ changes to $\sqrt{2}$ times of its initial value on dilution

Match each process given in LIST–I with one or more effect(s) in LIST–II. The correct option is

(A) P → 4; Q → 2; R → 3; S → 1
(B) P → 4; Q → 3; R → 2; S → 3
(C) P → 1; Q → 4; R → 5; S → 3
(D) P → 1; Q → 5; R → 4; S → 1

ANSWER KEYS

1. A, B, D	2. B, D	3. D	4. D	5. A, D	6. A, C
7. 5 or 6	8. 6.47	9. 126	10. 7	11. 495	12. – 14.6
13. 8500	14. – 11.62	15. C	16. D	17. B	18. D

SOLUTIONS

1. (A, B, D)

(A) The complex $[Co(en)(NH_3)_3(H_2O)]^{+3}$ is of the type $[M(AA)b_3c]$ and has two geometrical isomers:

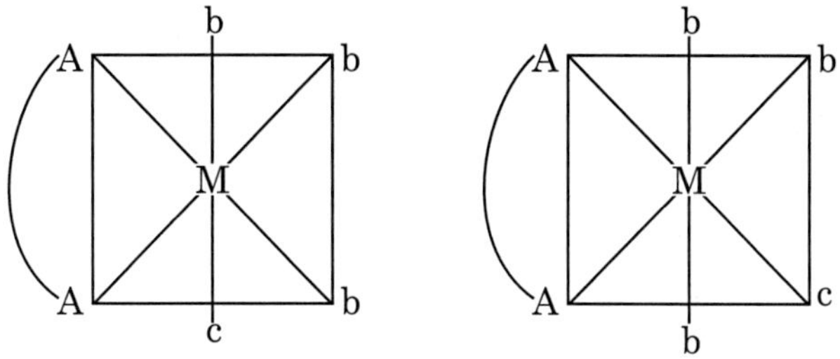

(B) If (en) is replaced by two cyanide ligand, the complex will be of the type $[Ma_3b_2c]$ and has three geometrical isomers:

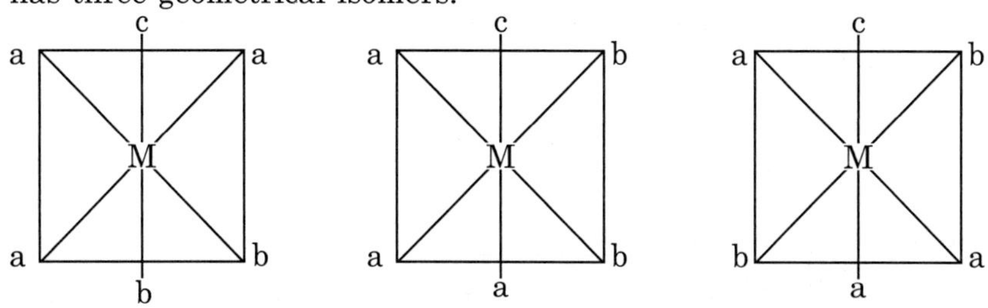

(C) The complex $[Co(en)(NH_3)_3(H_2O)]^{3+}$ has d^6 configuration (t_{2g}^6) on central metal with three strong field ligands. Hence, it is dimagnetic in nature.

(D) The complex $[Co(en)(NH_3)_3(H_2O)]^{3+}$ has lesser CFSE (Δ_0) value than that of $[Co(en)(NH_3)_4]^{3+}$. Hence, the complex $[Co(en)(NH_3)_3(H_2O)]^{+}$ absorbs longer wavelength for *d*-*d* transition.

2. (B, D)

(A) Both Cu^{2+} and Mn^{2+} give green colour flame in flame test and hence they cannot be distinguished.

(B) Cu^{2+} belongs to group-II of cationic radical and will give precipitate of CuS in acidic medium.

(C) Cu^{2+} and Mn^{2+} both form precipitate in basic medium.

(D) $Cu^{2+}/Cu = +\,0.34$ V; $Mn^{2+}/Mn = -\,1.18$ V.

3. (D)

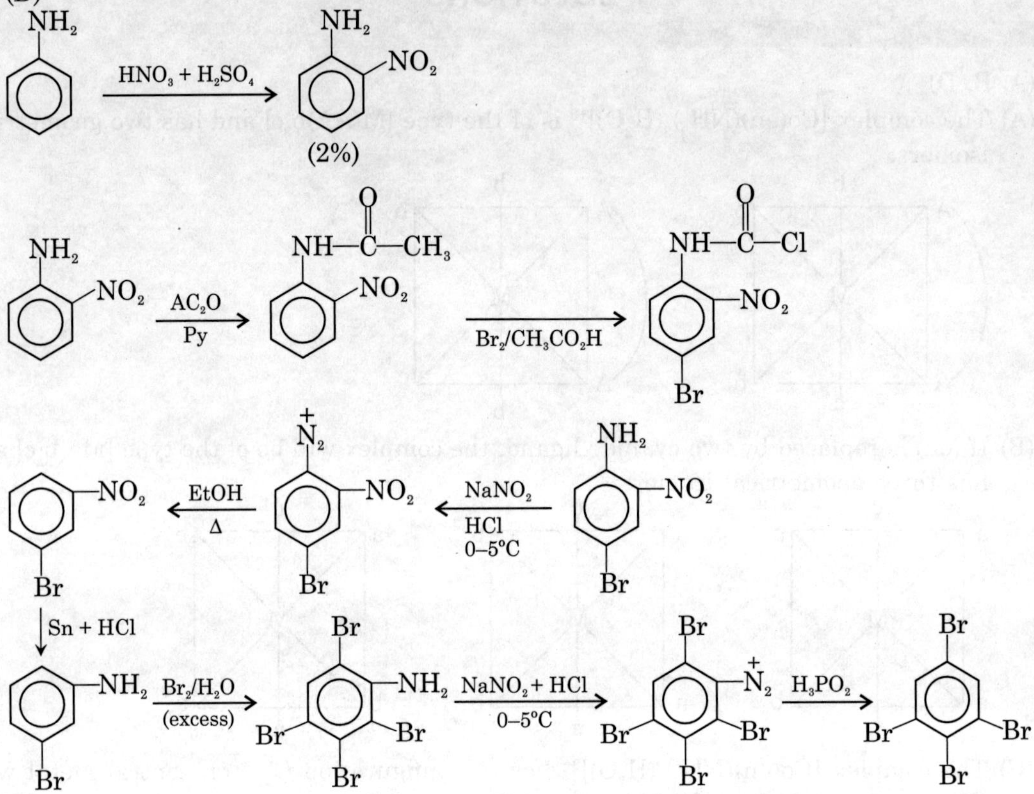

4. (D)

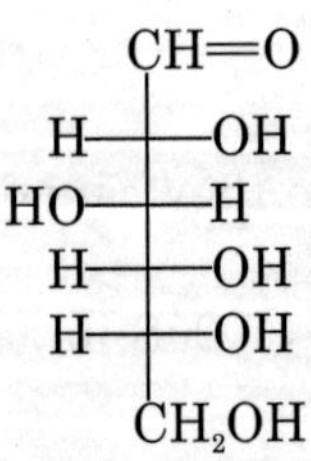

D-glucose

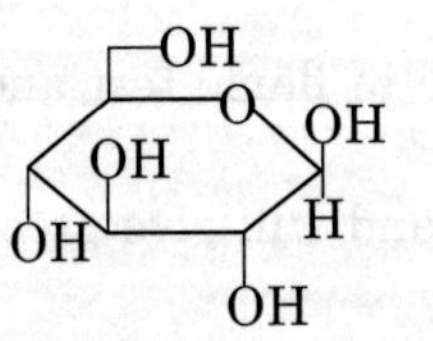

β-D-glucopyranose

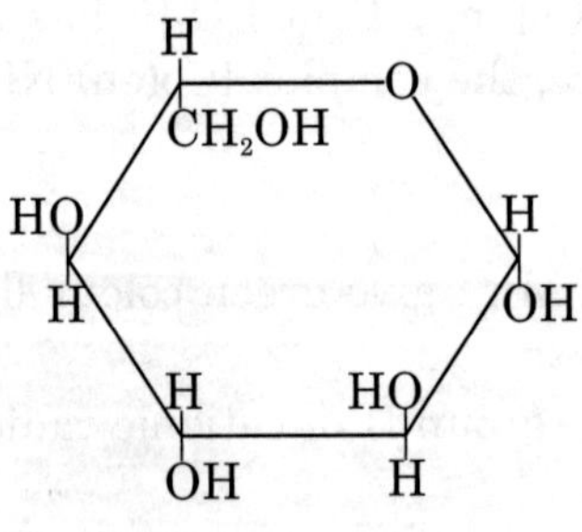

β-L-glucopyranose

5. (A, D)

$$\begin{array}{llll} & A & \rightarrow 2B & + \ C \\ t = 0 & P_0 & - & - \\ t = t & P_0 - P & 2P & P \end{array}$$

$P_0 + 2P = P_t$

$$K = \frac{1}{t}\ln\frac{P_0}{P_0 - P} = \frac{1}{t}\ln\frac{P_0}{P_0 - \dfrac{(P_t - P_0)}{2}}$$

$$K = \frac{1}{t}\ln\frac{2P_0}{3P_0 - P_t} \quad \text{or} \quad Kt + \ln 2P_0 = \ln(3P_0 - P_t)$$

$$t_{1/3} = \frac{1}{K}\ln\frac{P_0}{P_0/3} = \frac{1}{K}\ln 3 = \text{constant}$$

Rate constant does not depend on concentration.

6. (A, C)

$$A \rightleftharpoons P$$

Given: $T_2 > T_1$

$$\frac{\ln K_1}{\ln K_2} > \frac{T_2}{T_1}$$

$T_1 \ln K_1 > T_2 \ln K_2$

$-\Delta G_1^0 > -\Delta G_2^0$

$(-\Delta H^0 + T_1\Delta S^0) > (-\Delta H^0 + T_2\Delta S^0)$

$T_1\Delta S^0 > T_2\Delta S^0$

$\Delta S^0 < 0$

7. (5 or 6)

N_2O_3 = 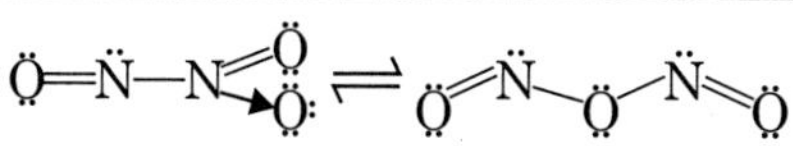

N_2O_5 = 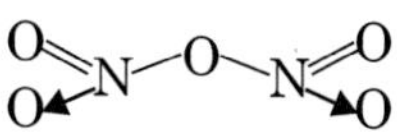

P_4O_6 = 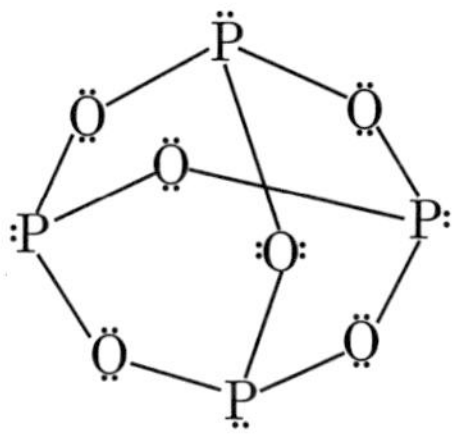

P_4O_7 = 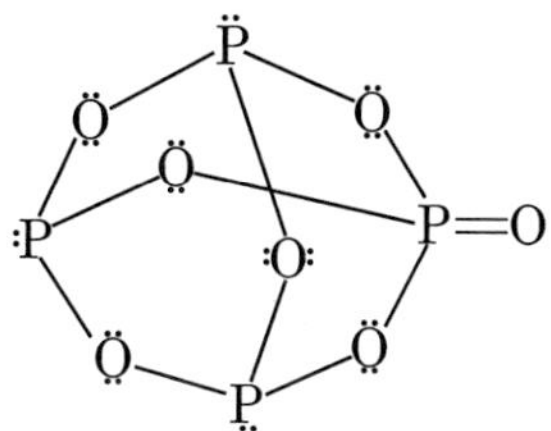

$H_4P_2O_5$ = 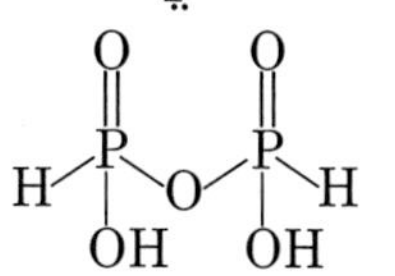

$H_2S_2O_3$ = 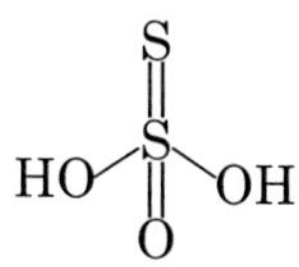

$H_5P_3O_{10}$ = 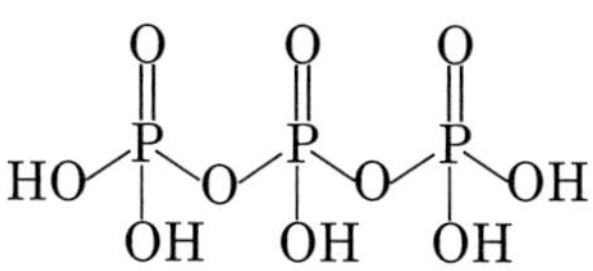

$H_2S_2O_5$ = 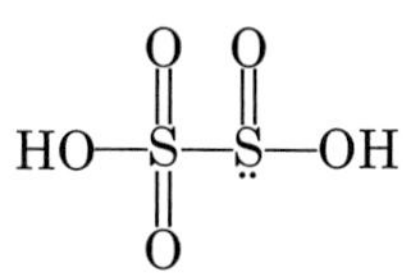

8. (6.47)

$$PbS + O_2 \rightarrow Pb + SO_2$$

$$\frac{1000}{32}\text{ mol} \qquad \frac{1000}{32} \times 207\text{ g}$$

mol of Pb = mol of $O_2 = \frac{1000}{32}$ mol

Hence, mass of Pb $= \frac{1000}{32} \times 207\text{ g} = \frac{207}{32}\text{kg} = 6.47\text{ kg}$

9. (126)

$$\underset{a\text{ mole}}{MnCl_2} + K_2S_2O_8 + H_2O \rightarrow \underset{a\text{ mole}}{KMnO_4} + H_2SO_4 + HCl$$

$$C_2O_4^{2-} + MnO_4^- \xrightarrow{H^+} CO_2$$

m_{eq} of $C_2O_4^{2-} = m_{eq}$ of MnO_4^-

$2 \times 0.225/90 = a \times 5$

$a = 1 \times [55 + 71] = 126$ mg

10. (7)

11. (495)

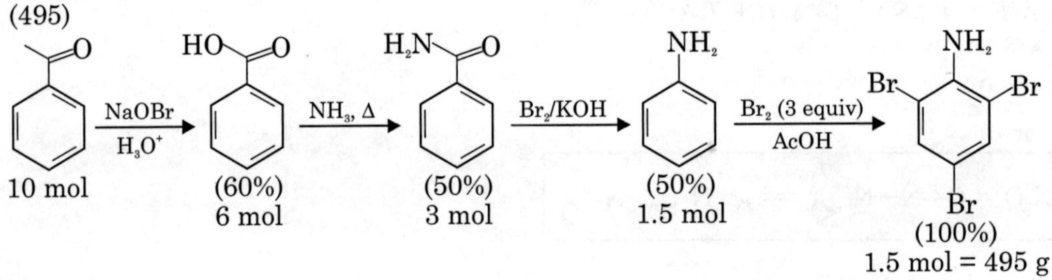

12. (– 14.6)

$2Cu(s) + \frac{1}{4}O_2(g) \rightarrow Cu_2O(s)$ $\quad\Delta G^0 = -78$ kJ

$H_2(g) + \frac{1}{2}O_2 \rightarrow H_2O(g)$ $\quad\Delta G^0 = -178\text{ kJ} \times (-1)$

Hence, $2Cu(s) + H_2O(g) \rightarrow Cu_2O(s) + H_2(g)$ $\quad\Delta G^0 = +100$ kJ

$\Delta G = \Delta G^0 + RT \ln Q$

$$0 = 100 + \frac{8}{1000} \times 1250 \ln \frac{p_{H_2}}{p_{H_2O}}$$

$$-\frac{100 \times 1000}{8} = 1250 \ln \frac{p_{H_2}}{\left(\frac{1}{100} \times 1\right)}$$

$\ln p_{H_2} = -14.6$

13. (8500)

$A(g) + B(g) \rightleftharpoons AB(g)$

$E_{ab} - E_{af} = 2RT$ or $\Delta H = -2RT$

$\frac{A_f}{A_b} = 4$

$$K_{eq} = \left(\frac{K_f}{K_b}\right) = \frac{A_f e^{-E_{af}/RT}}{A_b e^{-E_{ab}/RT}} = 4(e^2)$$

$\Delta G^0 = -RT \ln K = -2500 \times \ln(4 \times e^2) = -8500$ J/mol

Hence, absolute value of $\Delta G^0 = 8500$ J/mol

14. (– 11.62)

$$A(s)\left|A^{n+}(aq, 2M)\right\|B^{2n+}(aq, \underset{E_{cell}=0}{1M})\left|B(s)\right.$$

$\Delta H^\circ = 2\Delta G_0^0$

Cell reaction: $A \rightarrow A^{\Delta+} + ne^- \times 2$

$B^{2n+} + 2ne^- \rightarrow B(s)$

$$2A(s) + \underset{1M}{B}^{2n+}(aq) \rightarrow \underset{2M}{2A}^{n+}(aq) + B(s)$$

$$\Delta G = \Delta G^0 + RT \ln \frac{[A^{n+}]^2}{[B^{2n+}]}$$

$$\Delta G^0 = -RT \ln \frac{[A^{n+}]^2}{[B^{2n+}]} = -RT \ln \frac{2^2}{1} = -RT \ln 4$$

$\Delta G^0 = \Delta H^0 - T\Delta S^0$

$\Delta G^0 = 2\Delta G^0 - T\Delta S^0$

$$\Delta S^0 = \frac{\Delta G^0}{T} = -\frac{RT \ln 4}{T} = -8.3 \times 2 \times 0.7 = -11.62 \text{ J K}^{-1} \text{ mol}^{-1}$$

15. (C)

(1) $[FeF_6]^{4-}$

Fe^{2+} = [Ar] $3d^6$, Fe^{2+}

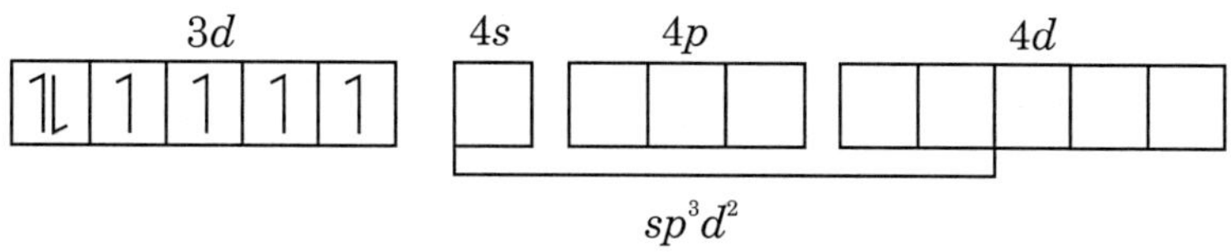

(2) $[Ti(H_2O)_3Cl_3]$

Ti^{3+} = [Ar] $3d^1$, Ti^{3+}

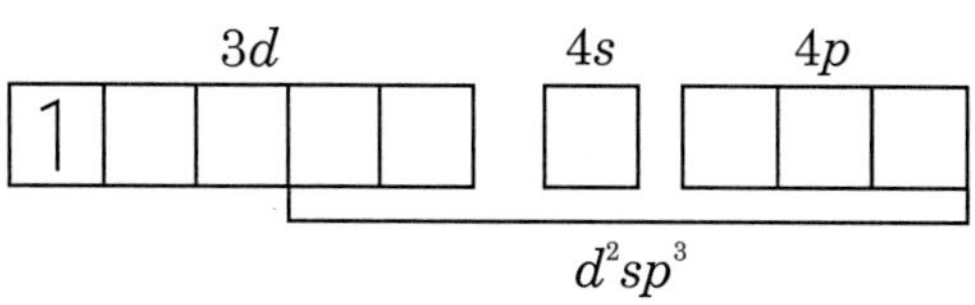

(3) $Cr(NH_3)_6]^{3+}$

Cr^{3+} = [Ar] $3d^3$, Cr^{3+}

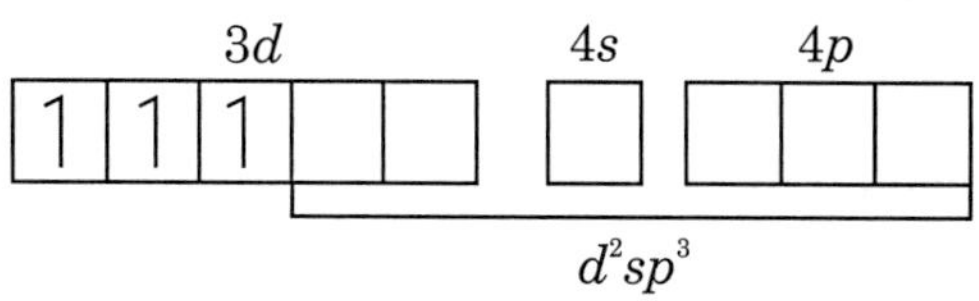

(4) $[FeCl_4]^{2-}$

Fe^{2+} = [Ar] $3d^6$, Fe^{2+}

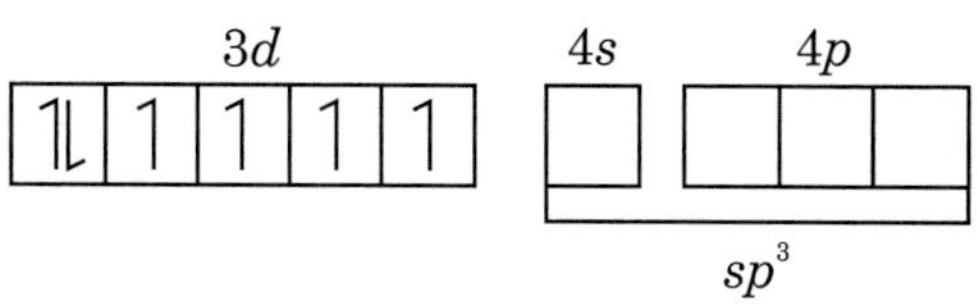

(5) $[Ni(CO)_4]$

Ni: $3d^8 4s^2$

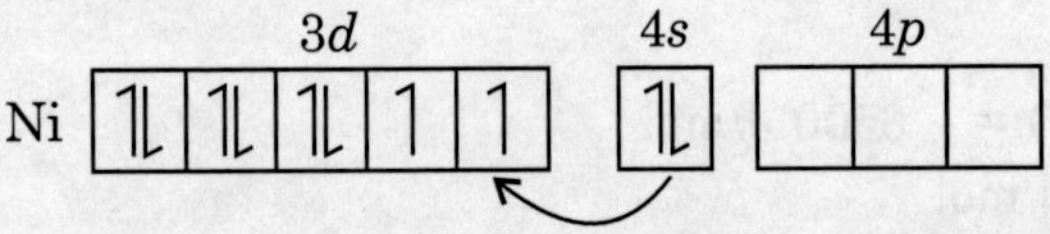

Back pairing of electrons due to the presence of strong field ligand CO

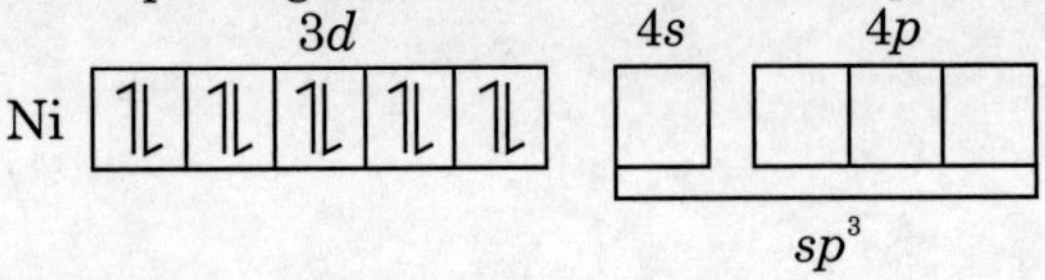

(6) $[Ni(CN_4)^{2-}$

Ni^{2+}: $3d^8$

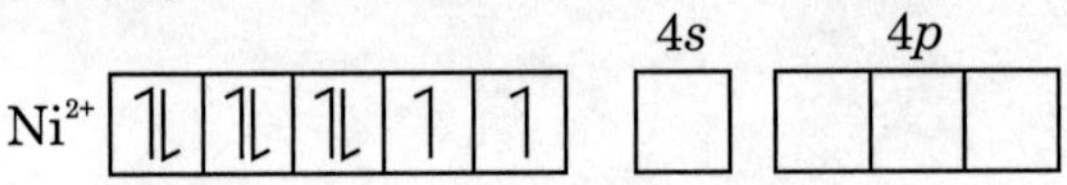

Electron pairing takes place due to the presence of strong field ligand CO.

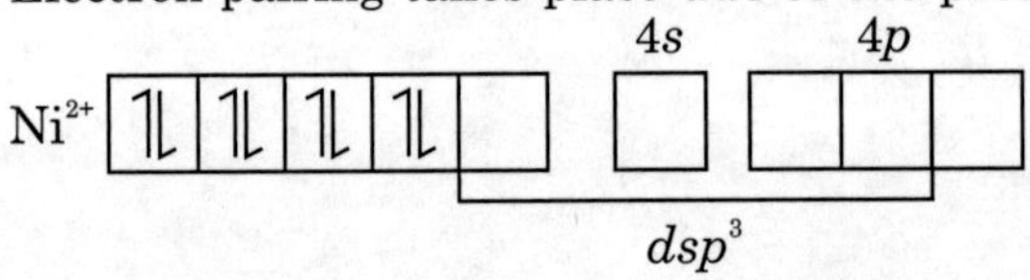

16. (D)

17. (B)

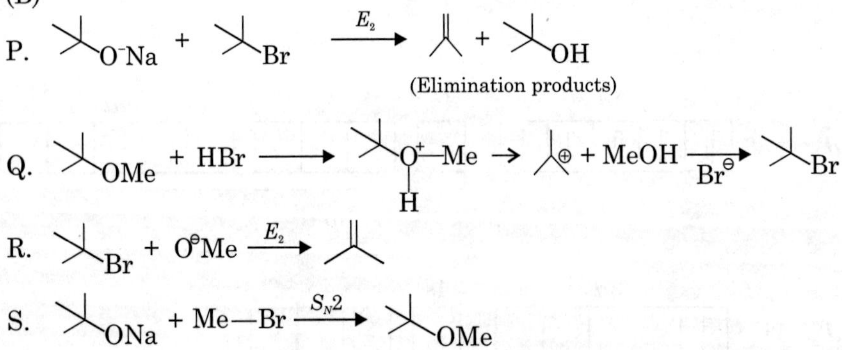

18. (D)

P. $\underset{0.1M,\ 20\ ml}{CH_3COOH} + \underset{0.1M,\ 10\ ml}{NaOH} \rightarrow CH_3COONa + H_2O$

$pH = pK_0$

$[H^+]$ will not change on dilution.

Correct match: P-1.

Q. $\underset{0.1M,\ 20ml}{CH_3COOH} + \underset{0.1M,\ 20ml}{NaOH} \rightarrow \underset{0.05M}{CH_3COONa} + H_2 O)$

$$[OH^-] = \sqrt{K_H C} = \sqrt{\left(\frac{k_w}{k_a} C\right)}$$

$$[H^+]_1 = \sqrt{\frac{k_w k_a}{C}}$$

$$\frac{[H^+]_2}{[H^+]_1} = \sqrt{\frac{C_1}{C_2}} = \sqrt{\frac{0.05}{0.025}} = \sqrt{2}$$

Correct match: Q-5

R. $\underset{0.1M,\ 20\ \text{ml}}{NH_4OH} + \underset{0.1M,\ 20\ \text{ml}}{HCl} \rightarrow \underset{0.05M}{NH_4Cl}$

$$[H^+] = \sqrt{K_H C}$$

$$\frac{[H^+]_2}{[H^+]_1} = \sqrt{\frac{C_2}{C_1}} = \frac{1}{\sqrt{2}}$$

Correct match: R-4

S. Because of dilution solubility does not change and hence $[H^+]$ = constant.

Appendix–I
Matching Type Questions

	X	Y
1.	(A) Heisenberg	(a) $\lambda = h/p$
	(B) de Broglie equation	(b) spectrum of H_2
	(C) Schrodinger	(c) stationary orbits
	(D) Rydberg	(d) uncertainty principle
	(E) Bohr	(e) wave theory
2.	(A) Soddy	(a) cathode rays
	(B) Thomson	(b) transuranium elements
	(C) Moseley	(c) molecular orbital theory
	(D) Seaborg	(d) atomic number
	(E) Mulliken	(e) isotopes
3.	(A) valence bond theory	(a) Gillespe and Nyholm
	(B) group displacement law	(b) Soddy and Rutherford
	(C) radioactive disintegration	(c) Aston and Soddy
	(D) isotopes	(d) Soddy, Fajans and Russel
	(E) VSEPR theory	(e) Heiteler and London
4.	(A) concept of hybridisation	(a) Guldberg and Waage
	(B) equation of state	(b) Thomson
	(C) law of mass action	(c) Sidgwick and Powel
	(D) electron	(d) Priestley
	(E) laughing gas	(e) van der Waals
5.	(A) kinetic theory of gas	(a) Sommerfield
	(B) idea of elliptical orbit	(b) Urey
	(C) nuclear fission	(c) Thomson
	(D) cathode rays	(d) Bernoulli
	(E) heavy water	(e) Otto Hahn and Strassman
6.	(A) isotopes	(a) $^{30}_{14}Si$, $^{31}_{15}P$
	(B) isobars	(b) CO, CN^-
	(C) isotones	(c) Fe_2O_3, Al_2O_3
	(D) isoelectronic	(d) $^{76}_{34}Sc$, $^{76}_{32}Ge$
	(E) isomorphs	(e) $^{35}_{17}Cl$, $^{37}_{17}Cl$

	Column I	Column II
7.	(A) isomers	(a) graphite, diamond
	(B) isotopes	(b) CH_3CH_2OH, CH_3OCH_3
	(C) isoelectronic	(c) $^{40}_{20}Ca$, $^{40}_{19}K$
	(D) isomorphs	(d) $^{16}_{8}O$, $^{18}_{8}O$
	(E) isobars	(e) $FeSO_4 \cdot 7H_2O$, $MgSO_4 \cdot 7H_2O$
	(F) allotropes	(f) Na^+, Mg^{2+} **(I.I.T. 1973)**
8.	(A) three electron bond	(a) AgI
	(B) covalent bond	(b) KI
	(C) ionic bond	(c) BDE = 2–10 kcal/mol
	(D) ionic, covalent and coordinate bonds	(d) NO
	(E) hydrogen bond	(e) NH_4I
9.	(A) superconductor	(a) antiferromagnetic
	(B) $Cu(CH_3COO)_2 \cdot 5H_2O$	(b) antiferroelectric
	(C) AgI	(c) paramagnetic
	(D) $PbZrO_3$	(d) diamagnetic
	(E) F-centre	(e) artificial rain
10.	(A) $YBa_2Cu_3O_7$	(a) antiferromagnet
	(B) MnO_2	(b) ferrimagnet
	(C) Fe_3O_4	(c) ferromagnet
	(D) CrO_2	(d) superconductor
	(E) CsCl	(e) Schottky defect
11.	(A) shape selective catalyst	(a) $k = A \exp(-E_a/RT)$
	(B) solar cell	(b) zeolite
	(C) Wilkinson catalyst	(c) Si
	(D) Arrhenius equation	(d) $(Ph_3P)_3RhCl$
	(E) hydrolysis of peptide	(e) pepsin
12.	(A) non-zero entropy at absolute zero	(a) a process in which volume remains constant
	(B) adiabatic process	(b) a process in which pressure remains constant
	(C) isochoric process	(c) a process in which temperature remains constant
	(D) isobaric process	(d) NO
	(E) isothermal process	(e) a process in which heat does not enter or leave the system
13.	(A) endothermic reaction	(a) a process which is carried out infinitesimally slowly
	(B) exothermic reaction	(b) $\Delta E = q - W$
	(C) reversible process	(c) $(\Delta H_2 - \Delta H_1)/(T_2 - T_1) = \Delta C_p$

(D) first law of thermodynamics	(d) ΔH is positive
(E) Kirchhoff's equation	(e) ΔH is negative

14.

(A) piezoelectricity	(a) ZnS
(B) pyroelectricity	(b) CsCl
(C) ferroelectricity	(c) application of heat
(D) Frenkel defect	(d) mechanical stress
(E) the 8 : 8 type of packing	(e) $BaTiO_3$

15.

(A) Aufbau	(a) Raoult's law
(B) de Broglie	(b) azeotropic mixture
(C) isotonic solutions	(c) electron has dual character
(D) ideal solutions	(d) 4*s* orbital is filled before 3*d* orbital
(E) non-ideal solutions	(e) osmotic pressure

16.

(A) number of mols of solute per kg of solvent	(a) conductance of ions
(B) number of mols of solute perlitre of solution	(b) *p*H scale
(C) Kohlrausch	(c) anode rays
(D) Sorensen	(d) molality
(E) Goldstein	(e) molarity

17.

(A) gold number	(a) cold cream
(B) water in oil emulsion	(b) vanishing cream
(C) oil in water emulsion	(c) negative sol
(D) $Fe(OH)_3$ sol	(d) protective colloid
(E) sulphur sol	(e) positive sol

18.

(A) Ziegler–Natta catalyst	(a) positive electron	(i) monoclinic S, orthorhombic S
(B) NaCl structure	(b) negative electron	(ii) -1.6×10^{-19} coulomb
(C) allotropes	(c) high pressure	(iii) 1.6×10^{-19} coulomb
(D) positron	(d) $(C_2H_5)_3Al + TiCl_4$	(iv) CsCl structure
(E) β-rays	(e) two or more forms of same element differing in physical properties	(v) hydrogenation of ethylene

19.

(A) Planck	(a) non-ideal solution	(i) $\Delta H_{mixing} = 0$
(B) plot of P_{total} *vs* x_A is non-linear	(b) ΔH unaffected	(ii) $\Delta V_{mixing} = 0$
(C) catalyst	(c) $E = h\nu$	(iii) 1.66×10^{-27} g
(D) 1 u	(d) ideal solution	(iv) ΔG unaffected
(E) Raoult's law	(e) 931 MeV	(v) 6.6×10^{-34} J s

20.

(A) The property dependent upon the amount of substance present in the system	(a) intensive property	(i) ice in a closed beaker
(B) The property independent of the amount of substance present in the system	(b) isolated system	(ii) ice in an open beaker
(C) a system which does not exchange mass and energy with the surroundings	(c) open system	(iii) ice in a thermos
(D) a system which exchanges energy with the surroundings and not mass	(d) extensive property	(iv) surface tension
(E) a system which exchanges energy and mass with the surroundings	(e) closed system	(v) entropy

21.

(A) a reaction which is spontaneous at all temperatures	(a) ABCABC... type of packing	(i) $\Delta G = +$ ve
(B) a reaction which is non-spontaneous at all temperatures	(b) AB AB ... type of packing	(ii) Cu
(C) cubic close packed arrangement	(c) $\Delta H = +$ ve	(iii) Zn
(D) hexagonal close packed arrangement	(d) anion vacancies	(iv) $\Delta G = -$ ve
(E) LiCl	(e) $\Delta H = -$ ve	(v) Schottky defect

22. (A) Bohr — (a) vapour pressure
(B) Roentgen — (b) colloids
(C) Tyndall — (c) chemical bond
(D) Raoult — (d) frequency condition
(E) Pauling — (e) X-rays **(M.L.N.R. 1997)**

23. (A) carbonium ions — (a) sp, sp^2, sp^3
(B) free radicals — (b) π bonds
(C) Baver and Nathan effect — (c) diamagnetic
(D) hybridisation of C atom — (d) paramagnetic
(E) sp, sp^2 hybridisations — (e) hyperconjugation

24. (A) relative energy of orbitals — (a) $sp > sp^2 > sp^3$
(B) bond angle — (b) $s > sp > sp^2 > sp^3 > p$
(C) bond length — (c) $sp^3 > sp^2 > sp$
(D) s-character — (d) $sp > sp^2 > sp^3$

(E) relative size of orbitals	(e) $sp^3 > sp^2 > sp$
25. (A) allotropes	(a) NO_3^-, CO_3^{2-}
(B) isoelectronic	(b) ${}^{76}_{32}Ge$, ${}^{77}_{32}Ge$
(C) isotopes	(c) ${}^{40}_{20}Ca$, ${}^{40}_{19}K$
(D) isotones	(d) ${}^{76}_{32}Ge$, ${}^{77}_{33}As$, ${}^{78}_{34}Sc$
(E) isobars	(e) O_2, O_3
26. (A) neutron	(a) Kohlraush
(B) molecular speed	(b) van der Waals
(C) intermolecular forces	(c) Maxwell
(D) conductance of ions	(d) Chadwick **(I.I.T. 1982)**
27. (A) ice	(a) insulator
(B) graphite	(b) ionic compound
(C) solid CO_2	(c) conducting covalent solid
(D) magnesiam oxide	(d) hydrogen bonded structure
(E) silica	(e) molecular crystal
(F) SiO_4^{4-}	(f) tetrahedra
28. (A) Bleaching agent	(i) Aluminium
(B) Smelling salt	(ii) Carbon
(C) Cryolite	(iii) Zinc
(D) Bell Metal	(iv) Ammonium carbonate
(E) Fluorspar	(v) Ammonium phosphate
(F) Fertiliser	(vi) Calcium
(G) Anthracite	(vii) Chlorine **(I.I.T. 1980)**
29. (A) Liquid air	(i) Deacon's process
(B) Na_2CO_3	(ii) Parken's process
(C) Nitric oxide	(iii) Claude process
(D) Silver	(iv) Ostwald process
(E) Chlorine	(v) Solvay process **(I.I.T. 1985)**
30. (A) Hg_2Cl_2	(i) Cassiterite
(B) $(NaPO_3)_6$	(ii) Lunar caustic
(C) NO_3^-	(iii) Producer gas
(D) SnO_2	(iv) Water softener
(E) $KCl{\cdot}MgCl_2{\cdot}6H_2O$	(v) Calomel **(I.I.T. 1982)**
31. (A) Al	(i) Calamine
(B) Cu	(ii) Cryolite
(C) Mg	(iii) Malachite
(D) Zn	(iv) Carnallite **(I.I.T. 1983)**
32. (A) Haber	(i) Activation energy
(B) Graham	(ii) Diffusion of gases

(C) Arrhenius (iii) Octet reule

(D) Lewis (iv) Ammonia synthesis **(I.I.T. 1983)**

33. (A) Spinel (i) $MgAl_2O_4$

(B) Feldspar (ii) $PbCO_3$

(C) Cerrusite (iii) $KAlSi_3O_8$

(D) Malachile (iv) $MgSO_4{\cdot}H_2O$

(E) Epsom salt (v) $Cu(OH)_2{\cdot}Cu_2CO_3$ **(I.I.T. 1983)**

34. (A) Self-reduction (i) Lead

(B) Carbon-reduction (ii) Silver

(C) Complex formation and displacement by metal (iii) Copper

(D) Decomposition of iodide (iv) Boron **(I.I.T. 2006)**

35. (A) $Bi_2O_3 \longrightarrow (BiO)^+$ (i) Heat

(B) $[AlO_2]^- \longrightarrow Al(OH)_3$ (ii) $PbCO_3$

(C) $SiO_4^{4-} \longrightarrow Si_2O_7^-$ (iii) $KAlSi_3O_8$

(D) $B_4O_7^{2-} \longrightarrow [B(OH)_3]$ (iv) $MgSO_4{\cdot}H_2O$ **(I.I.T. 2006)**

36. According to Bohr's theory of H-atom, E_n = total energy, K_n = kinetic energy of nth orbit, V_n = potential energy of nth orbit, r_n = radius of nth orbit.

(A) $V_n/K_n = ?$ (i) 0

(B) If radius of nth orbit $\propto E_n^x$, $x = ?$ (ii) – 1

(C) Orbital angular momentum (iii) – 2

(D) $\dfrac{1}{V_n} \propto Z^y$, $y = ?$ (iv) 1 **(I.I.T. 2006)**

37. (A) $[CuCl_4]^{2-}$ (i) sp^3d^2 hybridisation

(B) $[Fe(H_2O)_5(NO)^+]SO_4$ (ii) dsp^3 hybridisation

(C) $[Ni(CN)_4]^{2-}$ (iii) sp hybridisation

(D) $[Fe(CO)_5]$ (iv) dsp^2 hybridisation

(E) $[Ag(NH_3)_2]^+$ (v) sp^3 hybridisation

38. (A) $K_3[CoF_6]$ (i) $\mu = 0$

(B) $K_2[Ni(CN)_4]$ (ii) $\mu = 1.73$ B.M.

(C) $[VO(H_2O)_5]SO_4$ (iii) $\mu = 2.83$ B.M.

(D) $K_3[Mn(CN)_6]$ (iv) $\mu = 3.87$ B.M.

(E) $Na_3[FeF_6]$ (v) $\mu = 4.90$ B.M.

(F) $[Cr(NH_3)_6]Br_3$ (vi) $\mu = 5.92$ B.M.

39. (A) $[Zn(NH_3)_4]Cl_2$ (i) Yellow

(B) $CuSO_4{\cdot}5H_2O$ (ii) Green

(C) $[Ni(H_2O)_6]SO_4$ (iii) Pink

(D) $[Co(H_2O)_6]SO_4$	(iv) Colourless
(E) $[Fe(H_2O)_6]Cl_3$	(v) Blue

40.

(A) C_2H_5Br + alc. KOH	(i) cracking	
(B) preparation of alkanes	(ii) Friedel–Crafts reaction	
(C) phenol + $CHCl_3$ + NaOH	(iii) Reimer–Tiemann reaction	
(D) $CH_3COOC_2H_5$ + NaOH	(iv) Wurtz reaction	
(E) C_6H_6 + CH_3Cl + anhy.$AlCl_3$	(v) saponification	
(F) pyrolysis of alkanes	(vi) elimination reaction	**(I.I.T. 1982)**

41.

(A) dehydrohalogenation	(i) Kolbe reaction	
(B) electrolysis of sodium salt	(ii) alcoholic KOH	
(C) decarboxylation	(iii) addition reaction of ethylene	
(D) ozonolysis	(iv) sodalime	
(E) Williamson synthesis	(v) structure of alkene	
(F) dichloroethylene	(vi) ether	**(I.I.T. 1983)**

42.

(A) Lucas test	(i) phenol	
(B) neutral $FeCl_3$ test	(ii) glucose	
(C) Dye test	(iii) tertiary alcohol	
(D) Tollen's test	(iv) aniline	**(I.I.T. 1983)**

43.

(A) ethanol-methanol	(i) condensation reaction	
(B) phenol-methanol	(ii) homologous series	
(C) propanone sodium bisulphite	(iii) nucleophilic addition	
(D) carbon tetrachloride	(iv) fire extinguishing	**(I.I.T. 1981)**

44.

(A) Cyclopropane, Cl_2, light	(i) homolytic addition	
(B) propene, HBr, peroxide	(ii) eletrophilic substitution	
(C) benzene, HNO_3, H_2SO_4	(iii) free radical substitution	
(D) polymerisation	(iv) alcoholic KOH	
(E) dehydrohalogenation	(v) conversion of C_2H_2 to C_6H_6	**(I.I.T. 1981)**

45.

(A) phenol	(i) coloured glass	
(B) $Na_2S_2O_3$	(ii) antichlor	
(C) salicylic acid	(iii) refractory material	
(D) quicklime	(iv) antiseptic	
(E) CuO	(v) analgesic	**(I.I.T. 1985)**

46.

(A) freon	(i) antiseptic
(B) dacron	(ii) polymer of ethylene glycol and terephthalic acid
(C) Grignard reagent	(iii) refrigeration
(D) iodoform	(iv) CH_3MgX
(E) DDT	(v) insecticide

No.	Column I	Column II
47.	(A) phenetole	(i) anaesthetic
	(B) Williamson synthesis	(ii) estimation of alkoxy group
	(C) oxonium salt	(iii) reaction of either with inorganic acid
	(D) diethyl ether	(iv) reaction of alkyl halide with C_2H_5ONa
	(E) Zeisel method	(v) reaction of phenol with diethyl sulphate and NaOH
48.	(A) detergent	(i) Ni catalyst
	(B) hydrogenation	(ii) lubricant
	(C) bee wax	(iii) hydrocarbons
	(D) chromium soap	(iv) myricyl palmitate
	(E) paraffin wax	(v) sodium lauryl sulphonate

No.	Column I	Column II	Column III
49.	(A) Friedel–Crafts reaction	(a) oil	(i) alkenes
	(B) fermentation	(b) Lewis acid	(ii) soap
	(C) dehydrohalogenation	(c) copper(I) chloride	(iii) anhydrous $AlCl_3$
	(D) Sandmeyer	(d) yeast	(iv) chlorobenzene
	(E) saponification	(e) alcoholic alkali	(v) ethanol **(I.I.T. 1988)**
50.	(A) Fehling solution	(a) mesityle oxide	(i) reddish brown precipitate of Cu_2O
	(B) Tollen's reagent	(b) reducing agent	(ii) phorone
	(C) acetone	(c) Rochelle salt	(iii) ammoniacal $AgNO_3$
	(D) benzaldehyde	(d) 50% NaOH	(iv) urinary antiseptic
	(E) formaldehyde + NH_3	(e) urotropine	(v) $C_6H_5CH_2OH$ + C_6H_5COONa
51.	(A) Hell–Volhard–Zelinsky reaction	(a) aqueous CH_3COONa	(i) C_2H_5OH
	(B) Kolbe reaction	(b) $LiAlH_4$	(ii) α-chloroacid
	(C) CH_3COOH	(c) Cl_2 + P	(iii) carboxylic acid
	(D) Grignard reagent	(d) solid CO_2	(iv) ethane
	(E) decarboxylation	(e) sodalime	(v) removal of CO_2
52.	(A) acetamide	(a) H_2, $Pd/BaSO_4$ + S	(i) methylamine
	(B) ethyl acetate	(b) $LiAlH_2$	(ii) acetophenone
	(C) acetic anhydride	(c) Br_2 + NaOH	(iii) C_2H_5OH
	(D) benzene + acetic anhydride	(d) anhy. $AlCl_3$	(iv) CH_3CHO
	(E) acetyl chloride	(e) C_2H_5ONa	(v) acetoacetic ester
53.	(A) carbylamine	(a) primary amine	(i) diazonium salt
	(B) aromatic amine	(b) boiling aqueous HCl	(ii) $CHCl_3$ + alc. KOH
	(C) CH_3CN	(c) HN_2	(iii) NH_3 + CH_3COONa
	(D) CH_3CH_2CN	(d) boiling aqueous	(iv) butanone
	(E) CH_3NC	(e) CH_3MgBr, hydrolysis	(v) HCOOH + CH_3NH_3

	Column I	Column II	Column III
54.	(A) *m*-dinitrobenzene	(a) HNO_2	(i) less basic than aniline
	(B) *o*-nitroaniline	(b) + I group at ortho position	(ii) *m*-nitroaniline
	(C) *o*-toluidine	(c) $(NH_4)_2S$	(iii) more basic than aniline
	(D) *N*-phenylhydroxylamine	(d) ammoniacal $AgNO_3$	(iv) nitrolic acid
	(E) primary nitroalkane	(e) – I group at ortho position	(v) silver mirror
55.	(A) Hinsberg's reagent	(a) $K_2Cr_2O_7$, H^+	(i) *N*-nitrosoamine
	(B) secondary amine	(b) phenol	(ii) distinction between amines
	(C) Hofmann bromamide reaction	(c) HNO_2	(iii) primary amine
	(D) coupling reaction	(d) benzenesulphonyl chloride	(iv) aniline black
	(E) aniline	(e) Br_2 + KOH	(v) azo dye
56.	(A) Sabatier–Senderens reaction	(a) iodobenzene	(i) benzanilide
	(B) Schotten–Baumann reaction	(b) benzenediazonium chloride	(ii) CuBr/HBr
	(C) Sandmeyer reaction	(c) acetophenone	(iii) Cu powder, heat
	(D) Wolff–Kishner reaction	(d) aniline + benzoyl chloride	(iv) ethyne to ethane
	(E) Ullmann reaction	(e) H_2/Ni, 250–300°C	(v) NH_2NH_2, KOH, ethylene glycol, 200°C
57.	(A) Rosenmund reduction	(a) aromatic aldeyhde, aliphatic acid anhydride	(i) Na/C_2H_5OH
	(B) Reimer–Tiemann reaction	(b) $C_6H_5ONa + CO_2$	(ii) H_2, $Pd/BaSO_4$ + S
	(C) Perkin reaction	(c) acetonitrile	(iii) salicylic acid
	(D) Mendius reaction	(d) phenol, $CHCl_3$, NaOH	(iv) α, β-unsaturated acid
	(E) Kolbe–Schmidt reaction	(e) acetyl chloride	(v) salicylaldehyde
58.	(A) Claisen condensation	(a) Zn/Hg, HCl	(i) benzaldehyde
	(B) Clemmensen reduction	(b) aldehyde	(ii) Cu power, HCl
	(C) Cannizzaro reaction	(c) KCN	(iii) β-ketoester
	(D) Benzoin condensation	(d) benzenediazonium chloride	(iv) ketone to alkane
	(E) Gattermann reaction	(e) ethyl acetate, C_2H_5ONa	(v) 50% NaOH
59.	(A) Bergius process	(a) diene, dienophile	(i) ethylene
	(B) Fries reaction	(b) alkaline $KMnO_4$	(ii) distinction between alcohols
	(C) Diels–Alder reaction	(c) hyrogenation of coal	(iii) adduct

(D) Baeyer's reagent (d) conc. HCl + anhy. $ZnCl_2$ (iv) migration of acyl group
(E) Lucas reagent (e) phenyl acetate (v) petrol

60. (A) TNT (i) phenol-formaldehyde plastic
(B) teflon (ii) antiknock compound
(C) tetraethyl lead (iii) polyamide polymer
(D) nylon-66 (iv) corrosion resistant
(E) bakelite (v) explosive

61. (A) Liebig method (i) estimation of halogens
(B) Ziesel method (ii) estimation of N
(C) Carius method (iii) estimation of C, H
(D) Kjeldahl method (iv) detection of N, S, halogens
(E) Lassaigne's test (v) estimation of OCH_3 group

62. (A) *n*-heptane (i) cetane No. 0
(B) isooctane (ii) cetane No. 100
(C) hexadecane (iii) octane No. 0
(D) α-methylnaphthalene (iv) octane No. 100
(E) triplane (v) octane No. 124

63. (A) metamers (i) $CH_3COCH_2COCH_3$, $CH_3COCHC(OH)CH_3$
(B) functional isomers (ii) isopentane, neopentane
(C) position isomers (iii) CH_3COOCH_3, CH_3CH_2COOH
(D) chain isomers (iv) *n*-propyl chloride, isopropyl chloride
(E) tautomers (v) $C_2H_5OC_2H_5$, $CH_3OC_3H_7$

64. (A) $CHCl_3$ (i) addition polymer
(B) aspirin (ii) vitamin C
(C) PVC (iii) polymer of glucose
(D) starch (iv) antipyretic
(E) ascorbic acid (v) anaesthetic

65. (A) sublimation (i) H_2O_2
(B) steam distillation (ii) benzoic acid
(C) fractional crystallisation (iii) benzene and toluene
(D) fractional distillation (iv) Na_2SO_4, $Na_2Cr_2O_7$
(E) distillation under reduced pressure (v) turpentine oil

66. (A) caprolactum (i) antimalerial
(B) zwitterion (ii) protective action of colloid
(C) CFC (iii) fire extinguisher
(D) nitrous oxide (iv) amino acid
(E) ampicillin (v) green house effect
(F) CCl_4 (vi) nylon-66
(vii) general anaesthetic
(viii) broad spectrum anaesthetic
(ix) antipyretic **(Assam J.A.T. 1999)**

67. (A) Isobars — (i) C_nH_{2n-2} where $n = 1, 2, 3$
(B) Isomorphs — (ii) ${}^{40}_{20}Ca$ and ${}^{20}_{19}K$
(C) Alkynes — (iii) $FeSO_4{\cdot}7H_2O$ and $MgSO_4{\cdot}7H_2O$
(D) Isomers — (iv) C_nH_{2n-2} where $n = 3, 4$
(E) Cycloalkanes — (v) C_2H_5OH and CH_3OCH_3

(Bihar C.E.E. 1999)

68. (A) $R-CH=CH_2$ — (i) 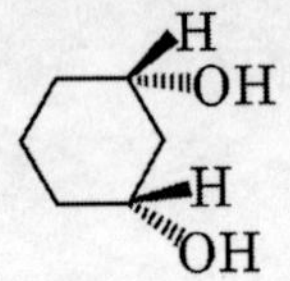

(B) 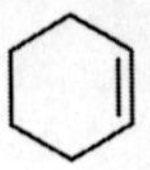 — (ii) 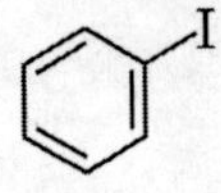

(C) $H-\overset{CH_3}{\underset{(CH_2)_3-CH_3}{C}}-Br$ — (iii) $HO-\overset{CH_3}{\underset{(CH_2)_3-CH_3}{C}}-H$

(D)  — (iv) 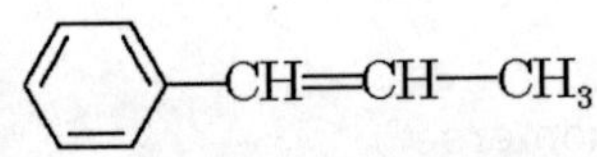

(E) 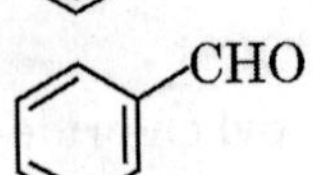 — (v) 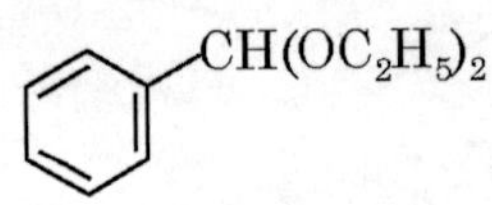

(F) 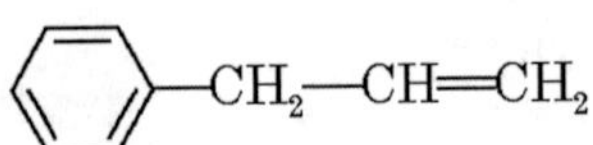 — (vi) $R-CHCl-CH_2-CCl_3$

Suggest a suitable reagent for conversion of each pair. **(Roorkee 2001)**

69. (A) Benzoic acid — (i) $K_a = 3.3 \times 10^{-5}$

(B) 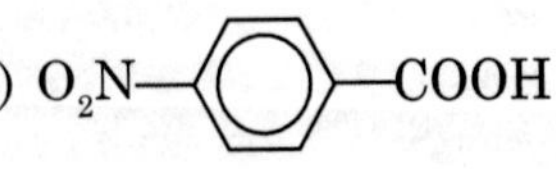 — (ii) $K_a = 6.3 \times 10^{-5}$

(C) 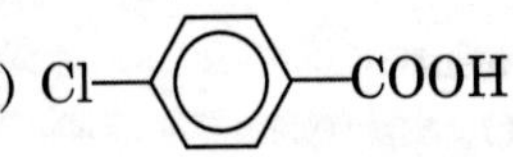 — (iii) $K_a = 30.6 \times 10^{-5}$

(D) 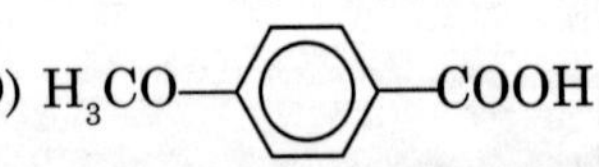 — (iv) $K_a = 6.4 \times 10^{-5}$

(E) 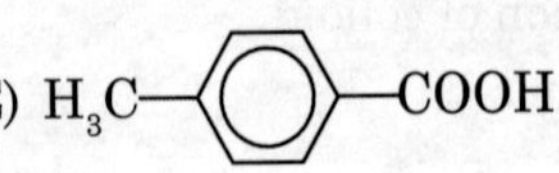 — (v) $K_a = 4.2 \times 10^{-5}$

(I.I.T. 2006)

70. (A) Compound X, B.P. = 100°C — (i) $K_b = 0.53$
(B) Compound Y, B.P. = 27°C — (ii) $K_b = 0.68$
(C) Compound Z, B.P. = 253°C — (iii) $K_b = 0.98$
(Assume the molecular masses of X, Y and Z same.) **(I.I.T. 2003)**
Hint: $K_b = RT_b^2/1000 l_v = RT_b^2\, M/\Delta H_v$ (since $l_v = \Delta H_v/M$)

71. (A) $PhCH_2CD_2Br$ on reaction with C_2H_5O gives PH—CH=CD_2 | (i) $E1$ reaction
(B) Ph—CH_2CH_2Br reacts faster than Ph—CD_2CH_2Br on reaction with $C_2H_5O^-$ in ethanol | (ii) $E2$ reaction
(C) Ph—CHD—CH_2—Br on treatment with $C_2H_5D/C_2H_5O^-$ gives Ph—CD=CH_2 as the major product. | (iii) $E1CB$ reaction
(D) $PhCHBrCH_3$ and $PhCHBrCD_3$ react with the same rate. | (iv) First order reaction

(I.I.T. 2006)

72. (A) Aufbau principle | (i) Atomic orbital with one radial node
(B) Pauli exclusion principle | (ii) $3d$ orbital is of higher energy than $4s$ orbital
(C) Hund's rule | (iii) Atomic orbital defined by $n=7$, $\ell=1$
(D) $3p$ orbitals | (iv) No two electrons in an atom can have all four quantum numbers identical
(E) d-orbitals | (v) Distribution of elections is symmetrical
(F) $7p$ orbitals | (vi) electron pairing is not possible until each orbital of the same subshell is occupied by a single election.

73. (A) XeF_2 | (i) 180°
(B) XeF_4 | (ii) 106°45′
(C) XeO_3 | (iii) 109.5°
(D) XeO_4 | (iv) 90°
(E) NH_3 | (v) 104°31′
(F) H_2O | (vi) 103°

74. (A) NH_4^+ | (i) Polar covalent bond
(B) S_8 | (ii) Coordinate bond
(C) HBr | (iii) Covalent bond
(D) $Al_2(CH_3)_6$ | (iv) Intermolecular H-bond
(E) HF | (v) Electron deficient

75. (A) Ni | (i) Covalent bond
(B) AgCl | (ii) Ionic bond
(C) SiC | (iii) van der Waals forces
(D) $NH_3(g)$ | (iv) coordinate bond
(E) H_3N—BF_3 | (v) metallic bond

76. (A) The charge carried by | (i) Strongest reducing agent
(B) For a galvanic cell $E_{cell} > 0$ | (ii) Liberation of O_2 at anode
(C) Li has the lowest standard electrode potential. | (iii) The cell is at equilibrium.

(D) Electrolysis of aqueous $CuCl_2$ using platinum electrodes
(E) Electrolysis of aqueous Na_2SO_4 using platinum electrodes

(iv) 1 Faraday
(v) Liberation of Cl_2 gas at anode

77. (A) $[Co(NH_3)_4(H_2O)_2]Cl_2$
(B) $[Pt(NH_3)_2Cl_2]$
(C) $[Co(H_2O)_5Cl]Cl$
(D) $[Ni(H_2O)_6]Cl_2$

(i) Geometrical isomers
(ii) Paramagnetic
(iii) Diamagnetic
(iv) Metal ion with +2 oxidation state

(I.I.T. 2007)

78. (A) Hydrogen gas (P = 200 atm, T = 273 K)
(B) Hydrogen gas (P = 0, T = 273 K)
(C) CO_2 (P = 1 atm, T = 273 K)
(D) Real gas with very large molar volume

(i) Compressibility factor = 1
(ii) Attractive forces are dominant
(iii) $PV = nRT$
(iv) $P(V - ab) = nRT$

(I.I.T. 2007)

79. (A) Cellulose
(B) Nylon-66
(C) Protein
(D) Sucrose

(i) Natural polymer
(ii) Synthetic polymer
(iii) Amide linkage
(iv) Glucoside linkage

(I.I.T. 2007)

80. (A) $CH_3CH_2CH_2CN$
(B) $CH_3CH_2OCOCH_3$
(C) $CH_3—CH═CH—CH_2OH$
(D) $CH_3CH_2CH_2CH_2NH_2$

(p) Reduction with Pd–C/H_2
(q) Reduction with $SnCl_2$/HCl
(r) Development of foul smell on treatment with chloroform and alcohalic KOH
(s) Reduction with diisobutylaluminium hydride (DIBAL-H)
(t) Alkaline hydrolysis

(I.I.T. 2009)

81. (A) B_2
(B) N_2
(C) O_2^-
(D) O_2

(p) Paramagnetic
(q) Undergoes oxidation
(r) Undergoes reduction
(s) Bond order $\geq$ 2
(t) Mixing of s and p orbitals

(I.I.T. 2009)

82. (A) 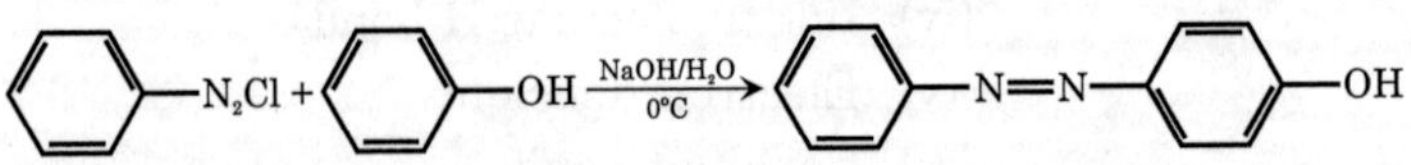

(B) 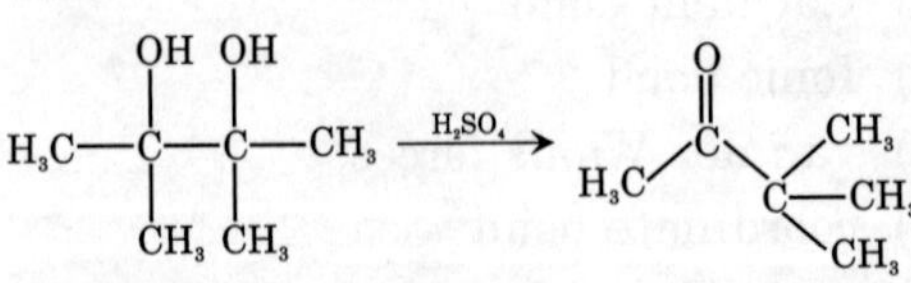

(C) 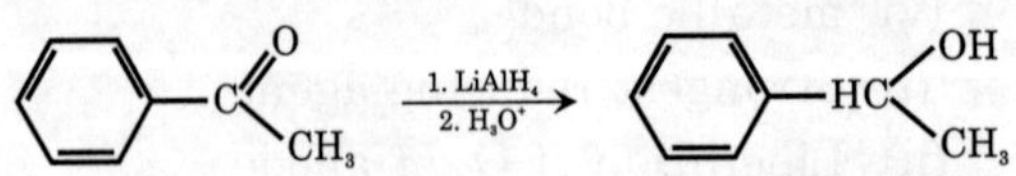

(p) Racemic mixture
(q) Addition reaction
(r) Substitution reaction

(D) 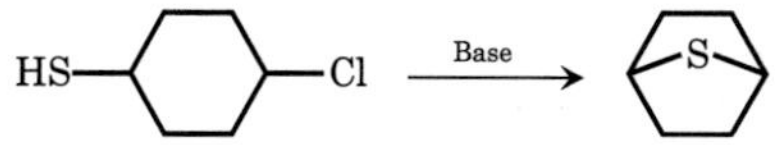

(s) Coupling reaction

(t) Carbocationintermediate **(I.I.T. 2010)**

83. (A) $(CH_3)_2SiCl_2$
(B) XeF_4
(C) Cl_2
(D) VCl_5

(p) Hydrogen halide formation
(q) Redox reaction
(r) Reacts with glass
(s) Polymerization
(t) O_2 formation **(I.I.T. 2010)**

84. (A) $CO_2(s) \longrightarrow CO_2(g)$
(B) $CaCO_3(s) \longrightarrow CaO(s) + CO_2(g)$
(C) $2H\cdot \longrightarrow H_2(g)$
(D) $P_{(\text{white, solid})} \longrightarrow P_{(\text{red, solid})}$

(p) phase transition
(q) allotropic change
(r) ΔH is positive
(s) ΔS is positive
(t) ΔS is negative **(I.I.T. 2011)**

85. (A) 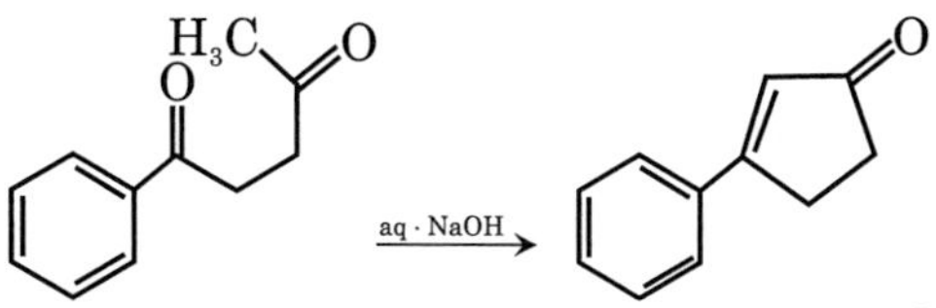

(B) 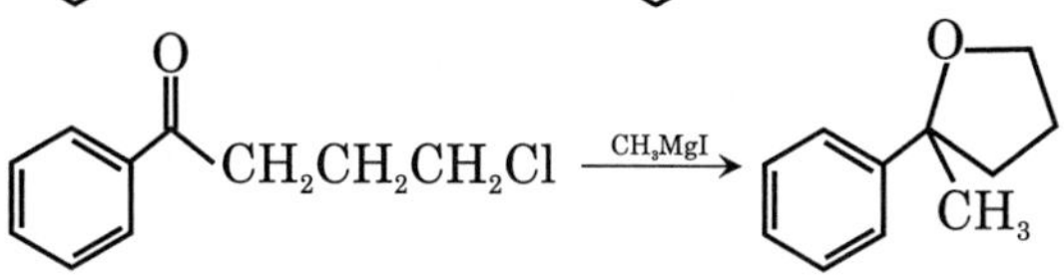

(C) 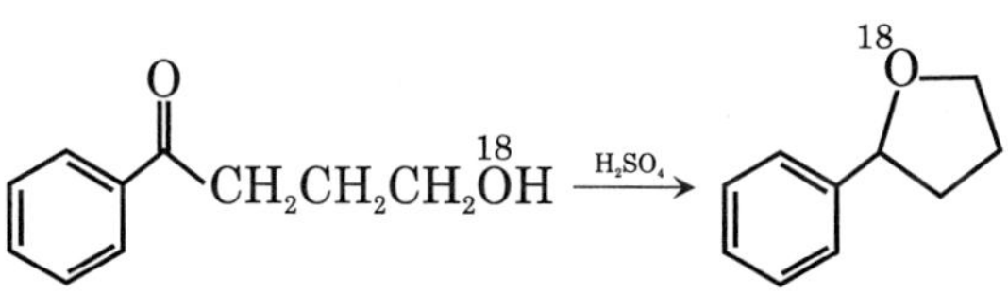

(D) 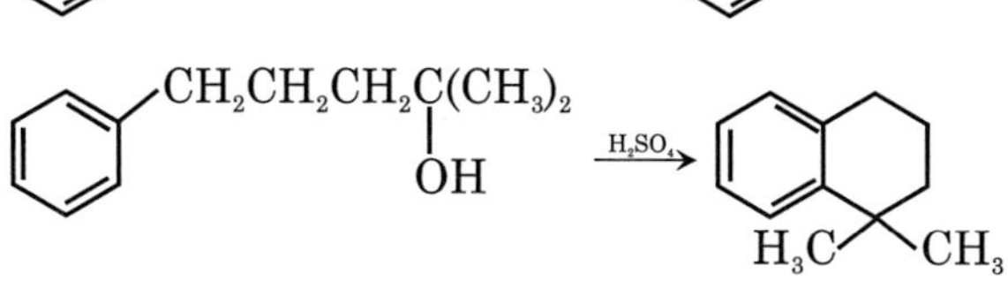

(p) Nucleophilic substitution
(q) Electrophilic substitution
(r) Dehydration
(s) Nucleophilic addition
(t) Carbanion **(I.I.T. 2011)**

86. (A) Cu + dil. HNO_3
(B) Cu + conc. HNO_3
(C) Zn + dil. HNO_3
(D) Zn + conc. HNO_3

(p) NO
(q) NO_2
(r) N_2O
(s) $Cu(NO_3)_2$
(t) $Zn(NO_3)_2$ **(I.I.T. 2009)**

87. (A) 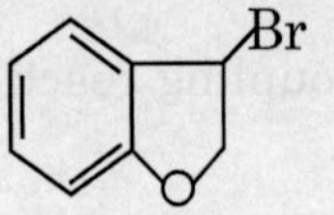 (p) Nucleophilic substitution

(B) 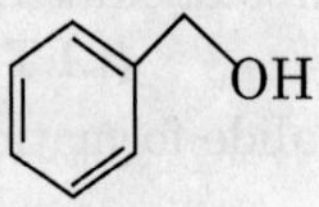 (q) Elimination

(C) 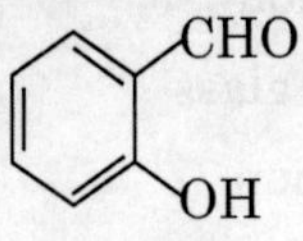 (r) Nucleophilic addition

(D) 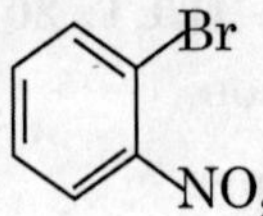 (s) Esterification with acetic anhydride

(t) Dehydrogenation **(I.I.T. 2009)**

88. P. $PbO_2 + H_2SO_4 \xrightarrow{?} PbSO_4 + O_2$ + other product 1. NO

Q. $Na_2S_2O_3 + H_2O \xrightarrow{?} NaHSO_4$ + other product 2. I_2

R. $N_2H_4 \xrightarrow{?} N_2$ + other product 3. Warm

S. $XeF_2 \xrightarrow{?} Xe$ + other product 4. Cl_2

Codes:	**P**	**Q**	**R**	**S**
(A)	4	2	3	1
(B)	3	2	1	4
(C)	1	4	2	3
(D)	3	4	2	1

(I.I.T. 2013)

89. P. 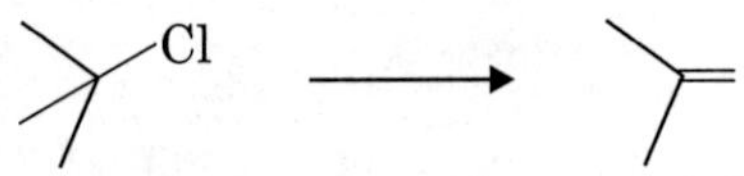 1. (i) $Hg(OAc)_2$; (ii) $NaBH_4$

Q. 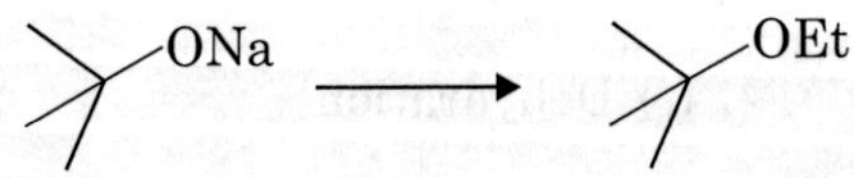 2. NaOEt

R. 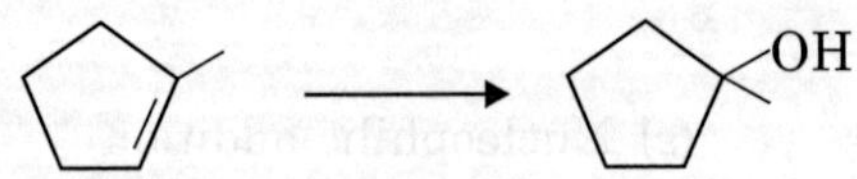 3. EtBr

S. 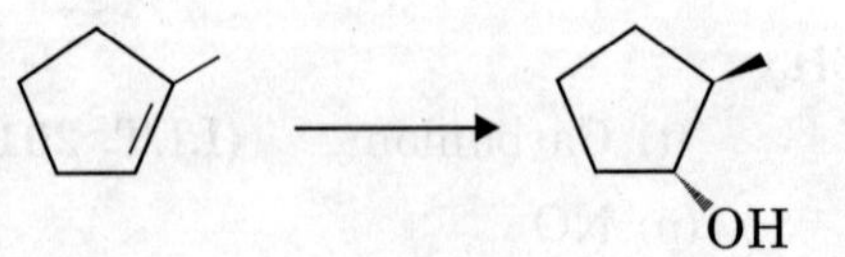 4. (i) BH_3; (ii) $H_2O_2/NaOH$

Codes:	**P**	**Q**	**R**	**S**
(A)	2	3	1	4
(B)	3	2	1	4
(C)	2	3	4	1
(D)	3	2	4	1

(I.I.T. 2013)

90. P. $(C_2H_5)_3N + CH_3COOH(aq)$

Q. $KI(aq, 0.1M) + AgNO_3(aq, 0.01M)$

R. $CH_3COOH(aq) + KOH(aq)$

S. $NaOH(aq) + HI(aq)$

1. Conductivity decreases and than increases
2. Conductivity decreases and then does not change much
3. Conductivity increases and then does not change much
4. Conductivity does not change much and then increases

Codes:	P	Q	R	S
(A)	3	4	2	1
(B)	4	3	2	1
(C)	2	3	4	1
(D)	1	4	3	2

(I.I.T. 2013)

91. Given:
E^0 (Fe^{3+}, Fe^{2+}) = + 0.77 V; E^0 (Fe^{2+}, Fe) = – 0.44 V; E^0 (Cu^{2+}, Cu) = + 0.34 V; E^0 (Cu^+, Cu) = + 0.52 V; E^0 ($O_2(g) + 4H^+ + 4e^- \rightarrow 2H_2O$) = + 1.23 V: Fe) = – 0.44 V; E^0 ($O_2(g) + 2H_2O + 4e^- \rightarrow 4OH^-$) = + 0.40 V; E^0 (Cr^{3+}, Cr) = – 0.74 V; E^0 (Cr^{2+}, Cr) = – 0.91 V

P. E^0 (Fe^{3+}·Fe)

Q. E^0 ($4H_2O \rightleftharpoons 4H^+ + 4OH^-$)

R. E^0 ($Cu^{2+} + Cu \rightarrow 2Cu^+$)

S. E^0 (Cr^{3+}, Cr^{2+})

1. – 0.18 V
2. – 0.4 V
3. – 0.04 V
4. – 0.83 V

Codes:	P	Q	R	S
(A)	4	1	2	3
(B)	2	3	4	1
(C)	1	2	3	4
(D)	3	4	1	2

(I.I.T. 2013)

92. P. $[Cr(NH_3)4Cl_2]Cl$

Q. $[Ti(H_2O)_5Cl](NO_3)_2$

R. $[Pt(en)(NH_3)Cl]NO_3$

S. $[Co(NH_3)_4(NO_3)_2]NO_3$

1. Paramagnetic and exhibits ionisation isomerism
2. Diamagnetic and exhibits *cis-trans* isomerism
3. Paramagnetic and exhibits *cis-trans* isomerism
4. Diamagnetic and exhibits ionisation isomerism

Codes:	P	Q	R	S
(A)	4	2	3	1
(B)	3	1	4	2
(C)	2	1	3	4
(D)	1	3	4	2

(I.I.T. 2014)

93. P. 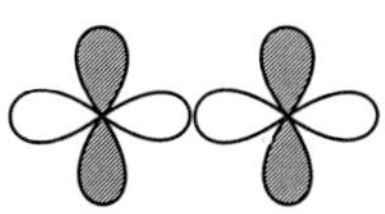

Q. 

1. *p-d* π antibonding
2. *d-d* σ bonding

R.  3. p-d π bonding

S.  4. d-d σ antibonding

Codes:	P	Q	R	S
(A)	2	1	3	4
(B)	4	3	1	2
(C)	2	3	1	4
(D)	4	1	3	2

(I.I.T. 2014)

94. Given: Possible pathways:

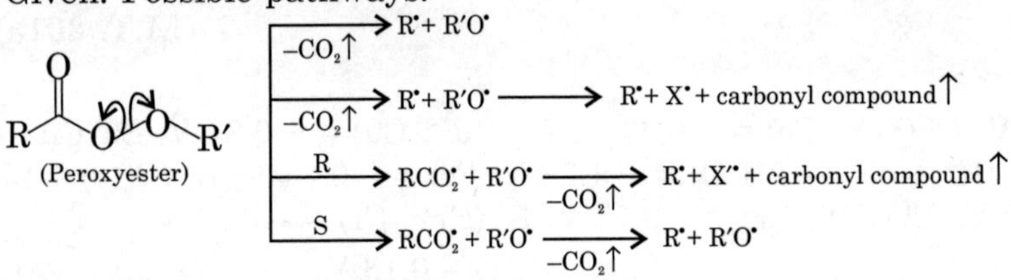

P. Pathway P — 1. 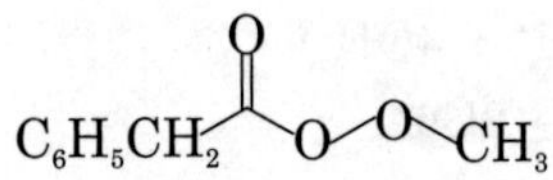

Q. Pathway Q — 2. 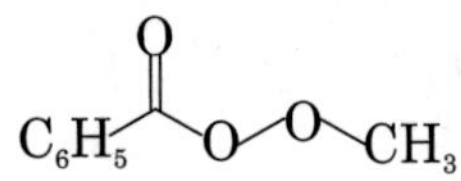

R. Pathway R — 3. 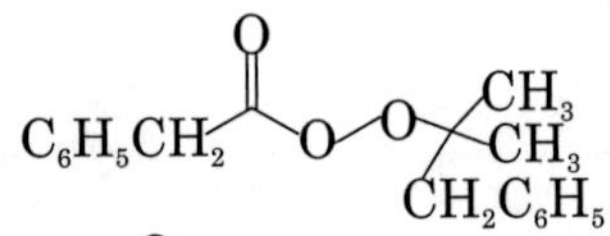

S. Pathway S — 4. 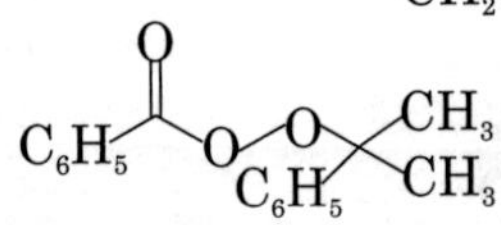

Codes:	P	Q	R	S
(A)	1	3	4	1
(B)	2	4	3	3
(C)	4	1	2	3
(D)	3	2	1	4

(I.I.T. 2014)

95. P. H—≡—H

1. Scheme I

? $\xrightarrow[\text{(ii) SOCl}_2\text{ (iv) NH}_3]{\text{(i) KMnO}_4\text{, HO}^-\text{, heat (ii) H}^+\text{, H}_2\text{O}}$ $C_7H_6N_2O_3$

Q. 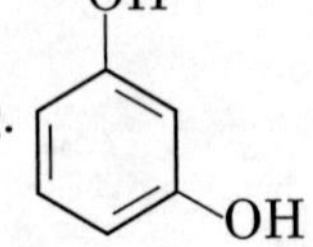

2. Scheme II

? $\xrightarrow[\text{(iv) HNO}_4\text{ (v) dil.H}_2\text{SO}_4\text{ heat (vi) HO}^-]{\text{(i) Sn/HCl (ii) CH}_3\text{COCl (iii) conc.H}_2\text{SO}_4}$ $C_6H_6N_2O_2$

R. 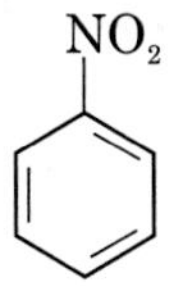

3. Scheme III

$$? \xrightarrow[\text{(iii) } H_2S,\ NH_3 \text{ (iv) } NaNO_2,\ H_2SO_4 \text{ (v) hydrolysis}]{\text{(i) red hot iron, 873 K (ii) fuming } HNO_2,\ H_2SO_4\text{, head}} C_6H_5NO_3$$

S. 

4. Scheme IV

$$? \xrightarrow[\text{(ii) conc.} HNO_3\text{, conc.} H_2SO_4 \text{ (iii) dil.} H_2SO_4\text{, heat}]{\text{(i) conc.} H_2SO_4\text{, } 60°C} C_6H_5NO_4$$

Codes:	**P**	**Q**	**R**	**S**
(A)	1	4	2	3
(B)	3	1	4	2
(C)	3	4	2	1
(D)	4	1	3	2

(I.I.T. 2014)

96. (A) $TiCl_3$ (catalyst) — (i) Wacker process
(B) $PdCl_2$ (catalyst) — (ii) Ziegler-Natta polymerization
(C) $CuCl_2$ (catalyst) — (iii) Contact process
(D) V_2O_5 (catalyst) — (iv) Deacon's process

(A) (A)–(ii), (B)–(i), (C)–(iv), (D)–(iii)
(B) (A)–(ii), (B)–(iii), (C)–(iv), (D)–(i)
(C) (A)–(iii), (B)–(i), (C)–(ii), (D)–(iv)
(D) (A)–(iii), (B)–(ii), (C)–(iv), (D)–(i)

(I.I.T. 2015)

97. (A) Carbonate ore — (P) Siderite
(B) Sulphide ore — (Q) Malachite
(C) Hydroxide ore — (R) Bauxite
(D) Oxide ore — (S) Calamine
(T) Argentite

(I.I.T. 2015)

98. (A) Freezing of water at 273 K and 1 atm — (P) $q = 0$
(B) Expansion of 1 mol of an ideal gas into a vacuum under isolated conditions — (Q) $w = 0$
(C) Mixing of equal volumes of two ideal gases at constant temperature and pressure in an isolated container — (R) $\Delta S_{sys} < 0$
(D) Reversible heating of $H_2(g)$ at 1 atm from 300 K to 600 K, followed by reversible cooking to 300 K at 1 atm — (S) $\Delta U = 0$
(T) $\Delta G = 0$

(I.I.T. 2015)

Directions (99-101): By appropriately matching the information given in the three columns of the given table, answer the questions that follow

Columns 1, 2 and 3 contain starting materials, reaction conditions and type of reactions, respectively		
Column–1	Column–2	Column–3
(I) Toluene	(i) $NaOH/Br_2$	(P) Condensation
(II) Acetophenone	(ii) $Br_2/h\nu$	(Q) Carboxylation
(III) Banzaldehyde	(iii) $(CH_3CO)_2O/CH_3COOK$	(R) Substitution
(IV) Phenol	(iv) $NaOH/CO_2$	(S) Haloform

99. For the synthesis of benzoic acid, the only correct combination is
(A) (III) (iv) (R) (B) (IV) (ii) (P) (C) (I) (iv) (Q) (D) (II) (i) (S)
(I.I.T. 2017)

100. The only correct combination in which the reaction proceeds through radical mechanism is
(A) (I) (ii) (R) (B) (II) (iii) (R) (C) (III) (ii) (P) (D) (IV) (i) (Q)
(I.I.T. 2017)

101. The only correct combination that gives two different carboxylic acids is
(A) (IV) (iii) (Q) (B) (III) (iii) (P) (C) (II) (iv) (R) (D) (I) (i) (S)
(I.I.T. 2017)

102. **List–I** **List–II**

P. dsp^2 1. $[FeF_6]^{4-}$
Q. sp^3 2. $[Ti(H_2O)_3Cl_3]$
R. sp^3d^2 3. $[Cr(NH_3)_6]^{3+}$
S. d^2sp^3 4. $[FeCl_4]^{2-}$
5. $Ni(CO)_4$
6. $[Ni(CN)_4]^{2-}$

(A) P → 5; Q → 4, 6; R → 2, 3; S → 1 (B) P → 5, 6; Q → 4; R → 3; S → 1, 2
(C) P → 6; Q → 4, 5; R → 1; S → 2, 3 (D) P → 4, 6; Q → 5, 6; R → 1, 2; S → 3
(I.I.T. 2018)

103. The desired product X can be prepared by reacting the major product of the reactions in LIST–I with one or more appropriate reagents in LIST–II. (Given: order of migratory aptitude: aryl > alkyl > hydrogen)

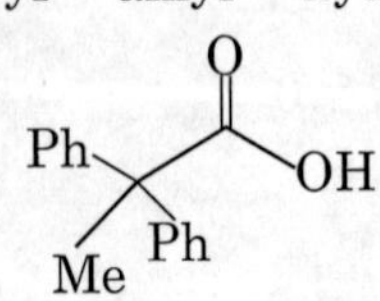

LIST–I		LIST–II
P.	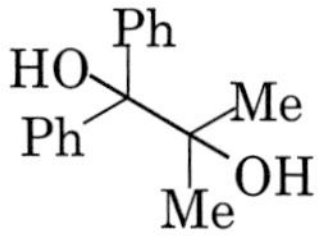 $+ H_2SO_4$	1. I_2, NaOH
Q.	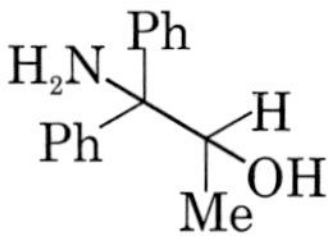 $+ HNO_2$	2. $[Ag(NH_3)_2]OH$
R.	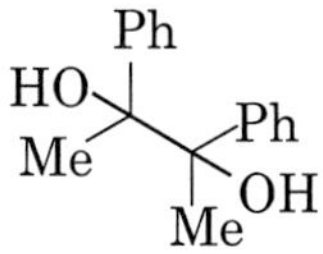 $+ H_2SO_4$	3. Fehling solution
S.	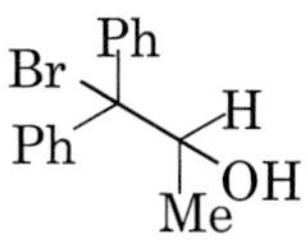 $+ AgNO_3$	4. HCHO, NaOH
		5. NaOBr

(A) P → 1; Q → 2, 3; R → 1, 4; S → 2, 4
(B) P → 1, 5; Q → 3, 4; R → 4, 5; S → 3
(C) P → 1, 5; Q → 3, 4; R → 5; S → 2, 4
(D) P → 1, 5; Q → 2, 3; R → 1, 5; S → 2, 3

(I.I.T. 2018)

104. LIST–I contains reactions and LIST–II contains major products.

LIST–I		LIST–II
P.	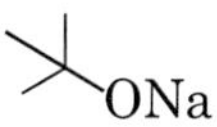 + 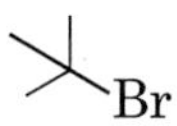 ⟶	1. 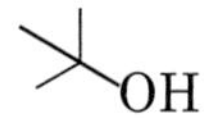
Q.	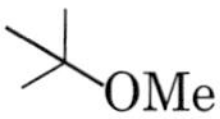 + HBr ⟶	2. 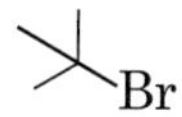
R.	 + NaOMe ⟶	3. 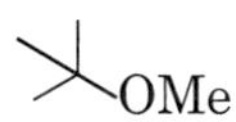
S.	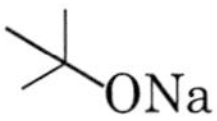 + MeBr ⟶	4. 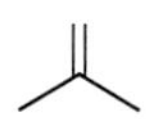
		5. 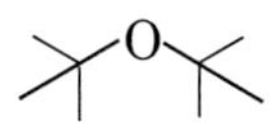

(A) P → 1, 5; Q → 2; R → 3; S → 4
(B) P → 1, 4; Q → 2; R → 4; S → 3
(C) P → 1, 4; Q → 1, 2; R → 3, 4; S → 4
(D) P → 4, 5; Q → 4; R → 4; S → 3, 4

(I.I.T. 2018)

105. Dilution processes of different aqueous solutions, with water, are given in LIST–I. The effects of dilution of the solutions on $[H^+]$ are given in LIST–II.

(Note: Degree of dissociation (α) of weak acid and weak base is << 1; degree of hydrolysis of salt << 1; $[H^+]$ represents the concentration of H^+ ions)

LIST–I	LIST–II
P. (10 mL of 0.1*M* NaOH + 20 mL of 0.1*M* acetic acid) diluted to 60 mL	1. the value of $[H^+]$ does not change on dilution
Q. (20 mL of 0.1*M* NaOH + 20 mL of 0.1*M* acetic acid) diluted to 80 mL	2. the value of $[H^+]$ changes to half of its initial value on dilution
R. (20 mL of 0.1*M* HCl + 20 mL of 0.1*M* ammonia solution) diluted to 80 mL	3. the value of $[H^+]$ changes to two times of its initial value on dilution
S. 10 mL saturated solution of $Ni(OH)_2$ in equilibrium with excess solid $Ni(OH)_2$ is diluted to 20 mL (solid $Ni(OH)_2$ is still present after dilution).	4. the value of $[H^+]$ changes to $\frac{1}{\sqrt{2}}$ times of its initial value on dilution
	5. the value of $[H^+]$ changes to $\sqrt{2}$ times of its initial value on dilution

(A) P → 4; Q → 2; R → 3; S → 1
(B) P → 4; Q → 3; R → 2; S → 3
(C) P → 1; Q → 4; R → 5; S → 3
(D) P → 1; Q → 5; R → 4; S → 1

(I.I.T. 2018)

ANSWER KEYS

1. A, d	B, a	C, e	D, b	E, c	
2. A, e	B, a	C, d	D, b	E, c	
3. A, e	B, b	C, d	D, c	E, a	
4. A, c	B, e	C, a	D, b	E, d	
5. A, d	B, a	C, e	D, c	E, b	
6. A, e	B, d	C, a	D, b	E, c	
7. A, b	B, d	C, f	D, e	E, c	F, a
8. A, d	B, a	C, b	D, e	E, c	
9. A, d	B, a	C, e	D, b	E, c	
10. A, d	B, a	C, b	D, c	E, e	
11. A, b	B, c	C, d	D, a	E, e	
12. A, d	B, e	C, a	D, b	E, c	
13. A, d	B, e	C, a	D, b	E, c	
14. A, d	B, c	C, e	D, a	E, b	
15. A, d	B, c	C, e	D, a	E, b	
16. A, d	B, e	C, a	D, b	E, c	
17. A, d	B, a	C, b	D, e	E, c	
18. A, d, v	B, c, iv	C, e, i	D, a, iii	E, b, ii	
19. A, c, v	B, a, ii	C, b, iv	D, e, iii	E, d, i	

20. A. d, v B, a, iv C, b, iii D, e, i E, c, ii
21. A, c, iv B, c, i C, a, ii D, b, iii E, d, v
22. A, d B, e C, b D, a E, c
23. A, c B, d C, e D, a E, b
24. A, a B, d C, c D, b E, e
25. A, e B, a C, b D, d E, c
26. A. d B, c C, b D, a
27. A, d B, c C, e D, b E, a F, f
28. A, vii B, iv C, i D, iii E, vi F, v G, ii
29. A, iii B, v C, iv D, ii E, i
30. A, vii B, iv C, v D, i E, vi F, ii G, iii
31. A, ii B, iii C, iv D, i
32. A, iv B, ii C, i D, iii
33. A, i B, iii C, ii D, v E, iv
34. A, i, iii B, i, iii C, ii D, iv
35. A, ii B, iv C, i D, ii, iii
36. A, iii B, ii C, i D, iv
37. A, v B, i C, iv D, ii E, iii
38. A, v B, i C, ii D, iii E, vi F, iv
39. A, iv B, v C, ii D, iii E, i
40. A, vi B, iv C, iii D, v E, ii F, i
41. A, ii B, i C, iv D, v E, vi F, iii
42. A, iii B, i C, iv D, ii
43. A, ii B, iii C, i D, iv
44. A, iii B, i C, ii D, v E, iv
45. A, iv B, ii C, v D, iii E, i
46. A, iii B, ii C, iv D, i E, v
47. A, v B, iv C, iii D, i E, ii
48. A, v B, i C, iv D, ii E, iii
49. A, b, iii B, d, v C, e, i D, c, iv E, a, ii
50. A, c, i B, b, iii C, a, ii D, d, v E, e, iv
51. A, c, ii B, a, iv C, b, i D, d, iii E, e, v
52. A, c, i B, e, v C, b, iii D, d, ii E, a, iv
53. A, a, ii B, c, i C, d, iii D, e, iv E, b, v
54. A, c, ii B, e, i C, b, iii D, d, v E, a, iv
55. A, d, ii B, c, i C, e, iii D, b, v E, a, iv
56. A, e, iv B, d, i C, b, ii D, c, v E, a, iii
57. A, e, ii B, d, v C, a, iv D, c, i E, b, iii

58. A, e, iii	B, a, iv	C, b, v	D, c, i	E, d, ii	
59. A, c, v	B, e, iv	C, a, iii	D, b, i	E, d, ii	
60. A, v	B, iv	C, ii	D, iii	E, i	
61. A, iii	B, v	C, i	D, ii	E, iv	
62. A, iii	B, iv	C, ii	D, i	E, v	
63. A, v	B, iii	C, iv	D, ii	E, i	
64. A, v	B, iv	C, i	D, iii	E, ii	
65. A, ii	B, v	C, iv	D, iii	E, i	
66. A, vi	B, iv	C, v	D, v, vii	E, viii	F, iii
67. A, ii	B, iii	C, v	D, i	E, iv	
68. A, vi	B, i	C, iii	D, ii	E, v	F, iv
69. A, ii	B, iii	C, iv	D, i	E, v	
70. A, ii	B, i	C, iii			
71. A, ii	B, ii	C, ii	D, i, iv		
72. A, ii	B, iv	C, vi	D, i	E, v	F, iii
73. A, i	B, iv	C, vi	D, iii	E, v	F, ii
74. A, ii	B, iii	C, i	D, v	E, iv	
75. A, v	B, ii	C, i	D, iii	E, iv	
76. A, iv	B, iii	C, i	D, v	E, ii	
77. A, i, ii, iv	B, i, iii, iv	C, ii, iv	D, ii, iv		
78. A, i, iv	B, iii	C, i, ii	D, i, iv		
79. A, i, iv	B, ii, iii	C, i, iii	D, iv		
80. A, p, q, s, t	B, s, t	C, p	D, r		
81. A, p, r, t	B, s, t	C, p, q	D, p, q, s		
82. A, r, s, t	B, t	C, p, q	D, r		
83. A, p, s	B, p, q, r, t	C, p, q, t	D, p		
84. A, p, r, s	B, r, s	C, t	D, p, q, t		
85. A, r, s, t	B, p, s	C, r, s	D, q, r		
86. A, p, s	B, q, s	C, r, t	D, q, t		
87. A, p, q, t	B, p, s, t	C, r, s	D, p		

88. D
89. A
90. A
91. D
92. B
93. C
94. A
95. C

96. A

97. (A) → (P, Q, S); (B) → (T); (C) → (Q, R); (D) → (R)

98. (A) → (R, T); (B) → (P, Q, S); (C) → (P, Q, S); (D) → (P, Q, S, T)

99. D

100. A

101. B

102. C

103. D

104. B

105. D

Appendix–II: Long Form of the Periodic Table

Atomic number → 6
Symbol → C
12.011 ← Atomic mass

GROUP NUMBER

IUPAC (1984)	1	2	3	4	5	6	7	8	9	10	11	12	13	14	15	16	17	18
Period number																		
1	1 H 1.0079																	2 He 4.0026
2	3 Li 6.941	4 Be 9.01218											5 B 10.811	6 C 12.011	7 N 14.0067	8 O 15.9994	9 F 18.9984	10 Ne 20.179
3	11 Na 22.98977	12 Mg 24.305											13 Al 26.98154	14 Si 28.0855	15 P 30.9734	16 S 32.066	17 Cl 35.453	18 Ar 39.948
4	19 K 39.0983	20 Ca 40.08	21 Sc 44.9559	22 Ti 47.88	23 V 50.9415	24 Cr 51.996	25 Mn 54.9380	26 Fe 55.847	27 Co 58.9332	28 Ni 58.69	29 Cu 63.546	30 Zn 65.38	31 Ga 69.72	32 Ge 72.59	33 As 74.9216	34 Se 78.96	35 Br 79.909	36 K 83.80
5	37 Rb 85.4678	38 Sr 87.62	39 Y 88.9059	40 Zr 91.22	41 Nb 92.9064	42 Mo 95.94	43 Tc (99)	44 Ru 101.07	45 Rh 102.905	46 Pd 106.42	47 Ag 107.87	48 Cd 112.40	49 In 114.82	50 Sn 118.69	51 Sb 121.75	52 Te 127.60	53 I 126.904	54 Xe 131.29
6	55 Cs 132.9054	56 Ba 137.34	57 La* 138.91	72 Hf 178.49	73 Ta 180.95	74 W 183.85	75 Re 186.206	76 Os 190.2	77 Ir 192.22	78 Pt 195.09	79 Au 196.967	80 Hg 200.59	81 Tl 204.383	82 Pb 207.19	83 Bi 208.9804	84 Po (209)	85 At (210)	86 Rn (222)
7	87 Fr (223)	88 Ra 226.0254	89 Ac** 227.028	104 Rf 261.11	105 Db 262.1	106 Sg (263)	107 Bh 262.12	108 Hs (265)	109 Mt (266)	110 Ds (269)	111 Rg (272.2)	112 Cn (285)	113 Uut (284)	114 Fl (289)	115 Uup (289)	116 Lv (291)	117 Uus	118 Uuo (294)

*Lanthanoids	58 Ce 140.12	59 Pr 140.907	60 Nd 144.24	61 Pm (145)	62 Sm 150.36	63 Eu 151.96	64 Gd 157.25	65 Tb 158.92	66 Dy 162.50	67 Ho 164.9304	68 Er 167.26	69 Tm 168.93	70 Yb 173.04	71 Lu 174.967
**Actinoids	90 Th 232.0381	91 Pa 231.0359	92 U 238.0289	93 Np 237.0482	94 Pu (242)	95 Am (243)	96 Cm (247)	97 Bk (247)	98 Cf (251)	99 Es (254)	100 Fm (257)	101 Md (258)	102 No (259)	103 Lr (260)

The element 117 is not known but is included to show its expected position.

Appendix–III
Electronic Configuration of Elements

Atomic number	*Element*	1	2		3			4				5				6				7			
		s	*s*	*p*	*s*	*p*	*d*	*s*	*p*	*d*	*f*	*s*	*p*	*d*	*f*	*s*	*p*	*d*	*f*	*s*	*p*	*d*	*f*
1	H	1																					
2	He	2																					
3	Li	2	1																				
4	Be	2	2																				
5	B	2	2	1																			
6	C	2	2	2																			
7	N	2	2	3																			
8	O	2	2	4																			
9	F	2	2	5																			
10	Ne	2	2	6																			
11	Na	2	2	6	1																		
12	Mg	2	2	6	2																		
13	Al	2	2	6	2	1																	
14	Si	2	2	6	2	2																	
15	P	2	2	6	2	3																	
16	S	2	2	6	2	4																	
17	Cl	2	2	6	2	5																	
18	Ar	2	2	6	2	6																	
19	K	2	2	6	2	6		1															
20	Ca	2	2	6	2	6		2															
21	Sc	2	2	6	2	6	1	2															
22	Ti	2	2	6	2	6	2	2															
23	V	2	2	6	2	6	3	2															
24	Cr	2	2	6	2	6	5	1															
25	Mn	2	2	6	2	6	5	2															
26	Fe	2	2	6	2	6	6	2															
27	Co	2	2	6	2	6	7	2															
28	Ni	2	2	6	2	6	8	2															
29	Cu	2	2	6	2	6	10	1															
30	Zn	2	2	6	2	6	10	2															
31	Ga	2	2	6	2	6	10	2	1														

Atomic number	Element	1	2		3			4				5				6				7			
		s	s	p	s	p	d	s	p	d	f	s	p	d	f	s	p	d	f	s	p	d	f
32	Ge	2	2	6	2	6	10	2	2														
33	As	2	2	6	2	6	10	2	3														
34	Se	2	2	6	2	6	10	2	4														
35	Br	2	2	6	2	6	10	2	5														
36	Kr	2	2	6	2	6	10	2	6														
37	Rb	2	2	6	2	6	10	2	6			1											
38	Sr	2	2	6	2	6	10	2	6			2											
39	Y	2	2	6	2	6	10	2	6	1		2											
40	Zr	2	2	6	2	6	10	2	6	2		2											
41	Nb	2	2	6	2	6	10	2	6	4		1											
42	Mo	2	2	6	2	6	10	2	6	5		1											
43	Tc	2	2	6	2	6	10	2	6	5		2											
44	Ru	2	2	6	2	6	10	2	6	7		1											
45	Rh	2	2	6	2	6	10	2	6	8		1											
46	Pd	2	2	6	2	6	10	2	6	10													
47	Ag	2	2	6	2	6	10	2	6	10		1											
48	Cd	2	2	6	2	6	10	2	6	10		2											
49	In	2	2	6	2	6	10	2	6	10		2	1										
50	Sn	2	2	6	2	6	10	2	6	10		2	2										
51	Sb	2	2	6	2	6	10	2	6	10		2	3										
52	Te	2	2	6	2	6	10	2	6	10		2	4										
53	I	2	2	6	2	6	10	2	6	10		2	5										
54	Xe	2	2	6	2	6	10	2	6	10		2	6										
55	Cs	2	2	6	2	6	10	2	6	10		2	6			1							
56	Ba	2	2	6	2	6	10	2	6	10		2	6			2							
57	La	2	2	6	2	6	10	2	6	10		2	6	1		2							
58	Ce	2	2	6	2	6	10	2	6	10	1	2	6	1		2							
59	Pr	2	2	6	2	6	10	2	6	10	3	2	6			2							
60	Nd	2	2	6	2	6	10	2	6	10	4	2	6			2							
61	Pm	2	2	6	2	6	10	2	6	10	5	2	6			2							
62	Sm	2	2	6	2	6	10	2	6	10	6	2	6			2							
63	Eu	2	2	6	2	6	10	2	6	10	7	2	6			2							
64	Gd	2	2	6	2	6	10	2	6	10	7	2	6	1		2							

Atomic number	*Element*	1	2		3			4				5				6				7			
		s	*s*	*p*	*s*	*p*	*d*	*s*	*p*	*d*	*f*	*s*	*p*	*d*	*f*	*s*	*p*	*d*	*f*	*s*	*p*	*d*	*f*
65	Tb	2	2	6	2	6	10	2	6	10	9	2	6			2							
66	Dy	2	2	6	2	6	10	2	6	10	10	2	6			2							
67	Ho	2	2	6	2	6	10	2	6	10	11	2	6			2							
68	Er	2	2	6	2	6	10	2	6	10	12	2	6			2							
69	Tm	2	2	6	2	6	10	2	6	10	13	2	6			2							
70	Yb	2	2	6	2	6	10	2	6	10	14	2	6			2							
71	Lu	2	2	6	2	6	10	2	6	10	14	2	6	1		2							
72	Hf	2	2	6	2	6	10	2	6	10	14	2	6	2		2							
73	Ta	2	2	6	2	6	10	2	6	10	14	2	6	3		2							
74	W	2	2	6	2	6	10	2	6	10	14	2	6	4		2							
75	Re	2	2	6	2	6	10	2	6	10	14	2	6	5		2							
76	Os	2	2	6	2	6	10	2	6	10	14	2	6	6		2							
77	Ir	2	2	6	2	6	10	2	6	10	14	2	6	7		2							
78	Pt	2	2	6	2	6	10	2	6	10	14	2	6	9		1							
79	Au	2	2	6	2	6	10	2	6	10	14	2	6	10		1							
80	Hg	2	2	6	2	6	10	2	6	10	14	2	6	10		2							
81	Tl	2	2	6	2	6	10	2	6	10	14	2	6	10		2	1						
82	Pb	2	2	6	2	6	10	2	6	10	14	2	6	10		2	2						
83	Bi	2	2	6	2	6	10	2	6	10	14	2	6	10		2	3						
84	Po	2	2	6	2	6	10	2	6	10	14	2	6	10		2	4						
85	At	2	2	6	2	6	10	2	6	10	14	2	6	10		2	5						
86	Rn	2	2	6	2	6	10	2	6	10	14	2	6	10		2	6						
87	Fr	2	2	6	2	6	10	2	6	10	14	2	6	10		2	6			1			
88	Ra	2	2	6	2	6	10	2	6	10	14	2	6	10		2	6			2			
89	Ac	2	2	6	2	6	10	2	6	10	14	2	6	10		2	6	1		2			
90	Th	2	2	6	2	6	10	2	6	10	14	2	6	10		2	6	2		2			
91	Pa	2	2	6	2	6	10	2	6	10	14	2	6	10	2	2	6	1		2			
92	U	2	2	6	2	6	10	2	6	10	14	2	6	10	3	2	6	1		2			
93	Np	2	2	6	2	6	10	2	6	10	14	2	6	10	4	2	6	1		2			
94	Pu	2	2	6	2	6	10	2	6	10	14	2	6	10	6	2	6			2			
95	Am	2	2	6	2	6	10	2	6	10	14	2	6	10	7	2	6			2			
96	Cm	2	2	6	2	6	10	2	6	10	14	2	6	10	7	2	6	1		2			
97	Bk	2	2	6	2	6	10	2	6	10	14	2	6	10	9	2	6			2			
98	Cf	2	2	6	2	6	10	2	6	10	14	2	6	10	10	2	6			2			
99	Es	2	2	6	2	6	10	2	6	10	14	2	6	10	11	2	6			2			

Atomic number	*Element*	1	2		3			4				5				6				7			
		s	*s*	*p*	*s*	*p*	*d*	*s*	*p*	*d*	*f*	*s*	*p*	*d*	*f*	*s*	*p*	*d*	*f*	*s*	*p*	*d*	*f*
100	Fm	2	2	6	2	6	10	2	6	10	14	2	6	10	12	2	6			2			
101	Md	2	2	6	2	6	10	2	6	10	14	2	6	10	13	2	6			2			
102	No	2	2	6	2	6	10	2	6	10	14	2	6	10	14	2	6			2			
103	Lr	2	2	6	2	6	10	2	6	10	14	2	6	10	14	2	6	1		2			
104	Rf	2	2	6	2	6	10	2	6	10	14	2	6	10	14	2	6	2		2			
105	Db	2	2	6	2	6	10	2	6	10	14	2	6	10	14	2	6	3		2			
106	Sg	2	2	6	2	6	10	2	6	10	14	2	6	10	14	2	6	4		2			
107	Bh	2	2	6	2	6	10	2	6	10	14	2	6	10	14	2	6	5		2			
108	Hs	2	2	6	2	6	10	2	6	10	14	2	6	10	14	2	6	6		2			
109	Mt	2	2	6	2	6	10	2	6	10	14	2	6	10	14	2	6	7		2			
110	Ds	2	2	6	2	6	10	2	6	10	14	2	6	10	14	2	6	8		2			
111	Rg	2	2	6	2	6	10	2	6	10	14	2	6	10	14	2	6	10		1			
112	Cn	2	2	6	2	6	10	2	6	10	14	2	6	10	14	2	6	10		2			
113	Uut†	2	2	6	2	6	10	2	6	10	14	2	6	10	14	2	6	10		2	1		
114	Fl	2	2	6	2	6	10	2	6	10	14	2	6	10	14	2	6	10		2	2		
115	Uup†	2	2	6	2	6	10	2	6	10	14	2	6	10	14	2	6	10		2	3		
116	Lv	2	2	6	2	6	10	2	6	10	14	2	6	10	14	2	6	10		2	4		
117*	Uus	2	2	6	2	6	10	2	6	10	14	2	6	10	14	2	6	10		2	5		
118	Duo†	2	2	6	2	6	10	2	6	10	14	2	6	10	14	2	6	10		2	6		

* Element yet to be discovered. † Official IUPAC name yet to be announced.

Appendix–IV
Atomic Masses of Elements

Name	*Symbol*	*Atomic number*	*Atomic mass*
Actinium	Ac	89	(227)
Aluminium	Al	13	26.9815
Americium	Am	95	(243)
Antimony	Sb	51	121.75
Argon	Ar	18	39.948
Arsenic	As	33	74.9216
Astatine	At	85	(210)
Barium	Ba	56	137.34
Berkelium	Bk	97	(247)
Beryllium	Be	4	9.0122
Bismuth	Bi	83	208.980
Bohrium	Bh	107	262.12
Boron	B	5	10.811
Bromine	Br	35	79.909
Cadmium	Cd	48	112.40
Calcium	Ca	20	40.08
Californium	Cf	98	(251)
Carbon	C	6	12.01115

Name	*Symbol*	*Atomic number*	*Atomic mass*
Cerium	Ce	58	140.12
Caesium	Cs	55	132.905
Chlorine	Cl	17	35.453
Chromium	Cr	24	51.996
Cobalt	Co	27	58.9332
Copernicium	Cn	112	(285)
Copper	Cu	29	63.54
Curium	Cm	96	(247)
Darmstadtium	Ds	110	(269)
Dubnium	Db	105	262.1
Dysprosium	Dy	66	162.50
Einstenium	Es	99	(254)
Erbium	Er	68	167.26
Europium	Eu	63	151.96
Flerovium	Fl	114	(289)
Fermium	Fm	100	(253)
Fluorine	F	9	18.9984
Francium	Fr	87	(233)
Gadolinium	Gd	64	157.25
Gallium	Ga	31	69.72
Germanium	Ge	32	72.59
Gold	Au	79	196.967
Hafnium	Hf	72	178.49
Hassnium	Hs	108	(265)
Helium	He	2	4.0026
Holmium	Ho	67	164.930
Hydrogen	H	1	1.00797
Indium	In	49	114.82
Iodine	I	53	126.904
Iridium	Ir	77	192.2
Iron	Fe	26	55.847
Krypton	Kr	36	83.80
Lanthanum	La	57	138.91
Lawrencium	Lr	103	(257)
Lead	Pb	82	207.19
Lithium	Li	3	6.939
Livermorium	Lv	116	(291)
Lutetium	Lu	71	174.97
Magnesium	Mg	12	24.312
Manganese	Mn	25	54.9380

Name	*Symbol*	*Atomic number*	*Atomic mass*
Meitnerium	Mt	109	(266)
Mendeleevium	Md	101	(256)
Mercury	Hg	80	200.59
Molybdenum	Mo	42	95.94
Neodymium	Nd	60	141.24
Neon	Ne	10	20.183
Neptunium	Np	93	(237)
Nickel	Ni	28	58.71
Niobium	Nb	41	92.906
Nitrogen	N	7	14.0467
Nobelium	No	102	(254)
Osmium	Os	76	190.2
Oxygen	O	8	15.9994
Palladium	Pd	46	106.4
Phosphorus	P	15	30.9734
Platinum	Pt	78	195.09
Plutonium	Pu	94	(242)
Polonium	Po	84	(210)
Potassium	K	19	39.102
Praseodymium	Pr	59	104.907
Promethium	Pm	61	(145)
Protactinium	Pa	91	(231)
Radium	Ra	88	(226)
Radon	Rn	86	(222)
Rhenium	Re	75	186.2
Rhodium	Rh	45	102.905
Roentgenium	Rg	111	(272.2)
Rubidium	Rb	37	85.47
Ruthenium	Ru	44	101.07
Rutherfordium	Rf	104	261.11
Samarium	Sm	62	150.35
Scandium	Sc	21	44.946
Seaborgium	Sg	106	263.118
Selenium	Se	34	78.96
Silicon	Si	14	28.086
Silver	Ag	47	107.870
Sodium	Na	11	22.9898
Strontium	Sr	38	87.62
Sulphur	S	16	32.064
Tantalum	Ta	73	180.94

Name	Symbol	Atomic number	Atomic mass
Technetium	Tc	43	(99)
Tellurium	Te	51	127.604
Terbium	Tb	65	158.92
Thallium	Tl	81	204.37
Thorium	Th	90	232.038
Thulium	Tm	69	168.93
Tin	Sn	50	118.69
Titanium	Ti	22	47.90
Tungsten	W	74	183.85
Ununpentium	Uup	115	(289)
Ununoctium	Uuo	118	(294)
Ununtrium	Uut	113	(284)
Uranium	U	92	238.03
Vanadium	V	23	50.942
Xenon	Xe	54	131.30
Yetterbium	Yb	70	173.04
Yttrium	Y	39	88.905
Zinc	Zn	30	65.37
Zirconium	Zr	40	91.22

Number in parentheses gives the mass number of the most stable isotope.

Appendix–V
Some Useful Constants and Conversion Factors

Constant	Symbol	Value
Acceleration due to gravity	g	980.665 cm s^{-2}
Atmospheric pressure	P	1.01325×10^6 dyne cm^{-2}
Avogadro number	N	6.0225×10^{23} mol^{-1}
Bohr radius	a_0	0.529×10^{-8} cm
Boltzmann constant	k	1.3805×10^{-6} erg deg^{-1} = 1.3805×10^{-23} JK^{-1}
Charge on electron	e	– 4.803×10^{-10} esu = – 1.602×10^{-19} C
Faraday	F	9.6487×10^4 C mol^{-1} = 82 mL atom/degree/mol
Gas constant	R	1.9872 cal deg^{-1} mol^{-1} = 0.082057 L atm deg^{-1} mol^{-1} = 8.314 J K^{-1} mol^{-1}
Mass of electron	m	9.1096×10^{-28} g = 5.4859×10^{-1} u
Mass of proton	–	1.6726×10^{-24} g = 1.007277 u
Mass of neutron	–	1.6749×10^{-24} g = 1.008665 u
Molar volume, ideal gas at STP	–	22.136 L
Planck constant	h	6.6256×10^{-27} erg s = 6.6256×10^{-34} J s
Rydberg constant	R_H	1.097373×10^5 cm^{-1}
Temperature	K	K = °C + 273.15°
Velocity of light in vacuum	c	2.99795×10^{10} cm s^{-1}
Velocity of α-particle	–	2.0×10^9 cm s^{-1}
Velocity of β-ray	–	2.36×10^{10} cm s^{-1}
Velocity of γ-ray	–	3.0×10^{10} cm s^{-1}

Conversion Factors

Unit	*Abbreviation*	*Equivalent/Definition*
amu	u	1.66×10^{-24} g = 1.66×10^{-31} J = 931.48 MeV
angstrom	Å	10^{-8} cm = 10^{-10} m = 0.1 nm
atmosphere	atm	760 mmHg = 760 torr 1.01325×10^{5} N m^{-2} = 1.01325×10^{5} Pa
becquerel	bq	1 dps
calorie	cal	4.184 J
cm^{-1}	cm^{-1}	1.986×10^{-23} J
curie	c	3.7×10^{10} dps
dyne	dyne	10^{-5} N
electron volt	eV	1.6022×10^{-19} J = 1.6022×10^{-12} erg = 2.3061×10^{4} cal mol^{-1} = 8066 cm^{-1}
entropy	S	4184 J kg^{-1} K^{-1}
erg	erg	10^{-7} J = 2.390057×10^{-8} cal
Faraday	F	96487 coulomb
joule	J	10^{7} erg = 0.2390057 cal
mmHg (torr)	mmHg	1.31579×10^{-3} atm
nanometer	nm	10^{-7} cm
rutherford	rd	10^{6} dps
triple point of H_2O		273.16 K
volt	V	amp ohm
watt	W	V amp

Appendix–VI
Eminent Scientists and Their Contributions

Name	*Contributions*
Alfred Rochow	Electronegativity scale
Arrhenius	Theory of electrolytic dissociation
Aston	Mass spectrograph
Avogadro	Avogadro hypothesis
Bartlett	Compounds of inert gases
Becquerel	Radioactivity
Bohr	Atomic structure
Boyle	Gas laws
Bronsted and Lowry	Acid-base concept
Brown	Brownian movement
Cavendish	Discovery of hydrogen

Name	*Contributions*
Chadwick	Discovery of neutron
Charles	Gas laws
Dalton	Atomic theory, law of partial pressures, law of multiple proportions
deBroglie	Wave nature of electron
Debye–Huckel	Theory of strong electrolytes
Dorn	Discovery of radon
Einstein	Theory of relativity, mass-energy relationship
Fajan	Polarisation of ions
Fajan, Russel and Soddy	Radiocative group displacement law
Faraday	Law of electrolysis
Gay-Lussac	Law of combining volumes
Goldberg and Waage	Law of mass action
Goldstein	Discovery of anode rays
Graham	Law of diffusion
Hann and Strassman	Nuclear fission
Helmholtz	First law of thermodynamics
Heisenberg	Uncertainty principle
Heitlor and London	Valence bond theory
Hund	Multiplicity rule
Hund and Mulliken	Molecular orbital theory
Irene Curie and F. Joliot	Artificial radioactivity
Kekule	Structure of benzene
Keyser	Adsorption
Kossel and Lewis	Electrovalency
Langmuir	Adsorption isotherm
Lavoisier	Laws of combination of mass
Lawrence	Cyclotron
Le Chatelier	Effect of *P*, *T* and C on equilibrium
Lewis	Convalency, acid-base concept
Libby	Radioactive dating
Lockyer and Frankland	Discovery of helium
Marie Curie	Discovery of radium and polonium
Maxwell	Kinetic theory of gases
Mendeleev	Periodic classification of elements
Milliken	Charge of electron
Moseley	Modern periodic law
Nernst	Electrode potential
Ostwald	Dilution law
Pauli	Exclusion principle
Pauling	Chemical bonding
Pauling and Slatter	Valence bond theory
Planck	Wave nature of light

Name	*Contributions*
Proust	Law of reciprocal proportions
Raman	Scattering of light by liquid molecules
Raoult	Relation between vapour pressure and mole fraction
Ramsay and Travers	Discovery of neon, krypton and xenon
Rayleigh and Ramsay	Discovery of argon
Ritcher	Law of definite proportions
Roentgen	Discovery of X-ray
Rutherford	Nuclear model of atom
Rutherford and Soddy	Radioactive disintegration
Schrodinger	Wave equation
Scheele	Discovery of nitrogen and chlorine
Schulze and Hardy	Coagulation of colloids by electrolytes
Seaborg	Transuranium elements
Sidgwick and Powell	Hybridisation
Smalley, Curt and Kroto	Buckminster fullerene
Soddy and Aston	Isotopes
Soddy and Fajan	Group displacement law
Sommerfield	Atomic model
Sorensen	*p*H
Sorel	Composition of ozone
Thenard	Discovery of hydrogen peroxide
Thomson	Discovery of electron, photoelectric effect
Tyndall	Scattering of light by colloidal particles
Urey	Discovery of D_2 and D_2O
van der Waals	Real gas equation
van't Hoff	Relation between rate constant, equilibrium constant and temperature
Werner	Theory of coordinate linkage
Wohler	Synthesis of first organic compound (urea)
Yukawa	Discovery of mesons

Appendix–VII
Some Important Ores/Minerals and Their Composition*

Name	*Composition*
1. Alumina	$\underline{Al}_2O_3$
2. Alunite	$K_2SO_4 \cdot \underline{Al}_2(SO_4)_3 \cdot 4\underline{Al}(OH)_3$
3. Azurite	$\underline{Cu}(OH)_2 \cdot 2\underline{Cu}CO_3$
4. Anhydrite	$\underline{Ca}SO_4$
5. Anglesite	$\underline{Pb}SO_4$

* The underlined metal may be manufactured from the ore/mineral.

Name	*Composition*
6. Argentite	$\underline{Ag}_2S$
7. Asbestos	$Ca\underline{Mg}_3(SiO_3)_4$
8. Bauxite	$\underline{Al}_2O_3\cdot 2H_2O$
9. Bengal salt petre	$\underline{K}NO_3$
10. Borax (tincal)	$Na_2\underline{B}_4O_7\cdot 10H_2O$
11. Calcite	$\underline{Ca}CO_3$
12. Calamine	$\underline{Zn}CO_3$
13. Carnallite	$KCl\cdot \underline{Mg}Cl_2\cdot 6H_2O$
14. Cassiterite	$\underline{Sn}O_2$
15. Cerussite	$\underline{Pb}CO_3$
16. Chalk (aragonite, marble, limestone)	$\underline{Ca}CO_3$
17. Chile salt petre	$\underline{Na}NO_3$
18. Chlorapatite	$\underline{Ca}_5(PO_4)_3Cl$
19. Cinnabar	$\underline{Hg}S$
20. Clay	$\underline{Al}_2O_3\cdot 2SiO_2\cdot 2H_2O$
21. Copper glance	$\underline{Cu}_2S$
22. Copper pyrites	$\underline{Cu}FeS_2$
23. Corrundum	$\underline{Al}_2O_3$
24. Cryolite	$Na_3\underline{Al}F_6$
25. Cuprite (ruby copper)	$\underline{Cu}_2O$
26. Diaspore	$\underline{Al}_2O_3\cdot H_2O$
27. Dolomite	$\underline{Mg}CO_3\cdot \underline{Ca}CO_3$
28. Epsomite (Epsom salt)	$\underline{Mg}SO_4\cdot 7H_2O$
29. Fluorapatite	$\underline{Ca}_5(PO_4)_3F$
30. Fluorspar	$\underline{Ca}F_2$
31. Galena	$\underline{Pb}S$
32. Greenockite	$\underline{Cd}S$
33. Gypsum	$\underline{Ca}SO_4\cdot 2H_2O$
34. Haematite (red)	$\underline{Fe}_2O_3$
35. Heavy spar	$\underline{Ba}SO_4$
36. Horn silver	$\underline{Ag}Cl$
37. Iron pyrites	$\underline{Fe}S_2$
38. Kieserite	$\underline{Mg}SO_4\cdot H_2O$
39. Lanarkite	$\underline{Pb}O\cdot \underline{Pb}SO_4$
40. Limonite (brown)	$\underline{Fe}_2O_3\cdot 3H_2O$
41. Magnesite	$\underline{Mg}CO_3$
42. Magnetite	$\underline{Fe}_3O_4$
43. Malachite	$\underline{Cu}(OH)_2\cdot \underline{Cu}CO_3$
44. Mica	$K_2O\cdot 3\underline{Al}_2O_3\cdot 6SiO_2\cdot 2H_2O$

	Name	*Composition*
45.	Phosphorite	$\underline{Ca}_3(PO_4)_2$
46.	Pitch blende	$\underline{U}_3O_8$
47.	Pollucite	$H_2Cs_4\underline{Al}_4(SiO_3)_9$
48.	Pyrargyrite (ruby silver)	$3\underline{Ag}_2S \cdot Sb_2O_3$
49.	Pyrolusite	$\underline{Mn}O_2$
50.	Rock salt	$\underline{Na}Cl$
51.	Schonite	$\underline{K}_2SO_4 \cdot MgSO_4 \cdot 6H_2O$
52.	Siderite	$\underline{Fe}CO_3$
53.	Sylvanite	$(\underline{Ag}, Au)Te$
54.	Sylvine	$\underline{K}Cl$
55.	Talc	$\underline{Mg}_2(Si_2O_5)_2 \cdot \underline{Mg}(OH)_2$
56.	Triphylite	$(\underline{Li}Na)_3(PO_4)_2 \cdot (FeMn)_3(PO_4)_4$
57.	Trona (natron)	$\underline{Na}H_2CO_3 \cdot NaHCO_3 \cdot 2H_2O$
58.	Whitherite	$\underline{Ba}CO_3$
59.	Willemite	$\underline{Zn}_2SiO_4$
60.	Zinc blende	$\underline{Zn}S$
61.	Zincite	$\underline{Zn}O$
62.	Zircon	$\underline{Zr}SiO_4$

Appendix–VIII
Some Important Compounds and Their Formulae*

	Compound	*Formula*
1.	Absolute alcohol	C_2H_5OH
2.	Alum	$M^I_2SO_4 \cdot M^{III}_2(SO_4)_3 \cdot 24H_2O$ $(M^I = Na^+, K^+, NH_4^+;\ M^{III} = Al^{3+}, Cr^{3+}, Fe^{3+})$
3.	Aqua fortis	HNO_3
4.	Anhydrone	$Mg(ClO_4)_2$
5.	Aspirin	*o*-$CH_3OCO—C_6H_4—COOH$
6.	Baking soda	$NaHCO_3$
7.	Baryta	BaO
8.	Baryta water	A solution of $Ba(OH)_2$
9.	Black oxide	MnO_2
10.	Bleaching power	$CaOCl_2$
11.	Blue vitriol	$CuSO_4 \cdot 5H_2O$
12.	Borax (tincal)	$Na_2B_4O_7 \cdot 10H_2O$

* For more compounds and their formulae, see Appendix-VI.

	Compound	Formula
13.	Borazole (inorganic benzene)	$B_3N_3H_6$
14.	Brine	A solution of NaCl
15.	Brown ring compound	$[Fe(H_2O)_5NO]SO_4$
16.	Butter of tin (oxymuriate of tin)	$SnCl_4 \cdot 5H_2O$
17.	Calgon	$Na_2[Na_4(PO_3)_6]$
18.	Calomel	Hg_2Cl_2
19.	Carbolic acid	C_6H_5OH
20.	Carborundum	SiC
21.	Caro's acid	H_2SO_5
22.	Caustic potash	KOH
23.	Caustic soda	NaOH
24.	Chloropicrin	$CCl_3—NO_2$
25.	Cream of tarter	KOOCCH(OH)CH(OH)COOH
26.	Chrome alum	$K_2SO_4 \cdot Cr_2(SO_4)_3 \cdot 24H_2O$
27.	Chrome yellow	$PbCrO_4$
28.	Chromyl chloride	CrO_2Cl_2
29.	Corrosive sublimate	$HgCl_2$
30.	Crystal carbonate	$Na_2CO_3 \cdot H_2O$
31.	Dead burnt	$CaSO_4$ (anhydrous)
32.	DDT	Dichlorodiphenyltrichloroethane
33.	Dry ice	Solid CO_2
34.	Ferric alum	$K_2SO_4 \cdot Fe_2(SO_4)_3 \cdot 24H_2O$
35.	Fisher salt	$K_3[Co(NO_2)_6]$
36.	Fremy salt	KHF_2
37.	Freon	CCl_2F_2
38.	Fullerene	An allotrope (C_{60}) of carbon
39.	Gammexane (BHC)	$C_6H_6Cl_6$
40.	Glauber salt	$Na_2SO_4 \cdot 10H_2O$
41.	Graphite	An allotrope of carbon
42.	Grape sugar	$C_6H_{12}O_6$ (fructose)
43.	Gun power	KNO_3 (75%) + S (12%) + charcoal (13%)
44.	Gypsom salt (gypsum)	$CaSO_4 \cdot 2H_2O$
45.	Inorganic rubber	$(PNCl_2)_n$
46.	Hair salt	$Al_2(SO_4)_3 \cdot 18H_2O$
47.	Hydrolith	CaH_2

	Compound	Formula
48.	Hypo	$Na_2S_2O_3·5H_2O$
49.	Jawelle water	A solution of KOCl
50.	Jeweller's rougue	Fe_2O_3
51.	Killed spirits	$ZnCl_2 + ZnO$
52.	King of chemicals (oil of vitriol)	H_2SO_4
53.	Laughing gas	N_2O
54.	Litharge	PbO
55.	Lithopone	$ZnS + BaSO_4$
56.	Lunar caustic	$AgNO_3$
57.	Marshal's acid	$H_2S_2O_8$ (perdisulphuric acid)
58.	Tear gas	$CCl_3—NO_2$
59.	Marsh gas	CH_4
60.	Microcosmic salt	$Na(NH_4)HPO_4·4H_2O$
61.	Milk of lime (slaked lime or lime water)	$Ca(OH)_2$
62.	Milk of magnesia	$Mg(OH)_2$
63.	Minium	Pb_3O_4
64.	Mohr salt	$FeSO_4·(NH_4)_2SO_4·6H_2O$
65.	Mosaic gold	SnS_2
66.	Muriatic acid	HCl
67.	Mustard gas	$ClCH_2CH_2SCH_2CH_2Cl$
68.	Nitre cake	$NaHSO_4$
69.	Oleum	$H_2S_2O_7$
70.	Oil of mirbane	$C_6H_5NO_2$
71.	Oil of vitriol	H_2SO_4
72.	Oil of winter green (methyl salicylate)	*o*-$HOC_6H_4COOCH_3$
73.	Oxone	Na_2O_2
74.	Pearl ash	K_2CO_3
75.	Perhydrol	30% H_2O_2
76.	Permutit (zeolite)	$Na_2O·Al_2O_3·2SiO_2·6H_2O$
77.	Pharoh's serpent	$Hg(SCN)_2$
78.	Philosopher's wool	ZnO
79.	Phosgene	$COCl_2$
80.	Plaster of Paris	$CaSO_4·0.5H_2O$
81.	Prussic acid	HCN
82.	Prussian blue	$Fe_4[Fe(CN)_6]_3$
83.	Quick lime (lime)	CaO

Compound	Formula
84. Realgar	As_2S_3
85. Red oxide	Pb_3O_4
86. Rinman's green	$[Co_2(ZnO_2)_3]$
87. Rochelle salt	NaOOCCH(OH)CH(OH)COOK
88. Rust	$Fe_2O_3 \cdot xH_2O$
89. Salt cake	Na_2SO_4 (anhydrous)
90. Scheele's green	$Cu_3(AsO_3)_2$
91. Smelling salt	$(NH_4)_2CO_3$
92. Soda ash	Na_2CO_3 (anhydrous)
93. Solammoniac	NH_4Cl
94. Spirit of wine (grain alcohol)	C_2H_5OH
95. Sugar of lead	$Pb(CH_3COO)_2$
96. Talc	$3\ MgO \cdot 4SiO_2 \cdot H_2O$
97. Tarter emetic	KOOCCH(OH)CH(OH)COO(SbO)
98. Teflon	$(C_2F_4)_n$
99. TEL	$(C_2H_5)_4Pb$
100. TNB	Trinitrobenzene (an explosive)
101. TNT	Trinitrotoluene (an explosive)
102. Turnbull's blue	$KFe[Fe(CN)_6]$
103. Ultramarine	$Na_5Al_3Si_4S_3O_{12}$
104. Vinegar	CH_3COOH (7-8%)
105. Washing soda	$Na_2CO_3 \cdot 10H_2O$
106. Water glass	Na_2SiO_3
107. White vitriol	$ZnSO_4 \cdot 7H_2O$
108. Wood spirit	CH_3OH
109. Zeisse's salt	$K[PtCl_3(C_2H_4)]$
110. Zinc white	ZnO

Appendix–IX
Some Important Compounds and Their Uses

1. Acetaldehyde: as an antiseptic in nose drops.
2. Acetaldehyde ammonia: as a rubber accelerator.
3. Acetanilide: as an antiseptic under the name antifebrin.
4. Acetic anhydride: as an acetylating agent, in the manufacture of aspirin, rayon and dyes.
5. Acetone: for storing acetylene, nail polish.
6. Alum: as a mordant in dye industry, as a styptic to stop bleeding, for sizing of paper, in tanning of leather and purification of water.
7. Alumina: as an abrasive, refractory material, for preparing cement, artificial gems, in chromatography.

8. Aluminum chloride: as a catalyst in Friedel–Crafts reaction.
9. Ammonia: as a cleaner for removal of grease in dry cleaning, in the manufacture of explosive (NH_4NO_3), fertilizers [$(NH_4)_2SO_4$, $CaNH_4(NO_3)_3$, $Ca(NH_4)_2(SO_4)_2$], $NaHCO_3$, HNO_3, artificial silk.
10. Aniline: in the synthesis of dyes and dye intermediates, sulpha drugs, antioxidants and accelerators.
11. Bleaching powder: in disinfection of water, for making wool unshrinkable.
12. Butter of tin (oxymuriate of tin): as a mordant in dyeing.
13. Calcium ammonium nitrate (CAN): as a fertilizer.
14. Calcium carbonate: as a building material (marble), flux in metallurgy, in paints, distempers, tooth powder, tooth paste, medicine for indigestion, manufacture of CO_2.
15. Calcium sulphate: in the manufacture of black board chalks, plaster of Paris, H_2SO_4, $(NH_4)_2SO_4$ and cement.
16. Carbon black: as filler in rubber industry, in printer's ink and black paints.
17. Carbon dioxide: in the manufacture of Na_2CO_3 (Solvay process), white lead, aerated water, as refrigerant (dry ice), fire extinguisher and agent for artificial respiration (carbogen).
18. Carbon monoxide: as a reducing agent, in the metallurgy of Fe and Ni, in the manufacture of producer gas and water gas.
19. Carbon tetrachloride: as solvent, fire extinguisher (pyrene), fumigant and anthelmentic (wormicide).
20. Chloretone: as sedative and hypnotic.
21. Chlorine: as germicide, disinfectant, bleaching agent, in the preparation of bleaching powder, NaOCl (antiseptic), metal chlorates, tear gas, mustard gas, $COCl_2$, DDT and gammexane.
22. Chloroform: as anaesthetic, preservative for anatomical specimens, in the preparation of chloretone (hypnotic and sedative) and chloropicrin (insecticide and tear gas).
23. Chloropicrin: as insecticide and tear gas.
24. Chloroprene: in the manufacture of synthetic rubber neoprene.
25. Copper metal: in jewellery, ornaments, copper coin, calorimeter, electrical wire and cable, for electrotyping and electroplating.
26. Copper sulphate: as insecticide, fungicide (Bordeaux mixture), in dyeing, calico printing, electrotyping, electroplating, for detection of moisture and preparation of Fehling solution.
27. Ether: as anaesthetic, solvent, in perfumes, Wurtz reaction, in the preparation of Grignard reagent.
28. Ethyl acetate: in perfumes.
29. Ethylene: as an anaesthetic.
30. Ethylene dibromide: as a petrol additive.
31. Ferric oxide (Fe_2O_3): as catalyst for conversion of CO to CO_2, polishing powder in jewellery, brown pigment.
32. Formaldehyde: in silvering of mirror, in the preparation of bakelite, urotropine, as a preservative for biological specimens [formalin (40% HCHO + 8% CH_3OH + 52% H_2O)].
33. Freon-12 (CCl_2F_2): as propellant and refrigerant.
34. Gammexane (BHC): as an insecticide.
35. Green vitriol ($FeSO_4 \cdot 7H_2O$): as insecticide, mordant in dyeing, in ink and tanning industry.
36. Heavy water (D_2O): as a moderator in nuclear reactor.
37. Hydrogen peroxide: as rocket propellant, germicide, antiseptic, oxidising agent, bleaching agent, for restoring white colour of lead painting blackened by H_2S.
38. Hydrogen sulphide: as laboratory reagent and reducing agent.
39. Hypo: in the metallurgy of Ag and Au, as a fixing agent in photography, antichlor, in volumetric analysis.

40. Iodine: in medicine for treatment of goiter, for increasing the yield of egg in poultry.
41. Iodoform: as an antiseptic.
42. Lead peroxide: as oxidizing agent, in match industry and storage battery.
43. Litharge: in paints, varnishes, in the preparation of lead glass and glazed pottery.
44. Lunar caustic: as a caustic in surgery, in ink, hair dye, for silvering of mirror.
45. Magnesium metal: as fuse in aluminothermit process, in photography as source of light, in the synthesis of Grignard reagent.
46. Magnesia: in Sorel's cement ($MgCl_2 \cdot 5MgO \cdot xH_2O$) for filling of teeth, as substitute for tiles, filler in rubber industry, insulator, antacid, basic lining in steel industry, refractory lining.
47. Magnesia alba ($xMgCO_3 \cdot yMg(OH)_2 \cdot zH_2O$): in tooth paste, tooth powder, silver polish, as laxative and antacid.
48. Magnesium carbonate: as filler in paper and rubber industry, as refractory material.
49. Magnesium sulphate: as filler in paper industry, purgative, for fire proofing of fabrics.
50. Mssicot (PbO, yellow): along with glycerol used as a cement for stone and glass.
51. Metaldehyde [$(CH_2O)_3$]: as a solid fuel in lamps.
52. Nitric acid: as oxidising agent, solvent for etching names and designs on copper, brass and bronze articles, in the preparation of dyes, perfumes, drugs, artificial silk, fertilizers [$Ca(NO_3)_2$, NH_4NO_3, $CaNH_4(NO_3)_3$], explosives (dynamite, nitroglycerine, picric acid, TNB, TNT, RDX (cyclomite), gun cotton).
53. Nitric oxide: as catalyst in lead chamber process, in the manufacture of HNO_3, for detection of oxygen.
54. Nitrous acid: in the synthesis of azo dyes.
55. Nitrous oxide (laughing gas): as anaesthetic, rocket propellant.
56. Oil of vitriol (H_2SO_4): as dehydrating agent, pickling agent, in lead storage batteries, in the manufacture of fertilizers, explosives, detergents and dyes.
57. Oxygen: as rocket fuel (liquid O_2), explosive (liquid O_2 + C), in oxy-acetylene and oxy-hydrogen flames for cutting and welding of metals, for artificial respiration [O_2 + He or O_2 + CO_2 (carbogen)].
58. Ozone: as oxidizing agent, bleaching agent, disinfectant for water, in the manufacture of camphor, artificial silk, $KMnO_4$, for detecting the position of double bond.
59. Paraldehyde [$(CH_2O)_n$]: as sporofic and hypnotic.
60. Permutit (zeolite, $Na_2O \cdot Al_2O_3 \cdot 2SiO_2 \cdot 6H_2O$): for purification of water.
61. Phenol: as antiseptic, in the preparation of aspirin, picric acid, bakelite and phenolphthalein.
62. Phosphorus: in match industry (red P), for making insecticide (AlP), rat poison (Zn_3P_2).
63. Phosphorus pentaoxide: as dehydrating agent.
64. Phenacyl chloride ($C_6H_5COCH_2Cl$): as lachrymator for dispersing mobs.
65. Potassium metal: as reducing agent, in photoelectric cells.
66. Potassium bicarbonate: as antacid, in baking powder.
67. Potassium carbonate: for washing wool, in the preparation of fusion mixture, hard glass and soft soap.
68. Potassium sulphate: as purgative, fertilizer, in the preparation of potash alum and glass.
69. Quick lime: as basic lining in steel industry, for softening of water, purification of coal gas, sugar, in the manufacture of cement, CaC_2, slaked lime, lime colours and bleaching power.
70. Silver metal: in medicines, silver plating, ornaments, coins, for filling of teeth (silver alloy).
71. Slaked lime [$Ca(OH)_2$]: in softening of water, white washing, preparation of sodalime, NaOH and bleaching powder.

72. Silver bromide: in photography.
73. Sodium metal: as coolant in nuclear reactor, thermometric liquid (Na–K alloy), in sodium vapour lamp, in the manufacture of $(C_2H_5)_4Pb$ and Na/Hg.
74. Sodium bicarbonate (baking soda): as antacid, in fire extinguishers, for the preparation of soft drinks, fruit salts, baking powder and Siedlitz powder.
75. Sodium bromide: as sedative.
76. Sodium carbonate (washing soda): in laundry, in the manufacture of soap, fusion mixture and glass.
77. Sodium hydroxide: for mercerizing cotton, adsorbent for CO_2, X_2, SO_2 and NO_2.
78. Sodium peroxide: as a source of O_2 (oxone), oxidising agent, bleaching agent (Na_2O_2 + HCl), air purifier.
79. Sodium sulphate (Glauber salt): as laxative, in the preparation of window glass and craft paper.
80. Stannous chloride: as reducing agent, mordant in dyeing, for the manufacture of purple of cassius.
81. Sulphur: in vulcanisation of rubber, skin ointments, fireworks, match industry, in the preparation of dyes and sulpha drugs.
82. Sulphur dioxide: as solvent (liquid SO_2), antichlor, bleaching agent.
83. Sulphur trioxide: for drying of gases, manufacture of H_2SO_4 and oleum.
84. Tetraethyl lead (TEL): as an antiknocking agent in petrol.
85. Tin metal: for making tin foil for wrapping, squeezable tubes for toothpastes and ointments.
86. Trilead (red lead, minium): as oxidising agent, in the manufacture of matches, paints, flint glass and silver mirror.
87. 2,2,4-Trimethylpentane (isooctane): petrol additive for decrease of knocking.
88. Urotropine (hexamethylenetetramine): as urinary antiseptic.
89. Vinyl acetate ($CH_2{=}CHOCOCH_3$): plastic industry.
90. Vinyl chloride ($CH_2{=}CHCl$): in the manufacture of PVC.
91. Westron and westrosol: as solvents.
92. White lead (basic lead carbonate, $Pb(OH)_2 \cdot 2PbCO_3$): as white paint.
93. White metal (alloy of Sn, Sb, Cu): used to make artificial jewellery.
94. Zinc metal: as reducing agent, in galvanisation, metallurgy of Pb, Ag and Au, in the preparation of Zn–Cu couple.
95. Zinc oxide (zinc white): as a white paint.
96. Zinc sulphide: as luminous paint in dials of watches, in X-ray screens, in the manufacture of lithopone (ZnS + $BaSO_4$).

Appendix–X
Some Important Reagents and Mixtures

1. Aqua regia	Conc. HNO_3 + conc. HCl (1:3) (used as a laboratory reagent)
2. Amatol	NH_4NO_3 (80%) + trinitrotoluene (20%) (used as an explosive)
3. Ammonal	Al powder + NH_4NO_3 (used as an explosive)
4. Baking powder	$NaHCO_3$ + sodium potassium tartrate

5. Baeyer reagent	Alkaline $KMnO_4$ solution (used for detecting ethylenic and acetylenic linkages)
6. Benedict solution	A solution of $CuSO_4 \cdot 5H_2O$ + NaOH + sodium citrate (used for detecting aldehydes)
7. Black ash	Na_2CO_3 + CaS (impure Na_2CO_3 obtained in Leblanc process)
8. Bordeaux mixture	A solution of $CuSO_4$ + lime (used to kill moulds and fungi on plants)
9. Carbogen	O_2 (90–95%) + CO_2 (5–10%) (used for artificial respiration)
10. Carbon oil	Vegetable oil + lime water (used for treatment of burns)
11. Coal gas	H_2 (47%) + CH_4 (32%) + CO (7%) + N_2 (4%) + C_2H_4 (3%) + C_2H_2 (2%) + CO_2 (1%) + other gases (4%) (used to produce reducing atmosphere in metallurgical operations)
12. Euchlorine	Cl_2 + ClO_2
13. Fenton reagent	H_2O_2 + few drops of $FeCl_3$
14. Fehling solution	$CuSO_4 \cdot 5H_2O$ + NaOH + sodium potassium tartrate (used for detecting aldehydes)
15. Freezing mixture	NaCl + ice (used for lowering temperature)
16. Fusion mixture	Na_2CO_3 + K_2CO_3 (used as a laboratory reagent)
17. Gobar gas	CH_4 + CO + H_2 (used as a domestic fuel)
18. Granite	Mica + rock + clay + sand
19. Gun powder	KNO_3 (75%) + S (12%) + charcoal (13%) (used as an explosive)
20. Ignition mixture	BaO_2 + Mg + Al (used in aluminothermit process)
21. Jones reagent	Chromic acid in aqueous acetone solution (used for oxidising 1° alcohol to aldihyde and 2° alcohol to ketone without affecting C=C bond)
22. Lindlar catalyst	Pd/$BaSO_4$, S (used for hydrogenation of alkyne to alkene)
23. Lithopone	ZnS + $BaSO_4$ (used as a white paint)
24. Lucas reagent	Conc. HCl + anhy. $ZnCl_2$ (used for distinguishing three types of alcohols)
25. Matte	Cu_2S + FeS
26. Methylated spirit	Rectified spirit (85–90%) + CH_3OH (10–15%) + pyridine + acetone (used as a solvent)
27. Milk of magnesia	An aqueous suspension of $Mg(OH)_2$ (used as an antacid)
28. Molish reagent	An ethanolic solution of α-naphthol (used for detecting carbohydrates)
29. Mortar	(slaked lime + sand) (1 : 3 in water)
30. Nessler reagent	A solution of K_2HgI_4 + KOH (used for detecting NH_4^+ ion)
31. Nitro chalk	NH_4NO_3 + $CaCO_3$ (used as a fertilizer)
32. Nitrolim	$CaCN_2$ + graphite (used as a fertilizer)
33. Nitrophos	$Ca(H_2PO_4)_2$ + $Ca(NO_3)_2$ (used as a fertilizer)
34. Oil gas	H_2 (50–55%) + CH_4 (25–30%) + CO (10–12%) + CO_2 (3%) (used in laboratory)
35. Power alcohol	(Petrol + C_2H_5OH) (4 : 1) + little benzene (used as a motor fuel)
36. Producer gas	N_2 (52–55%) + CO (22–30%) + H_2 (8–12%) + CO_2 (3%) (used as a fuel)
37. Purple of cassius	Colloidal sol of Au + $Sn(OH)_4$ (used for colouring of glass and pottery red)

38. Rectified spirit	C_2H_5OH (95.87%) + H_2O (4.13%) (used as a solvent)
39. Schiff reagent	An aqueous solution of rosaniline hydrochloride whose red colour has been discharged by SO_2 (used for detecting aldehydes)
40. Schweitzer reagent	$[Cu(NH_3)_4]SO_4$ (used in the manufacture of artificial silk)
41. Soda bleach	Na_2O_2 + HCl (used for bleaching of fabrics)
42. Sodalime	NaOH + $Ca(OH)_2$ (used in decarboxylation of carboxylic acids)
43. Sorel's cement	MgO + $MgCl_2$ (used as a substitute for tiles)
44. Sublimed white lead	PbO + $PbSO_4$ + ZnO (used as a white paint)
45. Superphosphate of lime	$Ca(H_2PO_4)_2$ + $CaSO_4$ (used as a fertilizer)
46. Thermite mixture	Al powder + metal oxide (used in metallurgy)
47. Tincture of iodine	I_2 + NaI + C_2H_5OH + water (used as an antiseptic)
48. Thomas slag	$Ca_3(PO_4)_2$, CaO (used as a fertilizer)
49. Water gas	H_2(51%) + CO(41%) + N_2(4%) + CO_2(4%) (used for manufacture of CH_3OH)

Appendix–XI
Some Important Facts

Most electronegative element	F
Least electronegative element	Cs
Element having highest ionisation energy	He
Element having lowest ionisation energy	Cs
Element having highest electron affinity	Cl
Element having lowest electron affinity	Noble gases
Most reactive non-mental	F
Most reactive metal	Fr
Lightest metal	Li (density = 0.534 g/mL)
Heaviest metal	Os (density = 22.69 g/mL)
Most electropositive element	Fr
Least electropositive element	F
Metal having lowest M. P.	Hg (M. P. = – 38.9°C)
Metal having highest M. P.	W (M. P. = 3410°C)
Non-metal having lowest M. P. and B. P.	He (B. P. = – 268.9°C)
Most stable element	Te ($t_{½} = 2 \times 10^{21}$ years)
Element having smallest atomic size	H
Element having largest atomic size	Cs
Smallest anion	H^-
Largest anion	At^-
Smallest cation	H^+
Largest cation	Cs^+

First man-made element	Tc
Most poisonous metal	Pu
Liquid non-metal	Bromine
Liquid metal	Hg, Ga, Cs, Fr
Best conductor among non-metals	Graphite
Least conductor among non-metals	S
Best conductor among metals	Ag
Least conductor among metals	Pb
Non-metal having highest M. P. and B. P.	Diamond
Hardest element	Diamond
Semi-metal elements	Si, Ge, As, Sb, Se, Te
Coinage metals	Cu, Ag, Au
Noble metals	Cu, Ag, Au, Rh, Ir, Ru, Os, Pd, Pt
Strongest oxidising agent	F_2
Strongest reducing agent	Li
Most malleable metal	Au
Most abundant element in the earth's crust	Oxygen
Rarest element in the earth's crust	At
Element exhibiting highest oxidation state	Ru (Ru^{8+}), Os (Os^{8+})
Polymorphic elements	O, S, P
Element having maximum number of naturally occurring isotopes	Sn (10)
Heaviest naturally occurring element	$^{228}_{92}U$
Element which sublimes on heating	Iodine
Element which is stored under water	Phosphorus
Element which is stored under kerosene	Na, K, Rb, Cs
Most abundant metal	Al
Most abundant element in the universe	Hydrogen
Element having maximum tendency for catenation	C
Element having highest tensile strength	B
Liquid radioactive element	Fr
Element which do not have neutron	$^{1}_{1}H$
Metal exhibiting strongest covalency	Ru, Os
Amphoteric non-metal	Si
Amphoteric metal	Zn, Al, Sn, Pb, Cr, As
Elements showing diagonal relationship	Li–Mg; Be–Al; B–Si
Non-metals having metallic lustre	Iodine, graphite, diamond
Heaviest solid non-metal	At
Lightest solid non-metal	B

Heaviest gaseous non-metal	Rn
Lightest gaseous non-metal	Hydrogen
Heaviest semi-metal	Te
Lighest semi-metal	Si
Last discovered element in the periodic table	Uuo (118)
Latest discovered element	Uup (115)
Volatile *d*-block elements	Zn, Cd, Hg
Most commonly used metal	Fe
Element with electronegativity next to fluorine	Oxygen
Total number of liquid element in the periodic table	3 (Br, Fr, Hg)
Total number of gaseous elements in the periodic table	12 (H_2, He, N_2, O_2, F_2, Cl_2, Ne, Ar, Xe, Kr, Rn, Uuo)
Total number of solid elements in the periodic table	102 (= 117 – 3 – 2)
Number of metals in the periodic table	93 (= 117 – 18 – 6)
Number of non-metals in the periodic table	18
Group containing maximum number of gaseous elements	Group 18
Most abundant gas	N_2
Largest group in the periodic table	Group 13
Element which constitutes 90% of the mass of sun	H_2
Recipient of first Nobel prize in chemistry	van't Hoff
Lowest refractive index	He (1.000036)
Highest refractive index	Diamond (2.4)
Coolant in nuclear reactors	D_2O
Neutral oxides of non-metals	CO, NO, N_2O, H_2O
Amphoteric oxides	ZnO, PbO, BeO, Al_2O_3, SnO, V_2O_5
Oldest known organic acid	CH_3COOH
First organic compound synthesised from its elements	CH_3COOH
Natural explosive	NCl_3
Artificial explosive	Dynamite
Dry ice	$CO_2(s)$
Dry bleacher	O_3
Some isomorphous substances	$FeSO_4 \cdot 7H_2O$, $MgSO_4 \cdot 7H_2O$, $ZnSO_4 \cdot 7H_2O$
Some efflorescent substances	$Na_2CO_3 \cdot 10H_2O$, $Na_2SO \cdot 10H_2O$, $MgSO_4 \cdot 7H_2O$, $Na_2S_2O_3 \cdot 5H_2O$, $FeSO_4 \cdot 7H_2O$
Some commonly used oxidising agents	H_2O_2, SO_2, SO_3, Cl_2, O_3, $KMnO_4$, $K_2Cr_2O_7$, HNO_3
Some commonly used reductants	H_2S, SO_2, H_2SO_3, $SnCl_2$, H_2, CO, $LiAlH_4$, C
Some commonly used bleaching agents	H_2O_2, SO_2, Cl_2, bleaching powder
Compound used to disperse mobs (tear gas)	Phenacyl chloride ($C_6H_5COCH_2Cl$)

Compounds used as war gases	Phosgene ($COCl_2$), mustard gas, lewisite
Liquid silver	Hg
Mosaic gold	SnS_2
White gold	Pt
Fool's gold	FeS_2
Liquid gold	Petrol
Rolled gold	An alloy of Cu and Al
Fulminating gold	$H_2N—Au═NH$
Modern periodic table is based upon concept put forward by	Moseley
Long form of periodic table was put forward by	Bohr
Most stable carbonate	Cs_2CO_3
Strongest alkali	CsOH
Strongest basic oxide	Cs_2O
Strongest hydrogen bond is present in	HF_2^- (BDE = 113 kJ mol^{-1})
Compound with highest ionic character	CsF
Compound with highest covalent character	H_2, O_2, N_2, Cl_2, Br_2, I_2
Softest form of carbon	Lamp black
Strongest acid	HSO_3F + 90% SbF_5 called magic acid
Most reactive form of P	White
Least reactive form of P	Red
Most stable allotrope of phosphorus	Black (due to layer structure)
Purest form of silica	Quartz
Recently discovered allotrope of cabon	Buckminsterfullerene
Noble gases which form compounds	Xe, Kr
Most recent element named by IUPAC	Lv (116)

Appendix–XII
Important Industrial Processes

Baeyer process	Manufacture of Al
Bessemer process	Manufacture of steel
Birkeland–Eyde process	Manufacture of HNO_3
Bosch process	Manufacture of H_2
Carter process	Manufacture of white lead (basic lead carbonate)
Castner process	Manufacture of Na
Castner–Kellner process	Manufacture of NaOH
Contact process	Manufacture of H_2SO_4
Cupellation process	Purification of Ag
Deacon process	Manufacture of Cl_2

Dow process	Manufacture of phenol
Down process	Manufacture of Na
Goldschmidt process	Thermite welding
Haber process	Manufacture of NH_3
Hasenclever process	Manufacture of bleaching powder
Hoop process	Purification of Al
Lead chamber process	Manufacture of H_2SO_4
Leblanc process	Manufacture of K_2CO_3
L. D. process	Manufacture of steel
MacArthur–Forest process (cyanide process)	Manufacture of Ag
Merck process	Manufacture of H_2O_2
Nelson process	Manufacture of NaOH
Open hearth process	Manufacture of steel
Ostwald process	Manufacture of HNO_3
Parke process	Manufacture of Ag
Pattinson process	Manufacture of Ag
Serpeck process	Manufacture of Al
Solvay ammonia soda process	Manufacture of Na_2CO_3 and $NaHCO_3$

Appendix–XIII
Catalysts Used in Some Industrial Processes

Process	*Catalyst*
Bergius process for manufacture of petrol from coal gas	Nickel oleate
Bosch process for manufacture of H_2	$Fe_2O_3 + Cr_2O_3$
Contact process for manufacture of H_2SO_4	V_2O_5 or Pt
Deacon process for manufacture of Cl_2	$CuCl_2$
Dehydrogenation of alcohol to aldehyde	Cu
Fischer–Tropsch process for manufacture of petrol from CO and H_2	Co + ThO_2 + MgO + keiselguhr
Friedel–Crafts process for alkylation of aromatic hydrocarbons	Anhydrous $AlCI_3$
Hydrogenation of alkenes	$(Ph_3P)_3RhCl$ (Wilkinson's catalyst)
Hydrogenation of alkynes to alkanes	Ni
Hydrogenation of alkynes to alkenes	Pd–$BaSO_4$, S
Hydrogenation of oils to fats	Ni
Lead chamber process for manufacture of H_2SO_4	NO
Manufacture of C_2H_5OH from starch	Diastase + maltase
Manufacture of C_2H_5OH from molasses	Invertase + zymase
Manufacture of CH_3OH from CO and H_2	ZnO + Cr_2O_3
Manufacture of CH_3COOH from CH_3CHO and O_2	V_2O_5

Manufacture of HCHO from CO and H_2	Cu
Manufacture of O_2 from $KClO_3$	MnO_2
Manufacture of phthalic acid from naphthalene	$HgSO_4$
Ostwald process for manufacture of HNO_3	Pt
Polymerisation of alkenes	H_3PO_4 + keiselguhr
Polymerisation of 1-alkenes	$(C_2H_5)_3Al + TiCl_4$ (Zeigler–Natta catalyst)

Appendix–XIV
Solubility of Some Common Inorganic Compounds in Water

Water Soluble

Acetate ($C_2H_3O_2^-$)	All acetates are soluble.
Bromide (Br^-)	All bromides are soluble except AgBr, $PbBr_2$, Hg_2Br_2 and $HgBr_2$.
Chloride (Cl^-)	All chlorides are soluble except AgCl, $PbCl_2$ and Hg_2Cl_2.
Nitrate (NO_3^-)	All nitrates are soluble.
Iodide (I^-)	All iodides are soluble except AgI, PbI_2, Hg_2I_2 and HgI_2.
Sulphate (SO_4^{2-})	All sulphates are soluble except Ag_2SO_4, $PbSO_4$, Hg_2SO_4, $CaSO_4$, $BaSO_4$ and $SrSO_4$.

Water Insoluble

Carbonate (CO_3^{2-})	All carbonates are insoluble except those of Group 1 elements and $(NH_4)_2CO_3$.
Hydroxide (OH^-)	All hydroxides are insoluble except those of Group 1 elements, $Ca(OH)_2$, $Ba(OH)_2$ and $Sr(OH)_2$.
Phosphate (PO_4^{3-})	All phosphates are insoluble except those of Group 1 elements and $(NH_4)_3PO_4$.
Sulphide (S^{2-})	All sulphides are insoluble except those of Group 1 and 2 elements and $(NH_4)_2S$.

Appendix–XV
Some Common Adsorbent for Gases

Gas	*Adsorbent*
O_2	Alkaline pyrogallol
O_3	Turpentine oil
N_2	Heated magnesium
NO	A solution of $FeSO_4$
CO, C_2H_2	Ammoniacal solution of Cu_2Cl_2
CO_2, SO_2, F_2, Cl_2, Br, I_2, NO_2, acidic oxides	NaOH or KOH solution
NH_3, $H_2O(g)$	Conc. H_2SO_4
$H_2O(g)$	$CuSO_4$ (anhydrous), $CaCl_2$ (anhydrous). P_4O_{10}, silica gel

Appendix–XVI
Some Common Gases and Their Characteristics

Gas	*Formula*	*Characteristics*
Ammonia	NH_3	Colourless, pungent odour, extremely soluble in water
Bromine	Br_2	Reddish brown, pungent and suffocating odour, toxic, sparingly soluble in water, fairly soluble in organic solvents
Carbon dioxide	CO_2	Colourless, odourless, toxic, fairly soluble in water
Carbon monoxide	CO	Colourless, odourless, toxic, sparingly soluble in water
Chlorine	Cl_2	Greenish yellow, choking and irritating smell, toxic, sparingly soluble in cold water and insoluble in hot water
Hydrogen chloride	HCl	Choking odour, corrosive, toxic, fairly soluble in water
Hydrogen cyanide	HCN	Slight odour of bitter almonds, very toxic
Hydrogen sulphide	H_2S	Odour of rotten egg, very toxic
Iodine	I_2	Violet, sparingly soluble in water, fairly soluble in organic solvents
Methane	CH_4	Colourless, odourless, inflammable
Nitric oxide	NO	Colourless, sparingly soluble in water, paramagnetic
Nitrogen dioxide	NO_2	Reddish brown, irritating odour
Nitrous oxide	N_2O	Colourless, sweet odour, laughing gas
Oxygen	O_2	Colourless, odourless, tasteless, sparingly soluble in water, paramagnetic, non-inflammable
Ozone	O_3	Pale blue, fishy smell, toxic, sparingly soluble in water, mercury loses its meniscus in contact with ozone
Sulphur dioxide	SO_2	Colourless, irritating odour

Appendix–XVII
Important Alloys and Their Composition

Alloy	*Composition*
Alclad	Al (95%) + Cu (4%) + Mg (0.5%) + Mn (0.5%)
Alnico	Fe (50%) + Al (20%) + Ni (20%) + Co (10%)
Aluminium bronge	Cu (90%) + Al (10%)
Babbitt metal	Sn (89%) + Sb (7.3%) + Cu (3.7%)
Bearing metal	Pb (98.66%) + Ca (0.7%) + Na (0.6%) + Li (0.04%)
Bell metal	Cu (78%) + Sn (22%)
Brass	Cu (60–80%) + Zn (20–40%)
Britania metal	Sn (86%) + Sb (12%) + Cu (2%)
Bronze	Cu (75-90%) + Sn (10–25%)
Constantan	Cu (60%) + Ni (40%)

Copper coin (brown)	Cu (95%) + Sn (4%) + Zn (1%)
Delta metal	Cu (60%) + Zn (38%) + Fe (2%)
Duralumin	Al (92.5–95.5%) + Cu (3.5–4.5%) + Mg (0.3–1.0%) + Mn (0.4–1.0%) + Si (0.3–1.0%)
Dutch metal	Cu (80%) + Zn (20%)
Elektron	Mg (95%) + Zn (4.5%) + Cu (0.5%)
Ferroaluminium	Fe (80%) + Al (20%)
Ferromanganese	Mn (70–80%) + Fe (20–30%)
Ferromdybdenum	Fe (35–40%) + Mo (55–65%)
Ferrovanadium	Fe (39.5–63.5%) + V (35–45%) + C (0.2–3.5%)
German silver	Cu (25–60%) + Zn (25–35%) + Ni (10–15%)
Gold 22 carat	Au (91.7%) + Ag (4.15%) + Cu (4.15%)
Gun metal	Cu (88%) + Sn (10%) + Zn (2%)
Invar	Fe (63.8%) + Ni (36%) + C (0.2%)
Magnalium	Al (70–90%) + Mg (10–30%)
Magnolia	Pb (85%) + Sb (15%)
Manganese bronze	Cu (55–65%) + Mn (16–17%) + Zn (20–30%) + Fe (0.5–2%)
Manganin	Cu (~ 84%) + Mn (~ 12%) + Ni (~ 4%) + Fe (trace)
Misch metal	Ce (50%) + La (25%) + Nd (10%) + Fe (4.5%) + (Ca, C, Si) (10.5%)
Monel metal	Ni (~ 67%) + Cu (~ 30%) + (Fe + Mn) (~ 3%)
Muntz metal	Cu (60%) + Zn (40%)
Nichrome	Ni (~ 60%) + Fe (~ 24%) + Cr (~ 16%) + C (0.1%)
Nickel coinage	Cu (80%) + Ni (20%)
Permalloy	Ni (78%) + Fe (21%) + C (1%)
Pewter	Sn (85%) + Cu (7%) + Bi (6%) + Sb (1.7%)
Phosphor bronze	Cu (79.7%) + Sn (10%) + Sb (9.5%) + P (0.8%)
Rhine metal	Sn (97%) + Cu (3%)
Rose metal	Bi (50%) + Pb (27.1%) + Sn (22.9%)
Solder	Sn (~ 50%) + Pb (~ 50%)
Spiegeleisen	Fe (80–85%) + Mn (15–20%) + C (trace)
Stainless steel	Fe (71–75%) + Cr (17–19%) + Ni (7–9%)+ C (~0.2%)
Tin foil	Sn (88%) + Pb (8%) + Cu (3.5%) + Sb (0.5%)
Tin sel	Sn (60%) + Pb (40%)
Silicon bronze	Cu (97%) + Si (3%)
Type metal	Pb (60%) + Sb (30%) + Sn (10%)
White metal	Sn (82%) + Sb (12%) + Cu (6%)
Wood metal	Bi (50%) + Pb (25%) + Sn (12.5%) + Cd (12.5%)
γ alloy	Al (92%) + Cu (4%) + Ni (2%) + Mg (2%)

Appendix–XVIII
Some Important Name Reactions

Name	Starting material	Reaction condition	Synthesis of
1. Aldol condensation	aldehyde having α-hydrogen atom	conc. NaOH	aldol and ketol
2. Beckmann rearrangement	aromatic ketoxime	180–200°C, PCl_5 or P_4O_{10} or SO_2Cl_2 or H_2SO_4	*N*-substituted acetamide
3. Benzoin condensation	aromatic aldehyde	KCN	benzoin
4. Cannizzaro reaction	aldehyde not having α-hydrogen atom	aq. KOH or alc. KOH	alcohol and acid
5. Carbylamine reaction	primary amine	$CHCl_3$ + NaOH	carbylamine (test for —NH_2 group)
6. Clemmensen reduction	aldehyde or ketone	Zn/Hg + conc. HCl	hydrocarbon
7. Claisen condensation	(aldehyde, ketone or ester containing α-hydrogen atom) + (aromatic aldehyde or ketone)	NaOH	unsaturated aromatic aldehyde or ketone
8. Coupling reaction	C_6H_5NNCl + phenol (or aniline)	NaOH (in case of phenol), HCl (in case of aniline)	azo dye (detection of—NH_2 and phenolic—OH group)
9. Diazotisation	aromatic primary amine	$NaNO_2$/conc. HCl, 0–5°C	C_6H_5NNCl
10. Diels–Alder reaction	α,β-unsaturated carbonyl compound + conjugated diene	100°C	cyclic addition product
11. Etard reaction	toluene	CrO_2Cl_2/CS_2	benzaldehyde
12. Fittig reaction	aryl halide	Na/dry ether	diphenyl
13. Friedel–Crafts reaction	C_6H_6 + RX (or RCOCl)	anhy. $AlCl_3$	alkyl and acyl benzene
14. Gattermann reaction	C_6H_5NNCl	Cu powder + HCl	halobenzene
15. Gattermann–Kotch reaction	C_6H_6 + CO + HCl	anhy. $AlCl_3$	aromatic aldehyde
16. Haloform reaction	CH_3CHO, methyl ketone, CH_3CH_2OH, $RCHOHCH_3$	$NaOH/X_2$ or NaOX	haloform (test for —$COCH_3$ group)
17. Hell–Volhard–Zelinsky reaction	carboxylic acid having α-hydrogen atom	Br_2/red P	α-halogenated carboxylic acid
18. Hofmann bromamide reaction	acid amide	Br_2/NaOH	primary amine
19. Hofmann mustard oil reaction	primary aliphatic amine + CS_2	$HgCl_2$	$CH_3CH_2—N{=}C{=}S$ (mustard oil)
20. Hunsdiecker reaction	Ag salt of carboxylic acid + Br_2	CCl_4, 80°C	alkyl or aryl bromide
21. Kolbe electrolytic reaction	alkali metal salt of carboxylic acid	electrolysis	alkane, alkene and alkyne

22. Kolbe–Schmidt reaction	C_6H_5OH	CO_2, 125°C, 4–7 atm, H^+	salicylic acid
23. Meerwein–Ponndorf–Verley reduction	ketone	$[(CH_3)_2CHO]_3Al$ + $(CH_3)_2CHOH$	secondary alcohol
24. Mendius reaction	alkyl or aryl cyanide	Na/C_2H_5OH	primary amine
25. Perkin condensation	aromatic aldehyde + $(CH_3CO)_2O$	CH_3COONa, H^+/H_2O	α,β-unsaturated aromatic acid
26. Reimer–Tiemann reaction	phenol	NaOH + $CHCl_3$ or CCl_4, dil. HCl	salicyaldehyde or salicylic acid
27. Rosenmund reduction	acid chloride	H_2, $Pd/BaSO_4$, S, boiling xylene	aldehyde
28. Sabatier–Senderens reaction	unsaturated hydrocarbon	raney Ni/H_2, 200–300°C	alkane
29. Sandmeyer reaction	C_6H_5NNCl	CuCl/HCl or CuBr/HBr or CuCN/KCN, heat	halo or cyanobenzene
30. Schieman's reaction	C_6H_5NNCl	HBF_4	Fluorobenzene
31. Schotten–Baumann reaction	(phenol or aniline or alcohol) + C_6H_5COCl	NaOH	benzoylated product
32. Stephen reaction	alkyl cyanide	$SnCl_2/HCl$	aldehyde
33. Ullmann reaction	iodobenzene + Cu	heat	diphenyl
34. Williamson synthesis	alkyl halide + sodium alkoxide or sodium phenoxie	heat	ether
35. Wurtz–Fittig reaction	alkyl halide + aryl halide	Na/dry ether	alkyl benzene

Appendix–XIX
Wavelengths and Frequencies of Electromagnetic Radiations

Radiation	*Wavelength (Å)*	*Frequency (Hz or s^{-1})*
Radiowaves	3×10^9 to 3×10^{13}	1×10^5 to 1×10^9
Microwaves	6×10^6 to 3×10^9	1×10^9 to 5×10^{11}
Infrared rays	6×10^6 to 7.6×10^3	5×10^{11} to 3.95×10^{14}
Visible rays	3800 to 7600	3.95×10^{14} to 7.9×10^{14}
Ultraviolet rays	150 to 3800	7.9×10^{14} to 2×10^{16}
X-rays	0.1 to 150	2×10^{19} to 3×10^{19}
Gamma rays	0.01 to 0.1	3×10^{19} to 3×10^{20}
Cosmic rays	0.00 to 0.01	3×10^{20} to infinite

Appendix–XX
Electrochemical Series

Half-reaction	*Standard electrode potential at 298 K (V)*	*Half-reaction*	*Standard electrode potential at 298 K (V)*
$Li^+ + e^- \longrightarrow Li$	– 3.05	$Sn^{2+} + 2e^- \longrightarrow Sn$	– 0.14
$K^+ + e^- \longrightarrow K$	– 2.93	$Pb^{2+} + 2e^- \longrightarrow Pb$	– 0.13
$Ba^{2+} + 2e^- \longrightarrow Ba$	– 2.91	$Fe^{3+} + 3e^- \longrightarrow Fe$	– 0.04
$Sr^{2+} + 2e^- \longrightarrow Sr$	– 2.89	$2H^+ + 2e^- \longrightarrow H_2$	0.00
$Ca^{2+} + 2e^- \longrightarrow Ca$	– 2.86	$Cu^{2+} + e^- \longrightarrow Cu^+$	+ 0.16
$Na^+ + e^- \longrightarrow Na$	– 2.71	$Sn^{4+} + 2e^- \longrightarrow Sn^{2+}$	+ 0.15
$Mg^{2+} + 2e^- \longrightarrow Mg$	– 2.36	$Bi^{3+} + 3e^- \longrightarrow Bi$	+ 0.20
$Al^{3+} + 3e^- \longrightarrow Al$	– 1.66	$Cu^{2+} + 2e^- \longrightarrow Cu$	+ 0.34
$Mn^{2+} + 2e^- \longrightarrow Mn$	– 1.18	$Cu^+ + e^- \longrightarrow Cu$	+ 0.52
$Zn^{2+} + 2e^- \longrightarrow Zn$	– 0.76	$I_2 + 2e^- \longrightarrow 2I^-$	+ 0.54
$Cr^{3+} + 3e^- \longrightarrow Cr$	– 0.74	$Fe^{3+} + e^- \longrightarrow Fe^{2+}$	+ 0.77
$U^{4+} + e^- \longrightarrow U^{3+}$	– 0.61	$Hg_2^{2+} + 2e^- \longrightarrow 2Hg$	+ 0.79
$S + 2e^- \longrightarrow S^{2-}$	– 0.48	$Ag^+ + e^- \longrightarrow Ag$	+ 0.80
$Fe^{2+} + 2e^- \longrightarrow Fe$	– 0.44	$Hg^{2+} + 2e^- \longrightarrow Hg$	+ 0.86
$In^{3+} + 2e^- \longrightarrow In^+$	– 0.44	$Br_2 + 2e^- \longrightarrow 2Br^-$	+ 1.09
$Cr^{3+} + e^- \longrightarrow Cr^{2+}$	– 0.41	$Pt^{2+} + 2e^- \longrightarrow Pt$	+ 1.20
$Cd^{2+} + 2e^- \longrightarrow Cd$	– 0.40	$Cl_2 + 2e^- \longrightarrow 2Cl^-$	+ 1.36
$Tl^{3+} + e^- \longrightarrow Tl^{2+}$	– 0.37	$Au^{3+} + 3e^- \longrightarrow Au$	+ 1.50
$Tl^+ + e^- \longrightarrow Tl$	– 0.34	$Ce^{4+} + e^- \longrightarrow Ce^{3+}$	+ 1.61
$Co^{2+} + 2e^- \longrightarrow Co$	– 0.27	$Co^{3+} + e^- \longrightarrow Co^{2+}$	+ 1.81
$V^{3+} + e^- \longrightarrow V^{2+}$	– 0.26	$Ag^{2+} + e^- \longrightarrow Ag^+$	+ 1.98
$Ni^{2+} + 2e^- \longrightarrow Ni$	– 0.23	$Au^+ + e^- \longrightarrow Au$	+ 1.69
$In^+ + e^- \longrightarrow In$	– 0.14	$F_2 + 2e^- \longrightarrow 2F^-$	+ 2.87

Appendix–XXI
Effect of Strong Heat on Compounds

Compound	*Products*	*Compound*	*Products*
$AlCl_3 \cdot 6H_2O$	Al_2O_3, HCl, H_2O	NH_4VO_3	V_2O_5, NH_3, H_2O
NH_4HCO_3	NH_3, CO_2, H_2O	BaO_2	BaO, O_2
NH_4Cl	NH_3, HCl	$Na_2B_4O_7 \cdot 10H_2O$	$NaBO_2$, B_2O_3
$(NH_4)_2Cr_2O_7$	Cr_2O_3, N_2, H_2O	$CaCO_3$	CaO, CO_2
NH_4NO_3	N_2O, H_2O	$Ca(NO_3)_2$	CaO, NO_2, O_2
NH_4NO_2	N_2, H_2O	$CuCO_3$	CuO, CO_2
NH_3	N_2, H_2	$Cu(NO_3)_2$	CuO, NO_2

Compound	Products	Compound	Products
$CuSO_4 \cdot 5H_2O$	CuO, SO_3	$LiNO_3$	Li_2O, NO_2, O_2
$CoCO_3$	CoO, CO_2	$Mg(NO_3)_2$	MgO, NO_2, O_2
$Co(NO_3)_2$	Co_2O_3, NO_2, O_2	$MgCO_3$	MgO, CO_2
$Ni(CO)_4$	Ni, CO	$MgCl_2 \cdot 6H_2O$	MgO, HCl
$Ni(NO_2)_3$	Ni_2O_3, NO, NO_2, O_2	$Na(NH_4)HPO_4$	$NaPO_3, NH_3, H_2O$
$ZnCl_2 \cdot 2H_2O$	ZnO, HCl	H_3PO_4	HPO_3, H_2O
$ZnSO_4 \cdot 2H_2O$	ZnO, SO_2, O_2	$H_2C_2O_4$	CO, CO_2, H_2O
$Fe_2(SO_4)_3$	Fe_2O_3, SO_3	$P_4 + S_8$	$P_4S_3, P_4S_5, P_4S_7, P_4S_{10}$
$FeSO_4 \cdot 7H_2O$	Fe_2O_3, SO_2, SO_3	Ag_2CO_3	Ag, CO_2, O_2
FeC_2O_4	FeO, CO, CO_2	Ag_2O	Ag, O_2
$Fe(OH)_3$	Fe_2O_3, H_2O	$AgNO_3$	Ag, NO_2, O_2
$FeCl_3$	$FeCl_2, Cl_2$	NaN_3	Na, N_2
$FeS_2 + O_2$	Fe_2O_3, SO_2	$NaHCO_3$	Na_2CO_3, CO_2, H_2O
$CaSO_4 \cdot 2H_2O$	CaO, SO_2, O_2	$Na_2HPO_4 \cdot 12H_2O$	$Na_4P_2O_7, H_2O$
$Ca(HCOO)_2$	HCHO, CaO, CO_2	$NaNO_3$	$NaNO_2, O_2$
$Ca(CH_3COO)_2$	CH_3COCH_3, CaO, CO_2	$Na_2CO_3 \cdot 10H_2O$	Na_2CO_3, H_2O
H_3PO_2	PH_3, H_3PO_4	$ZnCO_3$	ZnO, CO_2
PbO_2	PbO (black), NO_2, O_2	$Zn(NO_3)_2$	ZnO, NO_2, O_2
$2Pb_3O_4$	PbO (yellow), O_2	$KMnO_4$	K_2MnO_4, MnO_2, O_2
$Pb(CH_3COO)_2$	PbO, CO_2, CH_3COCH_3	$Mn(NO_3)_2$	MnO_2, NO_2
$Pb(NO_3)_2$	PbO (yellow), NO_2, O_2	$MnCO_3$	MnO, CO_2
HgO	Hg, O_2	MnO_2	Mn_2O_3, O_2
Hg_2Cl_2	$HgCl_2$, Hg	$K_2Cr_2O_7$	K_2CrO_4, Cr_2O_3, O_2
$Hg(NO_3)_2$	Hg, NO_2, O_2	$KClO_3$	KCl, O_2
Li_2CO_3	Li_2O, CO_2	Urea	Cyanuric acid, NH_3

Appendix–XXII
Colour Changes and *p*H Ranges of Some Important Acid-Base Indicators

Indicator	*p*H *range*	*Colour in acid solution*	*Colour in alkaline solution*	*Used in the titration of*
Methyl orange	3.1–4.4	pink	yellow	strong acid and weak or strong base
Methyl red	4.2–6.3	red	yellow	strong acid and weak base

Phenolphthalein	8.2–10.0	colourless	pink	strong or weak acid and strong base
Litmus	6.0–6.8	colourless	pink	strong acid and strong base
Bromothymol blue	6.0–7.6	yellow	blue	strong acid and strong base

Appendix–XXIII
Elements: Discoverers, Names and Symbols

Element	*Symbol*	*Atomic No.*	*Year of Discovery*	*Discoverer and Nationality*
Actinium	Ac	89	1899	A. Debierne (France)
Aluminium	Al	13	1827	F. Woehler (Germany)
Americium	Am	95	1944	A. Ghiorso (USA) R. A. James (USA) G. T. Seaborg (USA) S. G. Seaborg (USA)
Antimony	Sb	51	Ancient	
Argon	Ar	18	1894	Lord Raleigh (Great Britain) Sir William Ramsay (Great Britain)
Arsenic	As	33	1250	Albertus Magnus (Germany)
Astatine	At	85	1940	D. R. Corson (USA) K. R. Mackenzie (USA) E. Segre (USA)
Barium	Ba	56	1808	Sir Humphry Davy (Great Britain)
Berkelium	Bk	97	1950	G. T. Seaborg (USA) S. G. Thompson (USA) A. Ghiorso (USA)
Beryllium	Be	4	1828	F. Woehler (Germany) A. A. B. Bussy (France)
Bismuth	Bi	83	1753	Claude Geoffroy (France)
Bohrium	Bh	107	1976	USA
Boron	B	5	1808	Sir Humphry Davy (Great Britain) J. L. Gay-Lussac (France) L. J. Thenard (France)
Bromine	Br	35	1826	A. J. Balard (France)
Cadmium	Cd	48	1817	Fr. Stromeyer (Germany)
Calcium	Ca	20	1808	Sir Humphry Davy (Great Britain)
Californium	Cf	98	1950	G. T. Seaborg, S. G. Thompson,

Element	*Symbol*	*Atomic No.*	*Year of Discovery*	*Discoverer and Nationality*
				A. Ghiorso, K. Street, Jr. (USA)
Carbon	C	6	Ancient	
Cerium	Ce	58	1803	J. J. Berzelius (Sweden) William Hisinger (Sweden) M. H. Klaproth (Germany)
Caesium	Cs	55	1860	R. Bunsen, G. R. Kirchhoff (Germany)
Chlorine	Cl	17	1774	K. W. Scheele (Sweden)
Chromium	Cr	24	1797	L. N. Vauquelin (France)
Cobalt	Co	27	1735	G. Brandt (Germany)
Copernicium	Cn	112	1996	S. Hofmann *et al.* (Germany)
Copper	Cu	29	Ancient	
Curium	Cm	96	1944	G. T. Seaborg (USA) R. A. James (USA) A. Ghiorso (USA)
Darmstadtium	Ds	110	1994	P. Armbnuster, G. Munzenberg (Germany)
Dubnium	Db	105	1970	G. N. Flerov *et al.* (Russia)
Dysprosium	Dy	66	1886	Lecoq de Boisbaudran (France)
Einsteinium	Es	99	1952	A. Ghiorso (USA)
Erbium	Er	68	1843	C. G. Mosander (Sweden)
Europium	Eu	63	1896	E. Demarcay (France)
Fermium	Fm	100	1953	A. Ghiorso (USA)
Flerovium	Fl	114	2006	Kenton Moody *et al.* (USA)
Fluorine	F	9	1886	H. Moissan (France)
Francium	Fr	87	1939	Marguerite Perey (France)
Gadolinium	Gd	64	1880	J. C. Marignac (France)
Gallium	Ga	31	1875	Lecoq de Boisbaudran (France)
Germanium	Ge	32	1886	Clemens Winkler (Germany)
Gold	Au	79	Ancient	
Hafnium	Hf	72	1923	D. Coster (Dutch) G. von Hevesey (Hungary)
Helium	He	2	1868	P. Janseen (spectra) (France) Sir William Ramsay (isolation) (Great Britain)
Hassium	Hs	108	1984	P. Armbnuster, G. Munzenberg (Germany)
Holmium	Ho	67	1879	P. T. Cleve (Sweden)
Hydrogen	H	1	1766	Sir Henry Cavendish (Great Britain)

Element	*Symbol*	*Atomic No.*	*Year of Discovery*	*Discoverer and Nationality*
Indium	In	49	1863	E. Reich (Germany) T. Richter (Germany)
Iodine	I	53	1811	B. Courtois (France)
Iridium	Ir	77	1803	S. Tennant (Great Britain)
Iron	Fe	26	Ancient	
Krypton	Kr	36	1898	Sir William Ramsay (Great Britain) M. W. Travers (Great Britain)
Lanthanum	La	57	1839	C. G. Mosander (Sweden)
Lawrencium	Lr	103	1961	A. Ghiorso, T. Sikkeland, A. E. Larsh, R. M. Latimer (USA)
Lead	Pb	82	Ancient	
Lithium	Li	3	1817	A. Arfvedson (Sweden)
Livermorium	Lv	116	2006	Kenton Moody *et al.* (USA)
Lutetium	Lu	71	1907	G. Urbain (France) C. A. von Welsbach (Austria)
Magnesium	Mg	12	1808	Sir Humphry Davy (Great Britain)
Manganese	Mn	25	1774	J. G. Gahn (Sweden)
Mendeleevium	Md	101	1955	A. Ghiorso, G. R. Choppin, G. T. Seaborg, B. G. Harvey, S. G. Thompson (USA)
Meitnerium	Mt	109	1982	G. Munzenberg *et al.* (Germany)
Mercury	Hg	80	Ancient	
Molybdenum	Mo	42	1778	G. W. Scheele (Sweden)
Neodymium	Nd	60	1885	C. A. von Welsbach (Austria)
Neon	Ne	10	1898	Sir William Ramsay (Great Britain) M. W. Travers (Great Britain)
Neptunium	Np	93	1940	E. M. McMillan (USA) P. H. Abelson (USA)
Nickel	Ni	28	1751	A. E. Cronstedt (Sweden)
Niobium	Nb	41	1801	Charles Hatchett (Great Britain)
Nitrogen	N	7	1772	Daniel Rutherford (Great Britain)
Nobelium	No	102	1958	A. Ghiorso (USA)

Element	*Symbol*	*Atomic No.*	*Year of Discovery*	*Discoverer and Nationality*
				T. Sikkeland (USA)
				J. R. Walton (USA)
				G. T. Seaborg (USA)
Osmium	Os	76	1803	S. Tennant (Great Britain)
Oxygen	O	8	1774	Joseph Priestley (Great Britain)
				C. W. Scheele (Sweden)
Palladium	Pd	46	1803	W. H. Wollaston (Great Britain)
Phosphorus	P	15	1669	H. Brandt (Germany)
Platinum	Pt	78	1735	A. De Ullao (Spain)
			1741	Charles Wood (Great Britain)
Plutonium	Pu	94	1940	G. T. Seaborg (USA)
				E. M. McMillan (USA)
				J. W. Kennedy (USA)
				A. C. Wahl (USA)
Polonium	Po	84	1898	Marie Curie (France)
Potassium	K	19	1807	Sir Humphry Davy (Great Britain)
Praseodymium	Pr	59	1885	C. A. von Welsbach (Austria)
Promethium	Pm	61	1945	J. A. Marinsky (USA)
				L. E. Glendenin (USA)
				C. D. Coryell (USA)
Protactinium	Pa	91	1917	O. Hohn (Germany)
				L. Meitner (Austria)
Radium	Ra	88	1898	Pierre and Marie Curie (France)
Radon	Rn	86	1900	F. E. Dom (Germany)
Rhenium	Re	75	1925	W. Noddack (Germany)
				I. Tacke (Germany)
				Otto Berg (Germany)
Rhodium	Rh	45	1804	W. H. Wollaston (Great Britain)
Roentgenium	Rg	111	1995	S. Hofmann (Germany)
Rubidium	Rb	37	1861	R. W. Bunsen (Germany)
				G. Kirchhoff (Germany)
Ruthenium	Ru	44	1844	K. K. Klaus (Russia)
Rutherfordium	Rf	104	1965	USA
Samarium	Sm	62	1879	Lecoq de Boisbaudran (France)
Scandium	Sc	21	1879	L. F. Nilson (Sweden)
Seaborgium	Sg	106	1976	A. Ghiorso *et al.* (USA)

Element	*Symbol*	*Atomic No.*	*Year of Discovery*	*Discoverer and Nationality*
Selenium	Se	34	1817	J. J. Berzelius (Sweden)
Silicon	Si	14	1824	J. J. Berzelius (Sweden)
Silver	Ag	47	Ancient	
Sodium	Na	11	1807	Sir Humphry Davy (Great Britain)
Strontium	Sr	38	1808	Sir Humphry Davy (Great Britain)
Sulphur	S	16	Ancient	
Tantalum	Ta	73	1802	A. G. Ekeberg (Sweden)
Technetium	Tc	43	1937	C. Perrier (Italy)
Tellurium	Te	52	1782	E. J. Muller (Austria)
Terbium	Tb	65	1843	C. G. Mosander (Sweden)
Thallium	Tl	81	1861	Sir William Crookes (Great Britain)
Thorium	Th	90	1828	J. J. Berzelius (Sweden)
Thulium	Tm	69	1879	P. T. Cleve (Sweden)
Tin	Sn	50	Ancient	
Titanium	Ti	22	1791	W. Gergor (Great Britain)
Tungsten	W	74	1783	J. J. de Elehujar and F. de Elehuyar (Spain)
Uranium	U	92	1789	M. H. Klaproth (Germany)
			1841	E. M. Peligot (France)
Vanadium	V	23	1801	A. M. Del Rio (Spain)
			1830	N. G. Sefstrom (Sweden)
Xenon	Xe	54	1898	Sir William Ramsay (Great Britain) M. W. Travers (Great Britain)
Ytterbium	Yb	70	1907	G. Urbain (France)
Yttrium	Y	39	1843	C. G. Mosander (Sweden)
Zinc	Zn	30	1746	A. S. Marggraf (Germany)
Zirconium	Zr	40	1789	M. H. Klaproth (Germany)

Appendix–XXIV

Some Important Rearrangement Reactions in Organic Chemistry

1. Benzil-benzilic acid Rearrangement

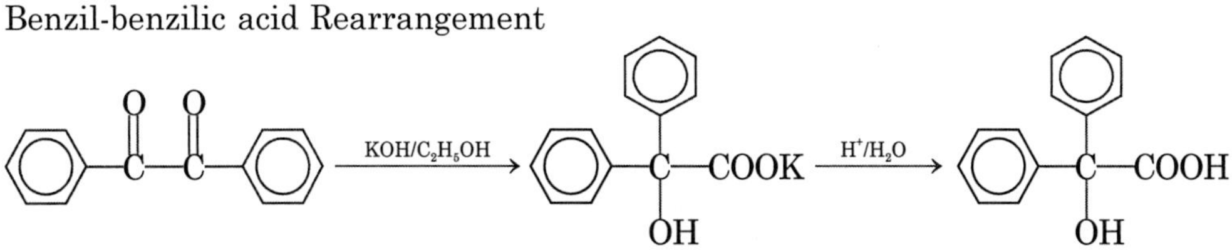

Benzil Potassium benzilate Benzilic acid

2. Benzidine Rearrangement

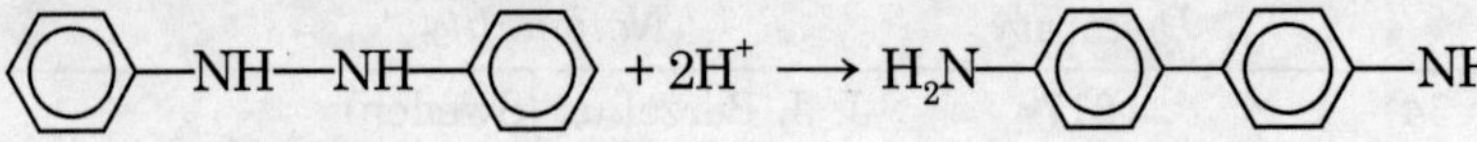

Hydrazobenzene Benzidine

3. Pinacol-pinacolone Rearrangement

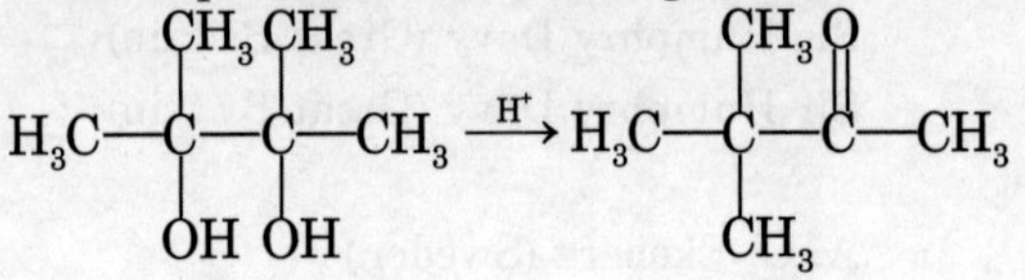

Pinacol Pinacolone

4. Claisen Rearrangement

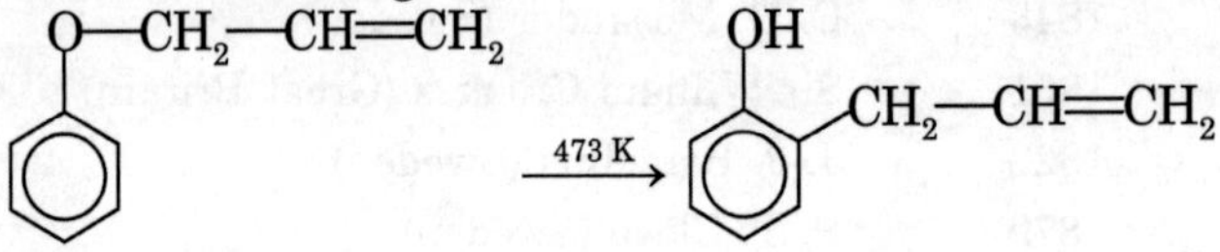

5. Beckmann Rearrangement

$$\begin{matrix} H_3C \\ H_5C_2 \end{matrix}\!\!>C{=}NOH \xrightarrow{\text{conc. } H_2SO_4} C_2H_5CONHCH_3 + CH_3CONHC_2H_5$$

6. Fries Rearrangement

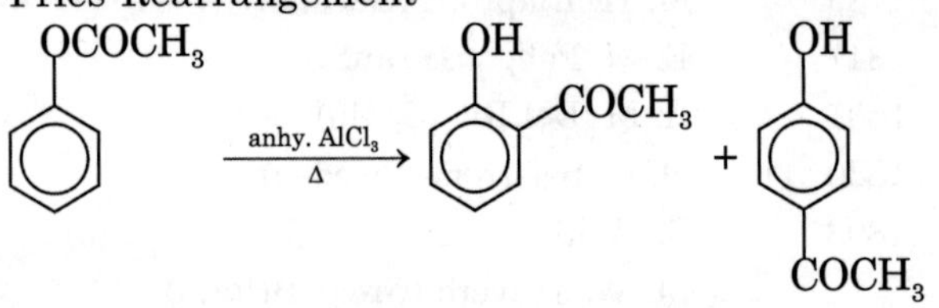

7. Friedel–Crafts Reaction

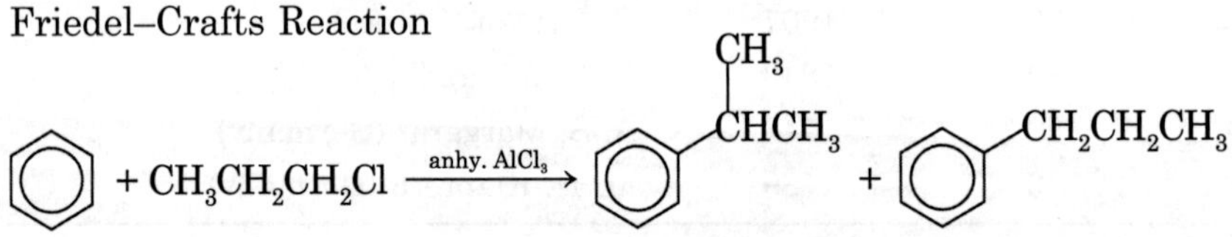

(major product) (minor product)

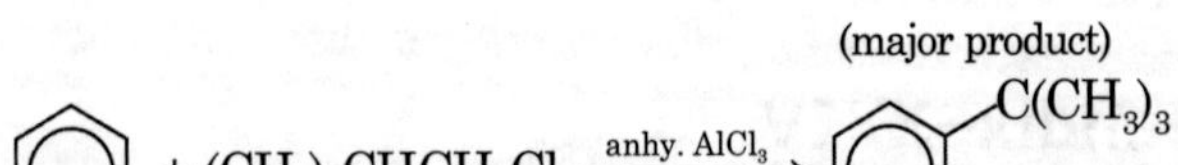

(only product)

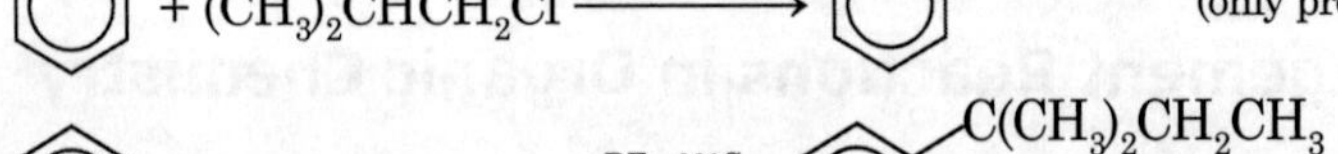

(only product)

8. Rearrangement, by Migration of Br, of 1° Free Radical into More Stable 2° or 3° Free Radical

$$CH_3CHBrCH_3 \xrightarrow{Cl_2} CH_3CHBr\dot{C}H_2 \xrightarrow{\text{rearr.}} CH_3\dot{C}HCH_2Br \xrightarrow{Cl_2} CH_3CHClCH_2Br$$

2-Bromopropane 1-Bromo-2-chloropropane

$$(CH_3)_3CBr \xrightarrow{Cl_2} (CH_3)_2CBr\dot{C}H_2 \xrightarrow{\text{rearr.}} (CH_3)_2\dot{C}CH_2Br \xrightarrow{Cl_2} (CH_3)_2CClCH_2Br$$

2-Bromo-2-methylpropane — 1-Bromo-2-chloro-2-methylpropane

9. Rearrangement of 1-Bromobutane to 2-Bromobutane in the presence of anhydrous $AlCl_3$ at 573 K

$$CH_3CH_2CH_2CH_2Br \xrightarrow{\text{anhy. } AlCl_3\text{, 573 K}} CH_3CH_2CHBrCH_3$$

1-Bromobutane — 2-Bromobutane

10. Hofmann Bromamide Reaction

$$CH_3CONH_2 \xrightarrow{Br_2/KOH} [CH_3CONHBr \xrightarrow[-H_2O]{OH^-} CH_3CO\ddot{\bar{N}}Br \xrightarrow{-Br^-}$$

Acetamide

$$CH_3C(=O)-\ddot{N}: \xrightarrow{\text{rearr.}} [O=C=N-CH_3] \xrightarrow{KOH} CH_3NH_2$$

Methylamine

11. Nucleophilic Substitution Reactions Involving S_{N^1} Mechanism

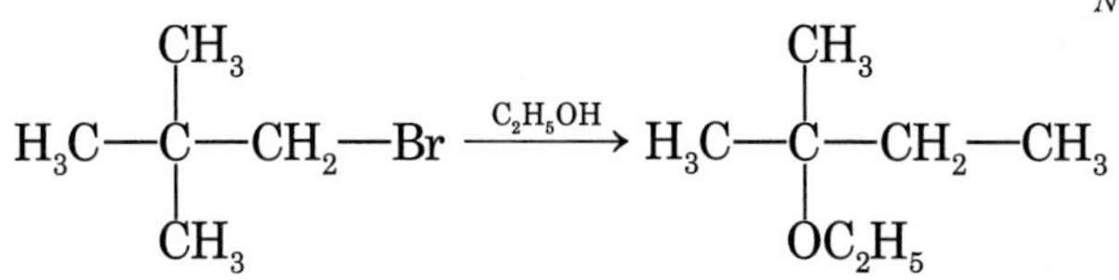

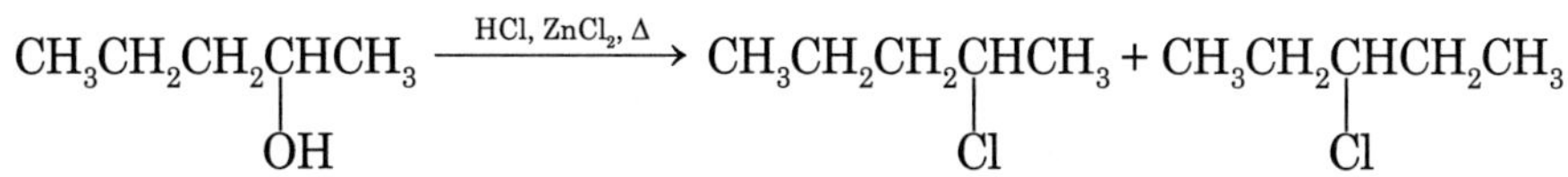

(major product) (minor product)

12. Elimination Reactions Involvnig $E1$ Mechanism

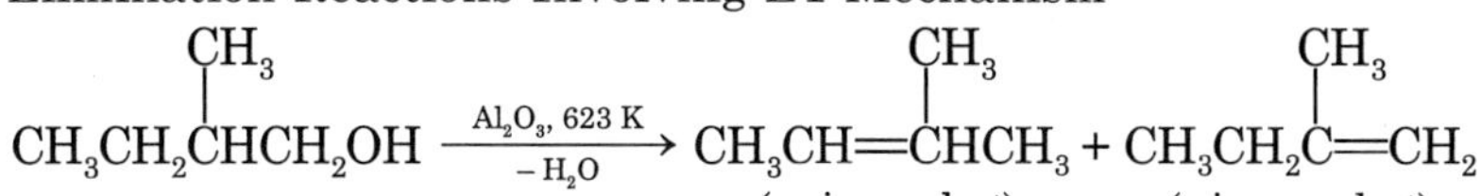

(major product) (minor product)

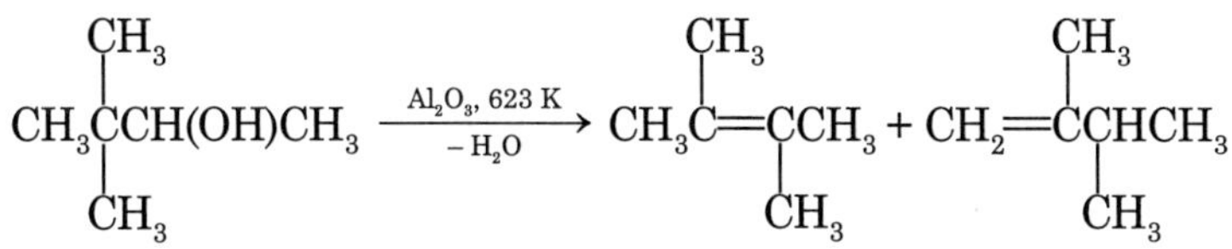

(major product) (minor product)

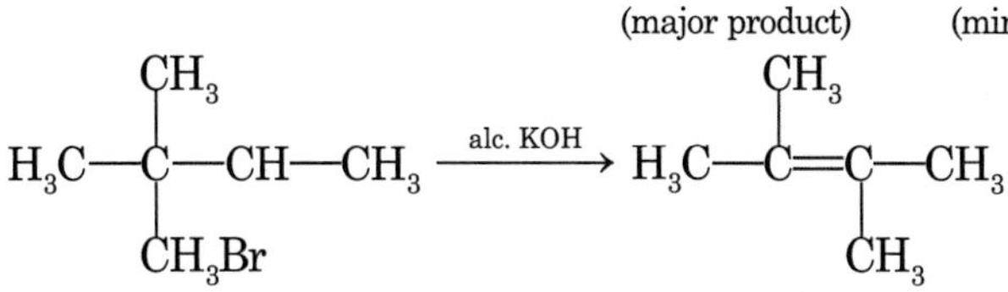

13. Electrophilic Addition Reaction

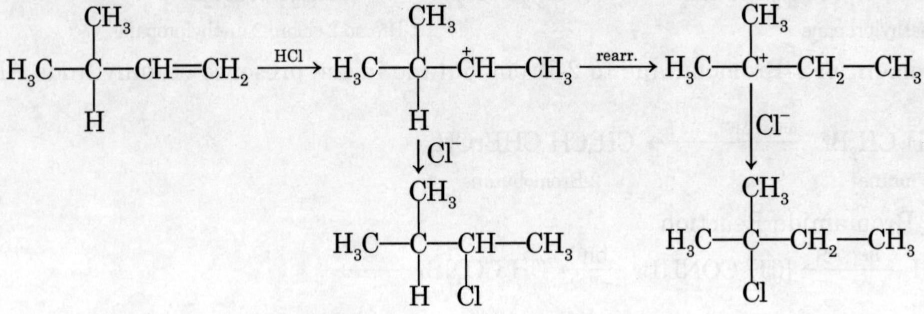

14. Hofmann–Martius Rearrangement

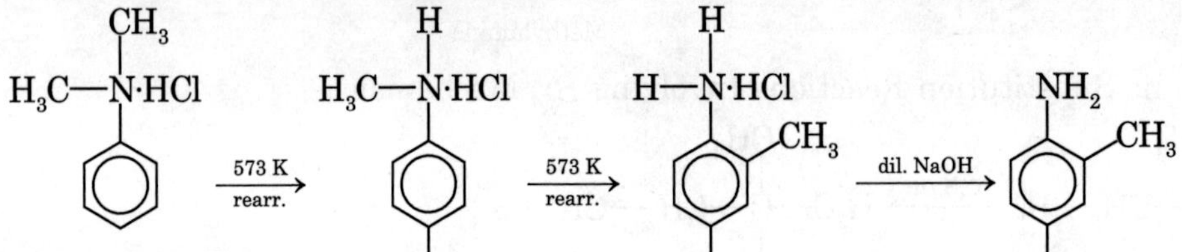

N, N-Dimethylamiline hydrochloride

2,4-Dimethylaniline

15. Rearrangement of Ammonium Cyanate to Urea on Heating

$$NH_4NCO \xrightarrow{\text{heat}} NH_2CONH_2$$

16. Rearrangement of Butane to 2-Methylpropane on Heating with Anhydrous $AlCl_3$

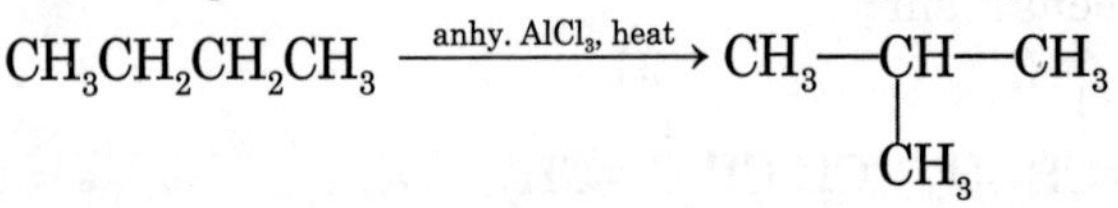

17. Rearrangement of But-1-ene to But-2-ene and 2-Methylpropene on Heating with Anhydrous $AlCl_3$

$$\underset{\text{But-1-ene}}{CH_3CH_2CH{=}CH_2} \xrightarrow{\text{anhy. } AlCl_3\text{, heat}} \underset{\text{But-2-ene}}{CH_3CH{=}CHCH_3} + \underset{\text{2-Methylpropene}}{(CH_3)_2C{=}CH_2}$$

18. Hydration of Alkenes

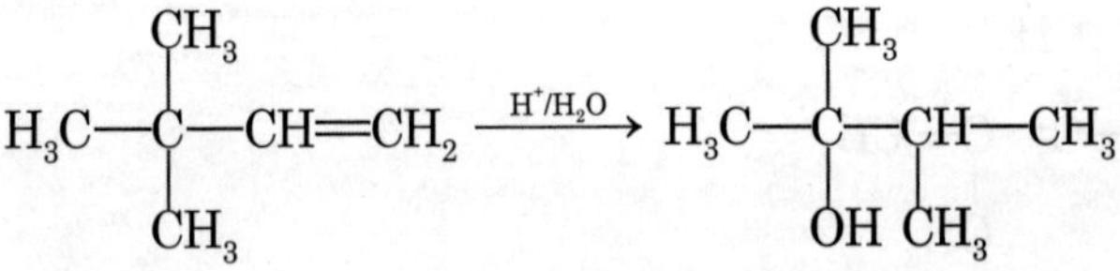

3,3-Dimethylbut-1-ene 2,3-Dimethylbutan-1-ol

Appendix–XXV
Conversion Chart of Aromatic Compounds

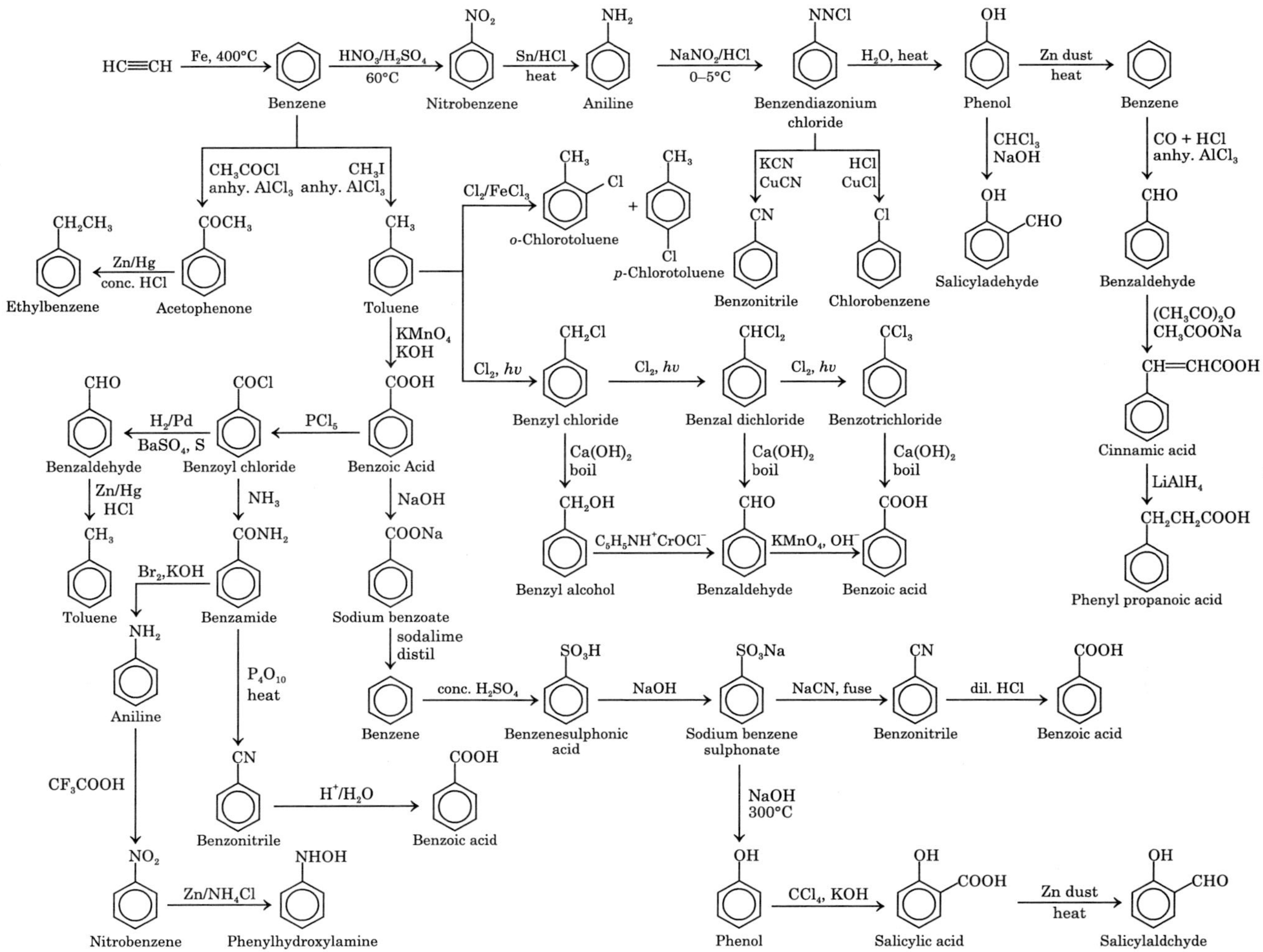

Appendix–XXVI

Colour of Some Transition and Non-transition Metal Compounds

	Compound	Colour
1.	Chromium(II) acetate	Red
2.	Chromium(III) acetate	Greyish-green
3.	Chromium(III) chloride hexahydrate	Violet
4.	Chromium(III) nitrate nonahydrate	Brown
5.	Chromium(III) oxalate monohydrate	Yellow
6.	Chromium(III) oxide	Green
7.	Cobalt(II) chloride	Black
8.	Cobalt(III) chloride hexahydrate	Pink
9.	Cobalt(II) hydroxide	Rose-red
10.	Cobalt(III) hydroxide	Blackish-brown
11.	Cobalt(II) oxide	Greenish-brown
12.	Cobalt(III) oxide	Blackish-grey
13.	Cobalt(II) sulphate hexahydrate	Red
14.	Cobalt(III) sulphate nonahydrate	Blue
15.	Hexaamminecobalt(III) chloride	Orange
16.	Copper(II) acetate monohydrate	Dark green
17.	Copper(I) chloride	White
18.	Copper(II) chloride dihydrate	Green
19.	Copper(I) cyanide	White
20.	Copper(II) cyanide	Yellowish-green
21.	Copper(II) ferrocyanide	Yellowish-green
22.	Copper(I) hydroxide	Yellow
23.	Copper(II) hydroxide	Blue
24.	Copper(II) nitrate	Blue
25.	Copper(II) oxalate	Blue
26.	Copper(I) oxide	Red
27.	Copper(II) oxide	Black
28.	Bis(acetylacetonato)copper(II)	Blue
29.	Copper(I) sulphate	Grey
30.	Copper(II) sulphate (anhydrous)	Green

31. Copper(II) sulphate pentahydrate	Blue
32. Copper(II) sulphide	Black
33. Diamminedichlorocopper(II) chloride	Yellow
34. Iron(II) chloride (anhydrous)	Yellow
35. Iron(II) chloride tetrahydrate	Bluish-green
36. Iron(III) chloride (anhydrous)	Blackish-brown
37. Iron(III) chloride hexahydrate	Brownish-yellow
38. Iron(II) ferrocyanide	Blackish-white
39. Iron(III) ferrocyanide	Dark blue
40. Iron(II) hydroxide	Pale green
41. Iron(III) nitrate hexahydrate	Pale violet
42. Iron(III) oxalate pentahydrate	Yellow
43. Iron(II) oxide	Black
44. Iron(III) oxide	Reddish-brown
45. Tris(acetylacetonato)iron(III)	Deep red
46. Iron(II) sulphate	Light green
47. Iron(III) sulphate hexahydrate	Yellow
48. Manganese(II) acetate	Pale red
49. Manganese(II) carbonate	Rose pink
50. Manganese(II) chloride tetrahydrate	Rose
51. Manganese(III) chloride	Black
52. Manganese(II) hydroxide	Whitish-pink
53. Manganese(II) nitrate hexahydrate	Rose-white
54. Manganese(II) oxalate dihydrate	Reddish-white
55. Manganese(II) oxide	Green
56. Manganese(IV) oxide	Black
57. Manganese(II) sulphate	Pale red
58. Manganese(III) sulphate	Green
59. Molybdenum(V) chloride	Greyish-black
60. Molybdenum(V) oxide	Violet blue
61. Nickel(II) carbonate	Light green
62. Nickel(II) chloride hexahydrate	Green

63. Bis(dimethylglyoximato)nickel(II)	Scarlet red
64. Nickel(II) nitrate hexahydrate	Green
65. Nickel(II) oxalate dihydrate	Light green
66. Nickel(II) oxide	Greyish-black
67. Nickel(II) sulphate hexahydrate	Blue
68. Nickel(II) sulphate heptahydrate	Green
69. Hexaamminenickel(II) chloride	Blue
70. Silver(I) nitrate	White
71. Silver(I) cyanide	White
72. Silver(I) oxide	Black
73. Silver(II) sulphide	Black
74. Tin(II) oxide	Black
75. Tin(II) sulphide	Greyish-black
76. Vanadium(V) chloride	Reddish-brown
77. Vanadium(V) oxide	Yellowish-red
78. Vanadium(IV) oxochloride	Green
79. Vanadium(V) oxochloride	Yellow
80. Oxovanadium(IV) sulphate	Blue
81. Zinc(II) oxide	White
82. Zinc(II) sulphide	White
83. Zinc(II) chloride	White
84. Zinc(II) ferrocyanide	White
85. Bis(acetylacetonato)zinc(II)	White
86. Zirconium(IV) oxide	Yellow
87. Zirconyl chloride hexahydrate	White

Appendix–XXVII
Logarithms

	0	1	2	3	4	5	6	7	8	9	1	2	3	4	5	6	7	8	9
10	0000	0043	0086	0128	0170						5	9	13	17	21	26	30	34	38
						0212	0253	0294	0334	0374	4	8	12	16	20	24	28	32	36
11	0414	0453	0492	0531	0569						4	8	12	16	20	23	27	31	35
						0607	0645	0682	0719	0775	4	7	11	15	18	22	26	29	33
12	0792	0828	0864	0899	0934						3	7	11	14	18	21	25	28	32
						0969	1004	1038	1072	1106	3	7	10	14	17	20	24	27	31
13	1139	1173	1206	1239	1271						3	6	10	13	16	19	23	26	29
						1303	1335	1367	1399	1430	3	7	10	13	16	19	22	25	29
14	1461	1492	1523	1553	1584						3	6	9	12	15	19	22	25	28
						1614	1644	1673	1703	1732	3	6	9	12	14	17	20	23	26
15	1761	1790	1818	1847	1875						3	6	9	11	14	17	20	23	26
						1903	1931	1959	1987	2014	3	6	8	11	14	17	19	22	25
16	2041	2068	2095	2122	2148						3	6	8	11	14	16	19	22	24
						2175	2201	2227	2253	2279	3	5	8	10	13	16	18	21	23
17	2304	2330	2355	2380	2405						3	5	8	10	13	15	18	20	23
						2430	2455	2480	2504	2529	3	5	8	10	12	15	17	20	22
18	2553	2577	2601	2625	2648						2	5	7	9	12	14	17	19	21
						2672	2695	2718	2742	2765	2	4	7	9	11	14	16	18	21
19	2788	2810	2833	2856	2878						2	4	7	9	11	13	16	18	20
						2900	2923	2945	2967	2989	2	4	6	8	11	13	15	17	19
20	3010	3032	3054	3075	3096	3118	3139	3160	3181	3201	2	4	6	8	11	13	15	17	19
21	3222	3243	3363	3284	3304	3324	3345	3365	3385	3404	2	4	6	8	10	12	14	16	18
22	3424	3444	3464	3483	3502	3522	3541	3560	3579	3598	2	4	6	8	10	12	14	15	17
23	3617	3636	3655	3464	3692	3711	3729	3747	3766	3784	2	4	6	7	9	11	13	15	17
24	3802	3820	3838	3856	3874	3892	3909	3927	3945	3962	2	4	5	7	9	11	12	14	16
25	3979	3997	4014	4031	4048	4065	4082	4099	4116	4133	2	3	5	7	9	10	12	14	15
26	4150	4166	4183	4200	4216	4232	4249	4265	4281	4298	2	3	5	7	8	10	11	13	15
27	4314	4330	4346	4362	4378	4393	4409	4425	4440	4456	2	3	5	6	8	9	11	13	14
28	4472	4487	4502	4518	4533	4548	4564	4579	4594	4609	2	3	5	6	8	9	11	12	14
29	4624	4639	4654	4669	4683	4698	4713	4728	4742	4757	1	3	4	6	7	9	10	12	13
30	4771	4786	4800	4814	4829	4843	4857	4871	4886	4900	1	3	4	6	7	9	10	11	13
31	4914	4928	4942	4955	4969	4983	4997	5011	5024	5038	1	3	4	6	7	8	10	11	12
32	5051	5065	5079	5092	5105	5119	5132	5145	5159	5172	1	3	4	5	7	8	10	11	12
33	5185	5198	5211	5224	5237	5250	5263	5276	5289	5302	1	3	4	5	6	8	9	10	12
34	5315	5328	5340	5353	5366	5378	5391	5403	5416	5428	1	3	4	5	6	8	9	10	11
35	5441	5453	5465	5478	5490	5502	5514	5527	5539	5551	1	2	4	5	6	7	9	10	11
36	5563	5575	5587	5599	5611	5623	5635	5647	5658	5670	1	2	4	5	6	7	8	10	11
37	5682	5694	5705	5717	5729	5740	5752	5763	5775	5786	1	2	3	5	6	7	8	9	10
38	5798	5809	5821	5832	5843	5855	5866	5877	5888	5899	1	2	3	5	6	7	8	9	10
39	5911	5922	5933	5944	5955	5966	5977	5988	5999	6010	1	2	3	4	5	7	8	9	10
40	6021	6031	6042	6053	6064	6075	6085	6096	6107	6117	1	2	3	4	5	6	8	9	10
41	6128	6138	6149	6160	6170	6180	6191	6201	6212	6222	1	2	3	4	5	6	7	8	9
42	6232	6243	6253	6263	6274	6284	6294	6304	6314	6325	1	2	3	4	5	6	7	8	9
43	6335	6345	6355	6365	6375	6385	6395	6405	6415	6425	1	2	3	4	5	6	7	8	9
44	6435	6444	6454	6464	6474	6484	6493	6503	6513	6522	1	2	3	4	5	6	7	8	9
45	6532	6542	6551	6561	6571	6580	6590	6599	6609	6618	1	2	3	4	5	6	7	8	9
46	6628	6637	6646	6656	6665	6675	6684	6693	6702	6712	1	2	3	4	5	6	7	7	8
47	6721	6730	6739	6749	6758	6767	6776	6785	6794	6803	1	2	3	4	5	6	6	7	8
48	6812	6821	6830	6839	6848	6857	6866	6875	6884	6893	1	2	3	4	5	5	6	7	8
49	6902	6911	6920	6928	6937	6946	6955	6964	6972	6981	1	2	3	4	4	5	6	7	8

Logarithms

	0	1	2	3	4	5	6	7	8	9	1	2	3	4	5	6	7	8	9
50	6990	6998	7007	7016	7024	7033	7042	7050	7059	7067	1	2	3	4	5	6	6	7	8
51	7076	7084	7093	7101	7110	7118	7126	7135	7143	7152	1	2	3	3	4	5	6	7	8
52	7160	7168	7177	7185	7193	7202	7210	7218	7226	7235	1	2	2	3	4	5	6	7	7
53	7243	7251	7259	7267	7275	7284	7292	7300	7308	7316	1	2	2	3	4	5	6	6	7
54	7324	7332	7340	7348	7356	7364	7372	7380	7388	7396	1	2	2	3	4	5	6	6	7
55	7404	7412	7419	7427	7435	7443	7451	7459	7466	7474	1	2	2	3	4	5	5	6	7
56	7482	7490	7497	7505	7513	7520	7528	7536	7543	7551	1	2	2	3	4	5	5	6	7
57	7559	7566	7574	7582	7589	7597	7604	7612	7619	7627	1	1	2	3	4	5	5	6	7
58	7634	7642	7649	7657	7664	7672	7679	7686	7694	7701	1	1	2	3	4	4	5	6	7
59	7709	7716	7723	7731	7738	7745	7752	7760	7767	7774	1	1	2	3	4	4	5	6	7
60	7782	7789	7796	7803	7810	7818	7825	7832	7839	7846	1	1	2	3	4	4	5	6	6
61	7853	7860	7868	7875	7882	7889	7896	7903	7910	7917	1	1	2	3	4	4	5	6	6
62	7924	7931	7938	7945	7952	7959	7966	7973	7980	7987	1	1	2	3	3	4	5	6	6
63	7993	8000	8007	8014	8021	8028	8035	8041	8048	8055	1	1	2	3	3	4	5	5	6
64	8062	8069	8075	8082	8089	8096	8102	8109	8116	8122	1	1	2	3	3	4	5	5	6
65	8129	8136	8142	8149	8156	8162	8169	8176	8182	8189	1	1	2	3	3	4	5	5	6
66	8195	8202	8209	8215	8222	8228	8235	8241	8248	8254	1	1	2	3	3	4	5	5	6
67	8261	8267	8274	8280	8287	8293	8299	8306	8312	8319	1	1	2	3	3	4	5	5	6
68	8325	8331	8338	8344	8351	8357	8363	8370	8376	8382	1	1	2	3	3	4	4	5	6
69	8388	8395	8401	8407	8414	8420	8426	8432	8439	8445	1	1	2	2	3	4	4	5	6
70	8451	8457	8463	8470	8476	8482	8488	8494	8500	8506	1	1	2	2	3	4	4	5	6
71	8513	8519	8525	8531	8537	8543	8549	8555	8561	8567	1	1	2	2	3	4	4	5	6
72	8573	8579	8585	8591	8597	8603	8609	8615	8621	8627	1	1	2	2	3	4	4	5	5
73	8633	8639	8645	8651	8657	8663	8669	8675	8681	8686	1	1	2	2	3	4	4	5	5
74	8692	8698	8704	8710	8716	8722	8727	8733	8739	8745	1	1	2	2	3	4	4	5	5
75	8751	8756	8762	8768	8774	8779	8785	8791	8797	8802	1	1	2	2	3	3	4	5	5
76	8808	8814	8820	8825	8831	8837	8842	8848	8854	8859	1	1	2	2	3	3	4	5	5
77	8865	8871	8876	8882	8887	8893	8899	8904	8910	8915	1	1	2	2	3	3	4	4	5
78	8921	8927	8932	8938	8943	8949	8954	8960	8965	8971	1	1	2	2	3	3	4	4	5
79	8976	8982	8987	8993	8998	9004	9009	9015	9020	9025	1	1	2	2	3	4	4	4	5
80	9031	9036	9042	9047	9053	9058	9063	9069	9074	9079	1	1	2	2	3	3	4	4	5
81	9085	9090	9096	9101	9106	9112	9117	9122	9128	9133	1	1	2	2	3	3	4	4	5
82	9138	9143	9149	9154	9159	9165	9170	9175	9180	9186	1	1	2	2	3	3	4	4	5
83	9191	9196	9201	9206	9212	9217	9222	9227	9232	9238	1	1	2	2	3	3	4	4	5
84	9243	9248	9253	9258	9263	9269	9274	9279	9284	9289	1	1	2	2	3	3	4	4	5
85	9294	9299	9304	9309	9315	9320	9325	9330	9335	9340	1	1	2	2	3	3	4	4	5
86	9345	9350	9355	9360	9365	9370	9375	9380	9385	9390	1	1	2	2	3	3	4	4	5
87	9395	9400	9405	9410	9415	9420	9425	9430	9435	9440	0	1	1	2	2	3	3	4	4
88	9445	9450	9455	9460	9465	9469	9474	9479	9484	9489	0	1	1	2	2	3	3	4	4
89	9494	9499	9504	9509	9513	9518	9523	9528	9533	9538	0	1	1	2	2	3	3	4	4
90	9542	9547	9552	9557	9562	9566	9571	9576	9581	9586	0	1	1	2	2	3	3	4	4
91	9590	9595	9600	9605	9609	9614	9619	9624	9628	9633	0	1	1	2	2	3	3	4	4
92	9638	9643	9647	9652	9657	9661	9666	9671	9675	9680	0	1	1	2	2	3	3	4	4
93	9685	9689	9694	9699	9703	9708	9713	9717	9722	9727	0	1	1	2	2	3	3	4	4
94	9731	9736	9741	9745	9750	9754	9759	9763	9768	9773	0	1	1	2	2	3	3	4	4
95	9777	9782	9786	9791	9795	9800	9805	9809	9814	9818	0	1	1	2	2	3	3	4	4
96	9823	9827	9832	9836	9841	9845	9850	9854	9859	9863	0	1	1	2	2	3	3	4	4
97	9868	9872	9877	9881	9886	9890	9894	9899	9903	9908	0	1	1	2	2	3	3	4	4
98	9912	9917	9921	9926	9930	9934	9939	9943	9948	9952	0	1	1	2	2	3	3	4	4
99	9956	9961	9965	9969	9974	9978	9983	9987	9991	9996	0	1	1	2	2	3	3	3	4

Appendix–XXVIII
Antilogarithms

	0	1	2	3	4	5	6	7	8	9	1	2	3	4	5	6	7	8	9
.00	1000	1002	1005	1007	1009	1012	1014	1016	1019	1021	0	0	1	1	1	1	2	2	2
.01	1023	1026	1028	1030	1033	1035	1038	1040	1042	1045	0	0	1	1	1	1	2	2	2
.02	1047	1050	1052	1054	1057	1059	1062	1064	1067	1069	0	0	1	1	1	1	2	2	2
.03	1072	1074	1076	1079	1081	1084	1086	1089	1091	1094	0	0	1	1	1	1	2	2	2
.04	1096	1099	1102	1104	1107	1109	1112	1114	1117	1119	0	1	1	1	1	2	2	2	2
.05	1122	1125	1127	1130	1132	1135	1138	1140	1143	1146	0	1	1	1	1	2	2	2	2
.06	1148	1151	1153	1156	1159	1161	1164	1167	1169	1172	0	1	1	1	1	2	2	2	2
.07	1175	1178	1180	1183	1186	1189	1191	1194	1197	1199	0	1	1	1	1	2	2	2	2
.08	1202	1205	1208	1211	1213	1216	1219	1222	1225	1227	0	1	1	1	2	2	2	2	3
.09	1230	1233	1236	1239	1242	1245	1247	1250	1253	1256	0	1	1	1	2	2	2	2	3
.10	1259	1262	1265	1268	1271	1274	1276	1279	1282	1285	0	1	1	1	2	2	2	2	3
.11	1288	1291	1294	1297	1300	1303	1306	1309	1312	1315	0	1	1	1	2	2	2	2	3
.12	1318	1321	1324	1327	1330	1334	1337	1340	1343	1346	0	1	1	1	2	2	2	2	3
.13	1349	1352	1355	1358	1361	1365	1368	1371	1374	1377	0	1	1	1	2	2	2	3	3
.14	1380	1384	1387	1390	1393	1396	1400	1403	1406	1409	0	1	1	1	2	2	2	3	3
.15	1413	1416	1419	1422	1426	1429	1432	1435	1439	1442	0	1	1	1	2	2	2	3	3
.16	1445	1449	1452	1455	1459	1462	1466	1469	1472	1476	0	1	1	1	2	2	2	3	3
.17	1479	1483	1486	1489	1493	1496	1500	1503	1507	1510	0	1	1	1	2	2	2	3	3
.18	1514	1517	1521	1524	1528	1531	1535	1538	1542	1545	0	1	1	1	2	2	2	3	3
.19	1549	1552	1556	1560	1563	1567	1570	1574	1578	1581	0	1	1	1	2	2	3	3	3
.20	1585	1589	1592	1596	1600	1603	1607	1611	1614	1618	0	1	1	1	2	2	3	3	3
.21	1622	1626	1629	1633	1637	1641	1644	1648	1652	1656	0	1	1	2	2	2	3	3	3
.22	1660	1663	1667	1671	1675	1679	1683	1687	1690	1694	0	1	1	2	2	2	3	3	3
.23	1698	1702	1706	1710	1714	1718	1722	1726	1730	1734	0	1	1	2	2	2	3	3	4
.24	1738	1742	1746	1750	1754	1758	1762	1766	1770	1774	0	1	1	2	2	2	3	3	4
.25	1778	1782	1786	1791	1795	1799	1803	1807	1811	1816	0	1	1	2	2	2	3	3	4
.26	1820	1824	1828	1832	1837	1841	1845	1849	1854	1858	0	1	1	2	2	3	3	3	4
.27	1862	1866	1871	1875	1879	1884	1888	1892	1897	1901	0	1	1	2	2	3	3	3	4
.28	1905	1910	1914	1919	1923	1928	1932	1936	1941	1945	0	1	1	2	2	3	3	4	4
.29	1950	1954	1959	1963	1968	1972	1977	1982	1986	1991	0	1	1	2	2	3	3	4	4
.30	1995	2000	2004	2009	2014	2018	2023	2028	2032	2037	0	1	1	2	2	3	3	4	4
.31	2042	2046	2051	2056	2061	2065	2070	2075	2080	2084	0	1	1	2	2	3	3	4	4
.32	2089	2094	2099	2104	2109	2113	2118	2123	2128	2133	0	1	1	2	2	3	3	4	4
.33	2138	2143	2148	2153	2158	2163	2168	2173	2178	2183	0	1	1	2	2	3	3	4	4
.34	2188	2193	2198	2203	2208	2213	2218	2223	2228	2234	1	1	2	2	3	3	4	4	5
.35	2239	2244	2249	2254	2259	2265	2270	2275	2280	2286	1	1	2	2	3	3	4	4	5
.36	2291	2296	2301	2307	2312	2317	2323	2328	2333	2339	1	1	2	2	3	3	4	4	5
.37	2344	2350	2355	2360	2366	2371	2377	2382	2388	2393	1	1	2	2	3	3	4	4	5
.38	2399	2404	2410	2415	2421	2427	2432	2438	2443	2449	1	1	2	2	3	3	4	4	5
.39	2455	2460	2466	2472	2477	2483	2489	2495	2500	2506	1	1	2	2	3	3	4	5	5
.40	2512	2518	2523	2529	2535	2541	2547	2553	2559	2564	1	1	2	2	3	4	4	5	5
.41	2570	2576	2582	2588	2594	2600	2606	2612	2618	2624	1	1	2	2	3	4	4	5	5
.42	2630	2636	2642	2649	2655	2661	2667	2673	2679	2685	1	1	2	2	3	4	4	5	6
.43	2692	2698	2704	2710	2716	2723	2729	2735	2742	2748	1	1	2	3	3	4	4	5	6
.44	2754	2761	2767	2773	2780	2786	2793	2799	2805	2812	1	1	2	3	3	4	4	5	6
.45	2818	2825	2831	2838	2844	2851	2858	2864	2871	2877	1	1	2	3	3	4	5	5	6
.46	2884	2891	2897	2904	2911	2917	2924	2931	2938	2944	1	1	2	3	3	4	5	5	6
.47	2951	2958	2965	2972	2979	2985	2992	2999	3006	3013	1	1	2	3	3	4	5	5	6
.48	3020	3027	3034	3041	3048	3055	3062	3069	3076	3083	1	1	2	3	4	4	5	6	6
.49	3090	3097	3105	3112	3119	3126	3133	3141	3148	3155	1	1	2	3	4	4	5	6	6

Antilogarithms

	0	1	2	3	4	5	6	7	8	9	1	2	3	4	5	6	7	8	9
.50	3162	3170	3177	3184	3192	3199	3206	3214	3221	3228	1	1	2	3	4	5	5	6	7
.51	3236	3243	3251	3258	3266	3273	3281	3289	3296	3304	1	2	2	3	4	5	5	6	7
.52	3311	3319	3327	3334	3342	3350	3357	3365	3373	3381	1	2	2	3	4	5	5	6	7
.53	3388	3396	3404	3412	3420	3428	3436	3443	3451	3459	1	2	2	3	4	5	6	6	7
.54	3467	3475	3483	3491	3499	3508	3516	3524	3532	3540	1	2	2	3	4	5	6	6	7
.55	3548	3556	3565	3573	3581	3589	3597	3606	3614	3622	1	2	2	3	4	5	6	7	7
.56	3631	3639	3648	3656	3664	3673	3681	3690	3698	3707	1	2	3	3	4	5	6	7	8
.57	3715	3724	3733	3741	3750	3758	3767	3776	3784	3793	1	2	3	3	4	5	6	7	8
.58	3802	3811	3819	3828	3837	3846	3855	3864	3873	3882	1	2	3	4	4	5	6	7	8
.59	3890	3899	3908	3917	3926	3936	3945	3954	3963	3972	1	2	3	4	5	5	6	7	8
.60	3981	3990	3999	4009	4018	4027	4036	4046	4055	4064	1	2	3	4	5	6	6	7	8
.61	4074	4083	4093	4102	4111	4121	4130	4140	4150	4159	1	2	3	4	5	6	7	8	9
.62	4169	4178	4188	4198	4207	4217	4227	4236	4246	4256	1	2	3	4	5	6	7	8	9
.63	4266	4276	4285	4295	4305	4315	4325	4335	4345	4355	1	2	3	4	5	6	7	8	9
.64	4365	4375	4385	4395	4406	4416	4426	4436	4446	4457	1	2	3	4	5	6	7	8	9
.65	4467	4477	4487	4498	4508	4519	4529	4539	4550	4560	1	2	3	4	5	6	7	8	9
.66	4571	4581	4592	4603	4613	4624	4634	4645	4656	4667	1	2	3	4	5	6	7	9	10
.67	4677	4688	4699	4710	4721	4732	4742	4753	4764	4775	1	2	3	4	5	7	8	9	10
.68	4786	4797	4808	4819	4831	4842	4853	4864	4875	4887	1	2	3	4	6	7	8	9	10
.69	4898	4909	4920	4932	4943	4955	4966	4977	4989	5000	1	2	3	5	6	7	8	9	10
.70	5012	5023	5035	5047	5058	5070	5082	5093	5105	5117	1	2	4	5	6	7	8	9	11
.71	5129	5140	5152	5164	5176	5188	5200	5212	5224	5236	1	2	4	5	6	7	8	10	11
.72	5248	5260	5272	5284	5297	5309	5321	5333	5346	5358	1	2	4	5	6	7	9	10	11
.73	5370	5383	5395	5408	5420	5433	5445	5458	5470	5483	1	3	4	5	6	8	9	10	11
.74	5495	5508	5521	5534	5546	5559	5572	5585	5598	5610	1	3	4	5	7	8	9	10	11
.75	5623	5636	5649	5662	5675	5689	5702	5715	5728	5741	1	3	4	5	7	8	9	10	12
.76	5754	5768	5781	5794	5808	5821	5834	5848	5861	5875	1	3	4	5	7	8	9	11	12
.77	5888	5902	5916	5929	5943	5957	5970	5984	5998	6012	1	3	4	5	7	8	10	11	12
.78	6026	6039	6053	6067	6081	6095	6109	6124	6138	6152	1	3	4	6	7	8	10	11	13
.79	6166	6180	6194	6209	6223	6237	6252	6266	6281	6295	1	3	4	6	7	9	10	12	13
.80	6310	6324	6339	6353	6368	6383	6397	6412	6427	6442	1	3	4	6	7	9	10	12	13
.81	6457	6471	6486	6501	6516	6531	6546	6561	6577	6592	2	3	5	6	8	9	11	12	14
.82	6607	6622	6637	6653	6668	6683	6699	6714	6730	6745	2	3	5	6	8	9	11	12	14
.83	6761	6776	6792	6808	6823	6839	6855	6871	6887	6902	2	3	5	6	8	9	11	13	14
.84	6918	6934	6950	6966	6982	6998	7015	7031	7047	7063	2	3	5	6	8	10	12	13	15
.85	7079	7096	7112	7129	7145	7161	7178	7194	7211	7228	2	3	5	7	8	10	12	-13	15
.86	7244	7261	7278	7295	7311	7328	7345	7362	7379	7396	2	3	5	7	8	10	12	13	15
.87	7413	7430	7447	7464	7482	7499	7516	7534	7551	7568	2	3	5	7	9	10	12	14	16
.88	7586	7603	7621	7638	7656	7674	7691	7709	7727	7745	2	4	5	7	9	11	12	14	16
.89	7762	7780	7798	7816	7834	7852	7870	7889	7907	7925	2	4	5	7	9	11	13	14	16
.90	7943	7962	7980	7998	8017	8035	8054	8072	8091	8110	2	4	6	7	9	11	13	15	17
.91	8128	8147	8166	8185	8204	8222	8241	8260	8279	8299	2	4	6	8	9	11	13	15	17
.92	8318	8337	8356	8375	8395	8414	8433	8453	8472	8492	2	4	6	8	10	12	14	15	17
.93	8511	8531	8551	8570	8590	8610	8630	8650	8670	8690	2	4	6	8	10	12	14	16	18
.94	8710	8730	8750	8770	8790	8810	8831	8851	8872	8892	2	4	6	8	10	12	14	16	18
.95	8913	8933	8954	8974	8995	9016	9036	9057	9078	9099	2	4	6	8	10	12	15	17	19
.96	9120	9141	9162	9183	9204	9226	9247	9628	9290	9311	2	4	6	8	11	13	15	17	19
.97	9333	9354	9376	9397	9419	9441	9462	9484	9506	9528	2	4	7	9	11	13	15	17	20
.98	9550	9572	9594	9616	9638	9661	9683	9705	9727	9750	2	4	7	9	11	13	16	18	20
.99	9772	9795	9817	9840	9863	9886	9908	9931	9954	9977	2	5	7	9	11	14	16	18	20